JOURNAL OF POLYMER SCIENCE: Polymer Symposium   No. 58

Orientation Effects
In Solid Polymers

Proceedings of the
5th Europhysics Conference on Macromolecular Physics
Budapest, Hungary
27–30 April, 1976
Held under the Sponsorship of the
European Physical Society
Hungarian Academy of Sciences
Section Chemical Sciences
Roland Eötuös Physical Society

Editor:

G. Bodor
Polymer Research Institute
H-1950 Budapest

an Interscience® Publication
published by JOHN WILEY & SONS

JOURNAL OF POLYMER SCIENCE: Polymer Symposia

G. Bodor has been appointed Editor for this Symposium by the Editorial Board of the *Journal of Polymer Science.*

Published by John Wiley & Sons, Inc., this book constitutes a part of the annual subscription of the *Journal of Polymer Science* and as such is supplied without additional charge to the subscribers. Single copies can be purchased from the Subscription Department, John Wiley & Sons.

Subscription price, *Journal of Polymer Science,* Vol. 16, 1978. $485.00. Postage and handling outside U.S.A.: $37.00. Please allow four to six weeks for processing a change of address. Back volumes, microfilm, and microfiche are available for previous years. Request price list from publisher.

Printed in the United States of America.

Preface

The papers collected in this volume were presented at the Fifth Europhysics Conference on Macromolecular Physics: Orientation Effects in Solid Polymers in Budapest on April 27–30, 1976.

This type of Conference is organized by the European Physical Society, Section Macromolecular Physics in each year since the foundation of this section. The subject of these Conferences is decided by the board of the section and it takes place at different places in Europe. Taking over the responsibility for the organization of this Conference, the members of the Organization Committee and myself were in the firm belief, that such conferences are helping a lot in the better understanding of different opinions, they are giving possibilities to meet people each other from different laboratories and that they make good for the development of the polymer physics.

To this Conference we invited some speakers from and outside of Europe. We are obliged to them for the participation.

The papers given at the Conference were ordered into five topics. In this volume they are reproduced in the order and in essentially the form in which they were presented. We were not able to include some papers, which were or will be published elsewhere, and there is one paper which was not presented because of the illness of the author, but we received the manuscript in due time.

I have to mention, that the participants of the Conference received an abstract volume, which contains the abstracts of these papers and the whole text of some lectures. This abstract volume is the Volume 1D of the Europhysics Conference Abstracts, published by the European Physical Society. The European Physical Society has the rights to the abstracts and the Journal of Polymer Science has the rights to the completed papers for general publication.

I am very much obliged to Mr. G. Thomas, Executive Secretary of the European Physical Society and to Prof. Ch. Overberger, Editor of the Journal of Polymer Science for their help in many ways, especially for the special arrangement of the rights mentioned above.

Many thanks to Prof. A. J. Kovacs, Chairman of the Section Macromolecular Physics and to the Hungarian colleagues for their efforts.

G. BODOR

Contents

THE MEASUREMENT OF MOLECULAR ORIENTATION IN POLYMERS BY SPECTROSCOPIC TECHNIQUES

I. M. WARD

University of Leeds, Leeds LS2 9JT, England

INTRODUCTION

The measurement of orientation in polymers by spectroscopic methods is of value for a number of different reasons. In a crystalline polymer it is complementary to X-ray diffraction measurements, in that it can provide information on the orientation of the noncrystalline regions. In an amorphous polymer it may well be the only method of obtaining detailed knowledge of the molecular orientation. In both crystalline and amorphous polymers it may be possible from spectroscopic data to define the changes in orientation and molecular conformation which take place during the stretching, rolling, or annealing processes which govern the structure of the final oriented polymer. Finally, there is the importance of molecular orientation in determining physical properties, especially mechanical properties. In this case the orientation and constitution of the noncrystalline regions may be as important or even more important than the crystalline regions. Spectroscopic methods have proved valuable in establishing links between mechanical anisotropy and molecular orientation, some examples of which will be given later in this paper.

It is not proposed to discuss here the detailed principles involved in each spectroscopic technique. Instead, we will consider which measures of orientation can be obtained and how these can be used in a general understanding of the material behavior. A more extensive account of each individual technique can be found elsewhere [1].

THEORY

The Definition of Orientation and the Determination of Orientation Functions

It is convenient to imagine that the partially oriented polymer can be regarded as an aggregate of units of structure. The orientation of a single unit can be described by three angles θ, ϕ, ψ. These define the three successive rotations required to bring into coincidence a set of cartesian axes in the unit, with a reference set of cartesian axes in the oriented polymer. The orientation situation in the polymer can then be described by an orientation distribution function $\rho(\theta,\phi,\psi)$.

A unique description of the molecular orientation may be extremely difficult to achieve, and as will be discussed later, it may not be possible to do so employing

Journal of Polymer Science: Polymer Symposium 58, 1–21 (1977)

a single spectroscopic technique. Nearly all the oriented polymer systems studied possess fiber symmetry, i.e., there is no preferred orientation in the plane perpendicular to the fiber axis or the initial draw direction. Drawing gives rise to preferred orientation of the molecular chain axis, which can now be described by an orientation distribution function $\rho(\theta)$, where θ defines the angle between the chain axis within one unit and the draw direction.

Wide angle X-ray diffraction does, in principle, allow the orientation distribution function to be obtained from the variation in intensity of suitable reflexions. Spectroscopic measurements, on the other hand, do not yield directly the orientation distribution function, but orientation averages usually called orientation functions. For example, infrared spectroscopy yields values for $\langle\cos^2\theta\rangle$, and Raman spectroscopy $\langle\cos^2\theta\rangle$ and $\langle\cos^4\theta\rangle$. At this stage it is appropriate to continue the discussion in terms of spherical harmonic functions

$$P_2(\cos\theta) = \tfrac{1}{2}(3\cos^2\theta - 1), \quad P_4(\cos\theta) = \tfrac{1}{8}(35\cos^4\theta - 30\cos^2\theta + 3), \text{ etc.}$$

The orientation distribution function is then given by

$$P(\theta) = \sum_{n=0}^{\infty} \left(n + \frac{1}{2}\right) \langle P_n(\cos\theta)\rangle \, P_n(\cos\theta)$$

and the orientation functions

$$\langle P_n(\cos\theta)\rangle = \int_0^{\pi} \rho(\theta)P_n(\cos\theta)\,\sin\theta\,\mathrm{d}\theta.$$

Table I summarizes the information obtainable from the techniques to be discussed in this paper. It follows from these considerations that spectroscopic measurements may not provide very detailed knowledge of $\rho(\theta)$. In our studies at Leeds University we have been primarily concerned with two aspects, (1) mechanical anisotropy (2) the understanding of mechanisms of deformation in drawing processes and the like.

TABLE I

Orientation Functions Obtained from Spectroscopic Measurements

Method	Orientation functions
Infra red	$\langle P_2(\cos\theta)\rangle$
Absorption dichroism	$\langle P_2(\cos\theta)\rangle$
Raman	$\langle P_2(\cos\theta)\rangle, \langle P_4(\cos\theta)\rangle$
Broad line NMR	$\langle P_2(\cos\theta)\rangle, \langle P_4(\cos\theta)\rangle$:Second moment
Polarized fluorescence	$(+P_6, P_8$:Fourth moment$)$
	$\langle P_2(\cos\theta)\rangle, \langle P_4(\cos\theta)\rangle$

Deformation Schemes

On the aggregate model the low strain mechanical anisotropy [2] also relates to the orientation functions $\langle P_2(\cos\theta)\rangle$ and $\langle P_4(\cos\theta)\rangle$. It was this which provided additional stimulus for developing broad line NMR, Raman spectroscopy and polarized fluorescence because these techniques yield $\langle P_4(\cos\theta)\rangle$ as well as $\langle P_2(\cos\theta)\rangle$. Moreover further knowledge of the distribution function is not required as far as the mechanical anisotropy is concerned.

It also transpires that a knowledge of $\langle P_2(\cos\theta)\rangle$ and $\langle P_4(\cos\theta)\rangle$ provides some understanding of the mechanism of deformation. The most extensively studied polymer deformation scheme is that proposed for a rubberlike network [3]. A simple model is to assume that the network chains consist of rotatable segments called random links. Each real chain is replaced by an equivalent random chain consisting of say N' random links. Following Kuhn and Grün [4], and later work by Treloar [5] and Roe and Krigbaum [6] $\langle P_2(\cos\theta)\rangle$ and $\langle P_4(\cos\theta)\rangle$ can be calculated as explicit functions of N' and the draw ratio λ. We have

$$\langle P_2(\cos\theta)\rangle = \frac{1}{5N'}\left(\lambda^2 - \frac{1}{\lambda}\right) + \frac{1}{25N'^2}\left(\lambda^4 + \frac{\lambda}{3} - \frac{4}{3\lambda^2}\right)$$

$$+ \frac{1}{35N'^3}\left(\lambda^6 + \frac{3\lambda^2}{5} - \frac{8}{5\lambda^3}\right)$$

using the Treloar expression for the inverse Langevin function and

$$\langle P_4(\cos\theta)\rangle = \frac{3}{175N'^2}\left(\lambda^4 - 2\lambda + \frac{1}{\lambda^2}\right)$$

$$+ \frac{216}{13475N'^3}\left(\lambda^6 - \frac{4\lambda^3}{5} - \frac{7}{5} + \frac{6}{5\lambda^3}\right) + \cdots$$

following Roe and Krigbaum, and taking only the equivalent approximation terms in the expansion.

Figure 1 shows these expressions, i.e., the orientation functions, as a function of λ for the case where $N' = 6$, which we will see is in the range estimated to be of interest for deformation of polyethylene terephthalate, which will be discussed in detail later. Two features of the nature of the curves in Figure 1 are worthy of attention. First, $\langle P_2(\cos\theta)\rangle$ is a rapidly increasing function of λ. This reflects the fact that the segmental orientation within the chain—the random link orientation—develops very slowly at first because the end-to-end distance is very much less than the fully extended chain length. Thus the orientation of the random link with respect to the draw direction initially lags much behind the orientation of the end-to-end vectors, i.e., the orientation of the lines joining the crosslink points. As the chains become extended with increasing deformation, the random links become more closely aligned with the end-to-end vectors and so their orientation accelerates. The second feature of these curves is that $\langle P_4(\cos\theta)\rangle$ is always very small in the region where the Gaussian approximation holds.

The optical birefringence gives a direct measure of $\langle P_2(\cos\theta)\rangle$ on the basis

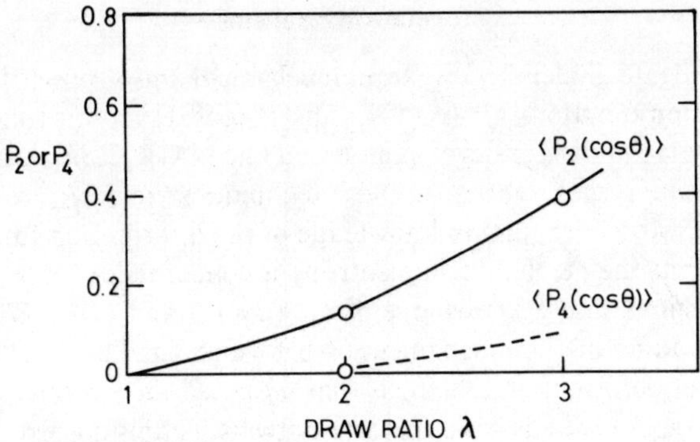

FIG. 1. $P_2(\cos\theta)$ and $P_4(\cos\theta)$ as a function of draw ratio λ for a rubber like network affine deformation scheme ($N' = 6$).

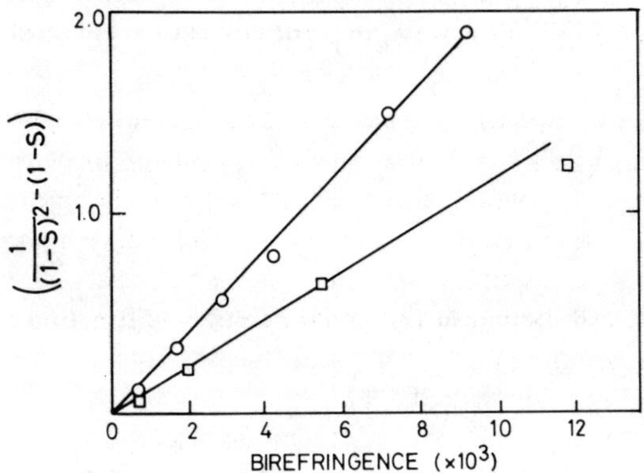

FIG. 2. Birefringence as a function of shrinkage S for spun PET fibers (after Pinnock and Ward, 1966).

of the aggregate model, and there is good experimental confirmation that the Kuhn and Grün model describes the behavior of rubbers. It also describes the deformation of amorphous polymers above the glass transition. Figure 2 shows data for polyethylene terephthalate (PET) based on a combination of birefringence and shrinkage measurements on melt spun fibers [7]. The shrinkage S is related to the extension ratio λ by the relationship $S = (\lambda - 1)\lambda$. This gives $\lambda = 1/(1 - S)$, which explains the use of the factor $\{1/(1 - S)^2 - (1 - S)\}$ as the ordinate in this figure. If the shrinkage force is also measured, the stress-optical coefficient can be used to calculate the number of monomer units per random link. The shrinkage force also provides a direct measurement of the number of chains per unit volume. We will show later how spectroscopic studies have given further understanding of the drawing behavior of PET based on such initial experiments.

It was early recognized that the deformation of polymers in the glassy state, or crystalline polymers in general, did not follow the affine deformation scheme of rubber elasticity. In particular the change in $\langle P_2(\cos\theta)\rangle$ (usually taken from measurements of the birefringence Δn, since $\Delta n = \langle P_2(\cos\theta)\rangle \Delta n_{max}$, where

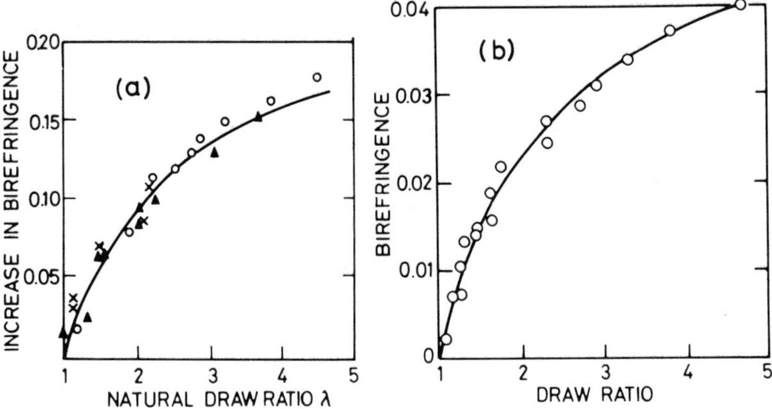

FIG. 3. Birefringence as a function of draw ratio for (a) cold draw PET and (b) low-density polyethylene.

Δn_{max} is the maximum birefringence at full orientation) contrasted immediately in that there was initially a sharp rise with increasing draw ratio at low draw ratios, turning over to a plateau at high draw, as shown in Figure 3 for cold-drawn PET and low density polyethylene [8, 9]. The deformation scheme which describes these results assumes that the polymer consists of an aggregate of transversely isotropic units whose symmetry axes rotate on drawing in the same manner as lines joining pairs of points in the bulk material, which deforms at constant volume [2]. This scheme takes the rotational part of the deformation of end-to-end vectors in the rubber network affine deformation scheme but ignores any change in length of the distances joining crosslink points. It was considered and rejected by Kuhn and Grün [4], and had been proposed earlier by Kratky [10] for crystallite orientation. It has been called pseudo-affine, and is particularly useful in predicting mechanical anisotropy for crystalline polymers and glassy polymers deformed by cold drawing, because it enables $\langle P_4(\cos\theta)\rangle$ as well as $\langle P_2(\cos\theta)\rangle$ to be calculated as a function of draw ratio. Following a previous publication [2] we have

$$\langle P_2(\theta)\rangle = \frac{1}{2}\left[\frac{2+k^2}{1-k^2} - \frac{3k\cos^{-1}k}{(1-k^2)^{3/2}}\right]$$

and

$$\langle P_4(\theta)\rangle = \frac{1}{8}\left\{\frac{35}{(1-k^2)^2}\left[1 + \frac{k^2}{2} - \frac{3k}{2(1-k^2)^{1/2}}\cos^{-1}k\right]\right.$$
$$\left. - \frac{30}{(1-k^2)}\left[1 - \frac{k\cos^{-1}k}{(1-k^2)^{1/2}}\right] + 3\right\}$$

where $k = \lambda^{-3/2}$. These expressions are shown in Figure 4, where both $\langle P_2(\cos\theta)\rangle$ and $\langle P_4(\cos\theta)\rangle$ calculated on this scheme are shown as a function of draw ratio. It is to be noted that $\langle P_4(\cos\theta)\rangle$ is comparable with $\langle P_2(\cos\theta)\rangle$, which is in complete contrast to the rubber network affine deformation scheme, where $\langle P_4(\cos\theta) \rangle$ is always very small. Measurements of $\langle P_4(\cos\theta) \rangle$ can therefore be of particular value in distinguishing different deformation schemes.

Broad Line NMR Measurements: Early Use of Aggregate Model

The first successful application [11] of broad line NMR to the determination of orientation in polymers was in low density polyethylene (LDPE). From an experimental viewpoint this was a good system to choose, because there is a large density of protons, it is possible to use reasonable power levels without saturation, and the intrinsic NMR anisotropy is high. The NMR anisotropy depends only on the spacial arrangement of the magnetic nuclei, in this case protons, and therefore relates to the crystal structure in a complicated manner. In general, it can be said that polymers with planar and regular structures, particularly where there are pairs of protons giving rise to the major internuclear interactions

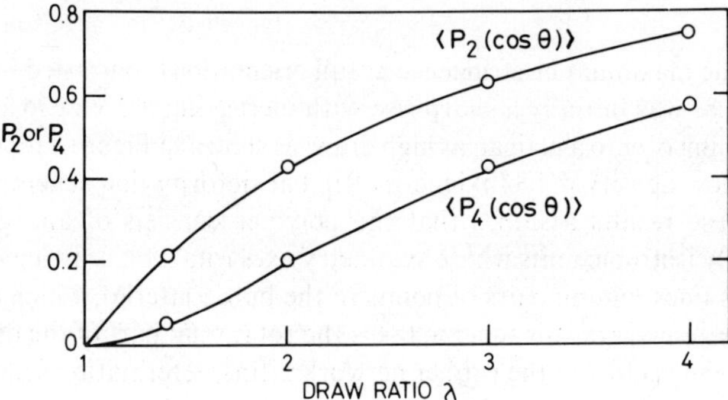

FIG. 4. $\langle P_2(\cos\theta) \rangle$ and $\langle P_4(\cos\theta) \rangle$ as a function of draw ratio λ for the pseudo-affine deformation scheme.

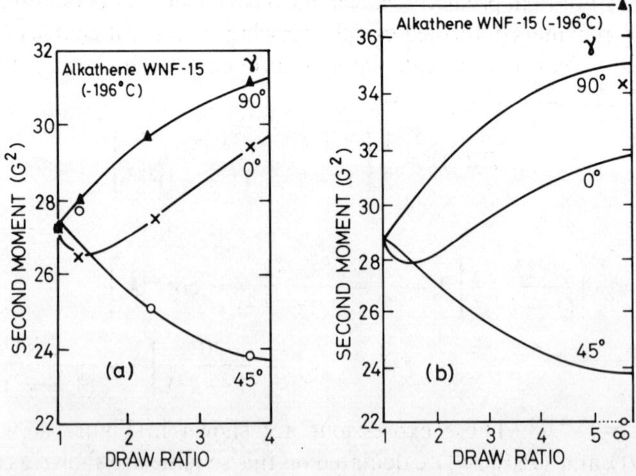

FIG. 5. (a) Observed variations of the second moment plotted against draw ratio. (b) Theoretical variation on pseudo-affine deformation scheme.

(such as LDPE, polyethylene terephthalate) will show higher anisotropy than those with crumpled or helical molecular configurations where there are many rather similar internuclear interactions (such as polymethylmethacrylate and polystyrene).

The theory of the NMR method is described in detail elsewhere [1, 11]. In the first experiments the second moment of the proton absorption line at $-196°C$ was determined as a function of the angle between the steady magnetic field and the initial draw direction, for a series of draw films. The results are shown in Figure 5, where they are compared with the predictions of an aggregate model which assumed (i) the sample was 100% crystalline—if you like composed of small groups of unit cells or crystalline grains of PE, (ii) that the mechanism of deformation was pseudo-affine. These results were sufficiently encouraging to suggest that a more effective use of the data might be to calculate orientation functions $\langle P_2(\cos\theta)\rangle$ and $\langle P_4(\cos\theta)\rangle$ directly, and use their values to predict the optical and mechanical anisotropy. The optical anisotropy relates directly to $\langle P_2(\cos\theta)\rangle$ and the best fit was achieved between the optical and NMR results for a constant level of 60% crystallinity, i.e., the anisotropy arises in both cases from orientation of the crystalline phase which forms 60% of the material. The fit to the mechanical anisotropy was very good, as shown in Figure 6.

A later paper [12] compared these results for $\langle P_2(\cos\theta)\rangle$ and $\langle P_4(\cos\theta)\rangle$ with those obtained from X-ray diffraction data and from measurements of the NMR fourth moment, which can also in principle yield values for $\langle P_6(\cos\theta)\rangle$ and $\langle P_8(\cos\theta)\rangle$. In fact the terms in $\langle P_8(\cos\theta)\rangle$ make such a small contribution to the fourth moment that it was considered satisfactory to put $\langle P_8(\cos\theta)\rangle =$ 0. Values for the other three coefficients and for the fourth moment of the amorphous phase, were chosen so as to minimize the root-mean-square deviation between the theoretical and experimental fourth moments. This procedure is analogous to that adopted for fitting the second moment data where values for

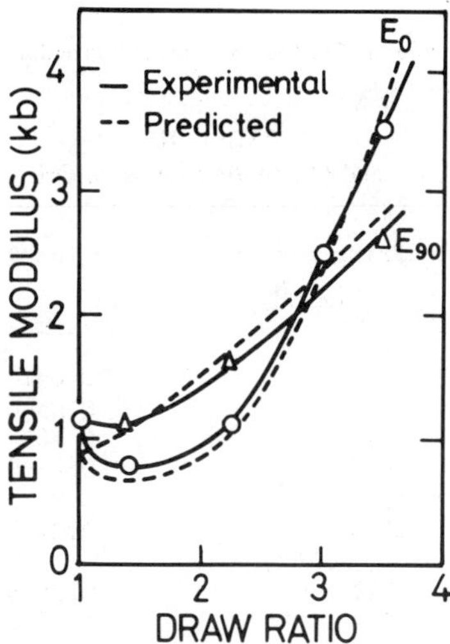

FIG. 6. Comparison of experimental and predicted variation of E_0 and E_{90} plotted against draw ratio for the two-phase model.

$\langle P_2(\cos\theta)\rangle$, $\langle P_4(\cos\theta)\rangle$ and the second moment of the amorphous phase were chosen to minimize the deviation between theoretical and experimental second moments. The collected results are shown in Table II. It is very satisfactory to note the approximate constancy of the second and fourth moments of the amorphous phase, and that second and fourth moment data yield comparable values for $\langle P_2(\cos\theta)\rangle$ and $\langle P_4(\cos\theta)\rangle$. Comparison with the X-ray diffraction values of $\langle P_n(\cos\theta)\rangle$ confirm that the NMR data are reasonably reliable with regard to $\langle P_2(\cos\theta)\rangle$ and $\langle P_4(\cos\theta)\rangle$ but cast doubt on the $\langle P_6(\cos\theta)\rangle$ value for the draw ratio 1.3 sample. There were clearly difficulties in obtaining the true best fit to fourth moment data at low degrees of anisotropy, which could not be resolved without a very considerable improvement in the accuracy of the data.

The low density polyethylene results were very encouraging with regard to the use of NMR as a technique for measuring molecular orientation. The

TABLE II
Orientation Distribution Coefficients Derived from Second and Fourth Moment Data

Draw ratio	$P_n(\cos\theta)$				$\langle\Delta H^m\rangle$ amorph		rms deviation
	n=2	n=4	n=6	n=8*	m=2 (G^2)	m=4 (G^4)	(G^4)
1.3a	0.19	−0.08	–	–	27.0	–	–
1.3b	0.19	−0.08	0.70	0	–	2000	17.8
2.3a	0.63	0.41	–	–	24.3	–	–
2.3b	0.73	0.47	0.25	0	–	1650	20.2
3.7a	0.78	0.82	–	–	24.3	–	–
3.7b	0.80	0.84	0.70	0	–	1650	10.3

[a] Based on second moment data.

[b] Based on fourth moment data.

[c] See text.

[d] The figures in the final column give the root-mean-square deviations between the calculated and experimental fourth moments.

TABLE III
Summary of $\langle P_2(\cos\theta)\rangle$ and $\langle P_4(\cos\theta)\rangle$ Results for Oriented PMMA and PVC

Polymer	Birefringence Δn $(\times 10^4)$	$\langle P_2(\cos\theta)\rangle$	$\langle P_4(\cos\theta)\rangle$
PMMA	2.2	0.0518	−0.0079
"	3.5	0.0546	−0.0306
"	5.7	0.1674	+0.0266
"	8.8	0.2350	−0.0065
"	13.2	0.3070	+0.0486
"	14.0	0.3134	−0.0184
	$(\times 10^3)$		
PVC	1.7	0.1415	0.0127
"	2.4	0.1869	0.0397
"	3.1	0.2362	0.0495

comparison with direct X-ray diffraction measurements showed that to a good approximation the NMR anisotropy arose from only the crystalline regions of the polymer. The NMR technique would, however, appear to be most valuable if it can determine *amorphous* orientation. For this reason, our subsequent efforts at Leeds University concentrated on two amorphous polymers, polymethylmethacrylate [13] (PMMA), and polyvinylchloride [14] (PVC) and polyethylene terephthalate [15] (PET) in which oriented samples of low crystallinity can be prepared.

The problem with amorphous polymers is that there is no crystalline lattice from which the lattice sums can be calculated. In PMMA and PVC, the equivalent lattice sums were calculated assuming that (1) the intramolecular proton interactions can be calculated on the basis of a model for the configuration of the chain (a helix in PMMA, a planar zig-zag in PVC), (2) the intermolecular proton interactions are isotropic.

The intermolecular interactions were not calculated on the basis of a model, but assumed to give a small isotropic contribution to the second moment. This was determined from the NMR data by a best fit procedure, and shown to be consistent for all samples. In PMMA a reasonably extensive set of data were obtained from samples whose birefringence and mechanical anisotropy had already been determined. Figure 7 shows that there is a good correlation between $\langle P_2(\cos\theta)\rangle$ found from NMR and birefringence, which is one test we can apply to the results. The predicted pattern of mechanical anisotropy, which uses both $\langle P_2(\cos\theta)\rangle$ and $\langle P_4(\cos\theta)\rangle$, also agreed well with that determined experimentally. In PVC only birefringence data were available, and again these correlated well with the $\langle P_2(\cos\theta)\rangle$ values.

Finally, in view of the relevance of $\langle P_4(\cos\theta)\rangle$ values to the deformation mechanism, it is interesting to summarize values of $\langle P_2(\cos\theta)\rangle$ and $\langle P_4(\cos\theta)\rangle$ for those polymers in Table III. It is notable that the $\langle P_4(\cos\theta)\rangle$ values are always comparatively small, as expected. In PMMA the actual values fluctuate around zero, which would be consistent with expectations based on deformation of a rubber network, accepting a certain amount of experimental error. In PVC, on the other hand the actual numbers are always positive and increase systematically with increasing $\langle P_2(\cos\theta)\rangle$. In fact, as remarked previously by examining $\langle \cos^2\theta \rangle$ and $\langle \cos^4\theta \rangle$ (ref. 1, p. 231) the results for PVC are more consistent with the pseudo-affine deformation scheme.

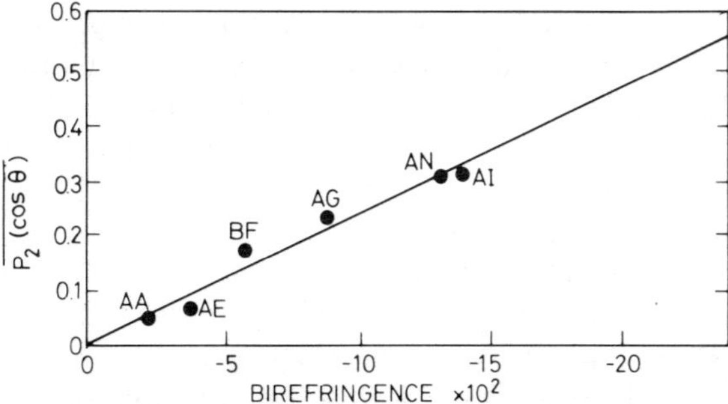

FIG. 7. Variation of $P_2(\cos\theta)$, obtained from NMR measurements, with the birefringence. Circles are experimental results while line is predicted from the aggregate theory.

These results for amorphous polymers suggested that it would now be of interest to study a crystalline polymer like PET where it was considered that evidence such as optical anisotropy showed that the noncrystalline regions could be highly oriented. A set of samples was examined [15], mostly one way drawn sheets, but including a transversely isotropic drawn rod. The second moment anisotropy was compared with optical measurements and X-ray data. It was necessary in the first instance to carry out a detailed examination of the second moment anisotropy for the case of general orthorhombic symmetry. It was shown that it is not possible to determine the distribution function $\rho(\theta,\phi,\psi)$ up to $l = 4$ completely from NMR second moment data, unless some simplifying assumptions are made with regard to the distribution in θ, ϕ or ψ (e.g., random ψ). The physical meaning is that since we measure second moments through only two parameters, say θ and ϕ, it is not possible to obtain information about the distribution in ψ. It was therefore decided to attempt an analysis of NMR data for sheets of low draw ratio where it is known that there is very little crystallization. It was assumed that there was an orthorhombic distribution of transversely isotropic structural units, which implies physically that only chain-axis orientation occurs during deformation. These assumptions had previously been found to give a very adequate prediction of the optical anisotropy in studies of deformation bands. However, the fit to the NMR data was poor, and this was also true for the uniaxially oriented rod. It was appreciated that these difficulties arise because of changes in molecular configuration which can occur on drawing PET. The aggregate model with its fixed set of lattice sums, in this case based on crystalline PET, breaks down because there are changes in the trans/gauche content of the polymer as well as changes in orientation. We will therefore now leave the use of the aggregate model and NMR isotropy until we have had a full discussion of the conformational changes. It will be shown that, knowing the conformational situation, it is possible to interpret the NMR anisotropy for the low draw sheets and the uniaxially oriented rod. The situations of preferred orientation, however, reveal a fundamental weakness in the NMR method, and again suggest that only by a combination of techniques will a satisfactory description of the orientation situation be achieved.

The Measurement of Orientation in PET

PET is an extremely suitable polymer for detailed comparative studies of molecular orientation for a number of reasons. First, it is available in a number of distinctly different physical states, from an isotropic amorphous glass to a highly oriented, highly crystalline fiber or sheet, with suitable intermediate situations. It is of particular value to be able to compare a highly oriented poorly crystalline sample with a sample of similar overall orientation (as determined by birefringence) but high crystallinity.

Secondly, PET can be readily fabricated into transparent oriented sheets, sufficiently good optically for measurements of optical dichroism, polarized fluorescence, infrared and Raman spectroscopy. Finally, some progress has already been made towards unravelling the deformation behavior of the polymer

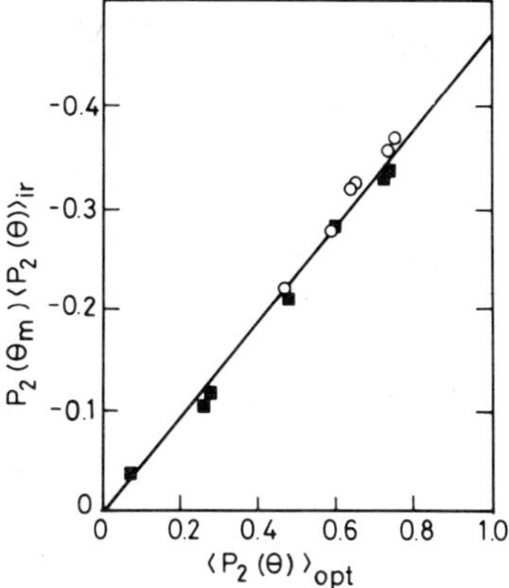

FIG. 8. Values of $P_2(\theta_m)\langle P_2(\theta)\rangle_{ir}$ for the 895 cm^{-1} band as a function of $\langle P_2(\theta)\rangle_{opt}$. O, Series (i); ■, series (ii).

using stress–optical measurements [7], and comparison of the optical and me-chanical anisotropy with the model deformation schemes discussed above.

Infrared measurements yield information on molecular orientation in a very direct manner. If the dipole transition moment vector makes an angle θ_m with the chain axis, and we choose a system where there is fiber symmetry, the in-frared data provide values for the product $P_2(\cos\theta_m)\langle P_2(\cos\theta)\rangle$ multiplied by a constant, which can be determined either by an independent calibration or from considerations of the structure. In PET it is valuable to examine bands associated with benzene ring mode vibrations, and also with the trans and gauche confor-mations of the glycol residue. The infrared data were calculated using the Lo-rentz–Lorenz internal field correction, which made small but significant dif-ferences to the numerical values obtained [16].

Two series of samples were examined, one series by drawing isotropic film to a draw ratio of 4:1 at different temperatures, and the second by drawing the same film to different draw ratios at 80°C [17]. There was excellent agreement between the infrared orientation functions for benzene ring bands and those obtained from optical measurements, as illustrated in Figure 8. This reflects the fact that optical anisotropy primarily measures overall orientation and is relatively insensitive to the trans/gauche conformational situation. It is therefore of greater interest to use the infrared method to examine the orientation and concentration of trans and gauche conformers for each sample. The results are summarized in Figures 9 and 10, where for convenience $\langle P_2(\theta)\rangle_{opt}$, the orien-tation function obtained from the optical measurements, is used as a measure of overall chain orientation. It can be seen from Figure 9 that the trans orien-tation function increases rapidly with increasing chain orientation, whereas the gauche orientation function is always small. Detailed examination of the results showed that to a good approximation

$$\langle P_2(\theta)\rangle_{opt} = f_{trans} \langle P_2(\theta)\rangle_{ir\ trans}$$

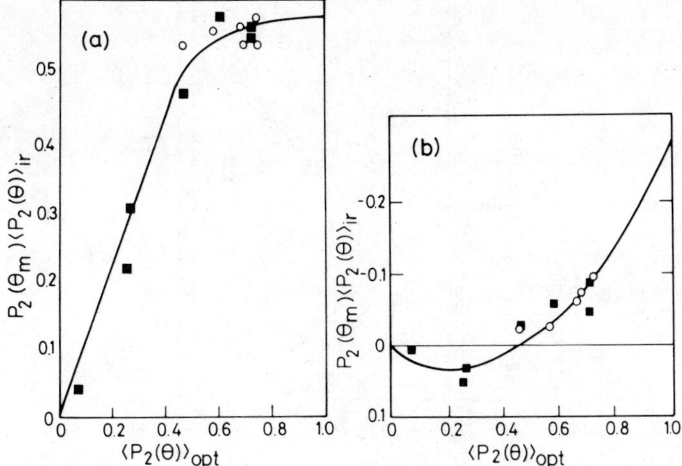

FIG. 9. Values of $P_2(\theta_m)\langle P_2(\theta)\rangle_{ir}$ for the 975 cm^{-1} trans band (a) and the 896 cm^{-1} gauche band (b) as a function of $\langle P_2(\theta)\rangle_{opt}$. O, Series (i); ■, series (ii).

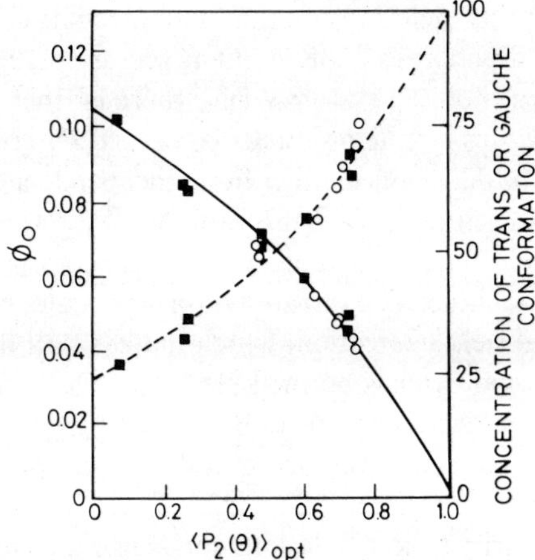

FIG. 10. Values of ϕ_0 and concentration for the trans conformation (975 cm^{-1} band) and the gauche conformation (896 cm^{-1} band) as a function of $\langle P_2(\theta)\rangle_{opt}$. - - -, trans; —, gauche. O Series (i) ■, series (ii).

where f_{trans} is the trans mass fraction. This equation suggests that the overall chain orientation arises from orientation of trans chain sequences only. We have recently confirmed this interpretation of the situation by a re-examination of the broad line NMR data on oriented PET [18]. If the model is correct the value of the coefficient B_{20}, which can be determined directly from the second moment anisotropy, should be a linear function of $\langle P_2(\theta)\rangle_{opt}$. Figure 11 shows that this expectation is fulfilled, the experimental points being close to the theoretical boundary for wholly trans oriented material.

A second remarkable feature of the results is the exact similarity of the data

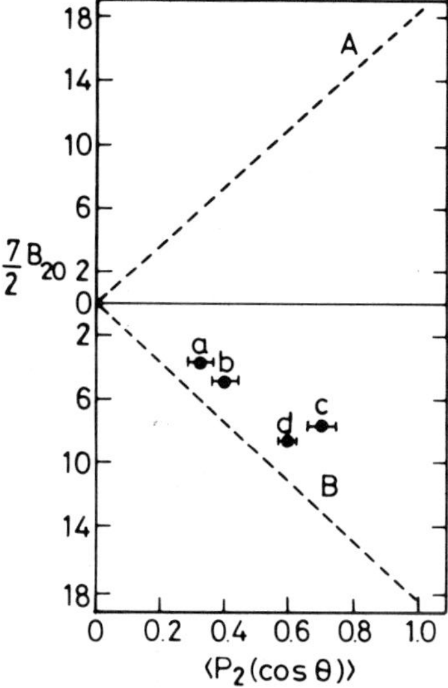

FIG. 11. $7/2 \, B_{20}$ plotted against $\langle P_2(\cos\theta)\rangle^{\text{overall}}$ together with the theoretical boundaries (A) for wholly gauche oriented material. Draw ratios (a) 2:1; (b) 2.5:1; (c) 5:1 for two-way drawn film; (d) 3.25:1 uniaxial drawn rod.

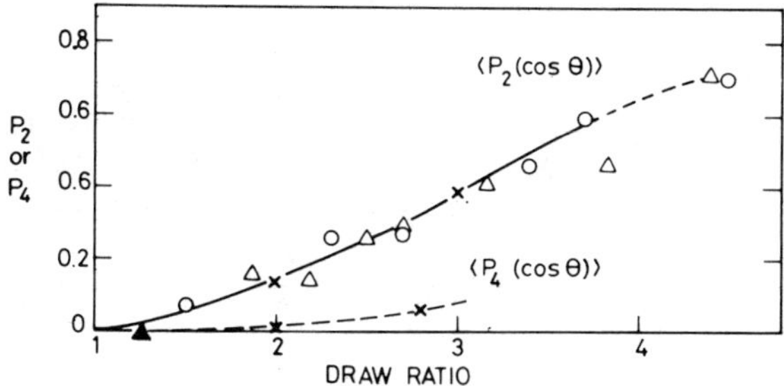

FIG. 12. Comparison of $\langle P_2(\cos\theta)\rangle$ determined by infrared 0 and Raman $\triangle$ spectroscopy with rubber network model X—also shown are $\langle P_4(\cos\theta)\rangle$ determined by Raman ▲ and calculated on the rubber network model X.

for the two series of samples, as shown in Figure 10, which suggests that there is a unique relationship between the concentration of trans and gauche conformers and the overall chain orientation. A simple explanation of this result is that both conformational content and the molecular orientation relate to the overall strain, irrespective of the draw conditions. This view is confirmed by a detailed examination of the data for drawing at 80°C. In Figure 12, values of $\langle P_2(\theta)\rangle$ for the chain orientation are plotted as a function of draw ratio together with points calculated on the basis of Treloar's exact expression for the random link network, using a value for N' of 6, which is in a good agreement with that obtained from the stress–optical data [7] mentioned previously. Also shown in Figure 12 are values of $\langle P_2(\cos\theta)\rangle$ obtained from Raman spectroscopy, using the line at 1616 cm^{-1}. Detailed vibrational assignments for PET and model compounds by Wilding [19] at Leeds University suggest that this line should be assigned to the benzene ring mode 8a (notation in reference 19). The Raman tensor has one of its principal axes (corresponding to α_3) parallel to the line of parasubstitution in the benzene ring, and one perpendicular to the plane of the ring, corresponding to α_1 or α_2. The Raman tensor is assumed to be cylindrically symmetrical ($\alpha_1 = \alpha_2$) and the angle between the α_3 direction and the chain axis calculated to be 19°12′ from crystal structure data. Incorporating data from random samples it is then possible to calculate $\langle P_2(\cos\theta)\rangle$ and $\langle P_4(\cos\theta)\rangle$ for the chain orientation [21, 22]. The results for $\langle P_2(\theta)\rangle$ confirm the values obtained for chain orientation from infrared and optical measurements. We have also shown some values for $\langle P_4(\cos\theta)\rangle$ in Figure 12. Although these are sufficiently small at very low draw ratios to be consistent with the rubber model, they do begin to move towards pseudo-affine values at comparatively low draw. If the accuracy of the data can be somewhat improved, there is clearly room for further detailed studies in this region.

Finally it was noted in the infrared studies that the trans/gauche ratio changed with draw ratio in the manner suggested by the end-to-end distance expansion formula of Abe and Flory [23]. Results for the samples drawn at 80°C (Figure 13) confirm the general validity of Flory's proposals, and it would be valuable to carry this work further and calculate the proportionality constant on the basis of estimated internal energies of the different conformational states.

So far we have discussed the measurement of orientation in PET under particularly straightforward situations, usually where there is low crystallinity and often when the deformation takes place as for a rubber network. With falling draw temperature, the mechanism of deformation changes to pseudo-affine [as shown in Figure 3(a)] and although the theoretical modelling for this cold drawing was worked out [24], it is only recently that we have begun to explore the detailed mechanics of orientation, and work is proceeding.

The analysis of crystalline oriented PET has, of course, been the subject of many studies, especially the question of preferred orientation [25, 26]. Although the spectroscopic studies are valuable in many respects, they do not distinguish between molecules in a crystalline environment and those in an amorphous environment.

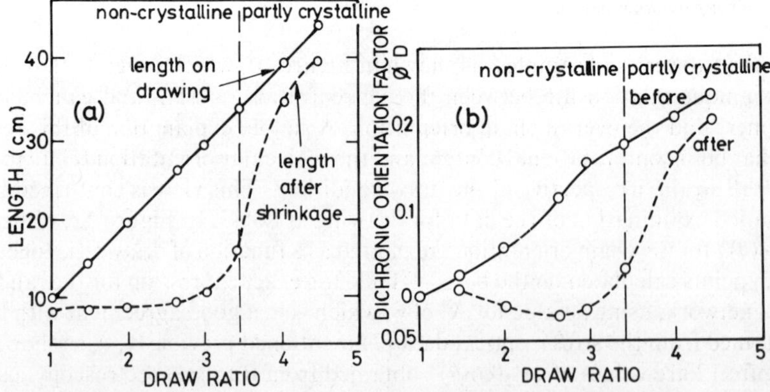

FIG. 13. Comparison of length of 10 cm monofilament (a) with dichroic orientation factor (b) before — and after - - - shrinkage.

There are no gauche conformations in the crystalline regions, but trans conformations can exist in both crystalline and noncrystalline regions [27, 28]. There is therefore merit in attempting to devise methods which will distinguish between the orientation of the crystalline and noncrystalline regions. At Leeds University we have explored the use of optical dichroism and polarized fluorescence, both of which examine the orientation of dye molecules introduced into PET at the melt spinning stage but which are hopefully only in the noncrystalline regions in the final drawn fiber or tape. The recent work relates to some much earlier studies of dichroism [29] and to the pioneering fluorescence investigations of Nishijima and co-workers [30].

The theoretical analysis of optical dichroism is analogous to that for infrared absorption. The dichroism measurements therefore yield values for $\langle P_2(\cos\theta)\rangle_{\text{dye}}$, the orientation of the dye molecules, where θ is the angle between the absorption direction and the draw direction. The earlier investigations [29] showed interesting correlations between shrinkage and the reduction in the dichroic orientation function (Figure 13). It was concluded that it is reasonable to assume that this is because both reflect primarily the disorientation of the noncrystalline regions of the fiber.

It was also found that the relationship between the orientation functions obtained from optical dichroism and from birefringence underwent a discontinuous change with the onset of crystallinity, i.e., for a series of drawn yarns at draw ratios greater than 3. This was again attributed to the dye molecules primarily reflecting the orientation of the noncrystalline regions, but no absolute orientation measurements were attempted.

In our most recent work [31, 32], we have made a detailed study of the polarized fluorescence method using a stilbene derivative at concentrations of 50–200 ppm in PET. The polarized fluorescence was first studied in a series of drawn samples of very little crystallinity. Excellent agreement was obtained between values of $\langle P_2(\cos\theta)\rangle$ obtained from polarized fluorescence and from optical dichroism measurements on the same system. The fluorescence orientation functions were also directly compared with those based on optical anisotropy, which we have shown above as an excellent measure of overall chain orientation in PET. Although the orientation functions obtained from these two techniques were uniquely related, the fluorescence measurements suggest that the dye molecules are more highly oriented than the PET chains. Figure 14 shows $\langle P_2(\cos\theta)\rangle_{\text{dye}}$ as a function of $\langle P_2(\cos\theta)\rangle_{\text{opt}}$. There are two possible lines of explanation of this result, which may indeed be two sides to the same explanation. First, we note that the shape of Figure 14 is very similar to that of Figure 9(a) where the trans orientation function was plotted as a function of $\langle P_2(\cos\theta)\rangle_{\text{opt}}$. It could well be that the dye molecules settle into regions where the molecular chains are in the extended trans conformation, i.e., the more highly ordered amorphous regions. Alternatively, similar values for $\langle P_2(\cos\theta)\rangle_{\text{dye}}$ are obtained if it is assumed that the dye molecules do not measure the segmental orientation in the chain, but take the average orientation of a number of segments along the length of the chain. Thus they are more highly oriented, in just the way that the vectors joining the crosslink points are more oriented than the random links, in the random link network model, as discussed above.

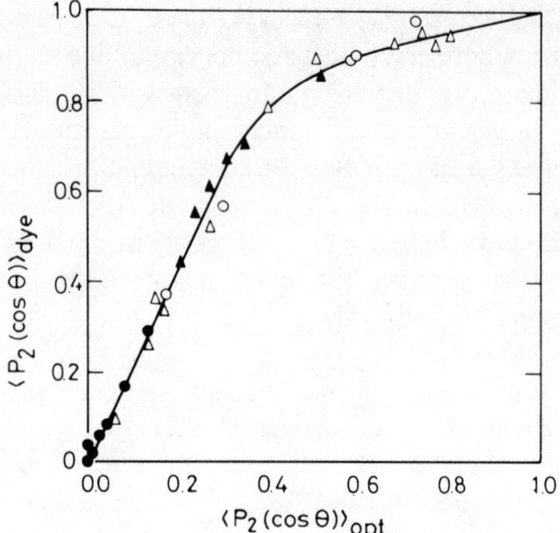

FIG. 14. $\langle P_2(\cos\theta)\rangle_{dye}$ compared with $\langle P_2(\cos\theta)\rangle_{opt}$ for samples of low crystallinity. △, samples drawn at 80°C to various draw ratios; ▲, samples drawn to draw ratio 2.66 at various temperatures between 65 and 90°C; ○ samples drawn at 80°C to various draw ratios; ● samples drawn and shrunk at 80°C.

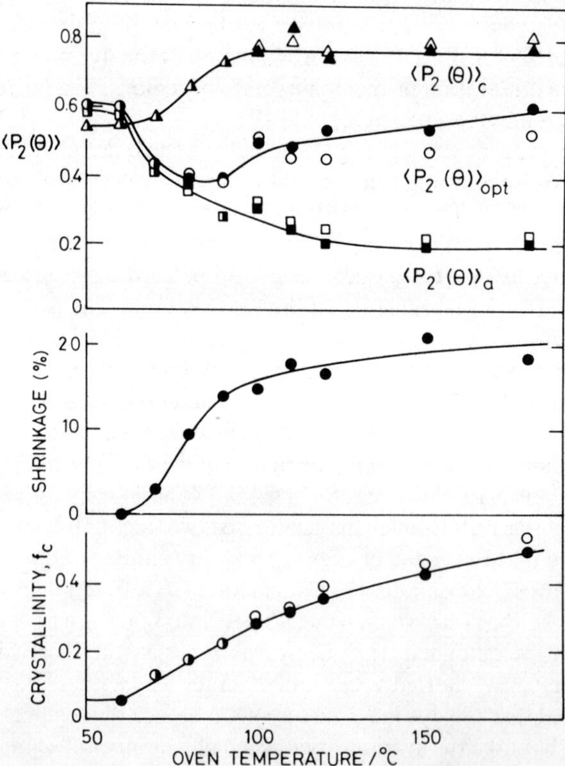

FIG. 15. $\langle P_2(\theta)\rangle_c$, $\langle P_2(\theta)\rangle_{opt}$, $\langle P_2(0)\rangle_a$, crystallinity and shrinkage plotted against oven temperature for samples initially drawn to $\lambda = 3.4$ at 80°C and subsequently shrunk freely for 10 min in an air oven.

The polarized fluorescence measurements were combined with optical measurements and X-ray diffraction data to obtain the overall chain orientation and the orientation of the crystalline regions as well as that of the noncrystalline regions, on the assumption that this is obtained uniquely from the polarized fluorescence data. The expected relationship between these three measures of orientation and the degree of crystallinity was confirmed, which justified the assumptions made in deriving $\langle P_2(\cos\theta)\rangle_a$ from the polarized fluorescence measurements. It was again shown that the orientation in drawn samples is consistent with stretching a rubber network—but more novel was the exploration of the shrinkage behavior. Some key results are summarized in Figure 15, where $\langle P_2(\cos\theta)\rangle$ for the various techniques are compared with the shrinkage and crystallinity of samples originally drawn to draw ratio 3.4 and subsequently allowed to shrink at a range of temperatures. It is to be noted that the shrinkage rises rapidly from 0 to 15% as the oven temperature rises from 60° to 90° and then much less rapidly with increasing temperature. The crystallinity, on the other hand, increases gradually with increasing temperature over the whole temperature range. $\langle P_2(\cos\theta)\rangle_{\mathrm{opt}}$ shows a clear minimum for oven temperatures near 80°C. It can be concluded from this that shrinkage is associated with a *decrease* in overall molecular orientation, whereas crystallization produces an increase in molecular orientation, which in these samples is the major effect above 100°C. For samples shrunk at 80°C which do not crystallize, it is clear that shrinkage is primarily associated with disorientation of the amorphous regions, as concluded from the dichroism studies. Statton [33] has proposed that shrinkage should be attributed to the refolding of chains which have been pulled out in the drawing process. These results, however, suggest that shrinkage is essentially disorientation of the amorphous regions, and that crystallization and refolding occur after this.

The fluorescence method is exactly analogous to Raman spectroscopy. We can therefore obtain values for $\langle P_4(\cos\theta)\rangle$ as well as $\langle P_2(\cos\theta)\rangle$. In our latest studies [32] we have examined values for $\langle P_4(\cos\theta)\rangle$ obtained from polarized fluorescence and X-ray diffraction measurements on samples after drawing and subsequent shrinkage. The results suggest that the overall molecular orientation is always close to the value expected for shrinkage occurring before crystallization, and the $\langle P_4(\cos\theta)\rangle$ results confirm those obtained from $\langle P_2(\cos\theta)\rangle$. We are at present examining the mechanical properties, especially the initial modulus of such samples, to see if the aggregate model for low strain behavior is applicable to a wide range of drawn and shrunk samples. The preliminary results are remarkably encouraging in this respect.

Molecular Orientation in Oriented Polymers of High Crystallinity, Including Ultra-High-Modulus Polymers

We have so far considered the development of molecular orientation in systems where continuum models for the deformation processes can be demonstrated to be applicable. The systems discussed include several amorphous polymers, together with PET where crystallization has occurred subsequent to the major orientation processes. In all these cases an aggregate model can be used to define the degree of orientation, even if the exact nature of the deformation process departs from the affine or pseudo-affine deformation schemes. An example of this is low density polyethylene where the deformation is nonaffine but the molecular orientation can still be described simply.

It is, however, also necessary to define molecular orientation in highly crystalline polymers such as high density PE and polypropylene (PP). In these polymers there are substantial morphological features—the breakdown of the spherulitic texture, the unfolding of folded chain molecules, etc. It was early recognized that such systems could not be adequately described by an aggregate model, certainly not beyond the initial stages of the deformation, i.e., as soon as lamellar breakdown occurs [34].

Broad line NMR can be of especial value in characterizing such materials, particularly ultra-high-modulus polymers, where this has been achieved by very high draw, with draw ratios of 30 and greater. It was first noted by Hyndman and Origlio [35] that drawn PE and PP showed a three component composite structure. There was a broad crystalline component, a narrow "amorphous" component and an intermediate component which was attributed to "strained amorphous material." We have found broad line NMR very useful in characterizing ultra-high-modulus high density PE [36]. In the first place, beyond draw ratios of ~15, the crystalline regions are virtually fully oriented. Moreover, infrared measurements of molecular orientation show that both the orientation functions and the concentration of gauche isomers reach constant values in this range within the accuracy of our measurements [37]. It is therefore very valuable to find that the NMR spectrum can provide quantitative information on the changes in the proportion of the rigid crystalline material, and changes in molecular orientation of the noncrystalline material. Second, it appears that the NMR components relate to molecules of different molecular mass. We have indeed proposed that the intermediate component corresponds to high molecular weight molecules which interconnect the crystalline regions and are sufficiently constrained to allow motion about the chain axis only. This component shows an increasing degree of molecular orientation and decreases in intensity with increasing draw ratio. Some typical results are shown in Figure 16. Finally, the intensity of the isotropic narrow component, attributed to the mobile fraction, correlates with the amount of low molecular weight material, and supports the proposal that this may act as a plasticizer in the drawing process.

Another spectroscopic technique which has proved useful in characterizing the ultra-high-modulus polymers is Raman spectroscopy. The longitudinal acoustic mode gives rise to a line very close to the exciting line whose frequency relates to the lamellar thickness and whose intensity and polarization reflect the proportion and orientation, respectively, of the lamellar material. With increasing draw ratio in the range 10–20, the lamellar texture is disrupted as the chains unfold. This can be monitored by the changes in the LA mode Raman line, as shown in Figure 17 [38].

The importance of these spectroscopic measurements on ultra-high-modulus polymers is that they enable quantitative understanding to develop of the

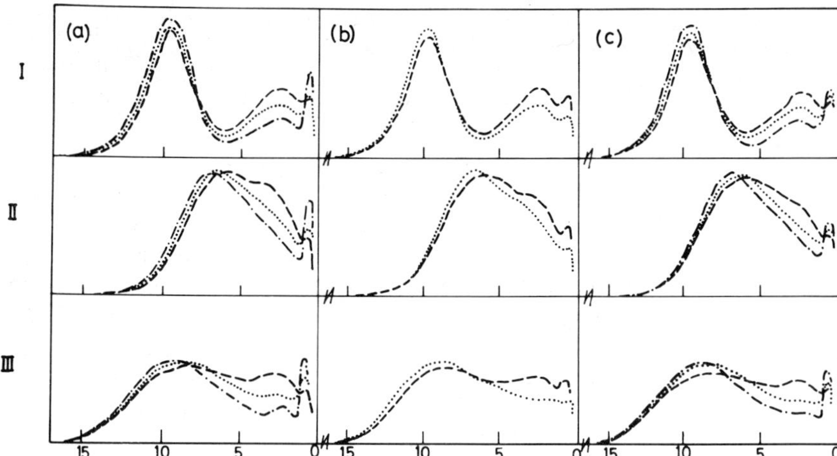

FIG. 16. Derivative wideline NMR spectra for the drawn Rigidex materials for the three orientations with respect to the static magnetic field $\gamma = 0°$, $45°$, and $90°$. (a) Rigidex 50; (b) Rigidex 25; (c) Rigidex 140–60. - - - draw ratio $\lambda = 11$; ······ draw ratio $\lambda = 19$; - · - · - draw ratio $\lambda = 30$; I, $\gamma = 0$; II, $\gamma = 45$; III, $\gamma = 90$. The abscissae are in G and the ordinates represent absorption derivative normalized to unit (integrated) intensity. Only the low field (left-hand) halves of the spectra are shown.

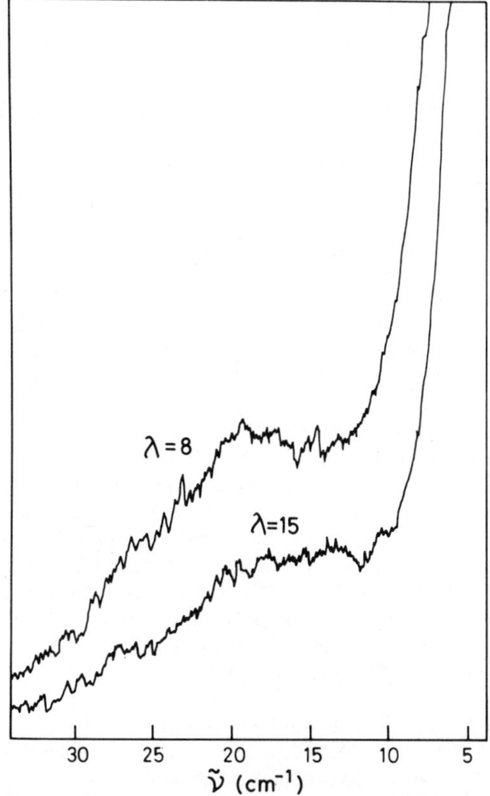

FIG. 17. Normalized intensities of the LA mode lines in the oriented linear PE (Rigidex 50 Grade) for draw ratio $\lambda = 8$ and 15.

structure/property relationships, as in the case of the much less well oriented amorphous polymers and the oriented crystalline polymers of lower crystallinity. The dynamic mechanical behavior of ultra-high-modulus high density PE has already been interpreted using the NMR data described above [39].

CONCLUSION

In this paper a broad survey of the use of spectroscopic techniques in characterizing oriented polymers has been attempted. It can be seen that the subject is now entering the truly quantitative phase for the first time. Considerable progress has been made in developing broad line NMR, infrared and Raman spectroscopy, absorption dichroism, and polarized fluorescence so that they can provide orientation functions for definitive attempts to establish (a) structure/property relationships for the oriented polymers, (b) detailed understanding of deformation mechanisms in fabrication processes. In all these cases spectroscopic measurements are at least complementary to X-ray diffraction measurements, and in some cases they are much more informative.

Finally, spectroscopic methods have much to contribute to the deformation of very crystalline polymers. The case of ultra-high-modulus high density PE has been highlighted as an area for exciting new developments in this respect.

REFERENCES

[1] *Structure and Properties of Oriented Polymers,* I. M. Ward, Ed., Applied Science Publishers, London, 1975, chaps. 1, 4, 5, 6.
[2] I. M. Ward, *Proc. Phys. Soc.* **80,** 1176 (1962).
[3] L. R. G. Treloar, *The Physics of Rubber Elasticity,* 3rd. Ed., Clarendon Press, Oxford, 1975.
[4] W. Kuhn and W. Grün, *Kolloid-Z.* **101,** 248 (1942).
[5] L. R. G. Treloar, *Trans. Faraday Soc.* **50,** 881 (1954).
[6] R. J. Roe and W. R. Krigbaum, *J. Appl. Phys.* **35,** 2215 (1964).
[7] P. R. Pinnock and I. M. Ward, *Trans. Faraday Soc.* **62,** 1308 (1966).
[8] S. W. Allison and I. M. Ward, *Br. J. Appl. Phys.* **18,** 1151 (1967).
[9] S. M. Crawford and H. Kolsky, *Proc. Phys. Soc.* **B64,** 119 (1951).
[10] O. Kratky, *Kolloid-Z.* **64,** 213 (1933).
[11] V. J. McBrierty and I. M. Ward, *Br. J. Appl. Phys.* (*J. Phys. D*) **1,** 1529 (1968).
[12] V. J. McBrierty, I. R. McDonald, and I. M. Ward, *J. Phys. D. Appl. Phys.* **4,** 88 (1971).
[13] M. Kashiwagi, M. J. Folkes, and I. M. Ward, *Polymer* **12,** 697 (1971).
[14] M. Kashiwagi and I. M. Ward, *Polymer* **13,** 145 (1972).
[15] M. Kashiwagi, A. Cunningham, A. J. Manuel, and I. M. Ward, *Polymer* **14,** 11 (1973).
[16] A. Cunningham, G. R. Davies, and I. M. Ward, *Polymer* **15,** 743 (1974).
[17] A. Cunningham, I. M. Ward, H. A. Willis, and V. Zichy, *Polymer* **15,** 749 (1974).
[18] A. Cunningham, A. J. Manuel, and I. M. Ward, *Polymer* **17,** 125 (1976).
[19] M. A. Wilding, Ph.D. Thesis, Leeds University, 1975.
[20] G. Varsanyi, *Vibrational Spectra of Benzene Derivatives,* Academic Press, New York, 1969, p. 152.
[21] J. Purvis, D. I. Bower, and I. M. Ward, *Polymer* **14,** 398 (1973).
[22] J. Purvis and D. I. Bower, *J. Polym. Sci., Polym. Phys. Ed.* **14,** 1461 (1976).
[23] Y. Abe and P. J. Flory, *J. Chem. Phys.* **52,** 2814 (1970).
[24] I. M. Ward, *Br. J. Appl. Phys.* **18,** 1165 (1967).
[25] C. F. Heffelfinger and R. C. Burton, *J. Polym. Sci.* **47,** 289 (1960).

[26] W. J. Dulmage and A. L. Geddes, *J. Polym. Sci.* **31,** 499 (1958).

[27] I. M. Ward, *Chem. Ind.* 905 (1956).

[28] D. Grime and I. M. Ward, *Trans. Faraday Soc.* **54,** 959 (1958).

[29] D. Patterson and I. M. Ward, *Trans. Faraday Soc.* **53,** 1516 (1957).

[30] Y. Nishijima, Y. Onogi, and T. Asai, *J. Polym. Sci.,* (*C*) **15,** 237 (1966).

[31] J. H. Nobbs, D. I. Bower, I. M. Ward, and D. Patterson, *Polymer* **15,** 287 (1974).

[32] J. H. Nobbs, D. I. Bower, and I. M. Ward, *Polymer* **17,** 25 (1976).

[33] W. O. Statton, J. L. Koenig, and M. Hannon, *J. Appl. Phys.* **41,** 1069 (1970).

[34] P. R. Pinnock and I. M. Ward, *Br. J. Appl. Phys.* **17,** 575 (1966).

[35] D. Hyndman and G. F. Origlio, *J. Polym. Sci.* **36,** 556 (1959).

[36] J. B. Smith, A. J. Manuel, and I. M. Ward, *Polymer* **16,** 57 (1975).

[37] A. J. Whitaker (unpublished work).

[38] D. I. Bower, M. A. Wilding, G. Capaccio, and I. M. Ward (unpublished work).

[39] J. B. Smith, G. R. Davies, G. Capaccio, and I. M. Ward, *J. Polym. Sci., Polym., Phys. Ed.* **13,** 2331 (1975).

METHODS OF CHARACTERIZATION AND MEASUREMENT OF MOLECULAR ORIENTATION

M. MAY

Technische Hochschule "Carl Schorlemmer" Leuna - Merseburg
Sektion Werkstofftechnik, DDR

INTRODUCTION

We issue from the fact that the physical and technical properties of high polymer materials can be greatly influenced by the manufacturing process. In this process one of the essential changes is caused by preferred orientation. Therefore, it is necessary to obtain an exact knowledge of orientations and their consequences, in order to be able to draw conclusions concerning the manufacturing process and the behavior of the material.

Some of these influences which, after stretching, result in a high anisotropy are represented in Figures 1(a)–1(c), and 2(a)–2(c).

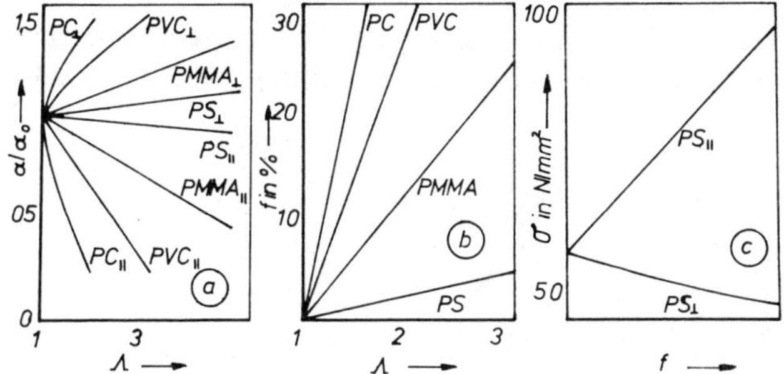

FIG. 1. (a) Anisotropie of polymers after stretching parallel ‖ and perpendicular ⊥ to the direction of stretching. Change of the coefficient of thermal expansion. (b) Effect on the degree of orientation. (c) Bending strength parallel and perpendicular to the direction of orientation in polystyrene.

FIG. 2. Orientation and stress crazing in an injection moulded polystyrene-plate. (a) Degree of orientation. (b) Direction of orientation. (c) Picture of orientation birefringence. (d) Susceptability to stress crazing by ball indentation method.

Journal of Polymer Science: Polymer Symposium, 23–42 (1977)

These few examples may suffice to demonstrate the great interest of materials engineering in the knowledge of orientation and texture in general, as well as the fact that these influences are far more important in high polymer materials as compared with metals.

ROUTINE MEASUREMENTS

A series of routine measuring methods were developed which at present are generally known. They are employed in technology, yielding useful results, but are often of a limited applicability, or their measuring techniques cannot be mastered exactly.
This will be illustrated by some examples:

(a) In amorphous high polymers, the degree of orientation f_x according to X-ray pictures is determined by an abridged method from the intensity distribution $I = f(\psi)$ at $\vartheta = $ constant, and ψ from 0° to 360°, based on the amorphous scattering curve $I = f(\vartheta)$, $f_x = (1 - \overline{h}/h)\cdot 100\%$ (Fig. 3). Such a determination may lead to completely erroneous results if the basic character of the texture of the sample investigated is unknown; if the texture is known it would have to be normalized to a certain angle of tilt φ.

(b) Extensive statements concerning the texture can be obtained most favorably by a pole figure. This is done by tilting the sample additionally by

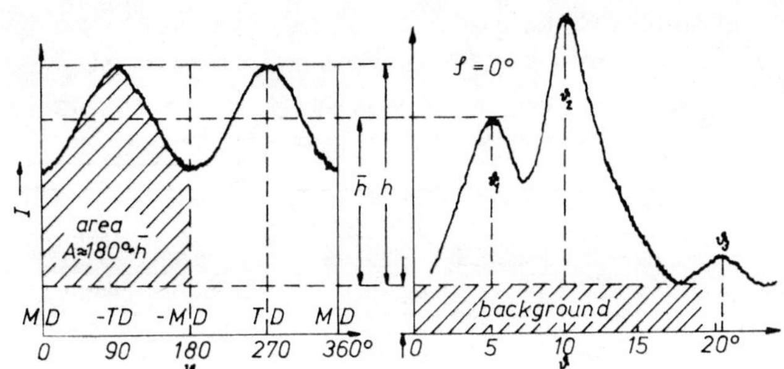

FIG. 3. Determination of the degree of orientation from the amorphous scattering curve $I = f(\vartheta)$ of the azimuthal intensity distribution $I = f(\psi)$.

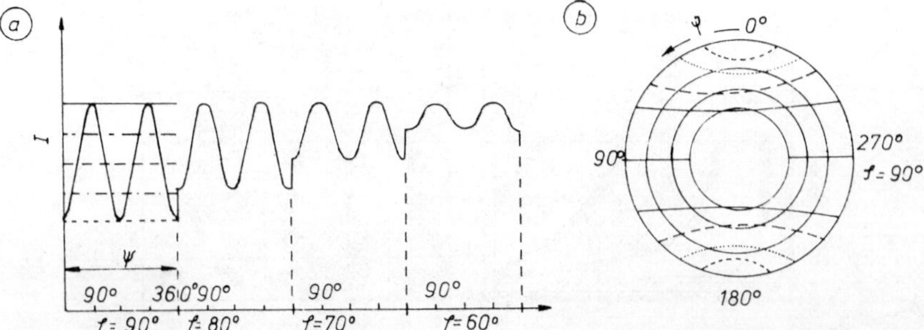

FIG. 4. Plotting of a pole figure: (a) diagram of texture and $I = f(\psi,\rho)\vartheta = $ const.; (b) transfer of the intensity values to the stereographic projection.

an angle φ. [ϑ (ϑ remaining constant, so that the function $I = f(\varphi,\psi)$ is plotted at ϑ = constant (Fig 4)].

Figure 5 [7] shows two pole figures of polystyrene obtained from the same place of the specimen. However, 5(a) was measured with ϑ_1 and 5(b) with ϑ_2. Compared with the direction of flow FD, the diagrams provide unequivocal evidence that 5(a) with ϑ_1 = 4.8° must be a peak of distance of molecules and 5(b) with ϑ_2 = 9.7° must be a peak of substituents. In this case, the statement is more unequivocal in 5(a) than in 5(b).

In Figure 5 it can already be recognized that the peak of substituents as a base of measurement does not stand an exact measurement and more exact investigations are necessary.

STRUCTURAL FOUNDATIONS

By use of the measurement results represented and the literature, relations between the amorphous scattering curve and the molecular structure, or the structure of the molecular assembly, can be proved to exist. In general, it is supposed that ϑ_1 = 4.8° can be considered as a distance of molecules and ϑ_2 = 9.7° as a distance of substituents and that both can be used for the determination of orientation.

Such a determination can be obtained by the Bragg equation and the Ehrenfest relation $r = 1.11$ to $1.25 \cdot d_{Bragg}$. It leads to results which in part are quite useful but, on the other hand, must be regarded as strongly doubtful for the use of certain peaks of the amorphous scattering curve. This is especially the case with polystyrene [7].

According to Figure 6, the following assignment has been possible thus far:

$$\vartheta_1 = 4.8°;\ d_1 = 9.3\ \mathring{A} = \text{distance of molecules}$$

$$\vartheta_2 = 9.7°;\ d_2 = 4.6\ \mathring{A} = \text{distance of substituents}$$

$$\vartheta_3 = 21°;\ d_3 = 2.2\ \mathring{A} = \text{next nearest neighbors}$$

$$\vartheta_4 = 41°;\ d_4 = 1.25\ \mathring{A} = \text{nearest neighbors}$$

In this representation, a certain necessary width in the scattering of the dis-

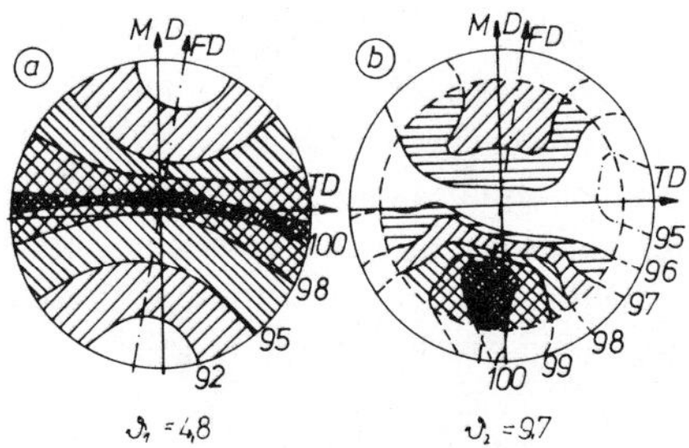

FIG. 5. Pole figures of an injection moulded plate of polystyrene.

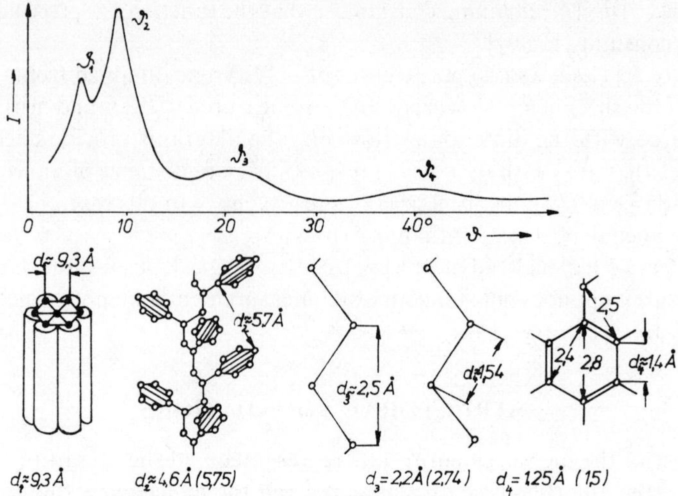

FIG. 6. Amorphous scattering curve of polystyrene and relations to the structure.

tances seems to be reasonable, the intermolecular distance at ϑ_1 being taken as relevant certain. However, in the range of ϑ_2, the main peak of the function $I = f(\vartheta)$, the statement of only one substituent peak of substituents will by no means be sufficient. At this distance, intermolecular distances can already be used.

Similar statements can be made with respect to the environment of ϑ_1. It results from the fact that it was necessary for the base of the determination of orientation by X-ray examination of amorphous high polymers to be strictly checked. We did this in two ways:

1. Elucidation of the correlations by means of the radial distribution function.

2. Introduction of a new method of measurement, the "oscillation curve," in order to complete the meaning of the radial distribution function.

RADIAL DISTRIBUTION FUNCTION AND AMORPHOUS SCATTERING CURVE

Fundamentals

The assignment of the maxima of the amorphous scattering curve to certain intermolecular and intramolecular distances, respectively, by means of the Bragg equation and the Ehrenfest relation is only a first approximation. As is shown in the literature [3,4], however, an exact statement about ρ as a distribution function of τ is obtained by the radial atomic density distribution function (RDF). The equation valid for these considerations

$$4\pi r^2 \rho_r - 4\pi r^2 \rho_0 = \left(\frac{2r}{\pi}\right) \int_0^\infty s \cdot i(s) \cdot \sin r \cdot s \cdot ds \tag{1}$$

with

$$i(s) = \frac{I_N}{f^2} - 1 \tag{2}$$

is obtained by inversion of the Debye equation (application of Fourier's integral theorem).

$$I = N \cdot f^2 \sum \frac{\sin \cdot sr_{mn}}{sr_{mn}} \qquad \text{Debye equation (3)}$$

In eq. 3, I is the scattering intensity of a noncrystalline configuration of N atoms of one specimen with the atomic factor f and the vector distance of two atoms r_{mn}, measured at a certain angle ϑ (equivalent $s = 4\pi \sin\vartheta/\lambda$). Thus, the mean atomic number per unit of volume

$$\rho_0 = \frac{\text{density} \cdot 6.02 \cdot 10^{23}}{\text{atomic weight} \cdot 10^{24}}$$

ϑ_r is the number of atoms per unit of volume at the distance r.

In eq. (2), I_N is the experimental coherent intensity in electron units. This means that prior to the transformation the corrected experimental intensity must be converted from arbitrary units into electron units, and the Compton scattering must be subtracted. There are two possibilities of representation. First $4\pi r^2 \rho_\tau - 4\pi r^2 \rho_0$ as a function of r which makes it possible to give an exact determination of the positions of the maxima, for example the mean distance to directly neighboring atoms, to second and third nearest atoms, respectively, or to the distances of the atoms of neighboring chains, and so on (Fig. 8).

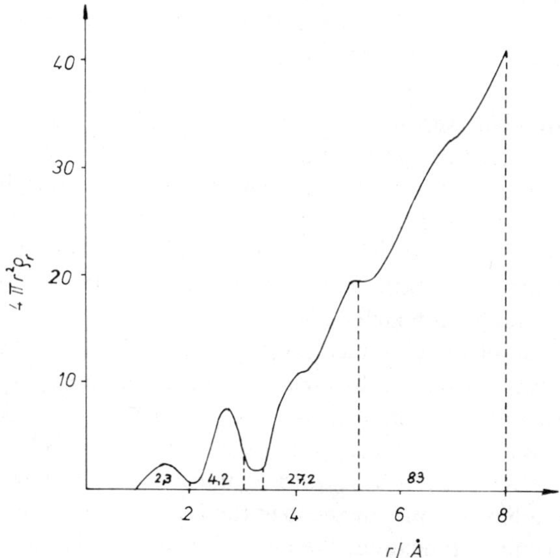

FIG. 7. Radial distribution function of atomic density (values according to output of computer).

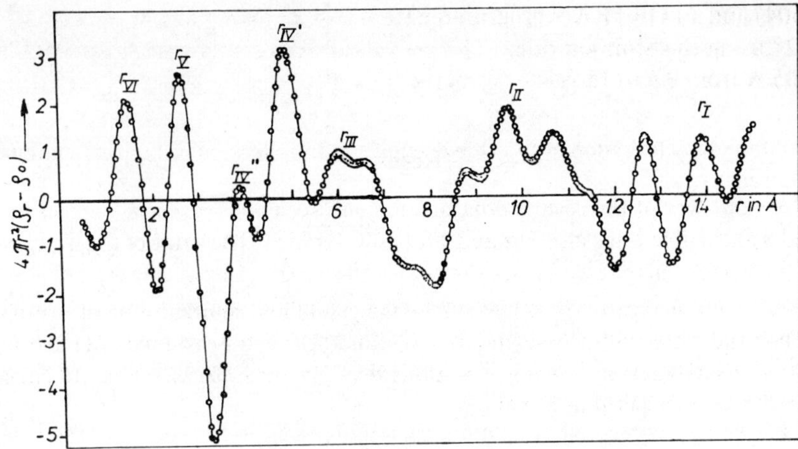

FIG. 8. Differential distribution function for polystyrene.

Second, the plotting of $4\pi r^2 \rho_\tau$ as a function of r, which apart from great r values, yields another important information stating that the area under the curve in the interval $\Delta\tau$ directly gives the number of atoms which are present in the spherical shell $4\pi r^2 \Delta\tau$ (Fig. 7). As has already been stated by Klug and Alexander [3], the area under the first peak, 2.25 for example, gives the number of direct neighbors which on the average would fall to each atom taken as a reference atom.

Conduct

The determination of the radial atomic density distribution function according to eqs. (1 and 2) was carried out in two steps. The first step includes the numerical calculation of function $i(s)$ from eq. (2) and the experimental scattering data.

The measurement were carried out with a crystal-monochromator and $CuK\alpha_1$-X-rays according to the Bragg–Brentano method by the aid of a diffractometer with step-by-step switch. The values were multiplied by the polarization factor, and the scattering in air determined experimentally was subtracted. The Compton scattering was corrected by calculation. For this purpose, the relation of the Compton scattering to the classical scattering was calculated and subtracted for each angle ϑ.

The conversion of the corrected intensity values from arbitrary units into electron units was carried out by adaption to the theoretical scattering curve (f^2 as a function of ϑ). The theoretical scattering curve was calculated for carbon (according to eq. (3), the H atoms can be neglected). The conversion factor $C = I$ in electron units/I in arbitrary units was determined from the ratio of the integral intensities and the ratio of the I and f^2 values between $\vartheta_{50°}$ and $\vartheta_{55°}$. The second step includes the numerical calculation of the integral in eq. (2).

The integration was carried out by the aid of an electronic computer (CDC

1604) and a FORTRAN programme from $\vartheta = 2°$ to $\vartheta = 55°$ in steps of $\Delta\vartheta = 0.2°$ using the Simpson rule. The $\tau-$ values were displayed in steps of $\Delta r = 0.05$ Å from 0.5 to 15 Å.

Results

A summary of the results obtained for polystyrene is given in Figures 6, 7, and 8 and Table I, see also Figure 12 for polystyrene. The value of the first peak of polystyrene may serve as an example for the values given in column 5 of Table I for the number of the neighboring atoms. The monomer unit includes eight C atoms, two of them having three direct neighbors, and six having two direct neighbors (that is $6.2 + 2.3/8$), and therefore on the average, 2.25 direct neighbors occur at $\bar{r}_{ideal} = 1.47$ Å.

The smoothing, which we did not carry out, gives on the one hand the possibility of a better perception of details in the course of the curve, and involves, on the other hand, the disadvantage that there are also disturbing secondary maxima which are not perceptible.

THE OSCILLATION CURVE

The results from the amorphous scattering curve and the radial distribution function yield very reliable statements about the position of intensity maxima conditioned by the structure. Statements concerning the structure, however, are possible only for the small distance values, and these are reliable only for the small distance values, and these are reliable only as long as the distances are intramolecular, which is also confirmed by the number of atoms per peak. The intensity maxima at smaller values of angles $\vartheta(\vartheta_1$ and ϑ_2; and ϑ_I; ϑ_{II}; ϑ_{III}; ϑ'_{IV}; respectively) cannot give any statements about the structure since the number of atoms per peak is very great and the assignment of neighboring atoms is totally uncertain.

However, in order to achieve a structural assignment to ϑ_{II}, ϑ_{III} and ϑ'_{IV} in

TABLE I

Results of the radial distribution function position of peak, atomic number, number of neighbours, calculated from the model for polystyrene

No	r-values (Å) (own results)	Number of atoms (own results)	Values of literature [2] r-values	Atomic numbers	Number of neighbors From model [2] average	total
VI	1.495	(0–2 Å): 2.3	1.51	2.1–2.15	(0–2 Å): 2.25	18
V	2.63	(2–3 Å): 4.2	2.35	3.67–4.23	(2–3 Å): 4.0	32
IV″	3.90	(3.4–5.2 Å): —	—	(3.4–5.2 Å): 22	164	
IV′	5.00	27.2	5.2	—		
III	6.30	(5.2–8 Å): 83	6.09–6.42	—	(5.2–8 Å): 87.5	700
II	10.70	—	10.1–10.7	—	—	—
I	15		14.7–14.9	—	—	—

[a] The calculation was broken off at 8 Å, because of the too large atomic numbers.

spite of these facts, we carried out further investigations on oriented specimens. At the same time we developed the measuring methods in such a way that statements about the character of the orientation can be obtained simultaneously with the amorphous scattering curve.

The technical solution [Fig. 9(a)] consisted in measuring the function $I = f(\vartheta)$ on a customary X-ray-diffractometer superimposed by a second function $I = f(\psi)$. In this case, the specimen rotates round the axis of the diffractometer at a speed which amounts of half of the angular speed of the countertube and simultaneously, it rotates round the normal of the specimen ($\psi = 0$ to $360°$). The course of both rotations is timed to one another.

We gave the diagram recorded by the recorder the term "oscillation curve" [Fig. 9(b)]. The measurements can be carried out by the transmission method as well as by the reflexion method. In our investigations it proved advisable to use the symmetrical transmission arrangement. For special orientations, however, we must proceed otherwise, and another angle of tilt φ must be adjusted as a constant value. According to Figure 9(a) and 9(b), an integration over the ϑ-interval which can be covered by the measurements is carried out, and the scattered intensity

$$\int_{\vartheta_{min}}^{\vartheta_{max}} \int_{\psi=0}^{\psi=2\pi} I_N(\vartheta,\psi)\varphi d\psi d\vartheta$$

is recorded (ϑ_{min} and ϑ_{max} were situated at $1.5°$ and $70°$, respectively).

The recorded diagram, the oscillation curve of Figure 9(b), yields the following information:

The amorphous scattering curve is distributed into regions of different widths of oscillations (amplitudes).

The regions of different widths of oscillation can be further characterized by corresponding phase shifts; these are plotted in the figure.

The regions which were found agree very well with the results of the radial distribution function, Figure 11.

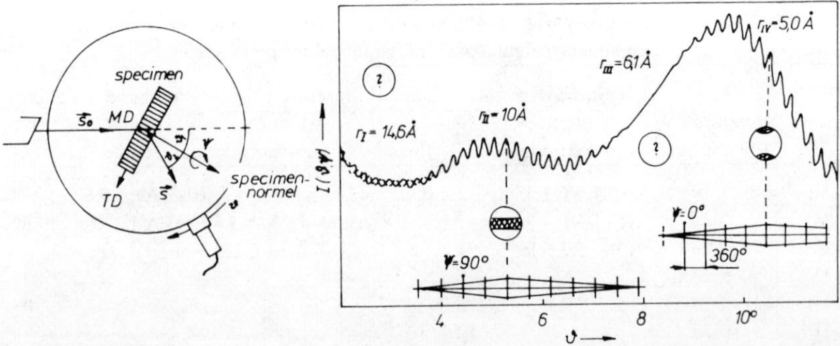

FIG. 9. Plot of the oscillation curve: (a) geometry of plotting $I = f(\vartheta, \psi)_{\varphi = \text{const.}}$; and (b) diagram of registration for polystyrene.

Since the main direction of the molecular orientation was known from measurements of the orientation, the phase shift could be related to it [Fig. 9(b)].

Thus the evaluation yields:

An univocal assignment of the peaks to extreme positions of the texture, for example, meridional or equatorial position, is possible. This is equivalent to periodic distances, preferentially in the longitudinal direction of the chains and in the perpendicular direction, respectively. On more exact evaluation, however, inclinations with respect to the main direction of orientation can be found. In case of doubt, it would be necessary to repeat the oscillation curve at another angle φ.

The exact assignment of the splitting of the peaks obtained from the radial distribution function, for example in polystyrene, the splitting of ϑ_2 into ϑ_{III} and ϑ_{IV}' according to the value of the distance, texture and structure.

In dependence on the parameters of manufacturing, the trend of the shift and the splitting of the peaks, respectively, can be explained, and based upon the direction of the shift and the change of the values of distance, respectively; conclusions concerning the manufacturing and the properties can be drawn (Fig. 10).

Thus in connection with the radial distribution function, the method of plotting the oscillation curve is suited very well to give exact statements about the structural assignment of the peaks of the amorphous scattering curve or to measure and define exactly the statements already known. With respect to the fixed value ϑ which is to be adjusted for the examination of the orientation and the determination of the degree of orientation, essentially new facts were found. For example, in polystyrene it is no longer advisable that the adjustment be made to the maximum of ϑ_{III}, ϑ_{IV}' and by no means in the direction of the slope of

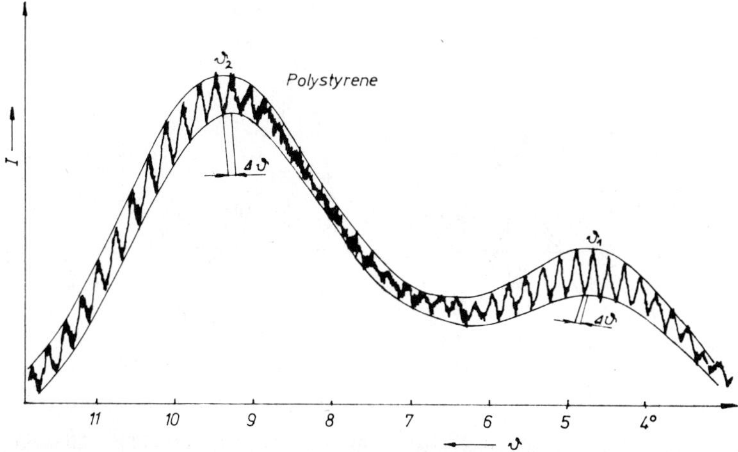

FIG. 10. Example of the evaluation of the oscillation curve with respect to the peak shift.

smaller ϑ values; this is done better and more reliably in the direction of the beginning slope to greater ϑ values.

RELATIONS BETWEEN OSCILLATION CURVE, RADIAL DISTRIBUTION FUNCTION AND STRUCTURE

Figure 11 shows the comparison of an oscillation curve of oriented polystyrene with the radial distribution function obtained by Wecker [2]. In order to be able to compare $I = f(\vartheta,\psi)$ directly with the $(4\pi r^2\rho_r - 4\pi r^2\rho_0)$-curve, the value $\vartheta \stackrel{\wedge}{=} \text{arc sin } [\lambda/2r\cdot(1, 11, \ldots 1, 25)]$ for the latter was plotted on the abscissa according to the Bragg equation and the Ehrenfest relation.

The comparison of both curves shows coincidence between the positions of the maxima of the radial distribution function and the regions which can be designated by a certain texture (phase shift according to Figure 9 and width of oscillation). In this case, both maxima r_{II} and r_{IV}' of the radial distribution function coincide roughly with the two regions of the greatest amplitude of the oscillation curve, the oscillations of which, however, underwent a phase shift of $\psi = 90°$. The results of both functions in connection with the calculated r values ($r_{II} \approx 10$ Å and $r'_{IV} \approx 5$ Å) permit an unequivocal structural assignment for these peaks to be made.

Further, it can be seen from the comparison that measurements of orientation and texture in polystyrene should not be carried out with adjustment of ϑ to the respective maximum of the amorphous scattering curve but to the maxima of $\vartheta_{II} \stackrel{\wedge}{=} r_{II} = $ peak of molecule distance and $\vartheta_{IV} \stackrel{\wedge}{=} r'_{IV} = $ distance of substituents, respectively.

Thus far, the evaluation of the secondary maxima r_I and r_{II} even after plotting the oscillation curve at an altered angle of tilt φ also did not permit a univocal structural assignment to be made. A special difficulty is caused by the fact that,

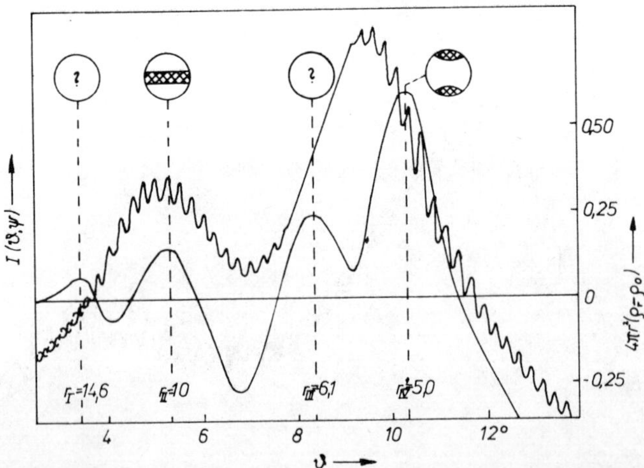

FIG. 11. Comparison of an oscillation curve $I = f(\vartheta,\psi)_{\varphi=0°}$ with the peaks of the difference radial distribution function.

with increasing orientation especially, the molecule chain distances perpendicular to the main orientation decrease to a greater degree, giving rise the Debye–Scherrer ring to deviate from the circular form and to become elliptic. This results in an additional asymmetry of the oscillation curve connected with different shifts of the lower and upper envelope curve. Here, it whould be pointed but once more that the comprehensive information of Figure 12 as well as of Table II would not have been possible without application of the oscillation curve. At least, a considerably more extensive program of investigation into orientation measurements would have been necessary.

DISCUSSION OF RESULTS

Figure 12 and Table II give a classification of our measurement results as well as conceptions of structure concerning the intensity maxima measured, the latter also including known statements from the literature. Table II shows that, with the exception of r_I all peaks can be assigned to structural elements and that the proposed structures are completed and secured by the information deduced from the oscillation curve. Thus far, there are no unanimous opinions as to the interpretation r_I and the assignment of a helical form of the molecular chain (as shown in Figure 12) is not univocal either. For example, this peak is not to be found in the radial distribution function of partially crystalline polystyrene [2] and moreover superimposition by small angle scattering already occurs in the region of this angle, $\vartheta \approx 2°$.

The splitting of the former ϑ-peak into ϑ_{III}, ϑ'_{IV}, and $r_{III} = 5.05$ Å, respectively, for isotactic regions and $r'_{IV} = 6.1 \ldots 6.6$ Å for syndiotactic regions has been discussed several times and is now proved sufficiently. The intensity of

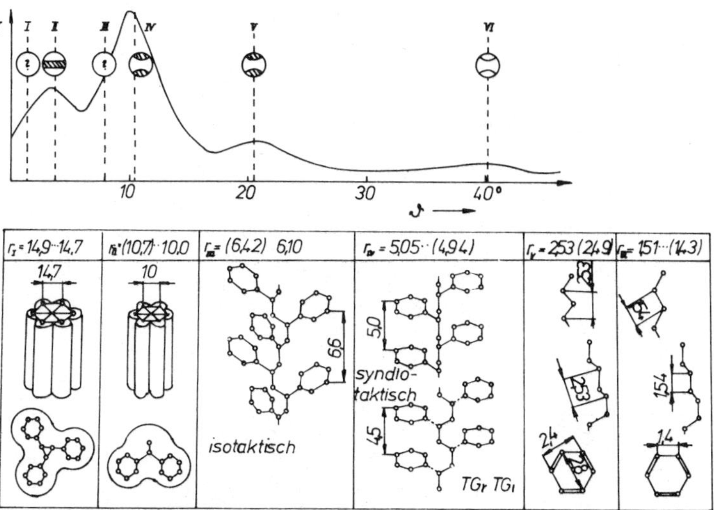

FIG. 12. Correlation between amorphous scattering curve and structural parameters for polystyrene.

TABLE II

Correlation between amorphous scattering and structural parameters of PS

r-value	Texture	Nature of distances	Tendency of the change of r [2] and own results	Most important structural element of the distance	Structural element with small contribution only
$r_{VI} =$ 1.5 Å	moderate, preferentially azimuthal	intrachain distances	not unequivocal due to overlapping	directly neighboring C-atoms in the chains ($d = 1.54$ Å)	directly neighboring C atoms in the phenylrings ($d = 1.4$ Å)
$r_V =$ 2.63 Å	moderate preferentially azimuthal	intrachain distances	not unequivocal due to overlapping	next nearest neighboring C-atoms in the chains ($d = 2.53$ Å)	second nearest neighboring C atoms in the phenylrings and diagonal distances ($d = 2.4$ Å°, u. d $= 2.8$ Å)
$r'_{IV} =$ 5 Å	markedly azimuthal	preferentially intrachain distance "Peak of substituents"	With increasing texture shift to smaller r-values	atoms from next nearest phenylrings in syndiotactic sequences	contributions to the total breath of the peak by distances phenylatoms chain atoms and phenylatoms-phenyl-atoms of the neighboring chain
$r_{III} =$ 6.3 Å	not discernible	intra- and interchain distances	with increasing isotactic proportions shift to greater r-values	atoms from third phenylrings in isotactic sequences (3 helix)	distances of parallelized phenylrings of neighboring chains contributions by distances phenylatoms-chain atoms of the same and of the neighboring chain
$r_{II} =$ 10.7 Å	markedly equatorial	interchain distances "Peak of chain distances"	with increasing isotactic proportions shift to greater r-values with increasing texture shift to smaller r-values	chain distances (preferentially syndiotactic sequences, and isotactic— $TG_r TG_i$—)	none
$r_I \approx 15$ Å		interchain distances and super structures	with increasing isotactic proportions shift to greater r-values; disappears in crystalline PS	chain distances of isotactic chains with helical confirmation	overlapping with small angle scattering superstructures greater than 15 Å°

this splitting depends on the proportions of the tacticities and in ϑ'_{IV} it is more or less distorted by different sterical configurations (see Figure 12).   Moreover, the proportions can be influenced by orientation and heat treatment.   Similar facts can be proved for the splitting of ϑ into ϑ_I and ϑ_{II} or r_I and r_{II}.

The peaks at r_{VI} = 1.51 Å and r_V = 2.53 Å which are univocally intramolecular do not show any shift of position dependent on manufacturing and heat treatment.

The scattering of the r values of the radial distribution function as well as the number of atoms per peak which can be calculated from the proportions of area prove that a dense series of atomic distances $\tau \pm \Delta\tau$ occur near τ.   They contribute to the formation of the relatively broad peaks [2], and a concrete distance in the molecule and the molecular aggregation, respectively, can be assigned to each value.

The comparison of radial distribution function and oscillation curve in Fig. 11 suggests the possibility of a very rapid determination of the base of measurements for the examination of the orientation and texture by plotting an oscillation curve (suitable ϑ value) when the structural assignment of the amorphous scattering curve is unknown.   This determination would certainly be more exact if it were done via the radial distribution function but in this case would be much more expensive.

Furthermore, the estimation of the permissible range of measurement $\vartheta \pm \Delta\vartheta$ for routine measurements can be carried out very rapidly by using the oscillation curve.   For example, this is the case in a region $\Delta\vartheta$ in which the width of oscillation related to the intensity of the randomly oriented basic material remains constant.

Figure 12 summarizes all possible structural correlations between the amorphous scattering curve and the measurements carried out on polystyrene which are known and were completed by us.   In order to obtain clear information, the adjustment should always be made to the values ϑ_{II} and ϑ'_{IV}, respectively, when the orientation is measured.

A special case is needed in the adjustment to ϑ'_{IV} since in this case a superimposition with ϑ_{III} may occur very easily which might cause a strong adulteration of the measurement results.   It is not advisable to include the peaks ϑ_V and ϑ_{VI} since they are of a low intensity.   As for the peaks ϑ_I and ϑ_{III}, the structural statements which are at present possible are not yet sufficient for an evaluation of the texture, and there is also the daugh of superimposition of intensity by ϑ'_{IV} and ϑ_{II}, respectively.

COMPARING INVESTIGATIONS OF POLYCARBONATE

The external shape of the scattering curve of polycarbonate (Figure 13) shows a course similar to that of polystyrene.   The maxima lie at ϑ_1 = 4.3° (w), ϑ_2 = 8.5° (vst), ϑ_3 = 13° (vw), ϑ_4 = 21° (m), and ϑ_5 = 41° (w).   The corresponding oscillation curve of the polycarbonate having undergone a treatment of cold-drawing to 150% of its original length is represented in Figure 14.   It can be seen at once that in polycarbonate the same cold-drawing causes a much

Table III.
Results from the RDF for polycarbonate and polystyrene.

	polycarbonate	polystyrene
	first neighbours	
peaks:	γ_5	γ_{VI}
d-value:		
experimental	1,49 Å	1,495 Å
theoretic	1,42 Å	1,470 Å
number of atoms:		
experimental	2,41	2,3
calculated	2,29	2,25

no texture in both polymers was observed

	second neighbours	
peaks:	γ_4	γ_V
d-value:		
experimental	2,55 Å	2,63 Å
number of atoms:		
experimental	4,91	4,2
calculated	–	4,0
texture:	uncertain	approximately azimuthal

	third neighbours and atoms from neighbouring chains	
peaks:	γ_3	$\gamma_{IV''}$
d-value:		
experimental	3,8 Å	3,9 A
number of atoms:		
experimental	7,43	–
texture:	equatorial ($\pm\ 15^{\circ}$)	azimuthal

peaks:	$\sim\vartheta$ 2''	$\sim\vartheta$ IV'
d-value: experimental	4,9 Å	5 Å distance of substituents
number of atoms: experimental	19,4	–
texture:	equatorial	azimuthal
comparison with crystall structure:	d_{200} = 5,05 Å chain distance	

peaks:	$\sim\vartheta$ 2'	ϑ III
d-value: experimental	6,3 ... 6,7 Å	6,3 Å distance of substituents
number of atoms: experimental	82	–
texture:	equatorial	random
comparison with crystall structure:	d_{020} = 5,95 A chain distance	

peaks:	ϑ 1	ϑ II
d-value: experimental	11,5 Å (chain distance ?)	10,7 Å chain distance
texture:	equatorial	equatorial
comparison with crystall structure:	d_{010} = 11,9 Å d_{100} = 10,1 Å	

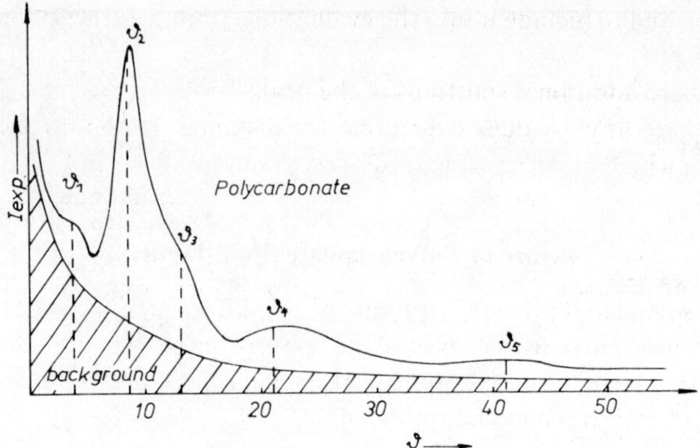

FIG. 13. Amorphous scattering curve for polycarbonate (experimental scattering curve).

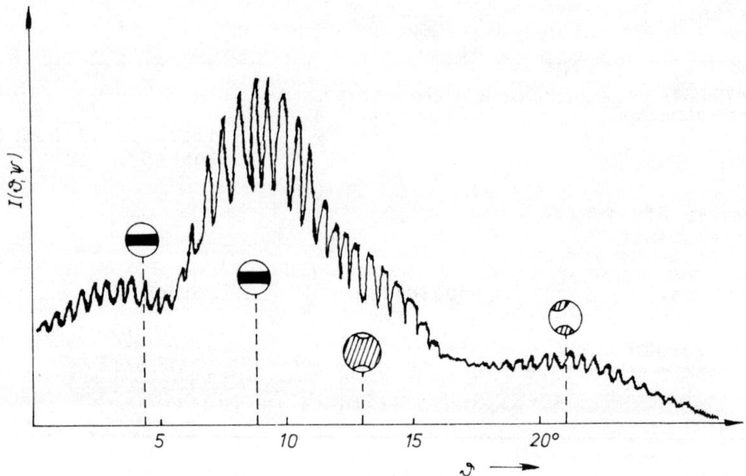

FIG. 14. "Oscillation curve" of cold drawn polycarbonate $I = f(\vartheta,\psi)$.

greater width of oscillation than it does in polystyrene (which agrees with Fig. 10). The greatest peak at $\vartheta_2 = 8.5°$ does not show azimuthal maxima but equatorial maxima in the oscillation curve. Therefore, this peak can be assigned to a molecular distance (in polystyrene ϑ_{III}, ϑ_{IV}' = peak of substitutents). By first approximation the third peak ϑ_3 at $\approx 13°$ may also be taken as a peak of molecular distance. In its orientation diagram, however, there are irregularities suggesting a superimposition with a distance value in the molecular direction.

In polycarbonate, there is no base of measurement which can be used for the determination of orientations in the longitudinal direction of the chains since, with the exception of the peak ϑ_4 at 21.5°, there are no distinct azimuthal reflexes. In this case, it is absolutely necessary to plot the radial atomic density distribution and to include it into the evaluation, aiming at the following questions:

Are there additional splittings of the peaks?

Are there any τ-values detectable for distances in the direction of the chains which are accessible for X-ray examination?

Texture of Polycarbonate (Pole-Figures)

In order to facilitate the discernibility of the differences in the single regions of the oscillation curve with respect to the texture, the phase shift, and the width of oscillation, the pole-figures of the peaks of ϑ_1, ϑ_2, ϑ_3, and ϑ_4 were plotted in the region of transmission and reflexion.

Contrary to polystyrene, the oscillation curve, Figure 14, and the pole-figures, Figure 15, show that with the exception of the pole-figure for the ϑ_3 peak at 21° which is not unequivocal (spacings of next nearest neighboring atoms) no pole-figure with distinct maxima of intensity occurs at the poles at MD and −MD. This means that no periodic distance in longitudinal direction of the chains contribute an appreciable amount of intensity to the peaks which can be recorded in the amorphous scattering curve.

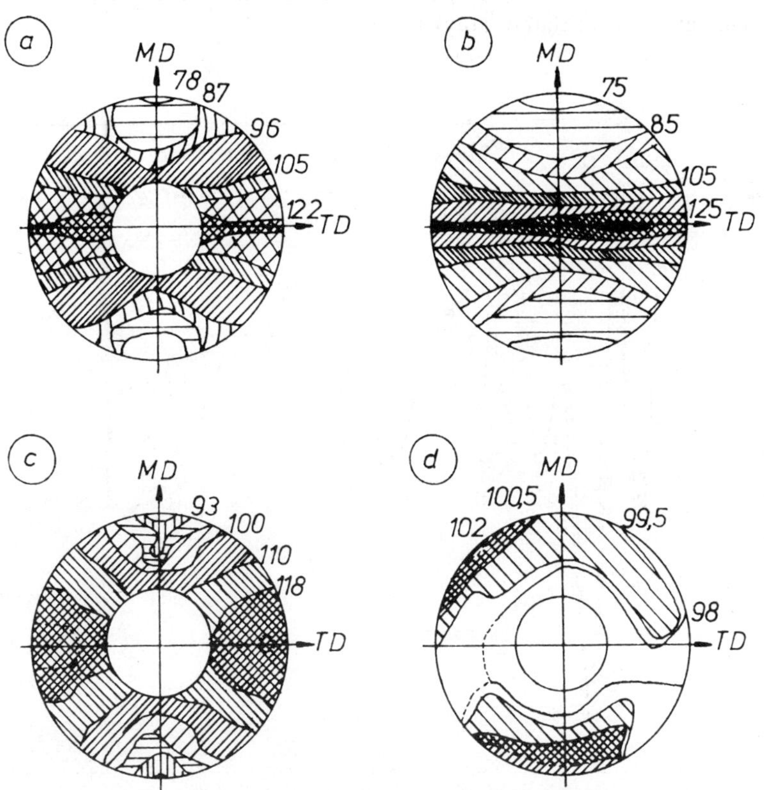

FIG. 15. Pole figures of polycarbonate.

Figure 15 represents the pole-figures of polycarbonate (having undergone a cold-drawing) for $\vartheta_1 = 4.3°$, $\vartheta_2 = 8.6°$, $\vartheta_3 = 13°$, and $\vartheta_4 = 21°$. As was already mentioned, that the width of oscillation in polycarbonate is very great, its relative proportion as related to the intensity of the nonoriented specimen $= 100\%$ is

(a) $\vartheta_1 = 4.3°$; $\Delta I_1 = \pm 22\%$ equatorial

(b) $\vartheta_2 = 8.6°$; $\Delta I_2 = \pm 25\%$ equatorial

(c) $\vartheta_3 = 13°$; $\Delta I_3 = \pm 17\%$ maxima at $\Delta\psi \approx 15°$ symmetrical to the equator and

(d) $\vartheta_4 = 21°$; $\Delta I_4 = \pm 1.6\%$ shifted unsymmetrically in azimuthal direction.

In this case it must be considered that the absolute intensity decreases from ϑ_2 via ϑ_1 to ϑ_3 and to ϑ_4. This suggests that contrary to polystyrene it is not possible to complete the examination of the texture in polycarbonate by another good X-ray reflex perpendicular to the first orientation.

Radial Distribution Function for Polycarbonate

The difference–radial distribution of atomic density as a function of r is represented together with that of polystyrene in Figure 16 in order to facilitate the comparison with that of polystyrene.

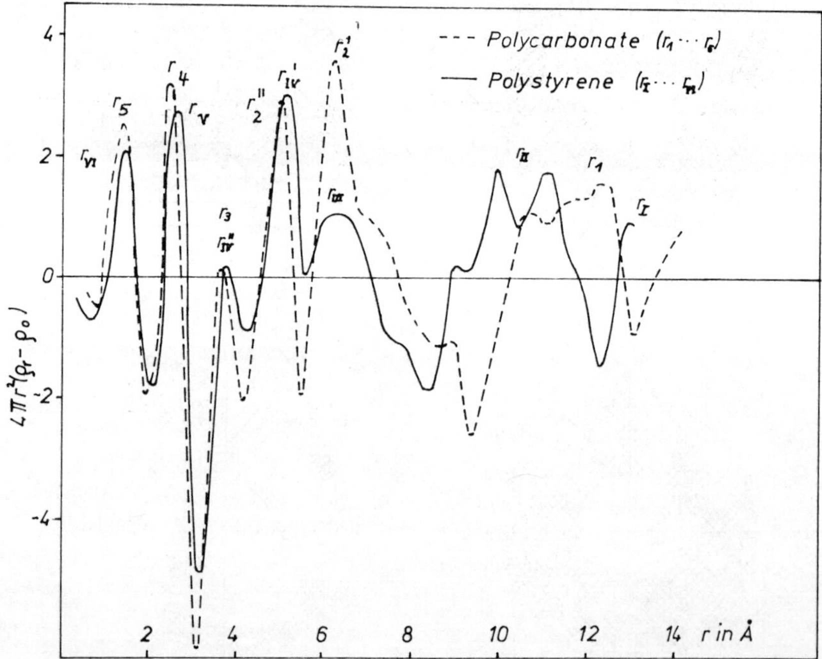

FIG. 16. Difference-radial distribution function of atomic density as a function of r: — polystyrene; ---polycarbonate

In some regions, both diagrams show great conformity which can be explained by the conformity of the basic structural elements of the macromolecule, especially in the case of small distances. This is the case especially with the last two peaks ϑ_V and ϑ'_{VI} in polystyrene and ϑ_4 and ϑ_5 in polycarbonate for next nearest and nearest neighbors, respectively.

Shift of Peaks

Referring to the example of polystyrene, it has already been indicated that the shift of peaks with increasing orientation is a measuring value which can be covered very well by the oscillation curve. In this case, the increasing order with increasing degree of orientation is measured by the decrease of the chain distances. The measurements known thus far were located within the margin of the measuring errors [4] so that in this respect a noticeable improvement could also be obtained by using the oscillation curve. As could be shown, the lower envelope curve can also be considered as constant and consistent with that of the unoriented material within the margin of error, even with increasing orientation. Therefore it is only necessary to measure the shift of the upper envelope curve with respect to the lower envelope curve and that of the unoriented material, respectively. In this case, the lines of heavy point are determined for both envelope curves, and the angular differences $\Delta\vartheta$ of both lines of heavy point are measured at different degrees of orientation, as is shown in Figure 17. With increasing degree of orientation up to $f_x = 8\%$, $\Delta\vartheta$ of polystyrene [Fig. 17(a)] increases proportionally, to the degree of orientation, the maximum shift being located at $\Delta\vartheta = 0.15°$. This corresponds to a change in the distances of the molecules by about 3%. This shift was also measured in polycarbonate [Fig. 17(b)] by the peak of chain distance $\vartheta_2 = 8.6°$.

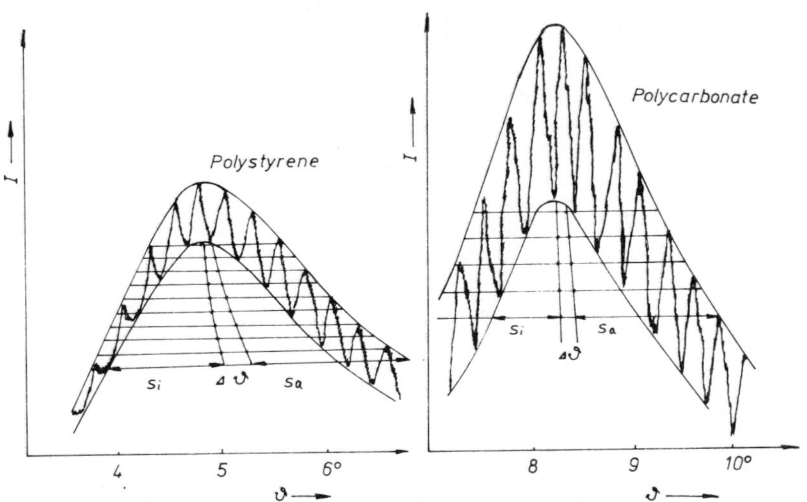

FIG. 17. Evalution of the oscillation curve for the determination of peak shift between upper and lower envelope curve: (a) in polystyrene and (b) in polycarbonate.

The degree of orientation of $f_x = 20\%$ being comparatively high in this case, a shift by $\Delta\vartheta = 0.14°$ was determined.   This corresponds to a shift by 1.7%.

Summary

It was shown that investigation of the orientation and texture of amorphous high polymers are of importance for technology.

The present measuring basis for X-ray orientation and texture measurements, the peaks of the amorphous scattering curve must be checked and improved accordingly.   There the "radial atomic density distribution function" and the "oscillation curve" introduced by us were an essential help.   With the example of polystyrene, the advance in knowledge was shown by applying these two measuring methods, and structure suggestions were made concerning the interrelation between the amorphous scattering curve and the polymer investigated. The measuring results obtained from a second material, polycarbonate, were discussed and compared with those of polystyrene.

Thus we paid attention to the special problems of the X-ray orientation measurement at amorphous high polymers, and hope to have made a contribution to an improved solution of the problem.

REFERENCES

[1] H. G. Kilian, and K. Boucke, *J. Polym. Sci.*, **58**, 311 (1962).

[2] S. Wecker, T. Davidson, J. G. Gohem, *J. Mater Sci.*, **7**, 1249 (1972).

[3] H. P. Klug, and L. E. Alexander, *X-Ray Diffraction Procedures*, John Wiley & Sons Inc., New York, 1954.

[4] R. N. Haward, *The Physics of Glassy Polymers*, Applied Science Pub., London, 1973.

[5] L. E. Alexander, *X-Ray Diffraction Methods in Polymer Science*, John Wiley & Sons, New York, 1969.

[6] T. E. Brady, and G. S. Y. Yen, *J. Polym. Sci. B*, **10**, 731 (1972).

[7] M. May, and Chr. Walther, *Plaste Kautsch*, **21**, 363 (1974).

PREFERRED ORIENTATION IN NONCRYSTALLINE POLYMERS

W. RULAND and W. WIEGAND
Fachbereich Physikalische Chemie-Polymere-D-3550 Marburg/Lahn, Germany

INTRODUCTION

The determination of the orientational anisotropy is an essential part of the characterization of polymer structures and their relationship to physical properties. There are a number of methods in use to determine orientation parameters (X-ray diffraction, light scattering, optical birefringence, IR dichroism, NMR, and fluorescence polarization) which are sensitive to the orientation of various components of the polymer structure (chain segments, crystallites, amorphous domains, lamellar structures, and spherulites) and which result in various types of orientation parameters. In general, these methods can be grouped into two classes, those which describe orientation in terms of angular distributions (e.g., pole figures) and those which result in the determination of weighted averages (e.g., moments) of these distributions. Angular distributions are obtained mainly by scattering methods (X-ray, electron, neutron, light scattering), whereas moments of these distributions are obtained by measurements of physical properties which can be related to tensors of various ranks, the rank determining the order of the moments. Since angular distributions contain generally more information than a limited number of moments of these distributions, especially since mostly only second moments and, less frequently, fourth moments are determined, we shall confine our discussion mainly to the problems related to the determination of angular distributions and, in particular, those occurring in the studies of preferred orientation of noncrystalline structures.

THEORETICAL

The orientation of the crystals in a polycrystalline material is conveniently described by the distribution $w(\alpha,\beta,\gamma)$ of the Euler angles α, β, γ which define the orientation of the crystal-fixed coordinate systems with respect to a sample-fixed system of axes. Diffraction methods result in the determination of the angular distributions (pole figures) $g_{hkl}(\varphi,\psi)$ of the normals of the plane of index (hkl) with respect to this sample-fixed coordinate system. It has been shown by Roe [1] that, in principle, an infinite number of plane-normal distributions is necessary to compute $w(\alpha,\beta,\gamma)$ using an inversion method based on series of

Journal of Polymer Science: Polymer Symposium 58, 43–58 (1977)

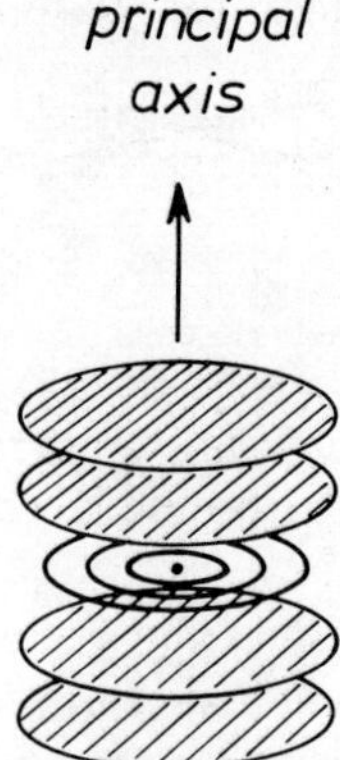

FIG. 1. Correlation function for a turbostratic (smectic) structure.

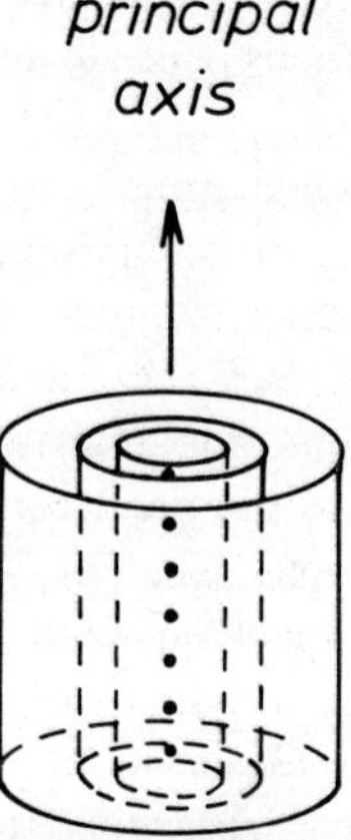

FIG. 2. Correlation function for a nematic structure.

spherical harmonics. The problem becomes, however, much simpler if one considers a system with cylindrical symmetry of the orientation of the crystallites about a sample-fixed principal axis (fiber structure) and a random rotation of the crystallites about a crystal-fixed principal axis (e.g., the main axis of the polymer chains). In this case, the orientation is completely defined by the distribution $g(\varphi)$ of the principal axes of the crystallites with respect to the principal axis of the sample.

Such an assumption is a convenient starting point to discuss the problems occurring in the determination of preferred orientation in noncrystalline materials.

When studying preferred orientation in a noncrystalline material by diffraction methods the main problem consists in defining the structural units, the orientation of which causes the anisotropy of the scattering. There are two types of short-range order for which a theoretical treatment of the preferred orientation is relatively easy: turbostratic (smectic) structures and nematic structures.

A turbostratic structure consists of a parallel stacking of layers with relatively

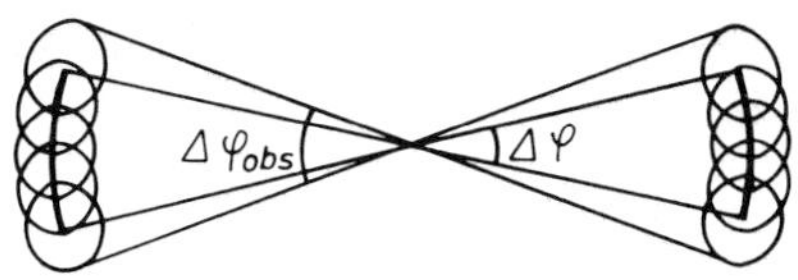

FIG. 3. Effect of the tangential width on the angular distribution of an (hkl) interference line.

perfect internal structure in which the layers are randomly rotated and translated with respect to each other so that no positional correlation exists between atoms belonging to different layers except the overall parallelism. The correlation function for such a structure is shown in Figure 1. The correlation function consists of a series of concentric rings in a plane through the origin and a periodic series of layers with constant surface density parallel to this plane. A nematic structure consists of a parallel stacking of rods with relatively perfect internal structure. The rods are randomly rotated and translated with respect to each other so that no positional correlation exists between atoms belonging to different rods except the overall parallelism. The correlation function for such a structure (considered to be rotated about its principal axis) is shown in Figure 2. The correlation function consists of a periodic series of narrow distributions on a straight line through the origin and a series of concentric cylinder walls with constant surface density.

The intensity distribution in reciprocal space of the turbostratic structure is similar to the correlation function of the nematic structure and vice versa, i.e., ($00l$) interference peaks and (hk) cylinder walls for the turbostratic structure, ($hk0$) interference rings and (l) layers in the case of a nematic structure. In the case of random orientation, the turbostratic structure shows symmetrical ($00l$) line profiles and asymmetrical (hk) line profiles of the type

$$I_{hk}(s) = 1/2\pi s\sqrt{s^2 - s_{hk}^2} \quad s \geq s_{hk} \tag{1}$$

(von Laue, [2]), the nematic structure shows symmetrical ($hk0$) line profiles and asymmetrical (l) line profiles of the type

$$I_l(s) = 1/2s \quad s \geq s_l \tag{2}$$

where $s = 2\sin\theta/\lambda$, s_{hk} and s_l are the absolute values of the reciprocal lattice vectors related to the internal structure of the layers and the rods, respectively. Finite size and perfection of the internal structure will, of course, introduce a supplementary broadening of $I_{hk}(s)$ and $I_l(s)$ which affects, however, essentially only the s values in the vicinity of s_{hk} or s_l.

If one considers a nonrandom orientation, there is obviously an essential difference between the effect of orientation on the intensity distributions of the ($kh0$) and ($00l$) lines on one hand, and the (hk) and (l) lines on the other.

The effect of preferred orientation on the ($hk0$) and ($00l$) lines is in principle the same as that on an interference peak of a three-dimensionally ordered structure. However, since the size and the perfection of the turbostratic and nematic domains are, in general, relatively small, the observed pole figures can contain a nonnegligible contribution from the finite width of the lines tangential to the radius vector s_{hkl} (see Figure 3). This effect has been discussed in an earlier paper (Ruland, [3]) where it was shown that it can be considered as a convolution of $g(\varphi)$ with a broadening function, and a general elimination procedure using Fourier series has been proposed and applied to a series of orientation measurements of turbostratic structures (carbon fibers). In order to assess the magnitude of the effect, let us assume the tangential line profile to be Lorentzian and the angular distribution of the center of the line to be given by

$$g(\varphi) = \frac{1 - q^2}{1 + q^2 - 2q \cos 2\varphi} \tag{3}$$

where q is an orientation parameter varying from 0 to 1 for an orientation varying from random to perfect orientation at $\varphi = 0$, and from 0 to -1 for an orientation varying from random to perfect at $\varphi = 90°$. Considering the orientation parameter

$$f = \tfrac{1}{2}(3\langle\cos^2\varphi\rangle - 1)$$

one can show that

$$f = 1 - \tfrac{3}{2}F(q)$$

with

$$F(q) = \frac{1 + q}{4\sqrt{q}\ \mathrm{artanh}\ \sqrt{q}} - \frac{(1 - q)^2}{4q}$$

For negative values of q, $F(q)$ is, to a good approximation, a linear function of q:

$$F(q) \simeq \tfrac{1}{3}(2 - q) \quad -1 < q < 0$$

which means that f is approximately $q/2$ for this type of orientation. For $q > 0$, the above approximation still holds reasonably well for $0 < q < 0.3$; for q values higher than 0.5 an approximation of the type

$$f = 1 - \frac{3}{2 \ln\ [4/(1 - q)]}$$

can be used.

The choice of a $g(\varphi)$ in the form of eq. (3) for calculating the effect of the tangential width on the observed $g(\varphi)$ has the advantage that this function can be considered as an infinite period series of Lorentzian distributions

$$g(\varphi) = \sum_n L(q,\varphi - n\pi)$$

for $0 < q < 1$ and

$$g(\varphi) = \sum_n L\left(q,\varphi - \frac{2n + 1}{2}\pi\right)$$

for $-1 < q < 0$ where

$$L(q,\varphi) = 2\left(\ln\frac{1}{|q|}\right)\frac{1}{\ln^2\left(\frac{1}{|q|}\right) + 4\varphi^2}$$

with an integral width

$$B_L(q) = \frac{\pi}{2} \ln \frac{1}{|q|}$$

If the line profile due to finite size and imperfection of the domains has the integral width $B(s)$ in $s = 2 \sin \theta / \lambda$, and if $B(s)$ is still small compared with s_{hkl}, its effect on $g(\varphi)$ can, to a first approximation, be considered as due to a broadening in φ with a width $B(\varphi) = B(s)/s_{hk0}$ or $B(s)/s_{00l}$, respectively. If this broadening is due to a Lorentzian distribution, the observed angular distribution is still of the type given in eq. (3) but has a smaller value of q given by the equation

$$q_{obs} = q \left(1 - \frac{2B_s}{\pi s_{hkl}} \right)$$

since

$$B_s \ll s_{hkl}.$$

For $-0.5 < f < 0.2$ this means, to a first approximation

$$f_{obs} \simeq f \left(1 - \frac{2B_s}{\pi s_{hkl}} \right), \tag{4}$$

and for $f \simeq 1$:

$$f_{obs} \simeq 1 - \frac{3}{2 \ln (2\pi s_{hkl}/B_s)}$$

For disordered structures of the kind considered here, $2B_s/(\pi s_{hkl})$ can be expected to be of the order of 0.1, the error in f is thus about 10% for f values below 0.2, increasing gradually to about 40% with respect to f, i.e., nearly 70% with respect to f_{obs}, when f tends towards unity (Fig. 4).

This shows, that the error can become considerable especially if the preferred orientation of s_{hkl} is along the primary axis, i.e., layers perpendicular to this axis in the turbostratic case and rods perpendicular to this axis in the nematic case. It is thus of interest to include the (hk) and the (l) interference in the determination of the preferred orientation since these interferences can be used to assess the tangential width of the $(00l)$ or $(hk0)$ lines, respectively, and their spacial distribution contains information on $g(\varphi)$, although in a more complicated way.

For (hk) interferences, this problem has been treated in an earlier paper (Ruland and Tompa, [4]). It was shown that the intensity distribution $I_{hk}(\mathbf{s})$ can be taken as a product of the isotropic distribution $I_{hk}(s)$ given in eq. (1) and a function $F(\sigma,\varphi)$

$$I_{hk}(\mathbf{s}) = I_{hk}(s) \cdot F(\sigma,\varphi) \tag{5}$$

in which $\sin \sigma = s_{hk}/s$ and

$$F(\sigma,\varphi) = 2 \int_0^\pi g(\beta)d\eta \tag{6}$$

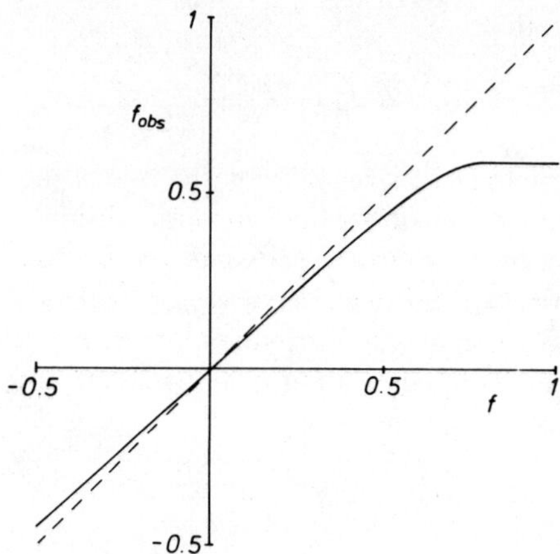

FIG. 4. The observed orientation parameter f_{obs} as a function of the correct parameter f for B_s/s_{hkl} = 0.16.

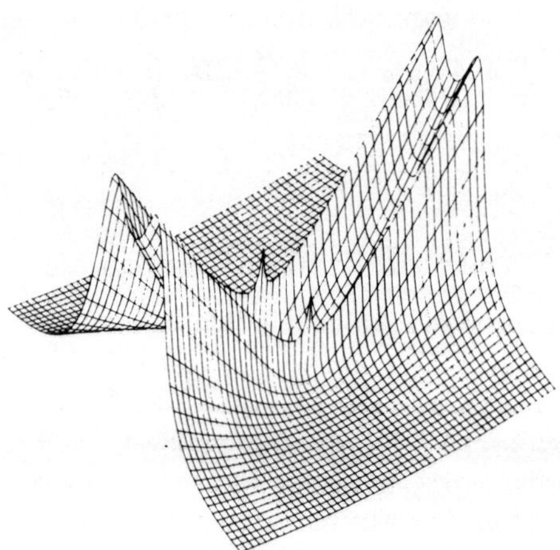

FIG. 5. The two-dimensional profile of the function $F(\sigma,\varphi)$ in the plane containing the principal axis ($\sin \sigma = s_{hk}/s$, $q = 0.8$).

where $g(\beta)$ is the normalized orientation distribution of the principal axis as a function of the angle β related to σ, η, and φ through

$$\cos \beta = \cos \varphi \cos \sigma + \sin \varphi \sin \sigma \cos \eta$$

The function $F(\sigma,\varphi)$ can be obtained in a closed form for orientation distributions of the type given in eq. (3).

Figures 5 and 6 show perspective plots of the two-dimensional profiles of $F(\sigma,\varphi)$ for $q = 0.8$ and -0.8, respectively, as functions of s in the plane containing the primary axis, the latter being parallel to the projection onto this plane of the lines running from the lower left to the upper right hand side of the diagrams. An inspection of the diagrams shows that pronounced maxima are observed at $\varphi = 90°$ for positive q values and $\varphi = 0°$ for negative q values, i.e., at right angles to the maxima of the corresponding $g(\varphi)$ functions. In contrast to the effect of preferred orientation on (hkl) interferences the (hk) interferences [and the (l)

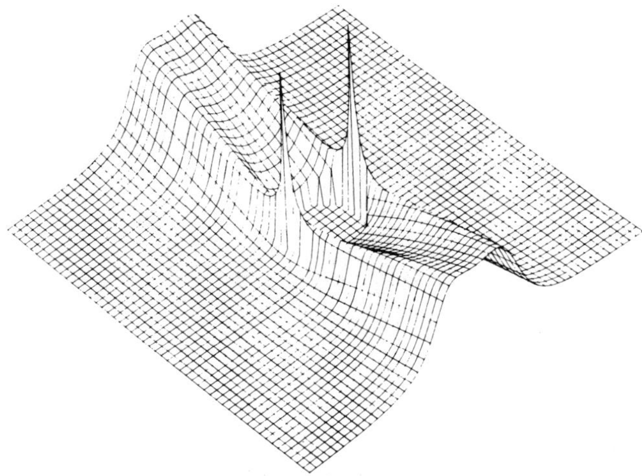

FIG. 6. The two-dimensional profile of the function $F(\sigma,\varphi)$ in the plane containing the principal axis ($\sin \sigma = s_{hk}/s$, $q = -0.8$).

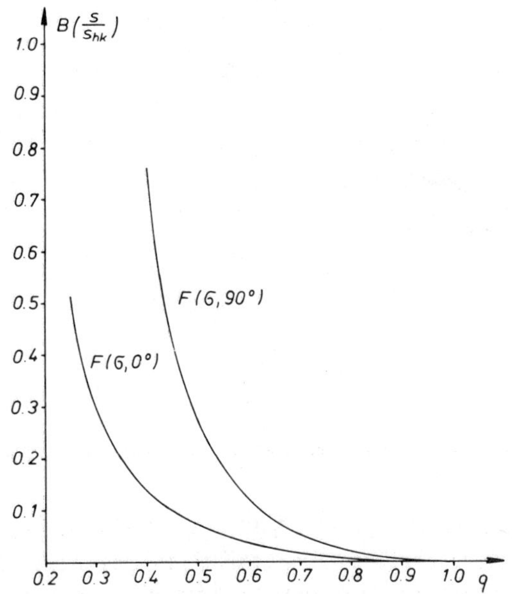

FIG. 7. Half-peak width in s/s_{hk} of $F(\sigma,0)$ and $F(\sigma,90°)$ as a function of $|q|$.

interferences, as will be demonstrated later] show a variation of the intensity in φ as well as in s. This means that the profile of an (hk) interference in s at constant φ is a function of $g(\varphi)$. The most simple relationships for such profiles are obtained for $\varphi = 0°$ and $\varphi = 90°$, respectively,

$$F(\sigma,0) = 2\pi g(\sigma) \tag{7}$$

$$F(\sigma,90°) = \frac{\sqrt{q}}{\text{artanh } \sqrt{q}} \frac{1}{[(1 + q)^2 - 4q \sin^2 \sigma]^{1/2}} \tag{8}$$

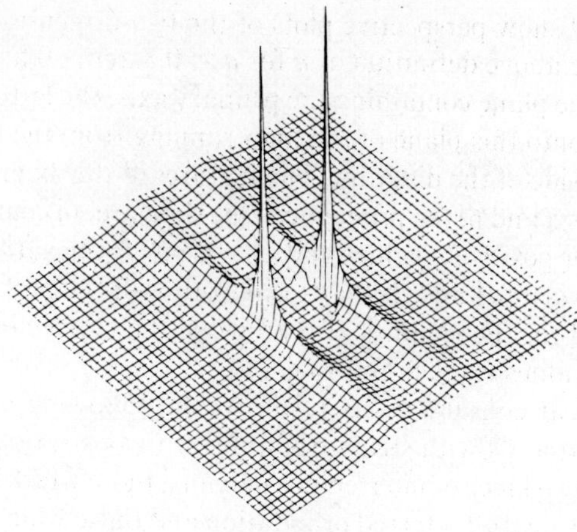

FIG. 8. The two-dimensional profile of the function $F(\sigma,\varphi)$ in the plane containing the principal axis ($\cos \sigma = s_l/s$, $q = 0.8$).

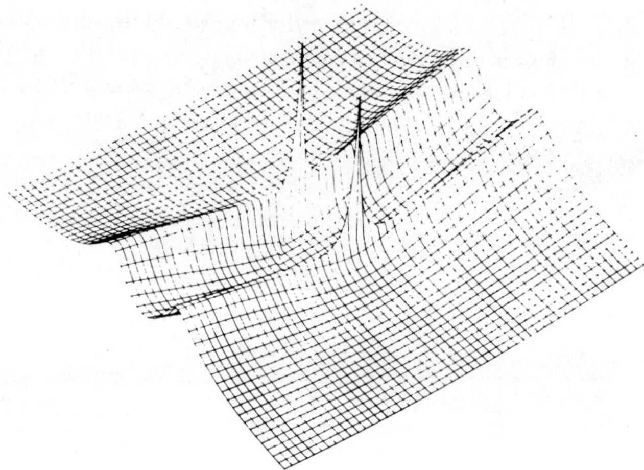

FIG. 9. The two-dimensional profile of the function $F(\sigma,\varphi)$ in the plane containing the principal axis ($\cos \sigma = s_l/s$, $q = -0.8$).

Figure 7 shows the variation of the half-peak width in s/s_{hk} of the line profiles in $F(\sigma,0)$ for negative q values and in $F(\sigma,90°)$ for positive q values. It appears that, in both cases, the widths change considerably with q and that the effect is larger in $F(\sigma,90°)$ than in $F(\sigma,0°)$.

If other line broadening effects can be eliminated these relationships should permit a determination of orientation parameters from a quantitative profile study of the (hk) lines.

For nematic structures one can show that the effect of preferred orientation on the (l) interferences leads to an expression similar to eq. (5)

$$I_l(\mathbf{s}) = I_l(s) \cdot F(\sigma,\varphi) \tag{9}$$

where $F(\sigma,\varphi)$ is the same function as in eq. (6) except that the definition of σ is changed to

$$\cos \sigma = s_l/s$$

Figures 8 and 9 show perspective plots of the two-dimensional line profiles of $F(\sigma,\varphi)$ with the above definition of σ for $q = 0.8$ and -0.8, respectively, as functions of **s** in the plane containing the primary axis, the latter being parallel to the projection onto this plane of the lines running from the lower left to the upper right hand side of the diagram. The features of the diagrams are similar to those of Figures 5 and 6, the pronounced maxima are, in contrast to the (hk) lines, at $\varphi = 0°$ for positive q values and at $\varphi = 90°$ for negative q values. The relationships between $|q|$ and the half-peak widths of the profiles in s given in Figure 7 are valid for the (l) lines also provided the change in the sign of the corresponding q values is taken into account.

These theoretical considerations result in the following conclusions: For noncrystalline materials with structural units of turbostratic or nematic type one has to expect two kinds of interference maxima, those which change intensity and line-profile in s with preferred orientation and those which change only in intensity. The latter can be evaluated in the same way as crystalline interferences provided the effect of the tangential width is taken into account. The asymmetry of the profile of the former and its change with preferred orientation can be used to characterize the type of short-range order and to obtain supplementary information on the degree of preferred orientation.

In those cases in which the type of short-range order cannot be decided from a study of the intensity distribution, a Fourier transformation of the total coherent scattering to obtain the orientation dependent pair correlation function is a useful although rather tedious procedure. Its evaluation would consist in determining the angular distribution of the interatomic distances; in favorable cases these may be separated into inter- and intramolecular distances and thus provide information on the orientation of molecular segments and their packing.

Apart from the determination of pole figures, scattering methods can be used to measure the anisotropy of the density fluctuation by studying the angular dependence of the diffuse small-angle scattering near the origin of reciprocal space. In a noncrystalline material the density fluctuations can be considered to be composed of a contribution from thermal motion (phonons) and from the "frozen-in" disorder. Both contributions can show an anisotropy. In the case of thermal motion this is due to the change of the group velocities of longitudinal waves with the propagation direction and thus directly related to the anisotropy of the dynamic modulus of the material. In the case of the "frozen-in" disorder an anisotropy of the diffuse small-angle scattering can be interpreted as due to an anisotropy of the short-range order.

A study of the temperature dependence of this effect can be used to separate the two components.

EXPERIMENTAL

A number of experimental results exist for the effect of preferred orientation on (hk) interferences of turbostratic structures. Such structures develop when organic material is thermally decomposed and small polycondensed aromatic

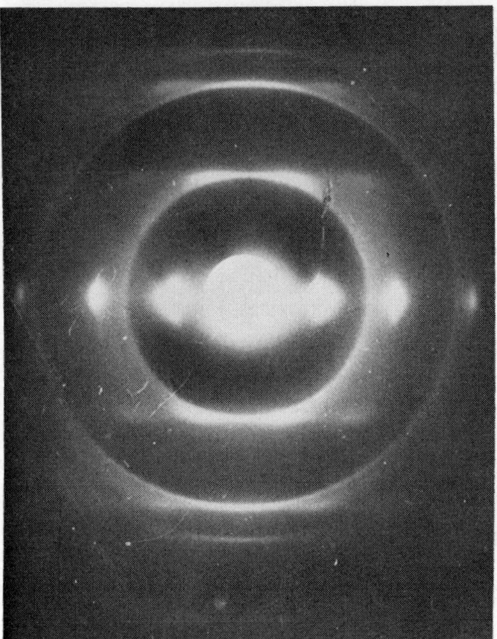

FIG. 10. Electron diffraction diagram of a highly oriented turbostratic structure (carbon fiber).

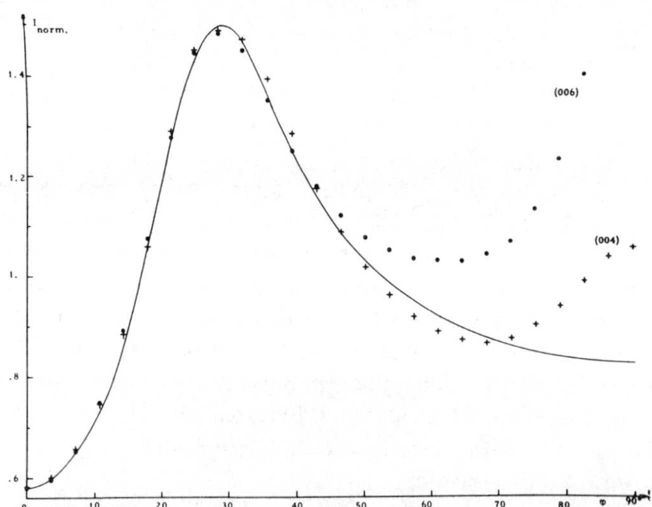

FIG. 11. Comparison of theoretical and experimental values for (hk) interferences measured at constant $s = 1.1s_{hk}$ as a function of φ.

ring systems are formed which develop a tendency towards parallel stacking. The structures obtained can show a preferred orientation of the layer normals in the direction of the principal axis (pyrolytic carbon, Guentert and Cvikevich, [5]) or perpendicular to it (carbon fibers, Ruland and Tompa, [4]; Fourdeux, Perret and Ruland, [6]). Figure 10 shows an electron diffraction diagram of a

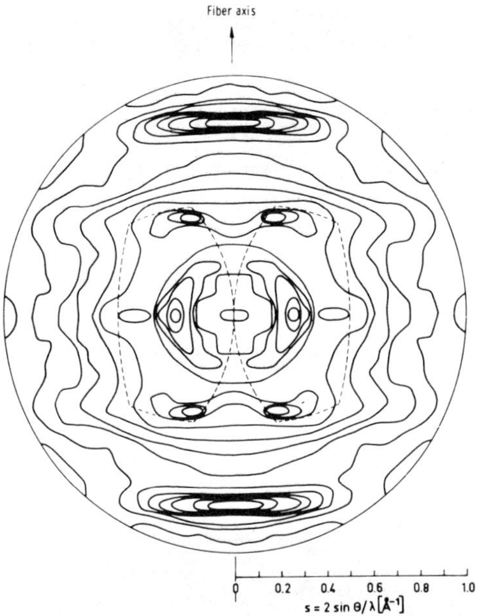

FIG. 12. Diffuse X-ray scattering intensity of polyamide 6 (isointensity lines).

FIG. 13. Diffuse X-ray scattering intensity of polyamide 66 (isointensity lines).

carbon fiber with high preferred orientation. The difference in the intensity distribution between the $(00l)$ lines (on the equator) and the (hk) lines is obvious. Figure 11 shows a plot of the intensity of a (10) and a (11) interference at constant $s = 1.1s_l$ as a function of φ together with a plot of the corresponding values of $F(\sigma,\varphi)$. The curves show a good agreement between the theoretical and experimental values outside the equatorial region in which the $(00l)$ intensities overlap with the (hk) intensities.

The effect of preferred orientation on nematic structures in polymers has not yet been studied in detail. This is mainly due to the fact that a neat distinction between nematic and amorphous is rather difficult. If bundles of molecules exist in the amorphous phase of polymers, the structure could be treated as nematic,

FIG. 14. Diffuse X-ray scattering intensity of polyamide 11 (isointensity lines).

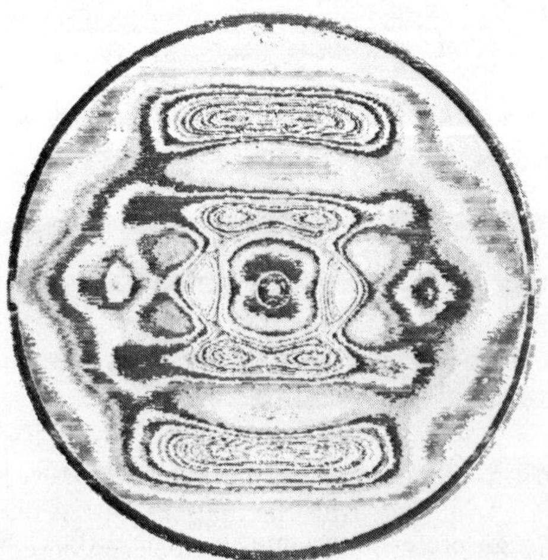

FIG. 15. Diffuse X-ray scattering of an alternating P(ETFE) copolymer (two-dimensional microdensitometer diagram).

provided the domains of anisotropic short-range order are large enough. The results of the neutron small-angle scattering studies on solid solutions of deuterated and undeuterated amorphous polymers seem to exclude the existence of bundle structures in atactic PMMA and polystyrene. However, polyamides, for example, are known to form structures of intermediate order between amorphous and crystalline which, in the oriented state, show intensity distributions very similar to those one would expect for nematic structures with preferred orientation. Figures 12–14 show equal-intensity lines for a series of polyamide fibers with diffuse intensity distributions which appear to be, at least qualitatively, composed of a function of the type given in Figure 8 and, of course, modulations due to the structure factor of the single chain. Similar intensity distributions are observed in the fiber diagrams of an oriented P(ETFE) alternating copolymer with about 10% sequential disorder. A two-dimensional microdensitometer plot is shown in Figure 15. The fact that the structure factor of the single extended molecule is not monotonic in the direction perpendicular to the chain axis makes it more difficult to extract the function $F(\sigma,\varphi)$ from the scattering intensity, in contrast to the case of the carbon fibers, however, this could, in principle, be possible.

A basic problem in the determination of the preferred orientation of the amorphous domains in semicrystalline polymers is the difficulty of finding an unambiguous way to separate the crystalline interferences from the amorphous ones. Figure 16 shows a series of intensity measurements in s at constant φ for a polyethylene film cold stretched about 300%. In the plots, two kinds of separation lines between the crystalline and the amorphous interferences are drawn which result in the two curves of the integrated intensity versus φ shown in Figure 17. The curve with the lower intensity at $\varphi = 90°$ is the result of a separation procedure in which care was taken that the total integrated intensity

$$\int \int I(s,\varphi) \sin \varphi d\varphi ds$$

of the crystalline and the amorphous interferences maintained the ratio calculated from the crystallinity of the sample measured by other methods, whereas the curve with the higher intensity at $\varphi = 90°$ corresponds to an unreasonably low degree of crystallinity. The orientation parameters f_{obs} related to the chain axes deduced from the two curves are +0.455 and +0.484, respectively, which become +0.540 and +0.575 after a correction using eq. 4 with $B_s/s \simeq 0.25$. The lower value of f can be considered the most probable; in combination with the value of $f = 0.936$ calculated for the crystalline domains it agrees well with the value measured for the optical birefringence.

Figure 18 shows, for the same sample, the variation of the diffuse small-angle scattering extrapolated to $s = 0$ as a function of φ at three different temperatures. The extrapolation has been carried out from the low-angle tail of the amorphous halo excluding the region in which the boundaries between crystalline and amorphous domains determine the small-angle scattering. The spherical averages of these functions are overall density fluctuations within the crystalline and amorphous domains which can be considered to be composed of a component due to structural disorder and a component due to thermal motion. At 4 K, the contribution of the latter component can be neglected so that only the structural

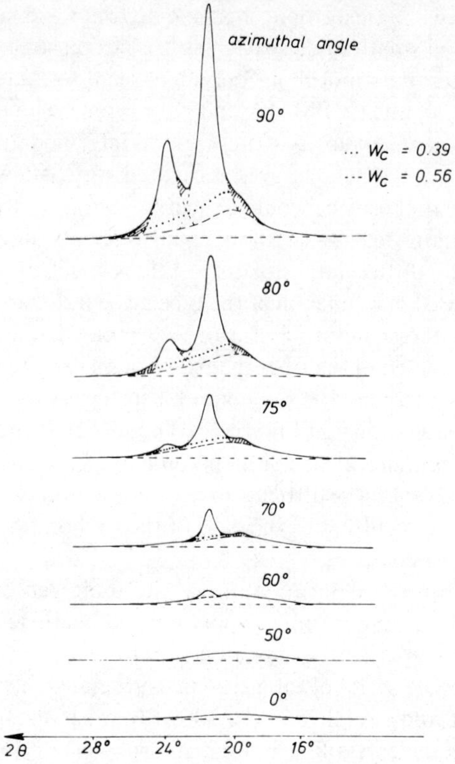

FIG. 16. X-ray scattering intensity of an oriented PE film as a function of s for constant φ.

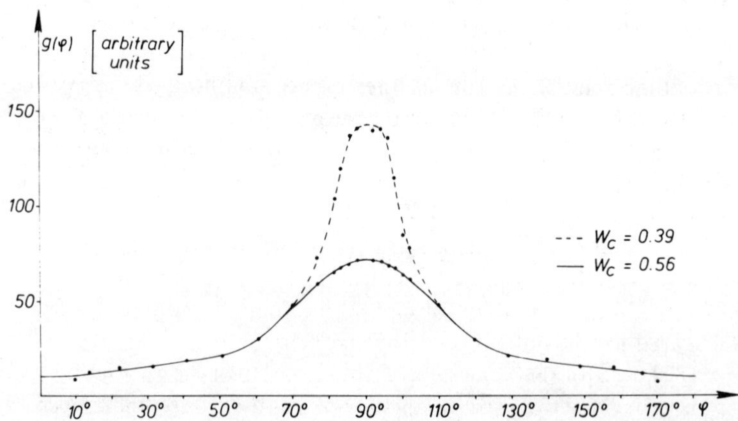

FIG. 17. Orientation function $g(\varphi)$ for the amorphous halo according to two different separation procedures.

disorder is determining the intensity values. It has been shown (Rathje and Ruland, [7]) that the amorphous domains produce the main contribution to these values, and since these domains are oriented one would expect a variation in φ if the short-range order in the amorphous domains is strongly anisotropic. Inspection of Figure 18 shows that this seems not to be the case. This result could

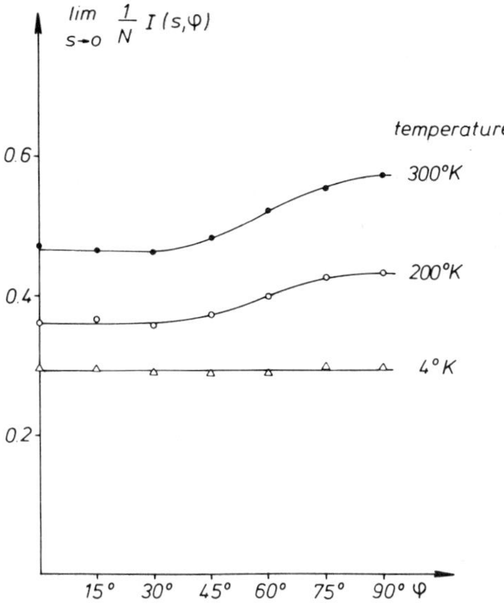

FIG. 18. Diffuse X-ray scattering of an oriented PE film extrapolated towards zero scattering angle as a function of φ for three different temperatures.

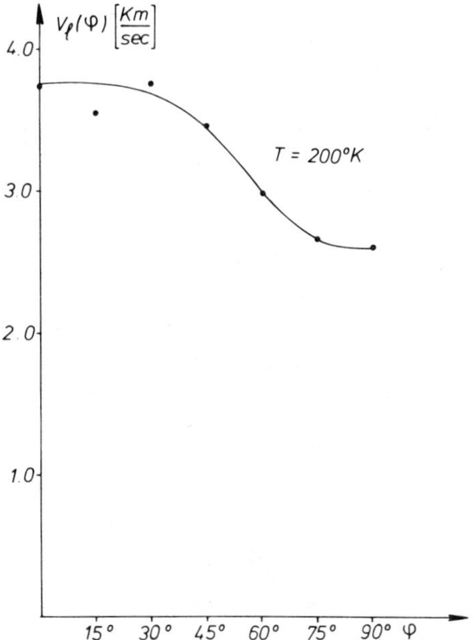

FIG. 19. Phonon velocities as a function of φ obtained from the temperature dependent part of

$$\lim_{s\to 0} (1/N)I(s,\varphi).$$

mean that either the short-range order is not of a bundle type, or if it is of this type, that the disorder in the packing is about as large as the disorder in the structure of the chains which form the rodlike structural units.

The anisotropy appearing at higher temperatures is related to the anisotropy

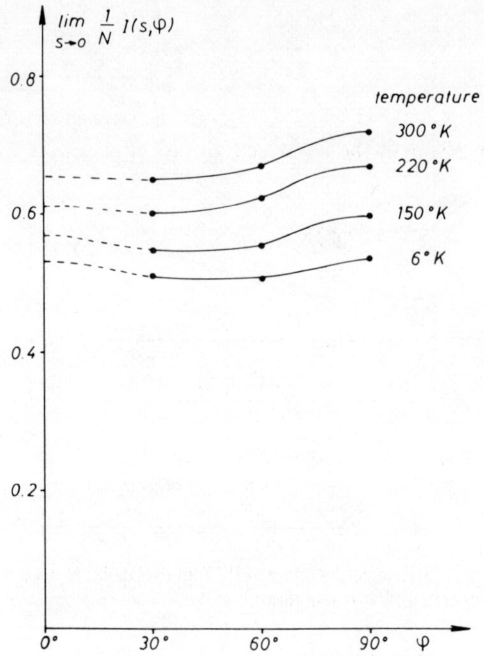

FIG. 20. Diffuse X-ray scattering of an oriented PET film extrapolated towards zero scattering angle as a function of φ for four different temperatures.

of thermal motion due to longitudinal vibration of long wavelength. It can be used to calculate the direction dependence of the group velocities of these vibrations which are shown in Figure 19.

Preliminary studies on an anisotropic sample of polyethyleneterephthalate (Fig. 20) show a not entirely vanishing anisotropy at very low temperatures which could be taken as an indication for small anisotropy in the short-range order. However, more detailed studies on a larger number of samples are necessary before more definite statements can be made.

REFERENCES

[1] R. J. Roe, *J. Appl. Phys.*, **36**, 2024 (1965).
[2] Laue, M. von, *Z. Kristallogr.*, **82**, 127 (1932).
[3] W. Ruland, *J. Appl. Phys.*, **38**, 3585 (1967).
[4] W. Ruland, and H. Tompa, *Acta Crystallogr.*, **A24**, 93 (1968).
[5] O. J. Guentert, and S. Cvikevich, *Carbon*, **1**, 309 (1964).
[6] A. Fourdeux, R. Perret and W. Ruland, *J. Appl. Crystallogr.*, **1**, 252 (1968).
[7] J. Rathje, and W. Ruland, *Coll. Polym. Sci.*, **254**, 358 (1976).

RECRYSTALLIZATION OF POLYMERS DURING DRAWING

H. ČAČKOVIĆ, J. LOBODA-ČAČKOVIĆ and
R. HOSEMANN

Fritz-Haber-Institut der Max-Planck-Gesellschaft, Teilinstitut für Strukturforschung, Berlin/Germany

SYNOPSIS

It is known that in branched polyethylene during cold-drawing a solid state diffusion of the single chains takes place during which the branches are trapped in the amorphous phase and an equilibrium state of highest possible crystallinity is attained if $\epsilon \leq 1.3\%$. This no longer can be achieved for larger ϵ values and the sizes $\overline{D}_{002}$ of the microparacrystallites (mPC) in fiber direction attain a definite smallest value of 60 Å. This is explained by a feedback of the lateral dimensions of the mPCs which on account of paracrystalline distortions, have sizes of 60–100 Å in a metastable state. The driving force of a minimum surface-to-volume ratio stabilizes therefore the $\overline{D}_{002}$ value and constraints the incorporation of more and more branches into the mPCs. Now the volume of the lattice cell increases monotonically with $\epsilon > 0.013$ and the melting point decreases according to Sanchez and Eby.

The superstructure of high polymers can be characterized as a network consisting of tie-molecules bound together by ordered domains in the range of 50 to 500 Å. They are the knots of this network. The drawing process results in a plastic deformation and destruction of these knots which reorder to new ordered regions. The ordered regions are the microparacrystallites (mPCs). In this work we will discuss only samples cold-drawn until the natural draw ratio. It could be shown by small and wide angle X-ray scattering (SAXS and WAXS) that in linear meltcrystallized polyethylene (PE) the average sizes of the mPCs are transformed by drawing into about twenty smaller ones which lie approximately like a string of pearls in the draw direction [1].

How this recrystallization takes place was investigated with respect to crystallinity, mPC size, and incorporation of defects with PE of various branching ratios ϵ from 0.05 to 3%. According to infrared measurements [2] PE is short chain branched with a $C_2H_5:C_4H_9$ ratio 2:1.

A statistical calculation based on the two-phase model leads to the relation between the crystallinity m/n and the branching ratio ϵ [3]:

$$m/n = e^{-m\epsilon} \tag{1}$$

m is the mean number of monomers in chain direction within one mPC measured from the 002 wide angle reflection; n is obtained from SAXS and gives the number of monomers in chain direction between the centers of adjacent mPCs (repeat distance). Three assumptions lead to eq. (1): (a) The branches are distributed randomly along the chains. (b) During drawing all branches segregate into the amorphous phase. (c) The sample tends to obtain the highest possible crystallinity during the drawing process.

59

Journal of Polymer Science: Polymer Symposium 58, 59–64 (1977)

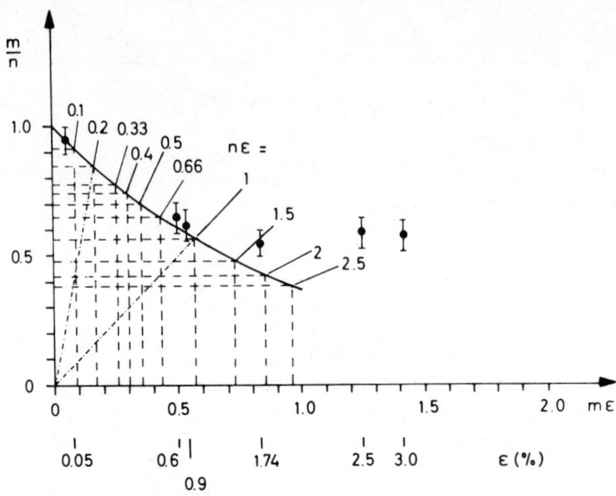

FIG. 1. Theoretical calculated "crystallinity" m/n as a function of $m\epsilon$ (curve) and the experimental points obtained from X-ray measurements. ϵ on the bottom indicates only the branching ratios of the six measured samples and is no scaling factor.

This theory defines function (1) until $m\epsilon = 1$. The experiments fully establish the correctness of all these assumptions (Fig. 1). The values m/n measured with SAXS and WAXS lie satisfactorily on the theoretical curve [eq. (1)] to $m\epsilon = 0.6$ and then are constant with $m/n = 0.55$. For lower values of $\epsilon \lesssim 1.3\%$, both the size $\overline{D}_{002}$ of the mPCs in chain direction and the repeat distance diminish with increasing ϵ [4]. If $\epsilon > 1.3\%$, $\overline{D}_{002}$ stays constant ~ 60 Å until $\epsilon = 3\%$. Obviously more and more branches are built now into the mPC in this ϵ range.

For branched meltcrystallized PE ($\epsilon = 5$ and 10%) Kilian and Müller [5] found, by differential thermal analysis (DTA), a size of $\overline{D}_{002} = 33$ Å for the smallest mPC in the sample. Taking into account the large polydispersity of the mPCs, these values agree satisfactorily with the WAXS result. Such small particle sizes were found also in other materials. For cold-drawn isotactic polypropylene, mPCs were measured with lateral diameters of 50 Å and heights of 40 Å [5]. In amorphous isotactic polystyrene, particle sizes of about 40 Å were found by electron microscopy [6]. By SAXS on polycaprolactam and polyvinylalcohol, particles of about 60 Å diameter were found [7]. Obviously these minimum possible average sizes of the mPCs in high polymers have thermodynamic reasons. Before discussing them we have measured the lateral sizes of branched PE too (Fig. 2). Both $\overline{D}_{110}$ and $\overline{D}_{200}$ decrease similar to $\overline{D}_{002}$ with increasing branching ratio ϵ and stay constant for $\epsilon \geq 1.3\%$.

The open question is why for $\epsilon > 1.3\%$ the perfect segregation of branches into the amorphous phase is finished and more and more branches are built into the mPC, so that the crystallinity stays constant with $m/n = 0.55$. A direct proof for increasing incorporation of branches into the mPCs with $\epsilon > 1.3\%$ can be observed by measuring the positions of the WAXS reflections, from which one can calculate the volume of a lattice cell. According to Figure 3, it stays constant till $\epsilon \sim 1\%$ and then increases monotonically with ϵ. Swann [8], Preedy [9], Baker et al. [10], and Wunderlich [11], found on the other hand that meltcrystallized

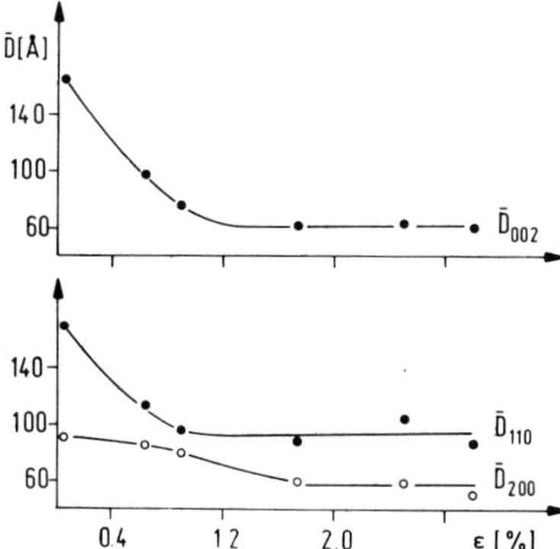

FIG. 2. The average mPC size in chain direction ($\overline{D}_{002}$) and orthogonal to it ($\overline{D}_{110}$ and $\overline{D}_{200}$) as a function of branching ratio ϵ.

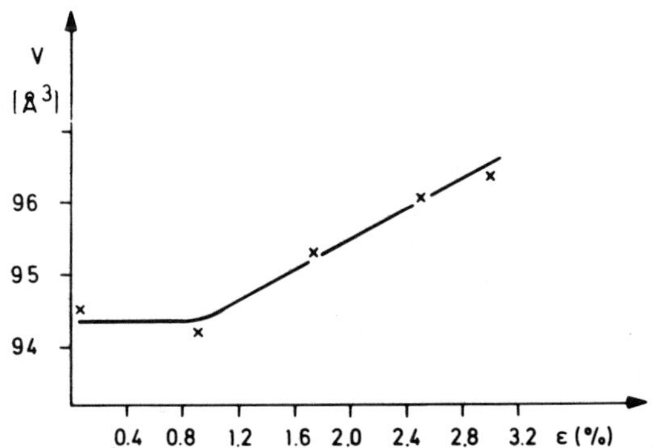

FIG. 3. The lattice cell volume V as a function of branching ratio ϵ.

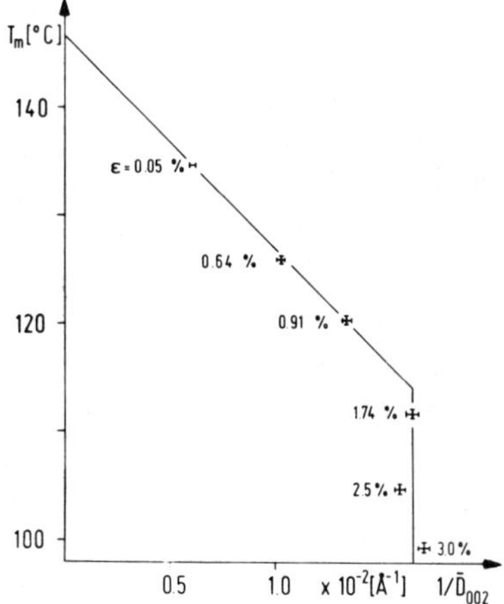

FIG. 4. Measured melting points T_m for drawn samples of various branching ratios ϵ (%) as a function of mPC height ($\overline{D}_{002}$).

branched PE has volumes increasing with ϵ beginning at $\epsilon = 0$. This is a direct proof that under stretching the branches are segregated into the amorphous phase totally until $\epsilon \cong 1\%$.

A further direct proof for this is the dependence of the melting point T_m on $\overline{D}_{002}$ and the concentration γ of defects. An equation exists from Lauritzen and Hoffmann [12], Sanchez and Eby [13] and Schultz [14]

$$T_m = T_m^0 \left[1 - \frac{2\sigma_e}{\Delta H_o} \frac{1}{\overline{D}_{002}} - \frac{\Delta H_d}{\Delta H_o} \gamma \right] \tag{2}$$

where ΔH_o is the heat of fusion per unit volume of the crystal with an "infinite" segment length at the equilibrium melting point T_m^0 and σ_e is the surface free energy normal to the chain direction per unit area. γ is the mole fraction of branches which are built into the mPCs and ΔH_d is the heat of transition associated with the incorporation of a defect. The measured results are plotted in Figure 4. $\epsilon \sim 1\%$, $\overline{D}_{002}$ decreases, as mentioned above, and the concentration γ of branches inside the mPCs is zero. It can be seen that the measured T_m decreases proportional to $1/\overline{D}_{002}$ and this confirms eq. (2) nicely. At $\overline{D}_{002} = 60$ Å, hence $1/\overline{D}_{002} \sim 0.0166$ Å^{-1}, T_m decreases further on with increasing ϵ because the concentration γ of branches in the mPCs now begins to grow. The extrapolation for an "infinite" crystal from our results on Figure 4 gives $T_m^0 = 146.6°C$ in good agreement with the result obtained by Flory and Vrij [15].

The experiments show that during drawing the chains suffer a solid-state diffusion and rearrange only until chain segments without branches are folded together to mPCs and until the crystallinity attains an optimal possible value. From statistical calculations this is theoretically possible until the number m of monomers within one fold distance is given by $1/\epsilon$. The crystallinity m/n does not reach the corresponding small value of 0.35 in reality, but remains constant with 0.55, if $m\epsilon$ has reached the value 0.6. The open question is why the crystallinity does not become smaller, if $m > 0.6$.

As mentioned above, eq. (1) is developed under the assumption of a total segregation and of reaching an equilibrium state with the highest possible crystallinity m/n. Such an equilibrium state exists, if $m < 1/\epsilon$ because branchless chain segments with too large m values hardly occur and a crystallinity m/n with such thick mPCs would therefore be very small. mPCs with too small m values need far too many backfolds which contribute to the amorphous phase and diminish m/n also. With increasing ϵ an equilibrium m must be therefore smaller than $1/\epsilon$ and m/n decreases according to eq. (1). Now a new thermodynamic driving force begins to work against the free energy driven force of the maximal crystallinity.

This is the free surface energy which tries to arrange a minimum possible surface-to-volume ratio. In the two-phase model, which was presupposed in the derivation of eq. (1), lateral boundaries of the crystalline phase do not exist but only folding surfaces. The lateral dimension of the crystalline domains is therefore unlimitedly large and this force therefore tries to increase m *ad infinitum*.

Taking into account the reality of lateral boundaries of the mPCs one obtains from Figure 2 lateral sizes $\overline{D}_{110}$ and $\overline{D}_{200}$ at larger ϵ values of 60–100 Å which are in the same order of magnitude as $\overline{D}_{002}$.

This leads to the conclusion that the surface free energy of the lateral boundaries is of the same order of magnitude as that of the fold surfaces.* $\overline{D}_{200}$ and $\overline{D}_{110}$ are determined by the paracrystalline distortions within the mPCs. They feed back to the size of $\overline{D}_{002}$ and m. They cause the deviation of m/n from the calculated kinetic curve at m ϵ values larger than 0.6 (Fig. 1) and prohibit a further adjustment of highest possible crystallinity m/n. As a consequence of this, an increasing percentage ϵ of branches must be assimilated from the mPCs which increase the volume of the lattice cell (Fig. 3) above $\epsilon = 1\%$. In this range $\overline{D}_{002}$ is constant and T_m decreases with increasing ϵ according to eq. (2) and Figure 4.†

We are grateful to the Deutsche Forschungsgemeinschaft for the kind support of our work. We thank Dr. F. J. Baltá-Calleja, Instituto de Estructura de la Materia, Madrid, for branching ratio measurements and Dr. P. H. Lindenmeyer for stimulating discussions.

REFERENCES

[1] R. Hosemann, J. Loboda-Čačković and H. Čačković, *Z. Naturforsch.*, **27a**, 478 (1972).

[2] F. J. Baltá-Calleja and A. Hidalgo, *Kolloid Z.Z. Polym.*, **229**, 21 (1969).

[3] H. Čačković, J. Loboda-Čačković, R. Hosemann, and D. Weick, *Colloid Polym. Sci.*, **252**, 812 (1974).

[4] H. G. Kilian and F. H. Müller, *Kolloid Z.Z. Polym.*, **192**, 34 (1963).

[5] J. Loboda-Čačković, R. Hosemann, H. Čačković, A. Ferrero, and E. Ferrancini, *Polymer*, **17**, 303 (1976).

[6] G. S. Y. Yeh, *J. Macromol. Sci., Phys.*, **B6** (3), 465 (1972).

[7] V. S. Kuksenko, O. D. Orlova and L. I. Slutsker, *Vysokomol. Soedin.*, **A15**, 2517 (1973).

[8] P. S. Swann, *J. Polym. Sci.*, **56**, 439 (1962).

[9] J. E. Preedy, *Br. Polym. J.*, **5**, 13 (1973).

[10] C. H. Baker and L. Mandelkern, *Polymer*, **1**, 71 (1966).

[11] B. Wunderlich and D. Poland, *J. Polym. Sci.*, **1A**, 357 (1963).

[12] J. I. Lauritzen, Jr. and J. D. Hoffman, *J. Res. Natl. Bur. Stand.*, **64A**, 73 (1960).

[13] I. C. Sanchez and R. K. Eby, *J. Res. Natl. Bur. Stand.*, **77A**, 353 (1973).

[14] J. M. Schultz, *Polym. Mat. Sci.*, Prentice-Hall, Englewood Cliffs, New Jersey, 1974, p. 190.

[15] P. J. Flory and A. Vrij, *J. Am. Chem. Soc.*, **36**, 3548 (1963).

[16] J. D. Hoffman, L. J. Frolen, G. S. Ross and J. I. Lauritzen, Jr., *J. Res. Natl. Bur. Stand.*, **79A**, 671 (1975).

[17] N. Morosoff, K. Sakaoku, and A. Peterlin, *J. Polym. Sci.*, **A2**, 10, 1221 (1972).

[18] H. Čačković, R. Hosemann, and W. Wilke, *Kolloid Z.Z. Polym.*, **234**, 1000 (1969).

* This would of course be exclusive for the energy required to make the folds and would correspond to the σ_{eo} suggested by Hoffman *et al.* [16]. σ_{eo} is only a small amount (10%–20%) of the whole surface energy including folds. In earlier works values of 40 erg/cm² were published [19, 20]. Recent calculations yield values near 100 erg/cm² [16]. Morosoff *et al.* [17] found similar equilibrium shapes of mosaic blocks in drawn and then annealed polypropylene. Čačković *et al.* [18] found on the other hand in PE single crystals that during annealing an equilibrium shape of the micropara-crystallites exists where $\overline{D}_{002}/\overline{D}_{310} \sim 0.55$.

† Preedy et al. found similar effects in high ethylene EPDM terpolymer [21]. The crystallinity increases strain induced from 15% to 73% if stretched from 15% to 225% and therefore more and more distorted chain segments are built into the mPCs.

[19] A. Peterlin and E. W. Fischer, *Z. Phys.,* **159,** 272 (1960).
[20] A. Peterlin, E. W. Fischer, and Chr. Reinhold, *J. Chem. Phys.,* **37,** 1403 (1962).
[21] J. E. Preedy, S. Wallis, and E. J. Wheeler, *Makromol. Chem.* (in press).

A NEW CONCEPT FOR UNDERSTANDING THE PLASTOELASTICITY OF POLYMERS

R. HOSEMANN and J. LOBODA-ČAČKOVIĆ

Fritz-Haber-Institut der Max-Planck-Gesellschaft, Teilinstitut für Strukturforschung, Berlin, Germany

SYNOPSIS

Chain molecules of partially crystalline polymers are built up as a three-dimensional network using one part of their chain segments. The other parts of segments of each molecule are embedded in microparacrystals (mPCs). It is pointed out that the random coil concept is nevertheless a good first approximation for the real conformation of a chain molecule. Its straight segments build up the mPCs. They act as the knots of a three-dimensional network and explain the phenomenon of the affine transformation and retransformation properties of these structures. On this basis, one can develop a rheology which is in accord with the molecular structure. Mechanical models can be applied, which, contrary to those of conventional methods, substantially contribute Kelvin and Maxwell elements to microdomains and not to volume elements. One microdomain contains at least 10^6 mPCs, including their surrounding network. The positions of the centers of the microdomains obey the rules of affine transformation. They consist of a large number of mPCs. Stretching experiments with linear polyethylene Lupolen 6001H (BASF) are discussed as an example. In the pseudo-Hooke's law range until elongations of $\gamma \sim 30\%$, statistically distributed mPCs in suitable positions are plastically deformed. At longer γ-values cooperative deformations of layers of mPCs set in and give rise to a necking process. Stretching is thermoreversible under a natural draw ratio because the relaxation time of the Maxwell elements is long enough compared with the time taken for the microdomains to travel through the neck. Their retardation time is calculated and has a minimum value of 2 min, 6 mm behind the beginning of the neck.

INTRODUCTION

Polymers in the solid state offer a lot of macroscopic properties which are unknown in solid-state physics of crystals or polycrystalline materials. It is therefore not surprising that all attempts must fail using the methods developed for the low molecular world. Models working with mechanical or electrical elements (springs, dashpots, electrical capacities, resistances, or induction coils) applied to volume elements of the solid must fail because polymers are colloidal systems with large inner surfaces. The rheology of polymers must be based on their structure. A direct way of understanding the special properties of polymers is available by a sophisticated analysis of diffraction phenomenon in wide angle (WAXS) and small angle (SAXS) domains combined with electron micrographs, nuclear magnetic resonance (NMR), and relaxation measurements.

Journal of Polymer Science: Polymer Symposium 58, 65–75 (1977)

It is the aim of this work to collect all those experimental data which are typical of polymers and of importance for an adequate rheology different from that of a polycrystalline material.

THE REALITY OF MICROPARACRYSTALLITES (mPCs)

The WAXS of partially crystalline polymers is quite different from that of polycrystalline material. It proves that microdomains exist in polymers proportional to their crystallinity with relatively well-arranged segments of the chain molecules [1]. Contrary to microcrystals they include point defects as a consequence of their conformational variety which are more or less randomly distributed. The free enthalpy of these so-called microparacrystals (mPCs) can therefore not be minimized. In this metastable thermodynamic state, liquidlike lattice distortions exist which can be described by the theory of paracrystals [2]. These distortions are the reason for the plasticity and the smallness of the mPCs and other properties unknown in crystals.

α^*-RULE

Collecting many WAXS data of synthetic polymers, biopolymers, catalysts and other colloids, one finds that the highest number N of netplanes of one mPC and the standard deviation

$$g = (d^2/\overline{d}^2 - 1)^{1/2} \tag{1}$$

of their netplane distances d are connected by the rule

$$g\sqrt{N} = \alpha^* \tag{2}$$

where α^* generally has values between 0.1 and 0.2 [3]. More or less statistically within one mPC, distributed point defects generate g values of some percent or less, and N is therefore of the order of 3 to 20 (or more), the mPCs have sizes of 10–100 Å (or sometimes more). This is the reason why all polymers and catalysts have extremely large inner surfaces unknown in the low molecular world and cannot crystallize whilst the g values are large enough. Otherwise they would lose their specific properties, a process known as aging in colloid chemistry [4].

MICRODOMAINS AND THE PHENOMENA OF AFFINE TRANSFORMATION AND RETRANSFORMATION

From rubber elasticity it is well known that a three-dimensional network of chain molecules exist. The molecules are fixed to each other at certain conjunctions by covalent bonds. If such a network is stretched, its entropy is diminished and it tends to return entropically to its old macroscopic shape if the temperature is above the glass transition. In general, such covalent bonds do not exist in nonirradiated semicrystalline polymers. Nevertheless these polymers also have a kind of memory and try to return to their old shape after deformation.

For instance polyethylene (PE), if stretched under a natural draw ratio (1:10 or more), retransforms back to its old macroscopic shape if heated above the melting point swimming in a glycerine bath [5]. This can only be understood by the fact that the mPCs act as knots of a three-dimensional network [6]. Careful studies prove that the deformation under stress and the retransformation above the melting point follow the rule of affine transformation which occurs down to dimensions of $\sim$10 μm [5]. Taking into account that the distance between the centers of neighboring mPCs is in the order of 100 Å one learns that each microdomain contains mPCs and their surrounding network of the order of 10^6 individuals.

RADIATION-INDUCED CROSSLINKS WITHIN THE MICROPARACRYSTALLITES

The important role of the mPCs as knots of a three-dimensional network is supported by the fact that the memory at the *unstretched* state is greatly strengthened if the *stretched* sample is irradiated with ^{60}Co radiation. This means that practically no crosslinks are generated between different molecules in the three-dimensional network because they would fix the stretched form. The crosslinks arise predominantly within the mPCs which are the knots of the rubberlike network of chain molecules. Now each knot consists, not only of one covalent bond between different chain molecules, but of thousands of van der Waals bond between the more or less well-aligned chain segments within one mPC and of some crosslinks when irradiated. This automatically explains the better memory of the undeformed state when irradiated in the stretched state. The high probability of generating crosslinks within the knots (not in the amorphous phase) can be explained by the fact that the knots are not microcrystals but micro*para*crystals in which mobile kinks are embedded and because of which a kind of solid state diffusion is possible. If two kinks of two adjacent chain segments meet each other at radiation-activated —$\dot{\text{C}}$H— groups, a crosslink can be easily formed within the mPCs because its conformation is quite similar to that of the two neighboring kinks [7].

RANDOM COIL MOLECULES AND mPCs

Neutron diffraction of mixtures of polyethylene with deuterated PE shows that the average diameter of single "coiled" chain molecules is practically the same in the meltcrystallized state as in high disoluted solvents of pe at the θ-point in xylene [8]. From this it is concluded that in the solid state the single molecules have the same random coiled shape as in solution. At first this seems to contradict the existence of mPCs, but by introducing the concept of the three-dimensional network with knots, one can combine both concepts: Each molecule lays randomly coiled through a series of mPCs, but within each mPC it is necessary that one or more chain segments must be well-aligned with segments of other molecules built into the same mPC (Fig. 1). This is the reason why partially crystalline polymers have knots even when not irradiated. Contrary to rubber, they

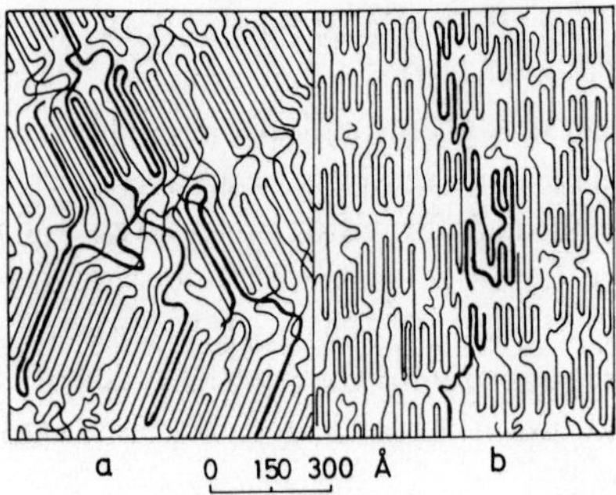

FIG. 1. Schematic drawing of the conformations of the chain molecules in semicrystalline polymers. Their averaged size is similar to the randomly coiled models in solutions at the θ-point. Contrary to these models a part of the monomers proportionally to the crystallinity runs through the mPCs and has therefore straight segments with some backfolds at the fold surfaces of the mPCs. Some molecules are fat-drawn: (a) meltcrystallized. The arrangement of the mPCs in spherolithes is not taken into account; (b) after natural draw ratio. Each mPC is separated into a couple of smaller ones aligned like pearls on a string.

have much larger volumes and must be deformed plastically when the sample is stretched. The entropy driving force of retransformation must therefore consume energy to deform the mPCs once again into shapes similar to those in the undrawn state. It is not surprising that reversibility exists only when the knots are warmed up to a large enough plasticity.

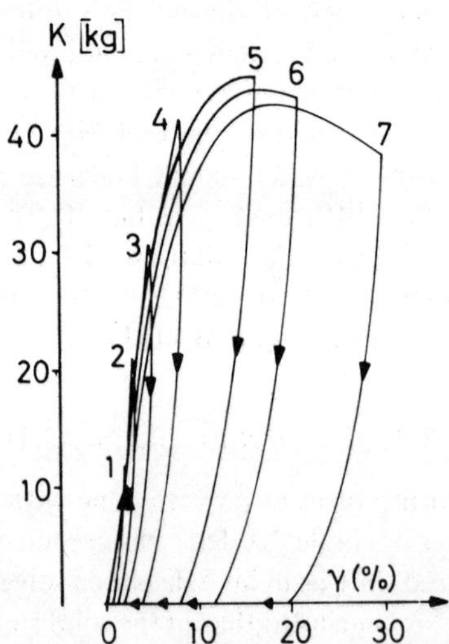

FIG. 2. Iterative drawing and relaxation of PE 6001 H (BASF) with increasing draw ratios of 2.5%, 5%, 7%, ... at room temperature. The rod with a 4 × 4 mm cross-section is 80 mm long. The drawing velocity is 10 mm/minute. Even beyond the yielding point 5 until $\gamma = 30\%$, the deformation of the rod is linear and follows the range of pseudo-Hooke's law, but becomes more and more tarry reversible.

THE RANGE OF PSEUDO-HOOKE'S LAW

Iterative stretching and relaxing with increasing γ-values (2.5%, 5%, 7%, 10%, ...) shows that the elasticity becomes smaller after each expansion and back shrinkage (Fig. 2). This is easily explained by an increasing number of plastically deformed mPCs. Those which are oriented within the stress field where twin generation [9], shearing along certain netplanes [10] or walking dislocation lines [11] can happen very easily, are initially plastically deformed. Since these mPCs are statistically distributed, the macroscopic shape of the rod shows a linear deformation. But in colloidal dimensions heterogeneous processes take place and Hooke's law is no longer applicable; very small γ-values ($\sim$0.1%) and short expansion times (high frequencies) excepted. If the stress is removed, the rod shrinks and practically reaches its old length within several minutes if $\gamma < 5\%$. Until $\gamma = 30\%$ the rod returns to its old length through more and more tarring because the statistically distributed deformed mPC mostly cannot be recast on account of the weak redriving force of the network. For instance, when stretched to $\gamma = 30\%$, it needs 7 min at 23°C to shrink back to $\gamma = 1.5\%$. The yield point 5, in Figure 2, indicates that the production of elongated mPCs has become so large that the lengthening of the rod with a constant velocity after passing the yield point, can be performed with smaller forces. At room temperature the stretching is reversible up to $\gamma = 30\%$ in the former macroscopic sense, but it is surely not so in a structural and thermodynamic sense because deformation heat is created and many mPCs have irreversibly changed their shape and structure.

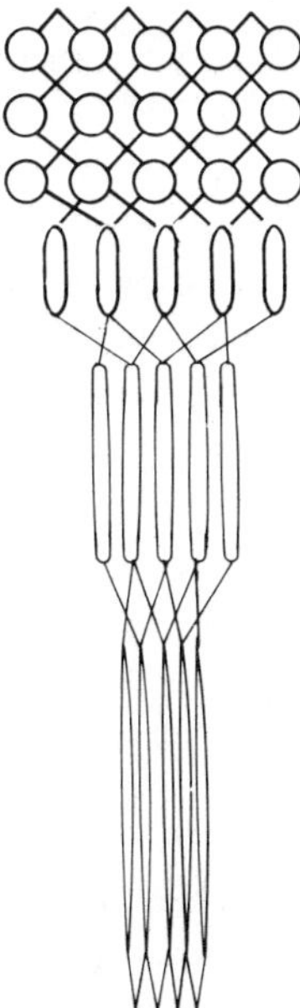

FIG. 3. Schematic drawing of the microdomains and the network within a neck. Each domain consists of about 10^6 mPCs and of the network tying them together.

NECKING PROCESS

We have seen that beyond the yield point the production of deformed mPCs becomes so large that only one fraction of the force is necessary to maintain the constant expansion velocity. Now the deformation changes to an unhomogeneous one. In certain parts of the network the mPCs are deformed so greatly that there the network is loosened. As a consequence a higher stress acts now between neighboring mPCs in a direction orthogonal to the strain direction, while it is smaller for mPCs in the strain direction. Because of this a cooperative plastic deformation of layers of mPCs begins orthogonal to the draw direction, in connection with growing microcrazes within those layers. The affine deformation of the position of the microdomains is nevertheless still fully established after passing the neck. A 6-mm-long part cut from the stretched rod, with planes cut orthogonal to the draw direction retransforms to a 0.6-mm-thick plane slice after being molten in a glycerine bath [5]. This proves that the layers of microdomains, which were orthogonal to the draw direction before stretching, still remain as plane layers orthogonal to the draw direction in the neck; this is drawn schematically in Figure 3. The stress field itself cannot be equal to all microdomains of one layer because the network in the outer part of the layer is deformed differently from the central part (see Fig. 3). As a consequence the WAXS and SAXS pattern of the different parts of one layer are different [12, 13]. With respect to macroscopic properties (such as temperature and plasticity) each layer n has a specific nature defined by its individual elongation γ_n [eq. (7)]. In spite of the deformation and the disintegration of its mPCs, which are different in the centers from the outer parts, it is logical to discuss these macroscopic properties as averages of all microdomains and their mPCs in one layer. The deformation heat enlarges the temperature within each layer and thus its plasticity also. The temperature and plasticity attain maximum values if the heat production by plastic deformation equals the heat lost by diffusion. This happens near the end of the neck, as shown below, where the mPCs have deformed so much that they are disintegrated into a couple of small mPCs. Natural draw ratio experiments with PE of different branching ratios ϵ prove that during this disintegration into a couple of smaller mPCs a forced solid-state diffusion of the single chains takes place where all branches are segregated out of the mPCs into the amorphous phase for values up to $\epsilon = 1.3\%$ [14]. At room temperature experiments with a 8 cm long rod of Lupolen H 6001 and a cross-section of 4 $\times$ 4 mm^2 showed that at the end of a tenfold elongation, each mPC is disintegrated, on average, into eighteen smaller mPCs. Here the temperature of a layer has become so small again that a further expansion takes place only very slowly. In our experiment, the neck, 15 mm from the front, had reached an elongation of $\gamma = 6.0$ which at 60 mm distance had increased slightly to $\gamma_e \sim 8.0$ (Fig. 6). Most of the necking process occurred in the first 15 mm. It is in this part that most of the heat of deformation is generated, as will be shown below.

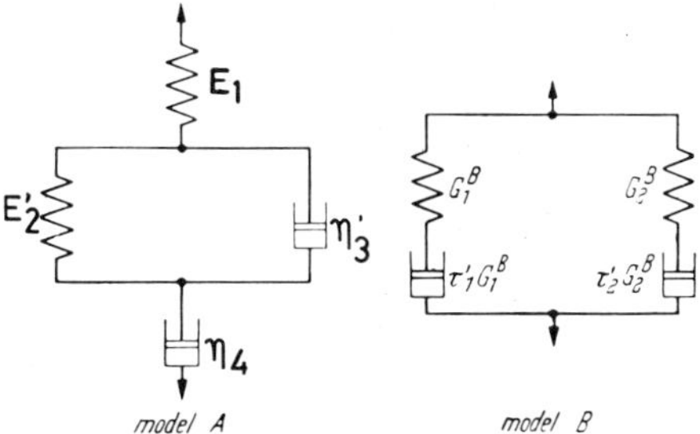

FIG. 4. Two mechanical models, each containing two dashpots and two springs. Both are equivalent in the rheology of low molecular solids. The stretching process of polymers can be described only by model A; model B is no longer equivalent to A.

MECHANICAL MODEL

Mechanical models with springs and dashpots are well known from the literature [15]. Figure 4 gives two models each with two springs and dashpots. Model A shows a spring of modulus E_1, a Kelvin(Voigt) unit with the constants E'_2 and η'_3 and a dashpot with the viscosity η_4 linked in series. The mechanical models of Figure 4 up until now are used to imitate properties of volume elements, assuming that all volume elements have the same properties. Model B then is nothing else than another representation of the mechanical behavior of model A. The quantities G_1^B, G_2^B, τ'_1 and τ'_2 can be expressed by the quantities E_1, E'_2, η'_3 and η_4 of model A.

The mechanical models can be applied nevertheless to the plastoelasticity of polymers if they are interpreted in quite another sense [16]. Taking into account the phenomenon of affine transformation and retransformation, where, according to Figure 2, all microdomains maintain their neighbors during deformation, one can substantially attribute one model A of Figure 4 to each domain. Now model A has the desired thermoreversible properties if the relaxation time

$$\tau_1 = \eta_4/E_1 \tag{3}$$

of the Maxwell element E_1, η_4 is large enough compared with the time where one layer n of microdomains passes through the necking zone. The retardation time

$$\tau_2 = \eta'_{3n}/E'_{2n} \tag{4}$$

of this assembly of microdomains of each layer n then comes into action solely. The dashpots which represent the average value of the plasticity of the mPCs of one microdomain and the springs E'_2 which represent the entropy driving force of the three-dimensional network around the mPCs are elongated about 1:10 at the end of the necking process. A thermoreversible retransformation is possible by heating the sample to an elevated temperature for a time $t \sim \tau_2$, if τ_2 is small enough compared with τ_1. Model B offers no thermoreversible properties. It is no longer a representation of the microdomains.

Figure 5 gives one of the well known generalized Voigt models which must

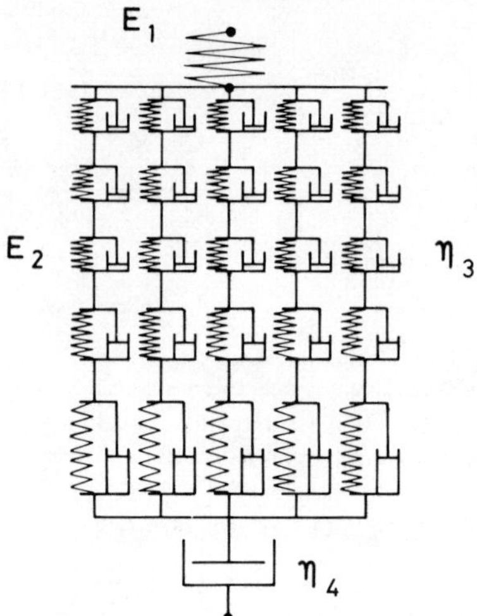

FIG. 5. Mechanical model of Figure 3 under the condition of natural draw ratio. For simplicity then only one Maxwell element can be put in series with them because the microdomains trace through the necking zone in a time which is small enough compared with the relaxation time of the Maxwell elements.

also now be understood in quite another sense: $n = 6$ layers of 5 microdomains per layer are drawn schematically representing the situation within one neck. Each layer has another draw ratio and another average of temperature and retardation time τ_2. For simplicity one can replace the Maxwell elements of each microdomain in Figure 5 by one Maxwell element in series with all Voigt elements, since under the condition of natural draw ratio the relaxation time τ_1 is large compared with the time of expanding a layer n tenfold.

The advantage of this model is that the Young moduli E can be replaced by spring forces E' (kp) of the microdomains and the viscosities η by quantities η' (kp sec) which are inversely proportional to the plasticity averaged over all mPCs of one layer. One then can calculate the deformation process in the neck, if one knows the temperature dependence of the plasticity of the mPCs and takes into account their heat production and heat capacity during plastic deformation and the heat diffusion to the environment. In the next section it is shown that the retardation time τ_2 can be calculated without knowledge of all these properties by analyzing solely the profile of the neck.

CALCULATION OF THE RETARDATION TIME τ_2 OF MICRODOMAINS IN A NECK

We will demonstrate with the help of the practical example mentioned above how one can calculate τ_2 from a profile analysis of the neck. If the neck is fully developed, both shoulders have attained dynamic equilibrium shapes and propagate to the ends of the rod of the initial cross-section D_0^2 with a constant velocity $\dot{d}_0$, which can be calculated: Let D_x be the final thickness of the rod after

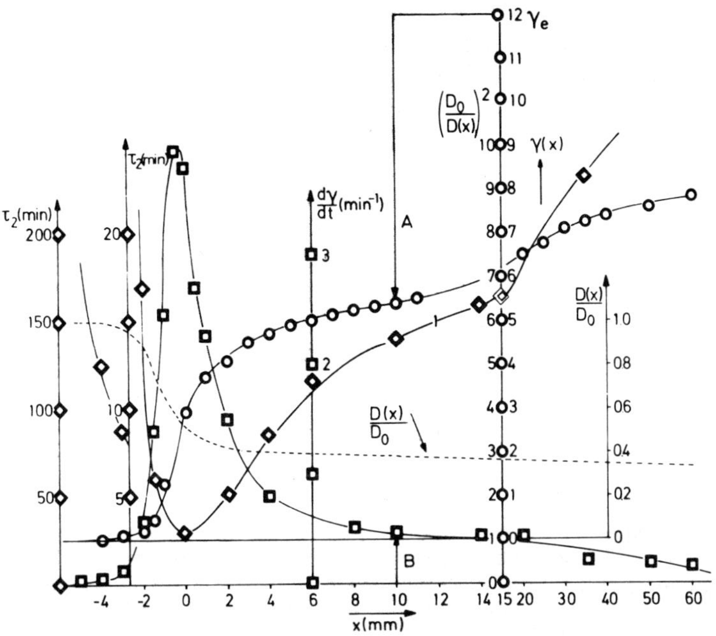

FIG. 6. The measured profile $D(x)/D_0$* of a neck in PE 6001 H produced under the conditions described in Figure 2, but without relaxation. The retardation time τ_2† is the quotient of the lengths A and B at any point x according to eq. (11). At $x = 0$ the minimum value lies of $\tau_2 = 2.5$ minutes, at $x = -6$ mm the necking process begins.

full elongation and Δl the elongation of the rod, then a good first approximation is

$$\Delta l \cdot D_0^2 = x_0(D_0^2 - D_x^2) \tag{5}$$

where x_0 is the distance between the half-widths $D^2 = \frac{1}{2}(D_0^2 + D_x^2)$ of the cross-sections of both shoulders. Assuming the neck is built up symmetrically at the middle of the rod with the initial length l_0 then the half-width of the neck has the same distance a from both ends of the rock

$$a = \frac{1}{2}(l_0 + \Delta l - x_0); \quad \dot{a} = \dot{\Delta l} \frac{D_x^2}{D_0^2 - D_x^2} \tag{6}$$

$\dot{a}$ is the velocity with which each shoulder of the neck moves towards the respective end of the rod.

We can now easily calculate the speed v_n of elongation γ_n of the single layers n within the neck, if we introduce its shape. Let us divide the unstretched rod into layers orthogonal to the stretching direction of the thickness d_0. During each second the neck may propagate by a distance $d_0 = \dot{a} \cdot 1$ sec. We now mark all these distances on the rod. If the cross-section of the nth layer is given by D_n^2 then the thickness d_n of the nth layer between adjacent time marks is given by

$$d_n/d_0 = 1 + \gamma_n = (D_0/D_n)^2 \tag{7}$$

* (---) and the local elongation γ (O O) with its true derivatives $d\gamma/dt$ (□ □).
† (◇ ◇).

where γ_n is the elongation of the nth layer. The function $D(x)/D_0$ measured from the neck profile along the rod is drawn in Figure 6 as a function of x; $x = 0$ is defined by the minimum value of τ_2 (see below). Furthermore we can calculate from the neck profile the quantity $d\gamma/dx$. Its time deviation $\dot{\gamma}$ is given by

$$\dot{\gamma} = \frac{d\gamma}{dx}\frac{dx}{dt} \tag{8}$$

dx/dt is nothing else than the distances between adjacent time necks. These are given by eq. (7). So we finally obtain

$$\dot{\gamma}_n = \frac{d\gamma}{dx}(1 + \gamma_n) \tag{9}$$

One now can calculate the retardation time τ_2 [eq. (4)] of the Kelvin elements with the help of all these measured quantities, if the time of passing the neck is small compared with the relaxation time of the Maxwell elements [eq. (3)]. This is realized in the above discussed experiments, because they are totally thermoreversible.

The force K_n acting on one microdomain of a layer n is given by its spring force E'_{2n} and plasticity $1/\eta'_{3n}$:

$$K_n = E'_{2n}\gamma_n + \eta'_{3n}\dot{\gamma}_n \tag{10}$$

hence

$$\tau_2 = \frac{\eta'_{3n}}{E'_{2n}} = \frac{\gamma_e - \gamma_n}{\dot{\gamma}_n} \tag{11}$$

γ_n is known from eq. (7), $\dot{\gamma}_n$ from eqs. (6), (8), and (9). γ_e is the final draw ratio which is attained for $x > 60$ mm and has a value between 8 and 12, as in Figure 6. At a distance of 6 mm from the beginning of the neck τ_2 has reached a minimum value of 2.5 min (2 min) for $\gamma_e = 12$ ($\gamma_e = 8$). In Figure 6 we put $x = 0$ here arbitrarily. At $x = -4$ mm, τ_2 has a value of 120 min at the beginning of the neck, a very large and not calculable value. At $x = 10$ mm with $\gamma = 4.5$, τ_2 has increased to 15 minutes and at $x \geqq 60$ mm far away from the neck to a value again immeasurably large.

DISCUSSION

The structural model of mPCs as knots of a three-dimensional network developed from WAXS and SAXS analysis gives a basis for a rheology which agrees with the structure of many polymers. Here we have solely discussed an application to samples of PE at room temperature, far above the glass transition point, where the plasticity of the mPCs is relatively large. Taking into account the heat production under plastic deformation and the plasticity of the mPCs, the heat capacity and the diffusion of heat, a theory obviously can be developed which can deduce the phenomena of deformation and shrinkage effects. It may be possible to refine this structural model to structures as PVC or PMMA near or below the glass transition, where the plasticity of the mPCs is nearly zero during the time of observation. Then the microcrazes play a predominant role for the technological properties of these samples and the Maxwell elements come into action because of thermo-irreversible processes.

REFERENCES

[1] R. Hosemann and J. Loboda-Čačković, *J. Polym. Sci.*, **53,** 159 (1975).
[2] R. Hosemann and S. N. Bagchi, *Direct Analysis of Diffraction by Matter,* North-Holland Publ. Comp., Amsterdam, 1962.
[3] R. Hosemann, *CRC Crit. Rev. Macromol. Sci.,* **1,** 351 (1972).
[4] R. Hosemann, *Progr. Colloid Polym. Sci.,* **60,** 213 (1976).
[5] R. Hosemann, J. Loboda-Čačković, and H. Čačković, *Z. Naturforsch.,* **27a,** 478 (1972).
[6] G. S. Y. Yeh, R. Hosemann, J. Loboda-Čačković and H. Čačković, *Polymer,* **17,** 309 (1976).
[7] R. Hosemann, H. Čačković, and J. Loboda-Čačković, *Makromol. Chem.,* **176,** 3065 (1975).
[8] J. Schelten, G. D. Wignall, D. G. H. Ballard, and W. Schmatz, *Colloid Polym. Sci.,* **252,** 749 (1974).
[9] M. Pietralla, Dissertation: "Die Funktion der Zwillingsbildung bei der Deformation teil-kristalliner Polymerer," Universität Ulm, 1974.
[10] W. Wilke and K. W. Martis, *Colloid Polym. Sci.,* **252,** 718 (1974).
[11] W. Pechhold and S. Blasenbrey, *Kolloid Z. Z. Polym.,* **241,** 955 (1970).
[12] N. Kasai and M. Kakudo, *J. Polym. Sci.,* **A2,** 1955 (1964).
[13] A. Peterlin and F. J. Baltá-Calleja, *Kolloid Z. Z. Polym.,* **242,** 1093 (1970).
[14] H. Čačković, J. Loboda-Čačković, and D. Weick, *Colloid Polym. Sci.,* **252,** 812 (1974).
[15] A. J. Stavermann and F. Schwarz, "Linear Deformation Behaviour of High Polymers," in, *Physik der Hochpolymeren,* A. H. Stuart, Vol. 4, Springer-Verlag, Berlin, 1956.
[16] R. Hosemann, J. Loboda-Čačković, and H. Čačković, *Colloid Polym. Sci.,* **254,** 783 (1976).

STUDY OF ORIENTATION IN AMORPHOUS POLYMERS BY THE CREEP THERMOMECHANICAL METHOD

P. FORGÁCS and P. HEDVIG

Research Institute for Plastics
H-1950 Budapest, Hungary

SYNOPSIS

Tensile creep of unplasticized and plasticized PVC and toughened polystyrene compounds were recorded as a function of the temperature with the UNIRELAX device developed in our laboratory and produced by Tetrahedron Associates, San Diego, California, U.S.A. The thermomechanical curves obtained this way were found to be governed by the creep process as well as by the recovery of the elastic deformation frozen-in by processing. Using small stress levels (10 g/mm^2) the thermomechanical curve was mainly determined by the recovery of the frozen-in deformation. Effects of thermal history on the creep thermomechanical curves were studied and interpreted. Using a mathematical formalism derived earlier for amorphous polymers, the curves were computer-processed in order to separate creep from the recovery of the frozen-in deformation and to obtain isothermal kinetic curves.

INTRODUCTION

Creep behavior of plastics is essentially determined by two processes of opposite direction: actual creep induced by the applied stress and shrinkage caused by the recovery of the frozen-in deformation which is always present unless special care in processing or annealing is taken. This recovery usually is very slow at room temperature; for full recovery, a time of the order of 10^{18} sec or more is needed, but investigation of its kinetics can be made quicker by elevating the temperature. Such measurements are easily made at swept temperature rather than isothermally at an elevated temperature.

In the present paper the shrinkage of several compounds of amorphous polymers is studied by recording extensional thermomechanical creep curves at very low load levels. A formalism based on the Boltzmann and time–temperature superposition principles is given, allowing a qualitative interpretation and quantitative transformation of thermomechanical data into long-term isothermal ones.

Journal of Polymer Science: Polymer Symposium 58, 77–84 (1977)

EXPERIMENTAL

A modified multichannel version of the UNIRELAX device constructed in this laboratory and produced by the Tetrahedron Associates Inc., San Diego, California was used for the measurements. This device is capable of recording, among other thermal and electrical parameters, extensional creep curves at constant and pulsed loads continuously from $-150°C$ up to $250°C$ with four parallel samples. The thermomechanical curves obtained this way were processed by an HP-65 programmable calculator.

Samples were processed under controlled conditions in a laboratory machine.

The results reported here are parts of a complex study involving morphological, calorimetric, dynamic mechanical, dielectric, and intermittent load thermomechanical investigations using the facilities of the UNIRELAX device.

THEORY

For quantitative interpretation of the results, the linear response approximation was used assuming the validity of the Boltzmann and time–temperature superposition principles. The process is separated into two parts: one part is assumed to have started at time $-t_1$ under a stress $\sigma(t)$ and finished at $t = 0$. This part is referred to as the history of the sample, which may include previous stretching or even processing. The second part of the process is considered to start at $t = 0$; in this period of time the stress is zero, it is referred to as the recovery part.

In the recovery region the extensional strain is expressed as [1]

$$\epsilon(t) = (D_r - D_u) \int_{-t_1}^{0} \int_{-\infty}^{\infty} L(ln\tau) \frac{\sigma(u)}{\tau} \exp\left(-\frac{t-u}{\tau}\right) dud(ln\tau) \\ + \frac{1}{\eta} \int_{-t_1}^{0} \sigma(u)du \quad (1)$$

where D_r and D_u respectively are the relaxed and unrelaxed extensional creep compliances, τ is the creep retardation time, $L(ln\tau)$ is the retardation time distribution function, η is the viscosity.

The temperature dependence of the creep retardation time is introduced into eq. (1) by defining an effective time [2, 3] as

$$t_{\text{eff}}(T) = \int_{T_0}^{T} \frac{1}{q} \frac{dT'}{a(T',T_r)} \quad (2)$$

where T is the actual, T_0 is the starting temperature, q is the rate of heating, $a(T',T_r)$ is the time temperature shift factor defined as

$$a(T,T_r) = \tau(T)/\tau(T_r) \quad (3)$$

where T_r is an arbitrary reference temperature. The temperature dependence of the shift factor is usually determined by the WLF equation, or can be measured separately. When the time is the independent variable

$$t_{\text{eff}}(t) = \int_{0}^{t} \frac{ds}{a[T(s),T_r]} \quad (4)$$

Assuming that the relaxed and unrelaxed compliances are independent of temperature it is found that

$$\epsilon(t) = \int_{-\infty}^{\infty} L(ln\tau) \exp\left(-\frac{t_{\text{eff}}}{\tau}\right) \epsilon_\sigma(\tau)d(ln\tau) + \int_{-t_1}^{0} \frac{\sigma(u)}{\eta(u)} du \qquad (5)$$

where

$$\epsilon_\sigma(\tau) = (D_r - D_u) \int_{-t_1}^{0} \frac{\sigma(u)}{\tau} \exp\left[-\frac{1}{\tau} \int_{u}^{0} \frac{ds}{a(T(s),T_r)}\right] du \qquad (6)$$

represents the partial elastic deformation corresponding to the retardation time τ.

At the beginning of the recovery region, i.e., at $t = 0$ from eq. (5) one obtains

$$\epsilon(0) = \int_{-\infty}^{\infty} L(ln\tau)\epsilon_\sigma(\tau)d(ln\tau) + \int_{-t_1}^{0} \frac{\sigma(u)}{\eta(u)} du \qquad (7)$$

The first term of eq. (7) is the elastic part of the deformation frozen-in at the end of the history period; this is recoverable, while the second term, the plastic deformation, is not.

From eqs. (5) and (7) the contraction of the sample in the recovery period is expressed as

$$\Delta\epsilon(t_{\text{eff}}) = \epsilon(0) - \epsilon(t)$$

$$= \int_{-\infty}^{\infty} L(ln\tau)\epsilon_\sigma(\tau)[1 - \exp(-t_{\text{eff}}/\tau)]d(ln\tau) \qquad (8)$$

This equation is applicable to the isothermal as well as thermomechanical case, according to eqs. (2) and (4).

Equation (8) makes it possible to calculate the contraction of the sample in the recovery period either at constant temperature or when the temperature is a known function of time.

RESULTS AND DISCUSSION

The validity of the given formalism has been checked by model experiments applying a constant stress $\sigma = \sigma_0$ uniaxially in the history period and allowing the material creep, then freezing in the deformation, and measuring the recovery at zero load while the temperature is scanned. In this case both the history and the recovery periods can be measured as well as calculated using the equations for the isothermal creep recovery experiments [1] by substitution of the real times by effective ones. In the history period the strain is given by

$$\epsilon(t) = \sigma_0 D_{T_r}(t_{\text{eff}})$$

where $D_{T_r}(t)$ is the extensional creep compliance at the reference temperature T_r.

In the recovery period

$$\epsilon(t) = \sigma_0[D_{T_r}(t_{\text{eff}}) - D_{T_r}(t^*_{\text{eff}})]$$

where t^*_{eff} is the effective time calculated by eq. (4), taking for lower limit of integration the moment when the stress was removed.

The temperature dependence of the shift factor as well as $D_{T_r}(t)$ were determined independently by the intermittent load technique described earlier [4].

A typical example of such an experiment is shown in Figure 1 for a typical plasticized polyvinylchloride compound. In the history period the sample was heated up to 60°C under an uniaxial stress $\sigma_0 = 37$ g/mm^2. The deformation was frozen in by a rapid cooling to −60°C. Then the temperature was increased again. At −33°C the stress was removed and the recovery was recorded. Figure 1 also shows the effective time scale for the reference temperature $T_r = 25$°C. Evidently the effective time changes quicker at higher temperatures and stays almost constant at lower temperatures.

Most of the discrepancies between the calculated and measured creep and recovery thermomechanical curves are due to the fact that the thermal dilatation was not taken into account.

Figure 2 shows the creep–thermomechanical curves at low load levels of samples cut out from an extruded unplasticized PVC profile at different directions. By such experiments the history period is the processing, so the actual stress is not known. The thermomechanical recovery curves represent the macroscopic manifestation of the frozen-in orientation of the sample viewed in one direction.

As in these experiments a small load is applied the recorded curves are composed of the recovery (shrinkage) and actual creep processes. Along the direction of the extrusion (II) a minimum strain is observed as in this orientation the value of $\epsilon_\sigma(\tau)$ [see eq. (6)] is much higher than that in the perpendicular orientation, leading to a more significant shrinkage.

By the approximation of the proposed linear formalism, a common effective time scale can be given for all directions. The shift factor values were measured for temperatures below 80°C, and were calculated by the WLF equation above this using the general constants [5] and taking $T_g = 75$°C. The effective time scale was obtained for $T_r = 60$°C by eq. (2) with $T_0 = 20$°C and a heating rate $q = 1.5$°C/min. It is seen that at 60°C significant deformation appears at time $t = 10^4$ sec.

Figure 3 shows the effect of storage below T_g (70°C) for 671 hr on the thermomechanical creep recovery curves of an extruded unplasticized PVC sample at two different load levels. It is seen that after prolonged storage below T_g the contraction decreases but still remains large, showing that recovery process is very slow. The strain minima are out of scale in this figure.

The shrinkage of stored samples is less as the frozen-in elastic deformation has already partly recovered. The storage immediately below T_g also may have an influence on the nonequilibrium state of the glassy polymer, resulting in a change in the temperature dependence of the shift factor [6].

Figure 4 shows a series of thermomechanical creep curves of impact polystyrene (Dow Styron 457) injection molded at different melt temperatures, as indicated. In this case, the history period was the injection molding process. The

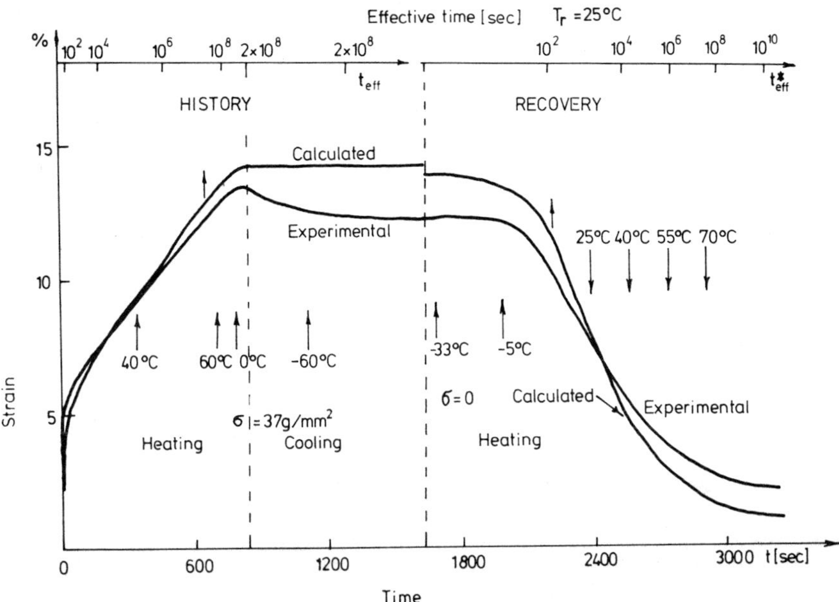

FIG. 1. Model experiment with 30% di-iso-octyl phthalate plasticized polyvinylchloride. History period: creep at a load of 37 g/mm² with increasing temperature (1.5°C/min), then rapid cooling. Recovery period: heating up at a rate of 1.5°C/min at zero load.

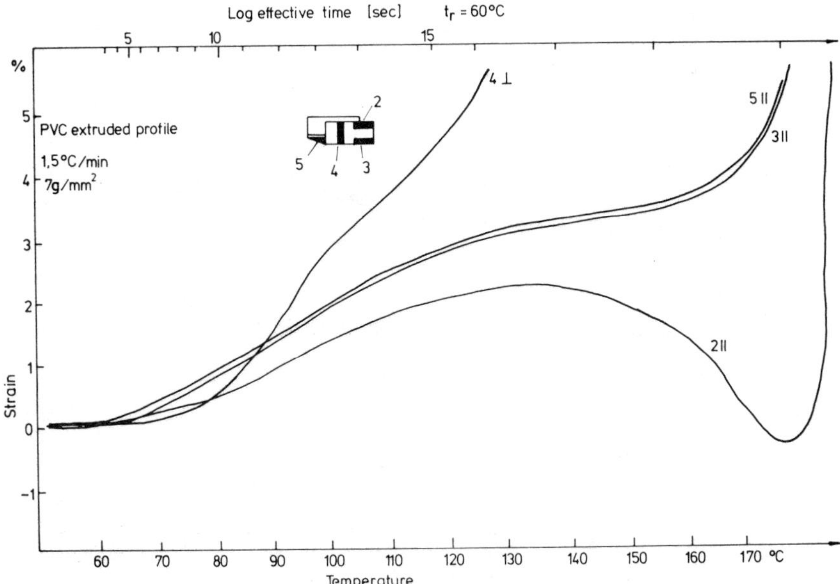

FIG. 2. Thermomechanical creep curves of samples cut out from an extruded polyvinylchloride profile at different directions as shown. Load 7 g/mm², speed 1.5°C/min.

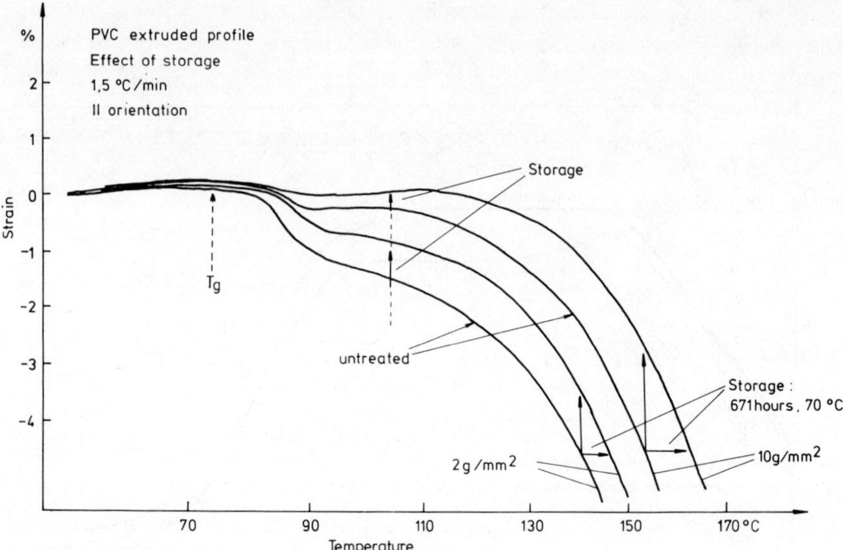

FIG. 3. Effect of storage at 70°C on the creep thermomechanical curves of extruded polyvinyl-chloride measured along the direction of the extrusion. Load 2 g/mm^2 and 10 g/mm^2, speed 1.5°C/min.

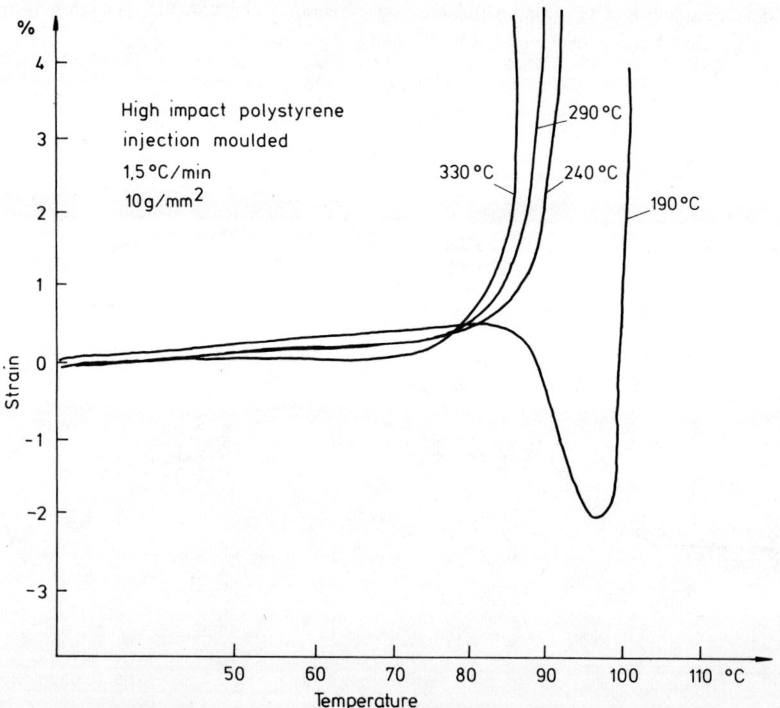

FIG. 4. Effect of the melt temperature on the creep thermomechanical curves of a high impact polystyrene compound (Dow Styron 457). Load 10 g/mm^2, speed 1.5°C/min.

recovery period shows significant frozen-in deformation for samples molded below 240°C, which decreases only at high temperatures. The samples in Figure 4 were all processed with a laboratory machine, the orientation of the measurement being along the direction of the injection.

By increasing the processing temperature, the elastic part of the strain decreases as the plastic part increases [see eq. (7)].

Unlike PVC, in polystyrene compounds recovery of the frozen-in orientation proceeds at a high rate immediately above the glass–rubber transition temperature. Comparison of Figures 2 and 4 shows that under similar experimental conditions the low load level "transition" from the recovery to actual creep is only by 10°C above T_g in polystyrene, while it is by about 80°–90°C above T_g in PVC. The recovery process in PVC is thus controlled by the 150°–180°C "liquid–liquid" type transition rather than the glass–rubber transition (75°C). The evidence of the existence of such a transition in PVC was obtained by dielectric depolarization spectroscopy [7] and by capillary viscometry [8].

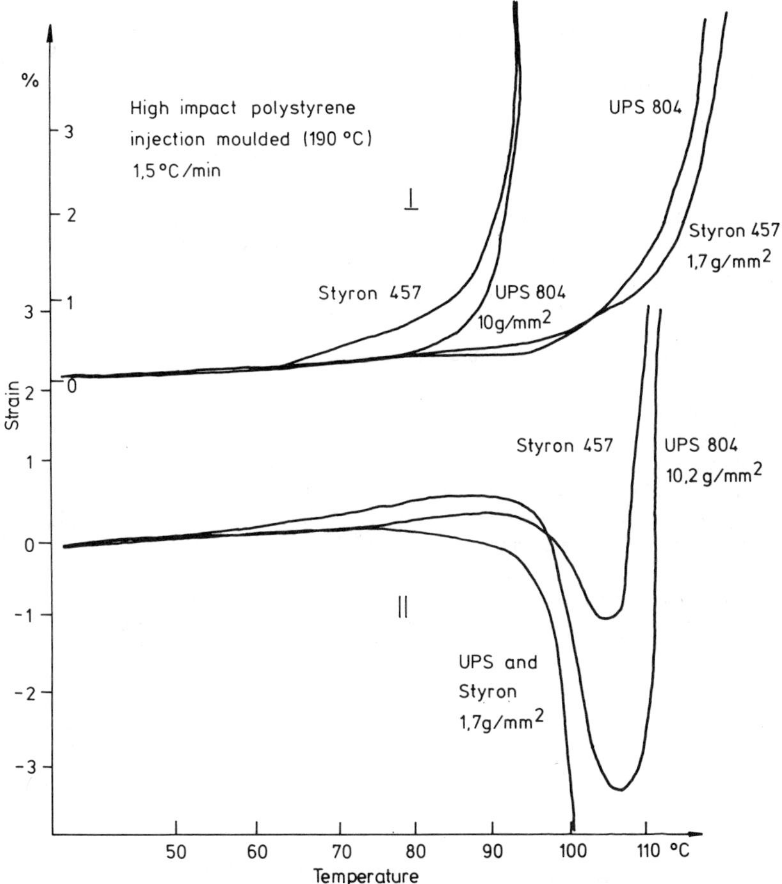

FIG. 5. Comparison of the thermomechanical curves of two high impact polystyrene compounds injection molded under similar conditions. Load 1.7 and 10 g/mm², speed 1.5°C/min. ⊥ (‖) mean perpendicular (parallel) orientations to the flow direction.

In Figure 5 a comparison of the thermomechanical creep curves of two different kinds of high impact polystyrene compounds are shown. The samples were prepared by injection molding at 190°C in the same way. The curves shown represent such runs when four samples were recorded simultaneously. The significant differences in the creep and shrinkage behavior found between the two compounds are due to the differences in their morphology.

CONCLUSIONS

The creep process as well as the recovery of the frozen-in deformation of plastics was described quantitatively by using linear response approximation. The validity of this approximation has been checked with model experiments. Low load level tensile creep thermomechanical curves were found to be sensitive to the structure and to the thermal history of the material, and the introduced computational formalism was found applicable for interpretation of such curves. The recovery process is, evidently, a macroscopic manifestation of the molecular or supermolecular (aggregate) orientation occurring during the history period. These effects should be studied separately on the molecular level. The phenomenological treatment, however, is of a great practical importance and seems to work at least at low deformation levels. By this formalism the kinetics of the recovery process can be studied and recovery isotherms can be constructed from the thermomechanical curve. Deviations from this formalism are expected due to the relaxation of the nonequilibrium glassy state below T_g [6]. Another source of the deviation is environmental stress crazing which might become effective in studying long-term recovery processes.

REFERENCES

[1] F. Bueche, *Physical Properties of Polymers,* Interscience Publichers, New York, 1962.
[2] I. L. Hopkins, *J. Polym. Sci.,* **28,** 631 (1958).
[3] P. Forgács, *Polym. Letters,* to appear.
[4] P. Forgács and P. Hedvig, VII. *Int. Congr. Rheol.* Göteborg, 1976.
[5] J. D. Ferry, *Viscoelastic Properties of Polymers.,* Wiley, New York, London, 1961.
[6] A. J. Kovács, R. A. Stratton, and J. D. Ferry, *J. Phys. Chem.,* **67,** 152 (1963); H. H. Meyer, P.-M. F. Mangin, and J. D. Ferry, *J. Polym. Sci.,* **A3,** 1785 (1965).
[7] P. Hedvig and E. Földes, *Angew. Makromol. Chem.,* **35,** 147 (1974).
[8] H. Münstedt, *BASF Prüflaboratorium, Internal Report.* Private communication (1974).

HYPERSOUND PROPAGATION IN ORIENTED POLYMETHYLMETHACRYLATE

S. M. LINDSAY and I. W. SHEPHERD
Physics Department,
Manchester University,
Manchester 13, U.K.

SYNOPSIS

The hypersonic elastic properties of oriented polymethylmethacrylate (PMMA) have been studied using high contrast ($>10^8$) multipass Fabry–Perot techniques. Four samples of different degrees of orientation (birefringence $\Delta n = -4, -9, -15, -24 \times 10^{-4}$) were prepared by drawing at 130°C. Measurements of longitudinal sound velocity, V_L, as a function of ϕ, the angle between the draw and propagation directions, show an anisotropy which scales with Δn and amounts to 6% in the sample having $\Delta n = 24 \times 10^{-4}$. At $\phi = 0$, $V_L = 2.80$ kilometers per second, (KMS^{-1}), the same as the value in an isotropic sample. The elastic constants C_{11} and C_{33} are measured as a function of Δn. Data as a function of sound wavelength, Λ, show dispersive effects in both V_L and attenuation $\alpha\Lambda$, for $\Lambda > 0.7\,\mu$. Over the range $0.7\,\mu < \Lambda < 1.3\,\mu$, $\alpha\Lambda$ increases from 0.1 to 0.5 and V_L increases by 8%. Attenuation is consistently higher for $\phi = 0$. The data are interpreted in terms of densification (up to 15%) accompanied by void formation on orientation, the structure having characteristic dimensions of approximately $1.5\,\mu$. The structure may be asymmetric with the longer dimension perpendicular to the draw direction. Additional Brillouin peaks may result from sound propagation in the voids ($V_L = 5.8$ KMS^{-1}) in which case the density is approximately 0.3 g cm^{-3}. Measurements of the Landau–Placzek ratio support these ideas, but evidence from electron micrographs is inconclusive. Dispersion measurements should in principle help determine details of the structure shape.

INTRODUCTION

In Brillouin scattering experiments light is inelastically scattered from thermal phonons to product a component of the scattered light shifted in frequency by an amount ν, given by [1]

$$\nu = \pm(2Vn/\lambda)\sin(\theta/2) \tag{1}$$

where θ is the scattering angle, λ the wavelength of the incident light in vacuo, V the phonon velocity, and n the refractive index of the scattering material. The $\pm$ signs refer to the two phonon wavevectors which satisfy the Bragg law. Attenuation of the phonon gives rise to broadening of the scattered component. Far from a relaxation the line profiles are Lorentzian with half-width at half-height Γ given by

$$\Gamma = \alpha V/2\pi \tag{2}$$

where α is the sound amplitude attenuation coefficient in nepers per unit length.

At any particular scattering angle θ, the phonon wavelength is given by

$$\Lambda = \frac{\lambda}{2n}\sin(\theta/2) \tag{3}$$

Journal of Polymer Science: Polymer Symposium 58, 85–96 (1977)
© by John Wiley & Sons, Inc.

With visible light sources, phonon wavelengths in the approximate range 0.2–5 μ may be studied.

Orientation effects in PMMA have been studied at a molecular level using Raman spectroscopy [2], and at a macroscopic level by ultrasonic and mechanical techniques [3]. Extensive studies have been made of the isotropic material using Brillouin scattering [4–6], however no hypersonic data have been published on the effects of orientation. Such studies have particular importance for both theoretical and practical reasons and are now possible following the development of high contrast multipass Fabry–Perot techniques.

(a) The phonon wavelengths, of the order 1 μ, are similar to the dimensions of structure in the samples, and Brillouin experiments are expected to provide a sensitive and direct probe for such structure. By contrast, effects on Raman and ultrasonic measurements are only indirect.

(b) Details of the anisotropy of oriented samples can easily be measured because the propagation direction of the phonon is accurately determined by the scattering geometry.

(c) Unlike electron microscopy, the technique does not require special sample preparation and thus the possibility of creating unwanted artefacts by thin film preparation or fracturing is avoided.

(d) Changes caused by very small degrees of orientation are easily studied because measurements can be made to high accuracy ($\sim\pm1\%$ in velocity and $\pm5\%$ in absorption) with a volume resolution of 10^{-13} M^3.

In the present work measurements of phonon velocities have been used to determine the anisotropy of the elastic constants of the oriented material. Only the longitudinal acoustic modes were observed (transverse modes have not been observed by Brillouin techniques in PMMA either because the ratio of the strain optical coefficients, P_{44}/P_{12}, is too low or because shear modes are strongly attenuated at these frequencies [4]). This restricts the measurements to a determination of just two of the five elastic constants. The stiffness matrix, C_{pq}, for a uniaxially oriented material is [7]

$$C_{pq} = \begin{pmatrix} C_{11} & C_{12} & C_{13} & 0 & 0 & 0 \\ C_{12} & C_{11} & C_{13} & 0 & 0 & 0 \\ C_{13} & C_{13} & C_{33} & 0 & 0 & 0 \\ 0 & 0 & 0 & C_{44} & 0 & 0 \\ 0 & 0 & 0 & 0 & C_{44} & 0 \\ 0 & 0 & 0 & 0 & 0 & C_{66} \end{pmatrix} \tag{4}$$

where the subscript 3 refers to the draw direction and $C_{66} = \tfrac{1}{2}(C_{11} - C_{12})$. The theory of wave propagation in an anisotropic medium has been considered by Musgrave [8–10] who shows that for a transversely isotropic material the velocities of propagation of waves traveling at an angle ϕ to the axis of symmetry are related to the stiffness constants through

$$[H - \tfrac{1}{2}C(1 - a_3^2)][H^2 - H\{a_3^2 h + a(1 - a_3^2)\}$$
$$+ a_3^2(1 - a_3^2)(ah - d^2)] = 0 \tag{5}$$

where

$$a_3 = \cos\phi, \qquad a = C_{11} - C_{44}$$

$$C = C_{11} - C_{12} - 2C_{44}, \; H = \rho V^2 - C_{44}$$

$$d = C_{13} + C_{44}, \; h = C_{33} - C_{44}$$

Solving eq. (5) for V_L, the velocity of longitudinal phonons, yields

$$\rho V_L^2 = C_{33} \qquad (6)$$

for propagation along the draw axis and

$$\rho V_L^2 = C_{11} \qquad (7)$$

for propagation perpendicular to the draw axis (ρ is the sample density). It is these two constants C_{11} and C_{33} which are determined in the present work.

Measurements of V_L and attenuation have also been made as a function of phonon wavelength, Λ. The observed dispersion is indicative of structure of size approximately 1.5 μ, and this is supported by measurements of the Landau–Placzek ratio. Additional components are observed in the Brillouin spectrum which may be caused by scattering from phonons propagating in regions of low density.

EXPERIMENTAL DETAIL

All samples used were made from 0.6 cm I.C.I. perspex sheet heated to $\sim 130°C$ and drawn to overall extension ratios of between 1.5 and 2.2 (including the unstrained portion in the lathe clamps). The samples were then left to cool to room temperature, sections cut from the drawn material, and birefringence measured by adjustment of the prisms in a Babinet compensator with the sample between cross polaroids [11]. This material is believed to be largely atactic with molecular weight $\sim 10^6$ [12]. The oriented material was then fashioned into discs of approximately 1 cm diameter which could be rotated in the plane of scattering. The light-scattering geometry is shown in Figure 1 where ϕ is the angle between the direction of phonon propagation and the draw axis, and θ is the scattering angle. The sample is calibrated by means of a laser beam reflected from a mirror located on the rotating sample holder. In this way the relative orientation of the

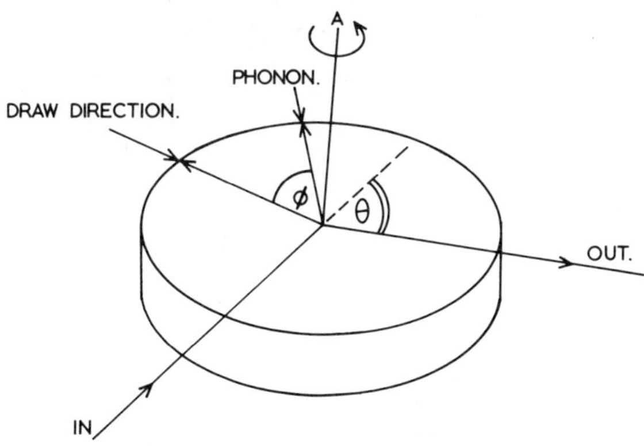

FIG. 1. The light-scattering geometry.

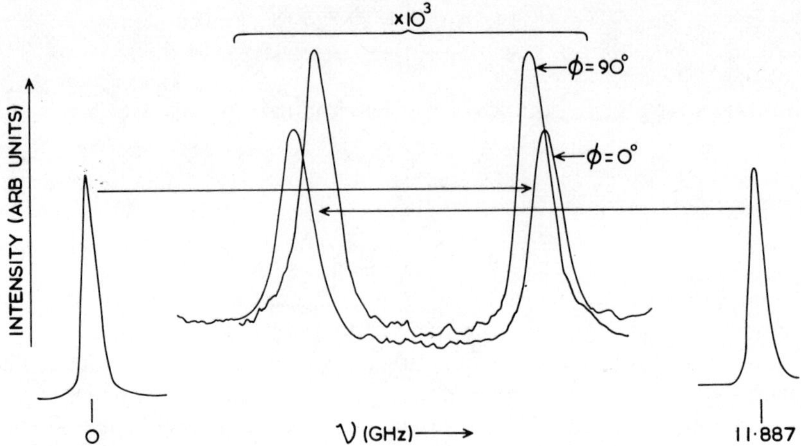

FIG. 2. Two juxtaposed Brillouin spectra from PMMA sample A with $\theta = 60°$, $\Phi = 0°$ and $90°$. The change in peak position and linewidth is evident. The orders are overlapped and the arrows indicate to which Rayleigh peak the Brillouin bands belong.

sample could be measured to better than $1°$. A serious source of potential error with this sample geometry is introduced into the internal angle θ by the uncertainty caused by refraction at the sample surfaces. To overcome this problem the rotating sample holder is mounted on an X, Y translation table and positioned so as to be normal to the input and output axes. This process is carried out using the back reflections of laser alignment beams. Using this system the overall error in velocity measurements as a function of θ was calculated to be $\sim\pm3\%$, and the relative error as a function of ϕ, $\sim\pm1.5\%$.

The light source was a coherent radiation 52G Argon ion laser focused to a $200\ \mu$ spot in the sample center. The scattered light was analyzed with the

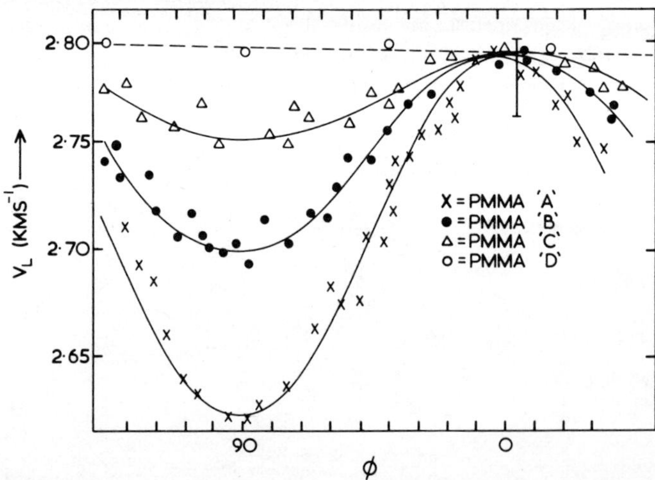

FIG. 3. Sound velocity dependence on draw direction for four samples of PMMA. The solid curves are fits to eq. (8). Birefringence values are A-24 $\times$ 10^{-4}; B-9 $\times$ 10^{-4}; C-4 $\times$ 10^{-4}; D amorphous.

triple-passed Fabry–Perot spectrometer described elsewhere [13, 14]. The multipass Fabry–Perot is essential in this work because of the large random scattering caused by inclusions in the sample. The free spectral range of the etalon was varied between 36.74 GHz and 8.02 GHz in order to resolve the phonons clearly. The finesse was never lower than 35. Linewidth measurements were taken with orders overlapped to take advantage of the higher instrumental resolution at larger etalon plate separations. The natural phonon widths are relatively small, and it is essential to correct for broadening caused by the finite aperture and the instrumental profile. Aperture effects are corrected using the procedure outlined by Danielmeyer [15] while instrumental broadening is corrected using the tables prepared for the multipass spectrometer by Lindsay, Burgess, and Shepherd [16]. The simultaneous application of these corrections incurs a certain error due to the non-Lorentzian nature of the observed profile in the multipass spectrometer [13], and the error bars indicated on the linewidth measurements represent the worst possible case.

Five samples were prepared with birefringences (Δn) varying from zero (annealed sample D) to -24×10^{-4} (sample A). The refractive index was measured as 1.495 ± 0.005 on an Abbé refractometer, and the density of all samples was 1.19 ± 0.07 g/cc.

RESULTS AND DISCUSSION

The Brillouin spectrum for sample A ($\Delta n = -24 \times 10^{-4}$) is shown in Figure 2 for both $\phi = 0$ and $90°$ with $\theta = 60°$. The orders are overlapped and the arrows indicate the Rayleigh peaks associated with each Brillouin component. It is clear from this figure that the velocity for $\phi = 90°$ is smaller than that for $\phi = 0$ by a significant amount, approximately 6%, and also that there is a small increase in attenuation along the draw direction. Values of V_L as a function of ϕ ($\theta = 90°$) for several samples were calculated from similar data using eq. (1) and are shown in Figure 3. Sample D is the isotropic material annealed at 140°C for 60 hours.

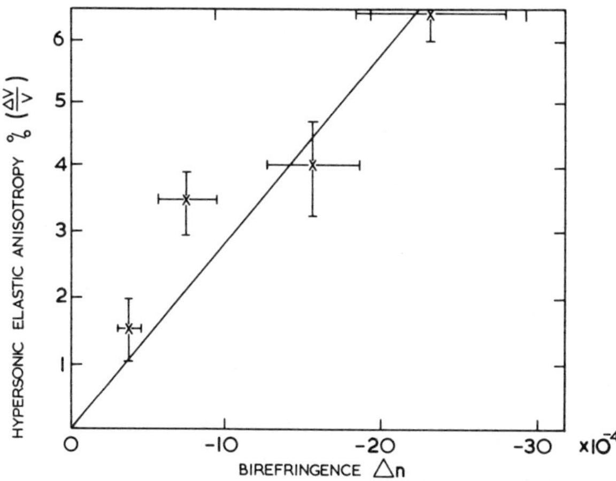

FIG. 4. The hypersonic elastic anisotropy plotted with sample birefringence.

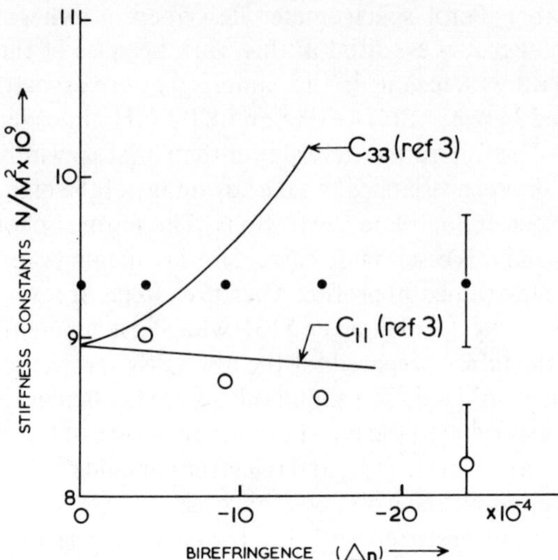

FIG. 5. Values of C_{33} (●) and C_{11} (O) derived from eqs. (6) and (7). The solid lines are the values given in [3].

Samples B and C have birefringences of -9×10^{-4} and -4×10^{-4} respectively. Measurements were also made on a fifth sample, E, with birefringence of -15×10^{-4}. These data fell between those of samples A and B, but are not shown on Figure 3 to avoid crowding. The solid curves are fits to the empirically chosen function

$$V_L(\phi) = V_0(1 - \Delta \sin^2\phi) \tag{8}$$

where V_0 is the velocity in the isotropic material and Δ a measure of the hypersonic elastic anisotropy. The anisotropy Δ and the birefringence, Δn, clearly correlate well, as shown in Figure 4. This implies that the hypersonic elastic properties are related to the quantity $\langle \cos^2\theta' \rangle$ where θ' is the angle between the draw direction and the axis of some typical molecular segment [17].

Using eqs. (6) and (7) C_{33} and C_{11} may be calculated. It is interesting at this point to compare the results of the present work with the measurements of Wright et al. [3] at 4 MHz ($\Lambda \sim 700\,\mu$). The solid lines on Figure 5 represent the ultrasonic values for C_{33} and C_{11}, the open circles the values we obtain for C_{11}, and the closed circles our values for C_{33}. Even though we cannot obtain values for the other elastic constants C_{12}, C_{13} and C_{44}, it can be seen that there is a clear discrepancy between eqs. (5) and (8). Using the values of Wright et al. [3] eq. (5) may be written (in the approximation that $H \gg C/2$)

$$\rho V_L^2 = C_{44} + h \cos^2\phi + a \sin^2\phi + (ah - d^2) \cos^2\phi \sin^2\phi \tag{9}$$

while for small Δ, eq. (8) may be rearranged to yield

$$\rho V_L^2 = C_{33} - 2C_{33}\Delta \sin^2\phi \tag{10}$$

There is clearly little correspondence between the two. The most striking difference is the weakening of the material perpendicular to the draw axis unac-

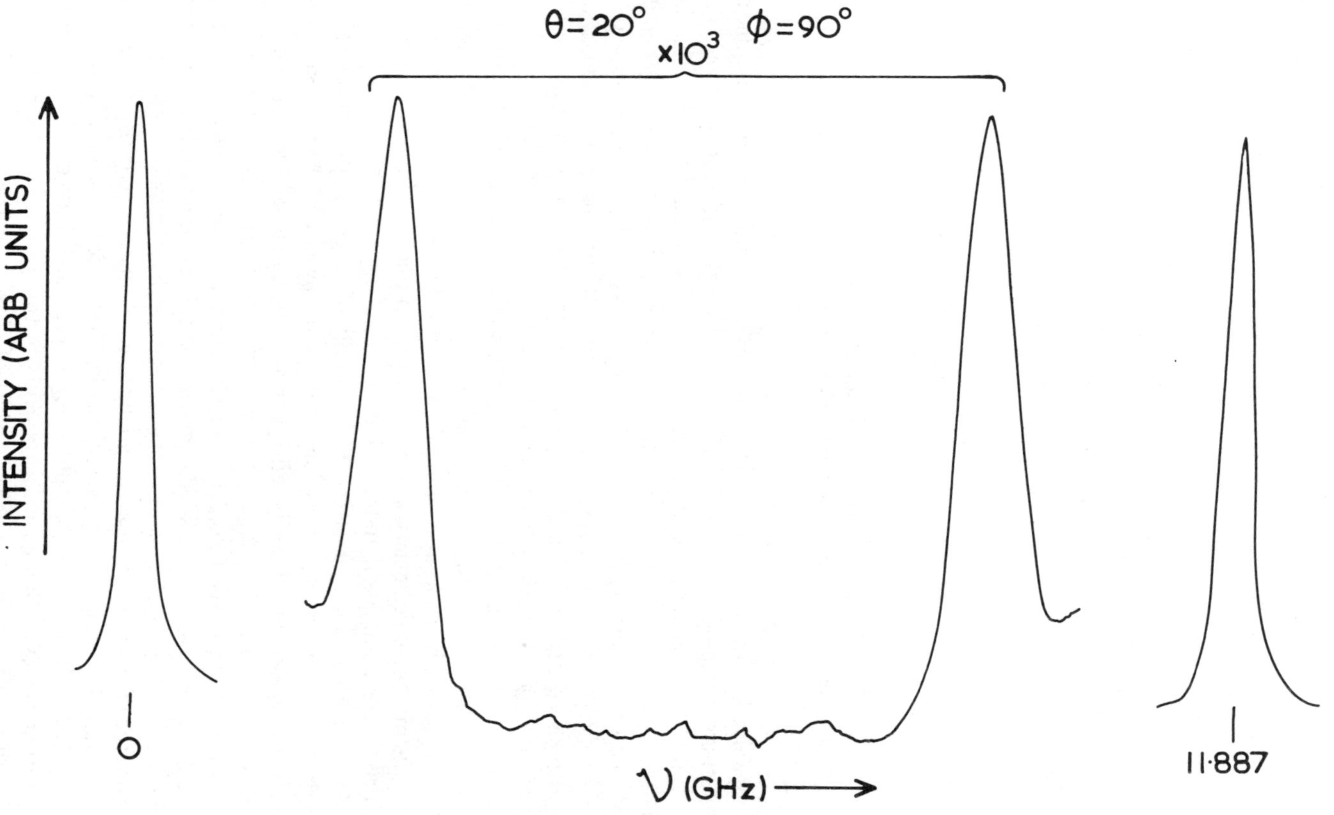

FIG. 6. Brillouin spectrum from PMMA sample A with $\theta = 20°$, $\Phi = 90°$.

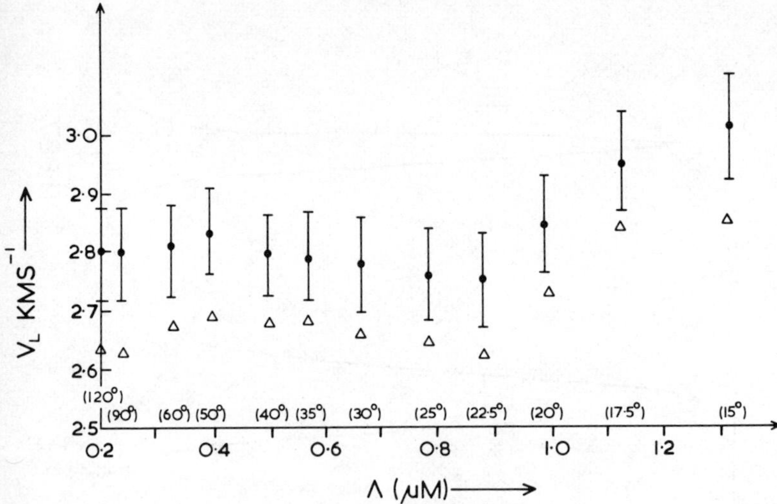

FIG. 7. Longitudinal velocity parallel (●) and perpendicular (△) to the draw direction plotted with phonon wavelength. For clarity, error bars are only shown on one set of data (PMMA Sample A).

companied by any strengthening in the draw direction. It is hard to imagine a molecular rearrangement that would give rise to these differences between ultrasonic and hypersonic elastic properties. We believe these effects may be explained by a densification of the material with an associated void formation. A 15% increase in density in the most highly oriented material would bring our values of C_{11} and C_{33} into approximate agreement with those of Wright et al. That such a densification must be accompanied by the formation of regions of rarefaction (voids) is clear from the fact that the macroscopic density is not a function of the degree of orientation. The formation of such voids in stressed polymers has, in any case, been established by many workers (see, for example, [18]). Measurements at ultrasonic frequencies [3] ($\Lambda = 700\ \mu$) are consistent with the macroscopic density which indicates that the linear dimensions of the structure are much less than this value.

The Brillouin technique offers the possibility of observing this structure directly. By varying the scattering angle θ down to low values, the phonon wavelength is increased and dispersive effects are observed. Figure 6 shows a spectrum from sample A with $\theta = 20°$, $\phi = 90°$ plotted with scale and resolution similar to Figure 2. In this case the orders are not overlapped. The line width is slightly increased relative to Figure 2 which implies a large increase in $\alpha\Lambda$, the absorption per cycle, because the phonon wavelength Λ is larger by a factor of approximately 3. Also the phonon velocity is substantially larger than at $\theta = 60°$. By decreasing the free spectral range to 8.02 GHz, phonons were clearly resolved down to $\theta = 15°$. Difficulties with the sample chamber geometry limited access to lower scattering angles, although measurements in this region become extremely difficult in any case due to the finite width of the phonon peaks, the extremely large random scattering, and the lower instrumental stability at the highest resolutions. The derived values of V_L over the range $120° \geqslant \theta \geqslant 15°$ are plotted

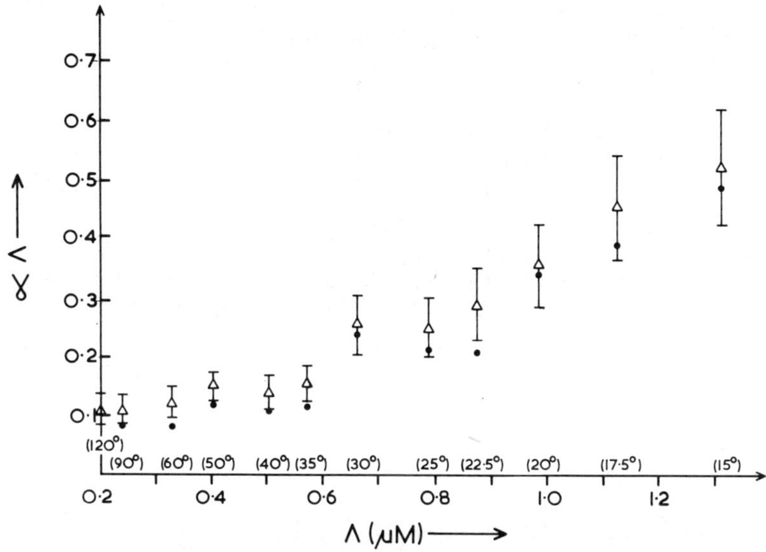

FIG. 8. Absorption per cycle, $\alpha\Lambda$, plotted with Λ for propagation parallel ($\triangle$) and perpendicular ($\bullet$) to the draw direction (PMMA sample A).

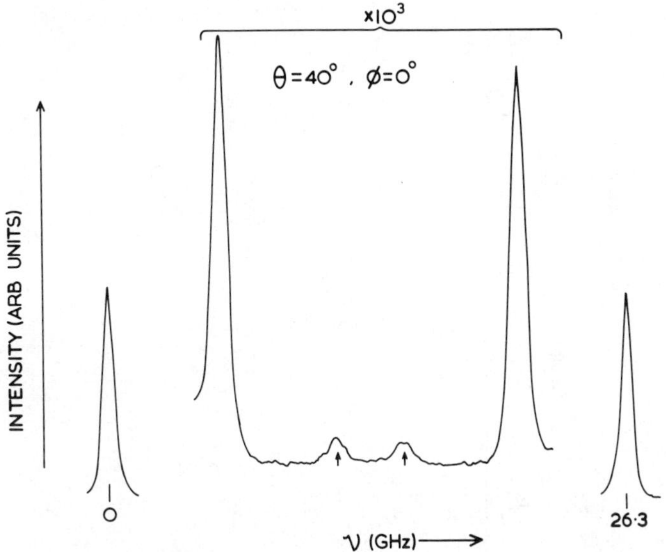

FIG. 9. Spectrum of PMMA 'A' with $\theta = 40°$, $\Phi = 0°$ with free spectral range increased to show extra components marked with arrows.

in Figure 7 against Λ as calculated from eq. (3). Values of $\alpha\Lambda$ calculated from eq. (2) and using $V_L = \nu\Lambda$ are plotted with Λ in Figure 8. In contrast, measurements on the annealed sample "D" yielded $V_L = 2.8 \pm 0.08$ KMS^{-1} and $\alpha\Lambda = 0.15 \pm 0.03$ over the entire range of Λ. This value of V_L is in good agreement with other work [4, 19] as is the absence of dispersion in $\alpha\Lambda$. However the absolute value of $\alpha\Lambda$ as measured here is larger by nearly a factor of 2 than the value of Jackson et al. [4]. We believe that the difference is due to a higher concentration of imperfections in our sample which contribute to the phonon scattering. This will be discussed later when measurements of the Landau–Placzek ratio are presented. The entire dispersion region for the oriented sample is not observed, and it is not therefore possible to draw firm conclusions about the dimensions of any structure. However the results indicate a significant amount of structure of linear dimensions approximately 1.5 μ.

At higher free spectral range, extra weak components in the spectrum are observed. They obey the Brillouin formula [eq. (1)] with $V_S = 5.8\ \mathrm{KMS}^{-1}$, and are shown in Figure 9. This value of velocity precludes transverse modes, and we believe these components may correspond to longitudinal phonons propagating in the rarefied regions. Assuming that the elastic properties of the polymer in the densified and void regions are similar, this value of velocity would indicate a density of $\sim$0.3 g/cc in the void region.

Without knowledge of the mechanism of phonon attenuation it is difficult to interpret the slight increase in attenuation of phonons having $\phi = 0$ relative to $\phi = 90°$ (Figs. 2 and 8). While phonon–phonon scattering may play a part in the attenuation mechanism [13] structural scattering is thought to dominate [4]. Although there are significant differences in the thermal conductivities

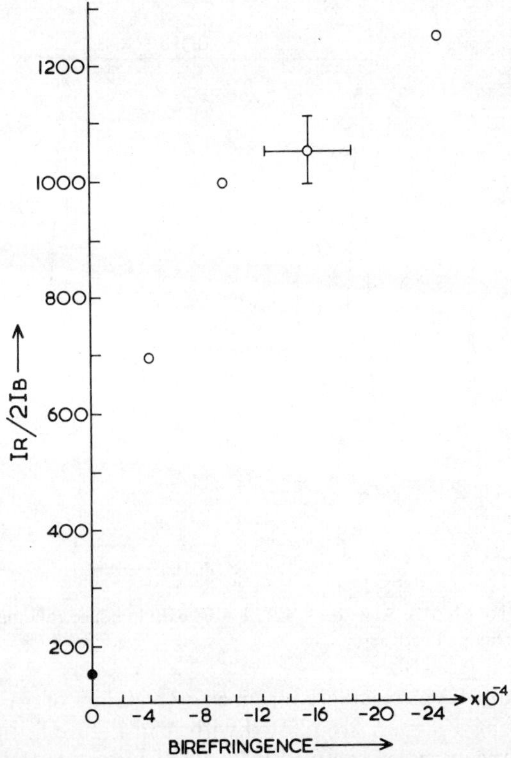

FIG. 10. The minimum Landau–Placzek ratio measured for five samples of PMMA plotted with birefringence.

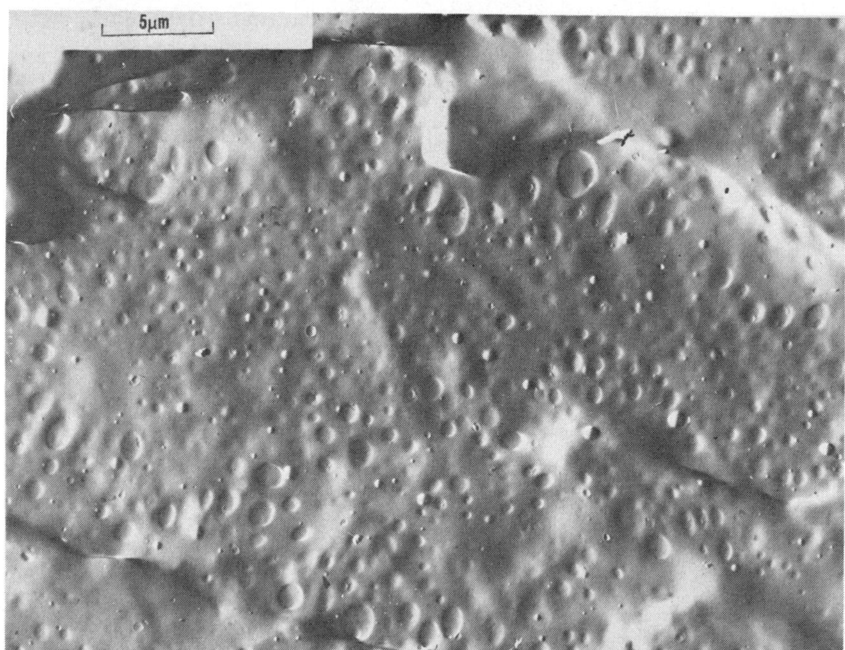

FIG. 11. Scanning electron microscope photograph [21] of PMMA sample A.

measured perpendicular and parallel to the draw axis in oriented polymers [20], the contribution of thermal conductivity to the line width is too small for this effect to be important. It may be that the void structure is asymmetric with the long axis perpendicular to the draw direction, in which case scattering would be enhanced for $\phi = 0$ relative to $\phi = 90°$ when Λ is close to, but smaller than, the structural dimensions. It is possible to speculate that such structure may play a role in accounting for the discrepancy between the anisotropy in the elastic properties and the anisotropy in the fracture properties [3].

Structure of approximately 1 μ size should enhance the random scattering of incident light, and, in order further to test the structural model proposed, measurements of the Landau–Placzek ratio (the ratio of total Rayleigh intensity to total Brillouin intensity) were made on all samples. The presence of large inclusions makes this measurement difficult, however the sample was carefully positioned in order to obtain a minimum value in each case. The results for the five samples investigated are plotted in Figure 10. It should be noted that values greatly in excess of those shown could be obtained by varying the sample position to maximize scattering from centers not associated with orientation. The value of $I_R/2I_B$ for the annealed sample is well above minimum values reported elsewhere [4], and the presence of the extra scattering centers that cause this discrepancy may well also account for the fact that $\alpha\Lambda$ is larger than the value reported by Jackson et al. [4]. In spite of these problems the qualitative feature of enhanced scattering with increased birefringence is beyond doubt and supports the model proposed.

Attempts were made to obtain electron microscope photographs of these materials, and Figure 11 is a typical result [21], showing structure of $\sim 1\ \mu$. However just such structure is induced by the fracture process used to expose the sample surface [18], and it is impossible to differentiate between such induced structure and intrinsic structure.

CONCLUSION

Brillouin measurements of sound velocity and attenuation as a function of propagation direction in oriented PMMA have been used to determine the anisotropy of the elastic properties at hypersonic frequency. The formation of voids is believed to play an important role; these voids occur in material that is densified by up to 15% at the highest draw ratios studied. The presence of extra components in the Brillouin spectra may result from scattering off phonons in the voids, in which case a void density of $0.3\ \mathrm{g\ cm^{-3}}$ is deduced. Measurements of sound velocity and attenuation as a function of sound wavelength show dispersion, indicating that the structure has approximate linear dimensions of $1.5\ \mu$, and that it is asymmetric with the longer dimension perpendicular to the draw direction. Measurements of the Landau–Placzek ratio support this model, but electron micrographs are inconclusive.

We thank Dr. P. Gaskell of Pilkington Glass Co., Dr. E. White of U.M.I.S.T. and S. J. Spells both for useful discussions and the help already referred to in the text. In addition we would like to acknowledge helpful discussions with Professor I. Ward and Dr. S. Burgess, and the assistance of A. Venebles in the construction of the sample holder. This work was supported in part by the S.R.C. Polymer Committee.

REFERENCES

[1]   L. Brillouin, *Ann. Phys.,* **17,** 88 (1922).

[2]   J. Purvis, and D. I. Bower, *Polymer,* **15,** 645 (1974).

[3]   H. Wright, C. S. N. Faraday, E. F. T. White, and L. R. G. Treloar, *J. Phys. D. Appl. Phys.,* **4,** 2002 (1971).

[4]   D. A. Jackson, H. T. A. Pentecost, and J. G. Powles, *Mol. Phys.,* **23,** 425 (1972).

[5]   A. B. Romberger, D. P. Eastman, and J. L. Hunt, *J. Chem. Phys.,* **51,** 3723 (1969).

[6]   E. A. Friedman, A. J. Ritger, and R. D. Andrews, *J. Appl. Phys.,* **40,** 4243 (1969).

[7]   R. F. S. Hearman, *Applied Anisotropic Elasticity,* Oxford, Clarendon Press, 1961.

[8]   M. J. P. Musgrave, *Proc. R. Soc. London Ser.,* **A226,** 339 (1954).

[9]   M. J. P. Musgrave, *Proc. R. Soc. London Ser.,* **A236,** 232 (1956).

[10]   M. J. P. Musgrave, *Rep. Prog. Phys.,* **22,** 77 (1959).

[11]   We are indebted to Dr. E. White and S. J. Spells for sample preparation and characterization.

[12]   C. R. Walsby (I.C.I. Plastics Division), private communication.

[13]   S. M. Lindsay, Ph.D. Thesis, University of Manchester (1976).

[14]   S. M. Lindsay, and I. W. Shepherd, *J. Phys. E. Sci. Instrum.,* **10,** 150 (1977).

[15]   H. G. Danielmeyer, *J. Acoust. Soc. Am.,* **47,** 151 (1970).

[16]   S. M. Lindsay, S. Burgess, and I. W. Shepherd, *Applied Optics,* **16,** 1404 (1977).

[17]   I. M. Ward, Ed., *Structure and Properties of Oriented Polymers,* Applied Sciences, London, 1975.

[18]   R. N. Haward, Ed., *The Physics of Glassy Polymers,* Applied Sciences, London, 1973.

[19]   J. R. Asay, D. L. Lamberson, and A. H. Guenther, *J. Appl. Phys.,* **40,** 1768 (1969).

[20]   W. Reese, *J. Macromol. Sci. Chem.,* **A3(7)** 1257 (1969).

[21]   We are indebted to Dr. P. H. Gaskell of Pilkington Bros. who took the SEM photographs.

THE ROLE OF INTERFIBRILLAR TIE MOLECULES IN DRAWING OF POLYMERS

V. A. MARIKHIN and L. P. MYASNIKOVA

A. F. Ioffe Physical Technical Institute of the Academy of Sciences of the USSR, Leningrad

SYNOPSIS

A structural-mechanical study of the orientation process in some flexible-chain semicrystalline polymers has been conducted. By their deformation properties, such polymers may be conditionally divided into two groups: the ultimate draw ratio is 20–25 for the first group and 6–10 for the second one, the transformation of the original structure into a microfibrillar one being completed at a draw ratio (λ) of 8–10 in the polymers of the first type and at a draw ratio of 2–4 in those of the second type. The low draw ratios characterizing the moment of a virgin neck propagation in the polymers of the second type are considered to be due to a large number of tie molecules in the original supermolecular structures. When microfibrils are formed, these molecules are supposed to be included between fibrils.

It is shown that a further increase of λ results from the microfibrillar slip. In the polymers of the first type, the slip is, however, relatively easy and is not accompanied by significant changes in the internal structure of microfibrils, because the interfibrillar tie molecules are almost absent in them. In the polymers of the second type, a significant change in the internal structure of microfibrils, associated with a large number of interfibrillar tie molecules, is observed (obliquity of crystallites and an increase of the long period attributed to an increase in the length of the amorphous intrafibrillar layer).

The microfibrillar slip leads to their more dense packing, which results in a sharp increase in the orientation load required for further drawing. This reduces significantly the lifetime of the specimens in high temperature drawing and leads to their fracture by a thermal fluctuation mechanism.

INTRODUCTION

To date, the high strength flexible-chain polymers are mainly manufactured with the help of orientational drawing. Many authors have got numerous experimental results both on the structural transformations and a change in the mechanical properties on orientation [1, 2, 3]. It has been shown that the tensile strength and deformation modulus grow significantly as the draw ratio increases [1, 4, 5]. Using the thermogradient heating technique [6], high strengths reaching 570 kg/mm^2 at liquid nitrogen temperature have been obtained (normally, the strength is approximately less by half at room temperature, i.e., 280 kg/mm^2). Even these high values of strength are, however, considerably lower than the theoretical values [7]. It is therefore quite natural that many investigators try to find the ways for further increase of the polymer strength using a well-developed manufacturing technique for production of fibers and

Journal of Polymer Science: Polymer Symposium 58, 97–107 (1977)

films from synthetic materials. However, here we encount difficulties associated first of all with the fact that drawing ceases after a definite λ is reached.

A number of papers [8, 9, 10] have been concerned with the discussion of some reasons for the orientational drawing termination, but we believe that this problem requires a deeper consideration.

With this in mind, we have carried out a structural–mechanical study of a number of crystallizing polymers oriented up to different degrees of λ. The results are reported in this paper.

EXPERIMENTAL

Materials

The study was carried out on films of linear polyethylene of different molecular weights (from 5.10^4 up to 1.10^6), branched polyethylene (MW $= 7.5 \times 10^4$), polyethyleneterephthalate, and polycaproamide. All the films, except linear PE, were industrial specimens. The linear PE specimens were pressed from powder and then quenched at $-95°C$ according to the technique described in [11]. The original unoriented films were 30–100 μ thick.

Mechanical Tests

A tensile strength of the specimens along the perpendicular to the orientation drawing direction was measured at liquid nitrogen temperature to avoid plastic deformation of the specimens at the moment of testing. The orientational drawing of polymers was conducted under the action of high temperatures during short times, the heating zone being moved along the specimen [6].

Structural Studies

The structure of polymer films stretched up to different degrees of elongation was studied with the help of small- and wide-angle X-ray diffraction. A device with a point collimation of the primary beam constructed on the basis of the DRON-1 X-ray apparatus was used. Wide-angle X-ray patterns were taken from one-layer films several microns thick. Small-angle X-ray patterns were recorded from wound specimens 300–500 μ thick.

RESULTS AND DISCUSSION

Table I presents the data on the ultimate draw ratios, which we have managed to achieve for the polymers examined. Let us first of all consider the results obtained for linear PE. It is clear from Table I that the maximum draw ratios can vary from $\lambda = 20$–23 for linear PE with molecular weight of 5×10^4 up to $\lambda = 9$–11 for PE with high molecular weights ($5 \times 10^5 \div 1 \times 10^6$), depending on the molecular weight of PE. Wide-angle X-ray patterns indicate (Figure 1, Ia, IIa, IIIa, IVa) that for all samples a high orientation of crystallites along the

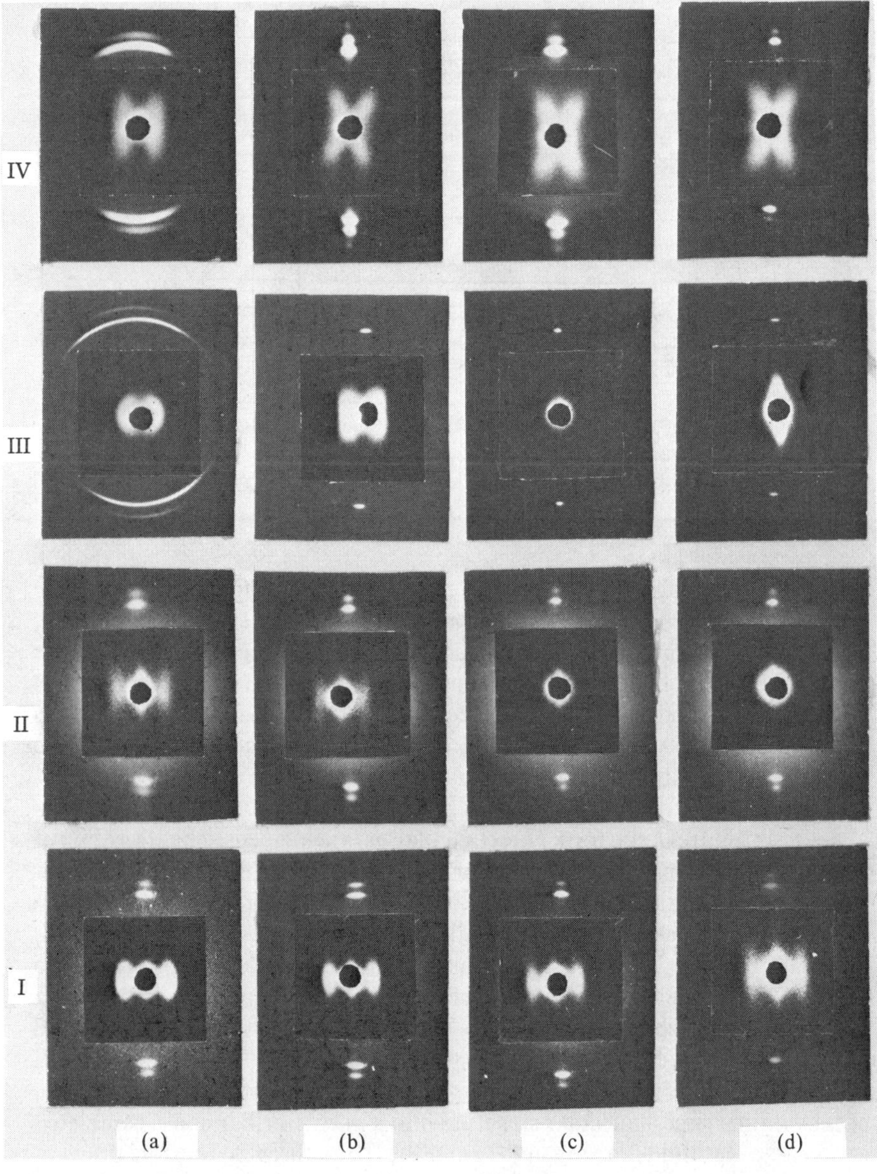

FIG. 1.

I	PE I	(a) $\lambda = 7.5$	(b) $\lambda = 12.5$	(c) $\lambda = 14.6$	(d) $\lambda = 22$
II	PE II	(a) $\lambda = 7.6$	(b) $\lambda = 13$	(c) $\lambda = 14.5$	(d) $\lambda = 19.5$
III	PE III	(a) $\lambda = 1.5$	(b) $\lambda = 5$	(c) $\lambda = 9$	(d) $\lambda = 11$
IV	PE IV	(a) $\lambda = 3$	(b) $\lambda = 5$	(c) $\lambda = 7$	(d) $\lambda = 9$

Wide- and small-angle patterns of drawn linear PE films for different draw ratios.

TABLE I

Ultimate Draw Ratios of Investigated Polymers Obtained for Definite Temperature of Drawing

Polymer	Draw ratio	$T\,^{\circ}C$
LPE ($M_w = 3\ 10^4$)	22	104
LPE ($M_w = 8\ 10^4$)	20	104
LPE ($M_w = 5\ 10^5$)	11	110
LPE ($M_w = 1\ 10^6$)	9	110
BPE	6,5	20
PET	8,7	230
NYLON-6	6,6	150

axis of drawing is observed long before the ultimate elongation takes place. The disorientation of crystallites is as small as 5–8°. Small-angle X-ray patterns are, however, significantly different: they are in the form of meridional streaks for linear PE I and PE II. This shape remains the same in a wide range of draw ratios up to the highest ones. For high molecular PE, the streak-type X-ray patterns observed at small λ are transformed into four-point or radial patterns as the maximum draw ratios are reached. It is known that the streak-type X-ray patterns are obtained from polymers with a microfibrillar structure, if the crystallites in microfibrils are in the form of rectangular parallelepipeds. The transition of streak-type patterns into four-point or radial ones is the evidence of the substantial change in the shape of crystallites: the rectangular crystallites become oblique ones. As one can see from small-angle X-ray patterns (Fig. 1), when λ, amounting to only 30–50% of the ultimate λ, is reached, the microfibrils in the linear PE films under investigation are already formed and then the drawing of the completely fibrillized films takes place. In this case, the crystallites in PE I and PE II do not change their shape with drawing, and the crystallites in PE III and PE IV become oblique as λ increases. The fact that the transformation of the original supermolecular structure into a microfibrillar one is completed at seven- to eightfold drawing in PE I and PE II and three- to fourfold drawing in PE III and PE IV have been also proved by other experimental techniques [11, 13, 14]. Thus, the fibrillized films of PE of any molecular weight can be drawn in two to three times.

What is the reason for a relatively high deformation ability of fibrillized films?

It is known that on plastic deformation of a microfibrillar structure, different deformation modes leading to an increase in the macroscopic dimensions of the specimen in the draw direction can occur. These modes are a rotation of otherwise unaltered crystallites, a change in the chain orientation within the crystallites, separation of the crystallites due to extension of the amorphous interlayer between them [15, 16, 17], and the microfibrillar slip [18].

In drawing of low molecular PE we have observed neither an increase of the long period, nor crystallite obliquity, nor a change in the chain inclination in the crystallites. The only change in the internal microfibrillar structure consists in a small increase of the longitudinal crystallite dimensions measured by the reduction in the half-width of the meridional reflection (002). It may be inferred from this fact that no one of the deformation modes mentioned above occur during the orientational drawing of the fibrillized films of such PE, except the microfibrillar slip.

As we have seen, at the same time the drawing of high molecular PE is accompanied by the crystallites' obliquity with rather a large angle of tilt to the draw direction. For the films of PE IV, the angle of obliquity is 40–45° (Fig. 1, IVa) even at the initial stages of drawing and it reaches 65–70° (Fig. 1, IVd) to the moment of the drawing completion. Moreover, the small-angle X-ray patterns taken from these samples indicate an increase of the long period from 300 up to 360 Å. It occurs due to extension of the amorphous interlayers between crystallites, since the longitudinal dimensions of the crystallites remain approximately constant (180 Å) in the whole range of elongations.

However, the estimation of the contribution of the crystallites' separation (50%) or their obliquity (20–30%) demonstrates that this contribution is too small to explain the observed elongation of fibrillized films during orientational drawing (from λ 3–4 up to λ 9–11). Therefore, in the case of high molecular PE we must also admit the possibility of microfibrillar slip.

Let us consider the character of the lateral contacts of microfibrils and estimate the possibility of their slip.

A number of authors suppose that microfibrils can be connected with each other by so-called interfibrillar tie molecules [8, 18]. Obviously, the number of such molecules, their conformation, and distribution by their lengths can significantly influence the slip.

Direct measurements of the number of interfibrillar tie molecules have not yet been carried out. An indirect estimate of their number can be derived from the measurements of strength of the oriented films transverse to the draw direction. We have conducted such measurements. The transverse strength of oriented PE specimens has turned out to be so small that practically it cannot be measured. It can be inferred from this fact that the interfibrillar tie molecules are either absent in these films or their quantity is very small. But the transverse strength of PE films of average and high molecular weights can be measured (see Fig. 2). It amounts to 3–4 kg/mm² for PE II and about 7 kg/mm² for PE III and PE IV. The microfibrils in the films of high molecular PE are, obviously, connected by a large number of interfibrillar tie molecules. As the draw ratio increases, the transverse strength of PE III and PE IV drops linearly, which makes one suppose that a fraction of molecules can rupture in the process of drawing.

Thus, the measurements of the transverse tensile strength of oriented specimens make one believe that the microfibrils in PE I and PE II films have a

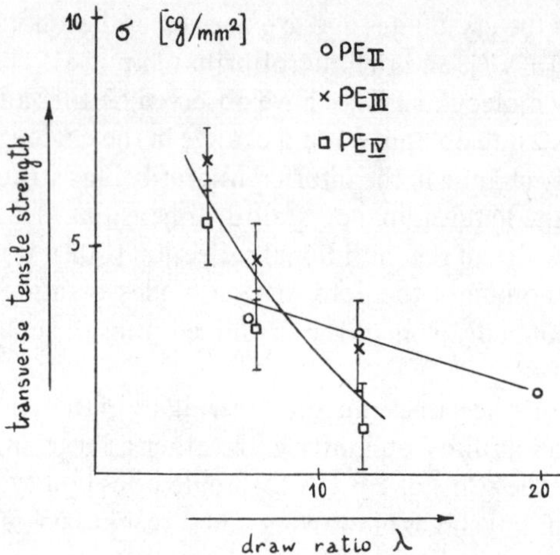

FIG. 2. Dependence of transverse tensile strength of drawn PE films on draw ratio.

weaker connection than the microfibrils in PE III and PE IV films. Moreover, their packing is not so close. This is evidenced by the presence of a zero layer line on small-angle X-ray patterns taken from these samples (Fig. 1, Ia, IIa). A zero layer line is absent in small-angle patterns for PE III and PE IV (Fig. 1, IIIa, IVa), which may be due to better packing of microfibrils.

In what manner does the microfibrillar slip occur during the orientational drawing of fibrillized films?

It is likely that, due to the absence of interfibrillar tie molecules, the microfibrils in PE I are displaced easily enough with respect to each other, since the microfibrillar slip does not produce any changes in their geometric parameters (such as cross section, long period, crystallite shape). The same picture is observed for PE II, though there are some interfibrillar tie molecules in it. The latter must, however, be long enough, because drawing of fibrillized films up to elongations preceding breakdown is not accompanied by scissions of the macromolecular chains [19]. During sliding motion, the tie molecules are either straightened or pulled out of microfibrils and are accommodated between fibrils [20]. It is difficult to say what process dominates here. In any case, the microfibrillar packing in PE II becomes better with drawing. This is indicated by a decrease in a zero layer line intensity in small-angle X-ray patterns (Fig. 1, IIb–d). The only change in the internal microfibrillar structure due to slip consists in an increase in the orientation degree ($\cos^2\theta_{am}$) of macromolecular segments in the amorphous intercrystallite layers [14]. Since the difference in the length of segments in the amorphous intercrystallite layers of these polymers is not large [21], the observed increase of $\overline{\cos^2\theta}_{am}$ cannot, obviously, lead to a significant growth of the long period. It is, therefore, not surprising that, within the accuracy of measurements ($\pm$10–15 A), we have observed no increase of the long period with drawing of films, though the orientation of chain segments in the amorphous regions becomes considerably better (from $\overline{\cos^2\theta}_{am} = 0.4$ up to 0.7). An orientation change is likely to result from shear displacement of microfibrils. This possibility has been noted, for example, in [8].

The microfibrillar slip in the orientational drawing of high molecular PE proceeds in a different manner. First of all, in this case the distribution of interfibrillar tie molecules by their lengths is, probably, nonuniform. The shortest chain segments, which are the first to extend due to slip, must rupture to provide a possibility of further slippage. Indeed, we have reported already [22] that the number of chain scissions significantly increases with drawing of fibrillized films of PE III and PE IV. The extended unruptured chains transform the orientation load to the microfibrillar crystallites, which can be cleaved into separate regions. The slip of these regions in the draw direction produces the obliquity of crystallites. The separation of crystallites is also likely to occur as a consequence of a strong connection of microfibrils with each other. The difference in the microfibrillar structure of the ultimately drawn films with a great and small number of tie molecules is schematically shown in Figure 3.

Thus, the presence of a large number of interfibrillar tie molecules changes substantially the process of the microfibrillar slip, activating additional deformation modes, and also causes the fracture seats in the specimen under drawing to appear.

What is the reason for the termination of drawing and breakdown of the polymer?

It has been supposed [8] that the termination of the orientational drawing is a consequence of a nearly exponential rise of the orientation load caused by a drastic growth of the taut interfibrillar tie molecules' number at the final stages of drawing. These molecules resist the axial displacement of microfibrils. Our results demonstrate that, indeed, the stress required for the orientational drawing increases sharply for all polymers under investigation, when the elongation preceding breakdown is reached (see Fig. 4). We believe, however, that this increase results not from the rise in the number of interfibrillar tie molecules, but from a better packing of microfibrils. As a result, the friction coefficient for sliding motion of microfibrils grows. As has been shown above, the number of interfibrillar molecules drops rather than increases with drawing, which is evidenced by a reduction in the transverse tensile strength of oriented samples (Fig. 2). This is also proved by the data on the accumulation of macromolecular scissions with drawing [19]. A closer packing of microfibrils can be derived from a decrease in the zero layer line intensity in small-angle X-ray patterns (Fig. 1, IIa–d). We think that a reduction in permeability of oriented films with drawing, reported in [8], is caused not by an increase in the number of interfibrillar tie molecules, but by a decrease in separation of microfibrils.

After all, a sharp rise of the orientation stress shortens the lifetime of the specimen under drawing. In accordance with the kinetic theory of strength [23] the lifetime is

$$\tau = \tau_0 l\,\frac{U_0 - \gamma\sigma}{KT}$$

where τ is the lifetime of the specimen under a stress σ, T is the temperature of drawing, U_0 is the activation energy for the fracture process, $\tau_0 = 10^{-13}$ sec, γ is a structural factor (0.07–0.1) and K is Boltzmann constant.

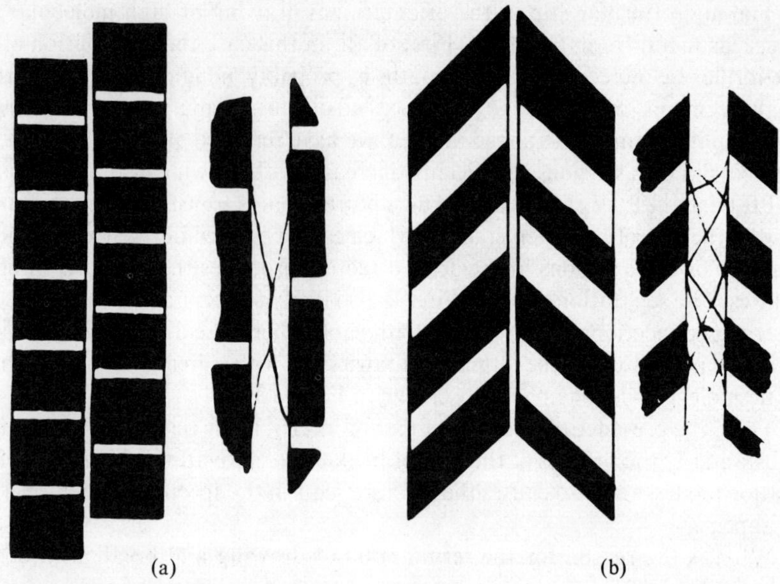

FIG. 3. The model of ultimately drawn PE films' structure. (a) PE I and PE II (a small number of interfibrillar tie molecules). (b) PE III and PE IV (a great number of interfibrillar tie molecules).

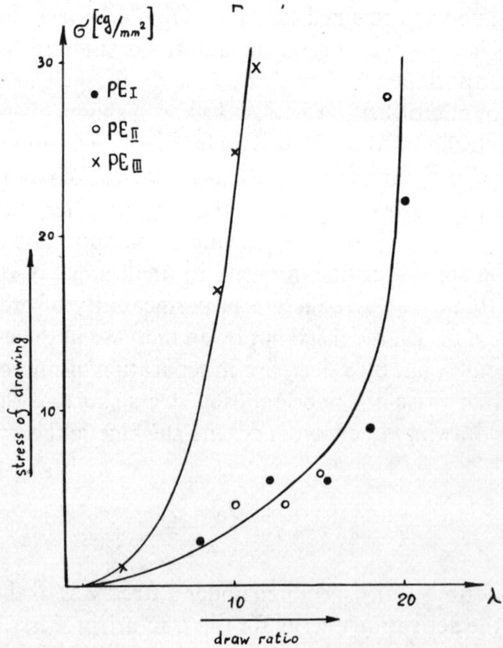

FIG. 4. Orientational stress–draw ratio curve for linear PE with different MW.

Figure 4 shows that $\sigma \geqslant 30$ kg/mm^2 for the ultimately drawn PE. The drawing was carried out at 100–110°C. The lifetime of the specimen under these conditions is only several seconds, which is comparable with the duration of drawing.

Thus, the main reason for the termination of the orientational drawing in polymers is a closer packing of microfibrils hindering a further axial displacement of microfibrils and leading to an increase in the orientation stress up to a critical value. In the specimens where the microfibrils are bridged by a great number of interfibrillar tie molecules, the former are brought together sooner and σ_{dr} increases more rapidly (compare the dependence of σ (λ) for different PE) (Fig. 4). When the microfibrils are loosely connected, only small efforts are required to displace them (see Fig. 5), and their bringing together occurs only at the final stages of drawing (σ begins to grow only at $\lambda = 18$). At the same time a steady increase of σ and ρ is observed (Figs. 4 and 5) in the systems where the microfibrils are closely packed.

It was of great interest to study the plastic deformation modes of the microfibrillar structure on the example of other polymers. For this purpose the drawing of branched PE, PETF, and PCA was carried out.

It is known that these polymers yield low draw ratios [5–7]. On the basis of

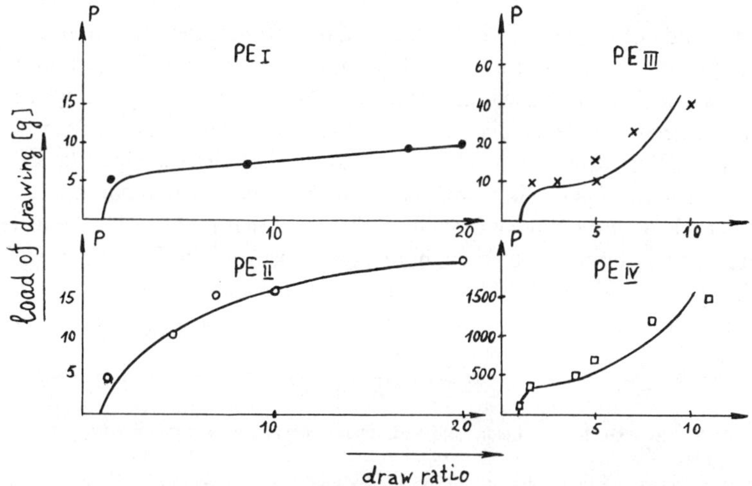

FIG. 5. Load–draw ratio curve for linear PE with different MW.

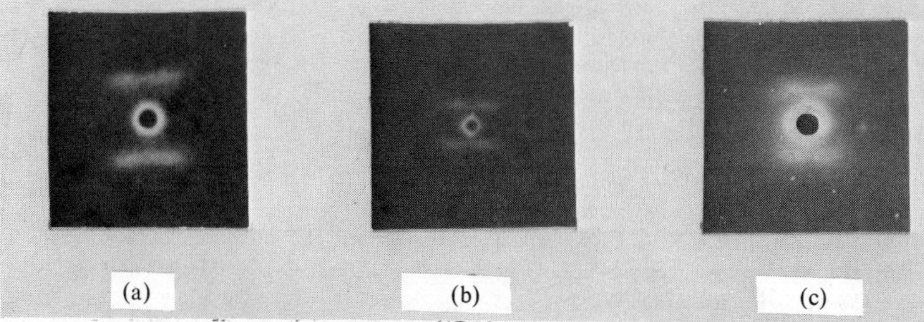

FIG. 6. Small-angle X-ray patterns: (a) branched PE; (b) polyethyleneterephthalate; (c) polycaproamide.

the data mentioned above, it can be expected that the transformation of the original structure into a microfibrillar one in these polymers will be completed at small draw ratios. A further elongation will proceed at the expense of the microfibrillar slip.

The X-ray patterns shown in Figure 6(a,b,c) demonstrate that the process of plastic deformation of the microfibrillar structure in these polymers is similar to that in high molecular PE. One can see that at ultimate draw ratios a distortion of the crystallite shape is observed, which can be derived from the four-point character of small-angle X-ray patterns. It should be noted that polycaproamide yields a four-point small-angle X-ray pattern only in the case of reorientation. The crystallite shape in a conventional unoriented film is not changed in drawing. This is likely to be due to high stability of crystallites resulting from strong intermolecular hydrogen bonds. Their shape can be changed only in the case when there is an extremely great number of interfibrillar tie molecules. Such a possibility is, probably, realized in the reorientation experiments. This assumption is qualitatively verified by a high transverse tensile strength of reoriented polycaproamide films ($\sigma = 25 \text{ kg/mm}^2$).

CONCLUSION

The study of the orientational drawing of different flexible-chain polymers leads to the conclusion that one of the main reasons for the termination of drawing is a sharp increase in the density of the microfibrillar packing, a growth of resistance to axial displacement, and, as a consequence, a drastic increase in the orientation stress. It reaches such values that the specimen breaks by a thermal fluctuation mechanism. The lifetime of the oriented polymer becomes comparable with the time of drawing. It is shown that interfibrillar tie molecules enable the microfibrils to be brought together more rapidly. The presence of these molecules strongly affects the character of the plastic deformation of the microfibrillar structure.

REFERENCES

[1] A. V. Savitskii, B. Ya. Levin, and V. P. Demitcheva, *Vysokomol. Soedin., A14,* **6,** 1286 (1973).

[2] S. N. Zhurkov, V. A. Marikhin, L. P. Myasnikova, and A. I. Slutsker, *Vysokomol. Soedin.,* 7, **6,** 1041 (1965).

[3] P. H. Geil, *Polymer Single Crystals,* Interscience Publishers, Division of John Wiley & Sons, New York, London, Sydney, 1964.

[4] G. Capaccio, T. J. Chapman, and I. M. Ward, *Polymer,* **16,** 469 (1975).

[5] P. J. Barham and A. Keller, *J. Mater. Sci.,* **11,** 1, 27 (1976).

[6] S. N. Zhurkov, B. Ya. Levin, and A. V. Savitskii, *Dokl. Acad. Nauk SSSR,* **186,** 132 (1969).

[7] K. E. Perepelkin, *Mekh. Polim.,* **6,** 845 (1966).

[8] A. Peterlin, *Colloid Polym. Sci.,* **253,** 809 (1975).

[9] V. V. Zhizhenkov, E. A. Egorov, and T. M. Petruchina, *Mekh. Polim.,* **3,** 387 (1973).

[10] L. I. Slutsker, A. V. Savitskii, L. E. Utevskii, I. M. Stark, and O. S. Lelinkov, *Vysokomol. Soedin., A13,* **12,** 2785 (1971).

[11] V. A. Marikhin, L. P. Myasnikova, and M. Sch. Tuchvatullina, *Mekh. Polim.,* **6,** 963 (1972).

[12] D. Ya. Tsvankin, Dokt. Dissertation, Moskva, Institut Elemento-organic. Soedin., 200 (1970).

[13] K. Adolphi, V. A. Marikhin, and L. P. Myasnikova, *Plaste Kautsch.*, **22,** Heft 3, 258 (1975).

[14] V. A. Marikhin, L. P. Myasnikova, V. A. Sutchkov, M. Sch. Tuchvatullina, and I. I. Novak, *J. Polym. Sci., Part C,* **N38,** 195 (1972).

[15] A. Cowking, J. G. Rider, I. L. Hay, and A. Keller, *J. Mater. Sci.,* **3,** 646 (1968).

[16] A. Keller and D. P. Pope, *J. Mater. Sci.,* **6,** 453 (1971).

[17] A. I. Slutsker, T. P. Sanphirova, A. I. Yastrebinskii, and V. S. Kuksenko, *J. Polym. Sci.,* **C16,** 4093 (1967).

[18] J. W. S. Hearle, *J. Polym. Sci.,* **28,** 432 (1958).

[19] V. E. Korsukhov, V. A. Marikhin, L. P. Myasnikova, and I. I. Novak, *J. Polym. Sci.,* **C42,** 847 (1973).

[20] V. A. Kosobukhin and A. D. Tchevytchelov, *Mekh. Polim.,* **5,** 771 (1973).

[21] V. A. Marikhin, L. P. Myasnikova, and N. L. Viktorova, *Vysokomol. Soedin.,* **17A,** 1302 (1976).

[22] K. Adolphi, V. A. Marikhin, and L. P. Myasnikova, *Plaste Kautsch.,* **21,** Heft 12, 902 (1974).

[23] V. R. Regel, A. I. Slutsker, and E. E. Tomashevskii, *Kinetic Nature of Solid Bodies Strength,* Nauka, ed., 600, Moskva, 1974.

UNI- AND BIAXIAL ORIENTATION OF POLYMER FILMS AND SHEETS

A. J. DE VRIES and C. BONNEBAT
Rhône-Poulenc Industries, Centre de Recherches de la Croix de Berny, 92160 Antony, France

J. BEAUTEMPS
Rhône-Poulenc Industries, Centre de Recherches des Carrières, 69190 Saint Fons, France

SYNOPSIS

Uni- and biaxial stretching of various polymer films has been studied under well-defined conditions of temperature and elongational strain rate in order to determine the relationship between stress and recoverable strain for both modes of deformation. The extent of molecular orientation has been investigated with the aid of stress optical methods: In amorphous polymers birefringence was found to be directly proportional to the frozen-in internal entropic stress irrespective of the latter's relationship with recoverable strain. In a first approximation, molecular orientation can be understood in terms of deformation of a rubberlike network with temporary junction points. The total internal stress in an oriented glassy polymer may be significantly larger than the entropic stress. Experimental methods based on retractive force measurements, able to distinguish between internal stresses of different nature, are described.

In uni- and biaxially drawn films of polyethylene terephtalate preferred planar orientation of (100) planes has been studied by means of X-ray diffraction and measurement of the three principal refractive indices. The intrinsic birefringence of completely oriented PET films has been estimated from these measurements with the aid of the Lorentz–Lorenz theory. The effect of draw ratio and temperature on the rate of crystallization and its consequences (increase of stretching stress and density, decrease of shrinkage) has also been studied.

Finally, the effect of molecular orientation on various mechanical properties: modulus, tensile strength, impact resistance, creep compliance, is discussed for both amorphous and semicrystalline polymers, with special emphasis on the predominant influence of amorphous phase orientation.

INTRODUCTION

It is widely recognized that molecular orientation resulting from deformation of a macromolecular material under appropriate conditions may have a profound effect on various properties of technological interest. Either uni- or biaxial molecular orientation are deliberately enhanced in certain industrial polymer processing techniques in order to improve the end-use properties of fibers and films, in particular. Since the induction of a considerable amount of molecular orientation inevitably leads to the appearance of a permanent state of anisotropy in the finished object, the choice of the most suitable orientation mode is mainly

Journal of Polymer Science: Polymer Symposium 58, 109–156 (1977)

governed by the final shape of the manufactured article and the type of mechanical constraints to which it will be subjected during service.

For this reason, objects whose thickness is small compared to the dimensions in the two other mutually perpendicular directions (films, sheets, but also hollow tubes and parisons) should be preferably oriented in a biaxial mode in order to prevent weakening in certain directions perpendicular to the thickness, unless the latter effect is desired in view of a specific technological application [1].

Because of its technological importance for the fiber manufacturing industry, uniaxial molecular orientation has been the subject of numerous investigations concerning, in particular, the characterization and measurement of the degree of orientation in relationship with the macroscopic deformation conditions, on the one hand, and with its effect on mechanical and other properties, on the other hand. A comprehensive account of the present state of knowledge in this field may be found in some recent publications [2, 3]. The major part of these investigations refers to the behavior of semicrystalline polymers, although it is generally admitted that in all cases an appropriate characterization of the molecular orientation in the amorphous phase is extremely important for a complete understanding of the mechanical properties.

The importance of amorphous phase orientation, already recognized in early investigations [4], has emphasized the need for characterization methods liable to distinguish between orientation distributions in both the crystalline and amorphous regions of a polymer. In this respect the extensive studies of Stein and co-workers [5] as well as those of Samuels [3] are of particular interest. Such a systematic approach making use of the results obtained by different experimental techniques may undoubtedly lead to a better understanding of the complex relationship between processing conditions, structure, morphology and properties, in spite of the fact that interpretation of the experimental results is very often based on a structural model containing specific features whose relevance is not necessarily justified.

Since orientation effects in the amorphous regions of a semicrystalline polymer are so important, it is still worthwhile to investigate in a detailed manner how molecular orientation is related to processing conditions in the case of completely amorphous polymers. Also, it should be kept in mind that the first stage of industrial drawing processes aimed at making a fiber or a film is most often performed on a mainly amorphous polymer and that, consequently, molecular orientation in the amorphous phase will precede and affect the formation of more or less perfectly oriented crystalline regions.

Moreover, essentially noncrystalline homo- and copolymers are widely used in the plastics processing industry for being subjected to thermoforming operations of various kinds (extrusion or calendering, followed by stretching, film blowing, vacuum or pressure forming, blow moulding) which may induce important amounts of uni- or biaxial molecular orientation in the finished object.

For these various reasons, the major part of the present paper will deal with the uni- and biaxial orientation of amorphous polymer films and sheets under well-defined thermal and mechanical conditions in order to determine the principal parameters common to all mechanical stretching operations, and which most significantly, as we believe, reflect the extent of molecular ordering and deformation with regard to its subsequent effect on mechanical properties. We will pay special attention to the differences between uni- and biaxial orientation as well as to some specific additional features in the case of uni- and biaxial orientation of a semicrystalline polymer (polyethylene terephthalate).

Some parts of this work are connected with a collaborative research program initiated by the IUPAC—Working Party on Structure and Properties of Commercial Polymers.

A detailed report on the first part of this collaborative study, entitled "The Effect of Molecular Orientation on the Mechanical Properties of Polystyrene" has been prepared for publication by T. T. Jones [6].

STRESS–STRAIN RELATIONSHIPS IN UNI- AND BIAXIAL EXTENSION

Experimental Methods

Imposed macroscopic strains are conveniently expressed by means of principal elongation ratios as follows:

Uniaxial extension

λ_1 is the elongation ratio in the direction of stretch, λ_2 the elongation ratio in the width direction, and λ_3 the elongation ratio in the thickness direction. In the absence of volume change, $\lambda_2 = \lambda_3 = \lambda_1^{-1/2}$.

Biaxial extension

λ_1 is either the greatest elongation ratio, or the elongation ratio in the first direction of stretch, λ_2 the elongation ratio in the second direction of stretch and λ_3 again the elongation ratio in the thickness direction. In the absence of volume change, $\lambda_3 = (\lambda_1\lambda_2)^{-1}$ and in the particular case of equal biaxial extension, $\lambda_1 = \lambda_2$, and hence, $\lambda_3 = \lambda_1^{-2}$.

By definition, uniaxial extension implies conservation of transverse isotropy, i.e., no preferential orientation in planes perpendicular to the draw direction $(\lambda_2 = \lambda_3)$. In oriented films, obtained under industrial conditions by means of unidirectional stretching, the orientation state is not necessarily uniaxial since the decrease in thickness during stretching may be much more important than the reduction in width [1].

In order to prevent, at least in the larger part of the specimen, any measurable amount of biaxial orientation, our laboratory experiments on uniaxial extension were always performed on test pieces of length-to-width ratio larger than 4. The identity of λ_2 and λ_3 in the final oriented specimens was checked by several width and thickness measurements in different parts of the sample.

Typically, specimens of uniform rectangular cross-section, having a length between 20 and 30 mm, a width of 5 mm and a thickness smaller than 1 mm were submitted to steady uniaxial extension until a maximum value of λ_1 in the order of 7 or 8 (unless rupture occurred at lower elongation ratios). The specimens were cut from calendered, extruded or molded sheets and films and clamped between the grips of a specially devised tensile tester equipped with an effective gas thermostat allowing the sample to be conditioned at various temperatures and to be maintained at a chosen fixed temperature during the elongation experiment. In this tensile tester the upper grip is rigidly connected to a force measuring device whereas the lower grip is traveling downward at a variable speed whose value is monitored by means of an electronic servo system so that the increase in specimen length (distance between the grips) per unit of time is, at any instant, proportional to the actual value of the specimen length, $l(t)$.

Under these conditions: $dl(t)/dt = \dot{\epsilon}l(t)$ where $\dot{\epsilon}$ denotes the time independent extensional strain rate, defined by:

$$\dot{\epsilon} = (dl/dt)/l = d \ln l/dt = d \ln \lambda_1/dt \tag{1}$$

Elongation ratio λ_1 is equal to $l(t)/l_0$ if l_0 is the initial specimen length. The force/length relationship is continuously recorded during the elongation experiment and subsequently converted to a stress–strain curve by dividing the measured force at any instant by the actual specimen cross-section, which is equal to the initial cross-section divided by λ_1. During its elongation the specimen may be visually observed through a slit-shaped window in the thermostat wall.

Biaxial extension experiments under well-defined conditions of strain rate are less easy to perform. In order to study the behavior in biaxial extension under conditions comparable to those of our uniaxial extension experiments, we have chosen the lateral bulging or inflation of circular shaped membranes, a type of experiment which has been applied in various laboratories to the investigation of rubbers [7, 8] and metals [9, 10] but more recently also to the study of thermoplastic polymers [11–15].

The experimental apparatus consists of a well-controlled environmental chamber, maintained at constant temperature in which a circular shaped flat specimen is gradually transformed into a spheroidal bubble under the influence of a variable gas pressure exerted on the lower face of the specimen, clamped at its perimeter between metal blocks of annular shape (Fig. 1). The inflation of the bubble, proceeding at a rate which depends on the gas pressure, the dimensions, and rheological properties of the specimen, may continue well beyond the attainment of hemispherical shape if the extensibility of the sample is high enough (Fig. 2). Although the bubble, even at high degrees of inflation, maintains its spheroidal shape, elongation ratios λ_1 (longitudinal direction) and λ_2 (latitudinal direction) at any point in the specimen, are in general unequal and different from one point to another. As first studied in detail in the case of rubber sheets by Treloar [7], both λ_1 and λ_2 generally increase gradually from the rim to the pole. Near the pole, however, both elongation ratios vary only little and are practically identical so that in this particular region a state of equal biaxial extension does exist. Typical experimental results obtained in biaxial extension

FIG. 1. Inflated bubble in biaxial extension.

FIG. 2. Effect of increasing degree of biaxial extension on bubble shape.

of polystyrene sheets in a range of temperatures above the glass transition region are shown in Figure 3. Similar results have been obtained with several other polymers under comparable conditions.

The elongation ratios at any point in the inflated bubble are determined by measuring the distances between concentric circles and between radii drawn on the specimen, as indicated in Figure 2. The condition of incompressibility: $\lambda_1\lambda_2\lambda_3 = 1$ is verified by means of thickness measurements:

$$e = \lambda_3 e_0 = e_0/\lambda_1\lambda_2 \tag{2}$$

if e is the local bubble thickness and e_0 the initial thickness of the flat specimen before inflation.

The inflation experiment may be used to investigate the relationship between stress and strain in equal biaxial extension at constant strain rate if the measurements are confined to the polar region of the bubble. In order to achieve a constant strain rate, the gas pressure required to inflate the bubble has to be continuously adjusted so that the distance between two points initially situated on a diameter of the flat specimen at equidistance from the center, increases during inflation as an exponential function of time. Because the polar region may be considered, to a good approximation, as a spherical cap, the arc length separating the two points, which is a direct measure of the latitudinal elongation ratio λ_2, will at any instant be equal to $\pi/2$ times the horizontal distance between the points. If this distance varies as an exponential function of time, the biaxial extensional strain rate in the polar region will be constant:

$$d \ln \lambda_2/dt = d \ln \lambda_1/dt = \dot{\epsilon} \text{ (constant)} \tag{3}$$

The continuous adjustment of gas pressure is monitored by means of an

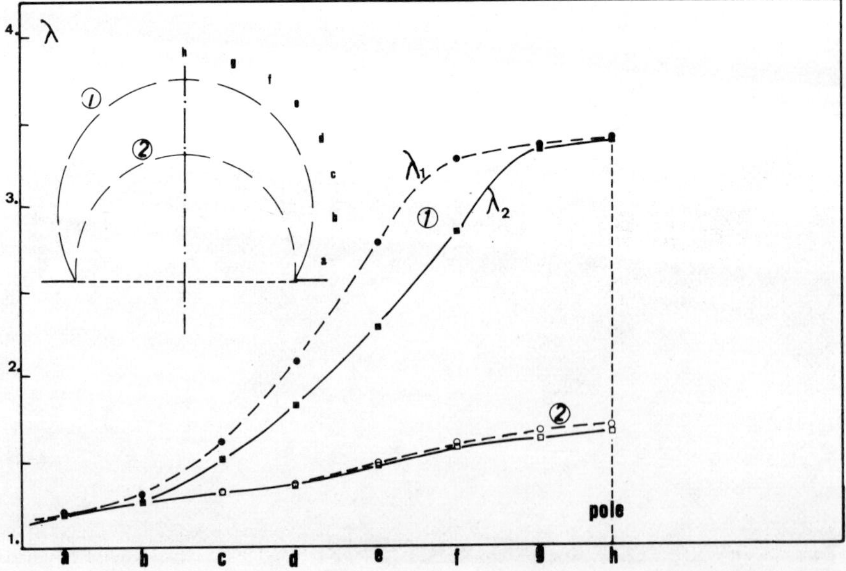

FIG. 3. Elongation ratios in an inflated bubble of polystyrene.

electronic servo-control system whose output signal is proportional to the difference between observed and pre-established elongation ratio λ_2, the latter being indicated by a programmer whereas the actual elongation ratio is measured with the aid of a specially devised strain gage sensor [15]. Gas pressure (p), elongation ratio in the polar region ($\lambda_2 = \lambda_1 \equiv \lambda$) and bubble height ($h$) are continuously recorded as a function of time, allowing the tensile stress (σ) in the polar region to be calculated with the aid of the classical equation:

$$\sigma = pR\lambda^2/2e_0 \tag{4}$$

where R, the radius of curvature in the polar region, is given by

$$R = (a^2 + h^2)/2h \tag{5}$$

if a is the radius of the flat circular specimen, equal to 75 mm in our experiments. Initial thickness values (e_0) were variable, but always smaller than 1 mm.

Results and Discussion

The extent of molecular orientation induced in amorphous polymers strongly depends on the temperature of deformation as well as on the extensional strain rate. The most appropriate temperature range is situated just above the glass transition region. This particular temperature range in which all polymers behave as viscoelastic rubbers, may be more or less broad, depending on the nature of the polymer and on the rate of deformation. The latter has to be high enough compared to the reciprocal characteristic relaxation time of the polymer network in order to avoid disorientation during stretching and subsequent cooling to temperatures below T_g.

All our experiments on uni- and biaxial extension have been performed in this temperature region and have confirmed the rubberlike behavior of amorphous polymers as shown by the fact that both uni- and biaxial strains were completely recoverable after stress release. However, in contrast with the behavior of ideal, crosslinked rubbers, the relationship between stress and strain for amorphous polymers such as polystyrene, polyvinylchloride and others, in the rubbery state, depends on strain rate and on temperature in a quite characteristic manner, as illustrated in Figures 4 and 5. As might be expected, for a given value of recoverable strain, the stretching stress increases with increasing strain rate and with decreasing temperature.

This behavior may be easily interpreted, at least qualitatively, in terms of formation and destruction of intermolecular secondary bonds accompanied by stretching and alignment in the direction of stress of molecular segments separated by successive intermolecular junction points. The former process will lead to energy dissipation and the latter one to buildup of elastic potential energy.

If the temperature is increased, energy dissipation becomes of greater relative importance compared to the storage of elastic energy because of the reduction in viscosity and relaxation times (average life times of network junctions). Lowering the strain rate has a similar effect because the corresponding increase in time of straining, required to attain a given value of strain, allows more disorientation (due to disappearance of network junctions) to occur.

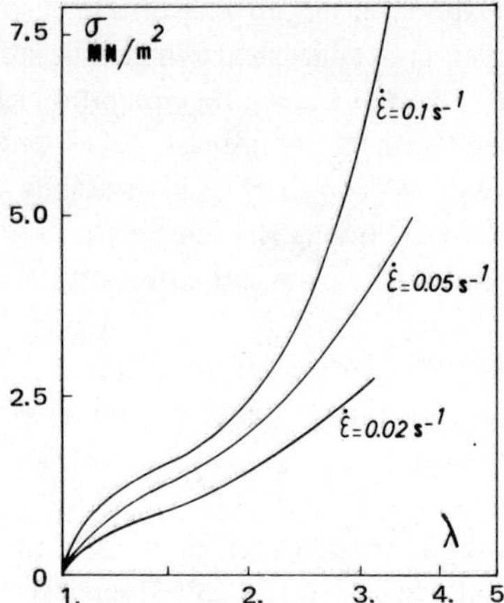

FIG. 4. Effect of strain rate on biaxial extension of polystyrene at 110°C.

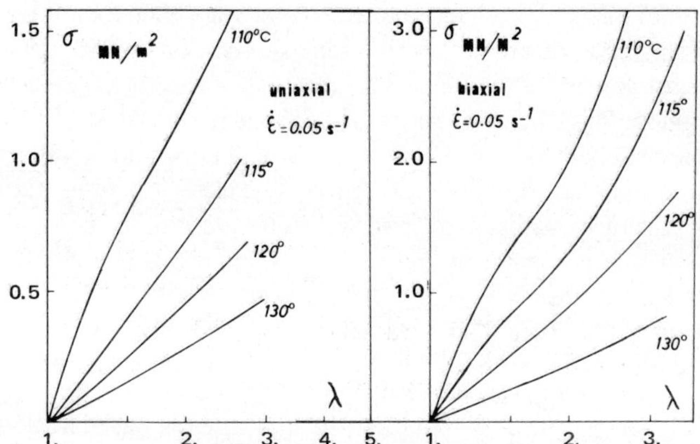

FIG. 5. Effect of temperature on uni- and biaxial extension of polystyrene.

It follows immediately from this argument that an increase in stretching stress (resulting from either a lower temperature or a higher strain rate) may expect to be associated with a greater extent of molecular orientation in the sample after it has been cooled to below T_g. A good correlation between stretching stress and frozen-in birefringence in the case of biaxially oriented polystyrene, for instance, has indeed been reported in the literature [16], but it is obvious that such correlations should, in general, depend on specific experimental conditions (e.g., rate of cooling) and hence, are of little use in the study of fundamental relationships between molecular orientation parameters and macroscopic, readily measurable characteristics.

Before discussing the latter point in greater detail in the following section, we want to emphasize the similarity between the observed stress–strain relations in uni- and biaxial extension. A typical example is shown in Figure 5 from which it is clear that for given values of temperature and strain rate, the general shape of the stress–strain curves is identical in both modes of deformation. However, under identical conditions the tensile stress in biaxial extension is to a good ap-

proximation twice as large as the tensile stress in uniaxial extension. Our results
obtained with different polymers seem to indicate that the exact value of this
ratio decreases somewhat with increasing temperature without deviating much
from 2 which is the theoretical value for "linear" elastic solids (but not for
rubbers except in the limit of infinitesimal strains) as well as for "linear" viscous
or viscoelastic liquids. Apparently, the deviation from linearity of the observed
stress–strain relationships does not seriously affect the value of the above ratio.
Cogswell and Moore [17] have recently reported that the creep behavior at room
temperature of a high molecular weight polyisobutylene in shear, uni- and biaxial
extension can be represented, in a first approximation, by a single curve if it is
assumed that for a given value of strain, the stresses in these three deformation
modes have a ratio 1:3:6.

From a practical point of view, the study of stress–strain relationships under
variable conditions of temperature and strain rate is of interest, not only for
defining the energy requirements of uni- and biaxial orientation processes, but
also for a better understanding of the occurrence of nonuniform deformations
and instability phenomena. The latter problem which has already been studied
in considerable detail for the biaxial extension of rubber sheets [18, 19], is also
of great importance for orientation processes of thermoplastic polymers in the
rubbery state. The nonuniform biaxial extension in the bubble inflation experiment, shown in Figure 3, does not only result from the particular boundary
conditions (zero latitudinal strain at the perimeter) but is essentially due to the
intrinsic mechanical behavior of the polymer under the chosen conditions.

The increase of stress for a given strain increment, in other words the slope
of the stress–strain curve, is of particular importance in this respect, because
it determines the deformation gradient along the bubble surface and consequently the bubble thickness profile. In order to reduce thickness gradients and
to avoid excessive thinning in biaxially extended sheets, obtained by thermoforming processes, conditions must be chosen so that the slope of the stress–strain
curve will be as high as possible [14, 15]. In the case of amorphous polymers this
may be achieved by lowering of the temperature as illustrated in Figure 6, which

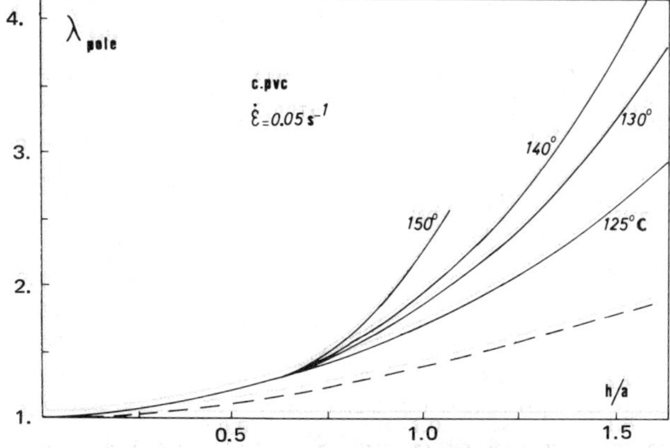

FIG. 6. Elongation ratio in polar region as a function of height-to-radius ratio of inflated bubble
of CPVC. Lower curve represents theoretical relationship for uniform biaxial extension.

shows that the elongation ratio in the thinnest part of a bubble (the polar region) increases less rapidly as a function of total deformation (characterized by the height-to-radius ratio of the bubble) when the temperature is lowered, although a uniform state of biaxial extension is never attained, even at the lowest temperature. The latter situation is indicated by the lower curve in Figure 6, representing the theoretical relationship for uniform deformation:

$$\lambda_1 = \lambda_2 = [1 + (h/a)^2]^{1/2} \tag{6}$$

This ideal situation may be approximately reached if the slope of the stress–strain curve becomes extremely high, for instance as a result from strain-induced crystallization. A practical example referring to the uniaxial extension of initially amorphous polyethylene terephthalate is shown in Figure 7.

If conditions of temperature and strain are appropriate for crystallization to occur, the slope of the stress–strain curve will considerably increase. This "strain hardening" effect due to crystallization starts for strain values which are smaller the lower the temperature. A biaxially oriented bubble inflated under the same conditions will present a relatively uniform thickness except for the region near the rim where the strain values are the smallest (Fig. 8).

The opposite effect, "strain softening," quite generally occurs in amorphous polymers below T_g but also in crystalline polymers such as polyethylene and polypropylene submitted to uni- or biaxial extension below the melting temperature. The stress–strain curves for uniaxial extension of isotactic polypropylene, given in Figure 9, show that in a certain range of strain values the slope may actually become negative after the so-called "yield point" has been reached, followed by a plateau zone in which the slope remains very small. In this region, the behavior is no longer homogeneous and the sample thins locally to a smaller cross-section with formation of a "neck." The thinner part of the specimen increases gradually its length at the expense of the thicker part until, finally, the cross-section becomes uniform again accompanied by a rapid increase of the slope of the stress–strain curve. The region of inhomogeneous extension is indicated in Figure 9 by means of dashed lines; the values for stress and strain in this region are only "apparent values" since they have been calculated on the assumption of uniform uniaxial extension. The nonhomogeneous behavior can be characterized by the ratio of cross-sections in the two parts of the specimen. This ratio has been plotted in Figure 10 as a function of total global deformation. In this particular example, nonhomogeneous behavior disappears for elongation ratios higher than about 5.

This peculiar deformation behavior of semicrystalline polymers below their melting point has been extensively discussed in the literature and is usually interpreted in terms of a transition from spherulitic to fibrillar crystalline structures [3, 20]. Such studies are of great value for a better understanding of deformation processes in polymers in relation with their molecular and supramolecular structure, but it should be kept in mind that a completely different microstructure (as found, e.g., in amorphous polymers below T_g) may give rise to similar macroscopic behavior. For this reason, it is extremely useful to investigate orientability and optimum conditions for polymer orientation processes, both

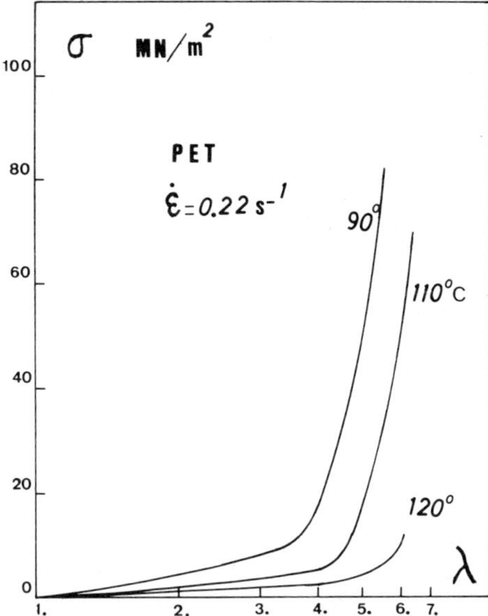

FIG. 7. Effect of strain-induced crystallization on uniaxial extension of PET at different temperatures.

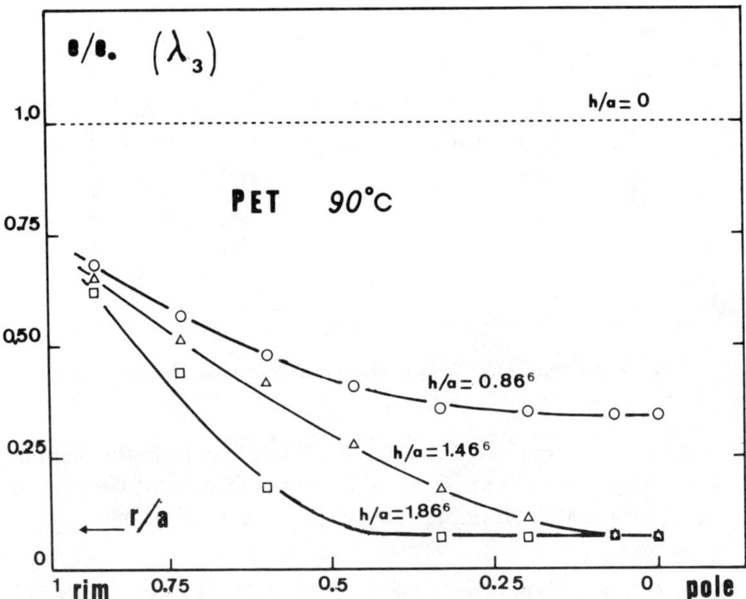

FIG. 8. Thickness variation in inflated bubbles of PET for different height-to-radius ratios.

subjects of great technological interest, without the aid of any specific structural model but only on the basis of observed mechanical behavior under well-defined conditions. Obviously, a thorough knowledge of the relationship between structure, morphology, and macroscopic mechanical behavior in the case of

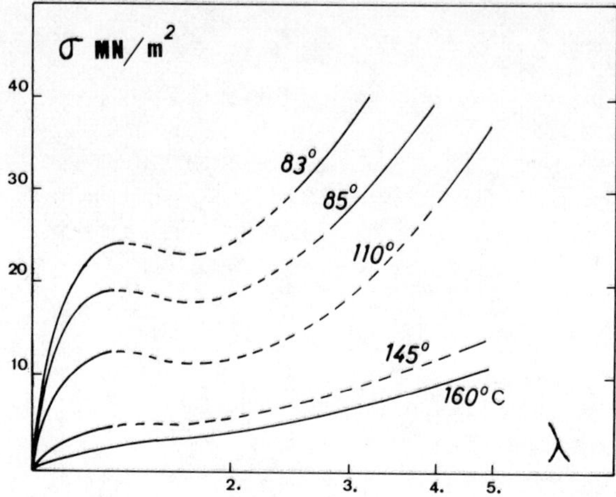

FIG. 9. Effect of temperature and elongation ratio on nonuniform uniaxial extension of isotactic polypropylene.

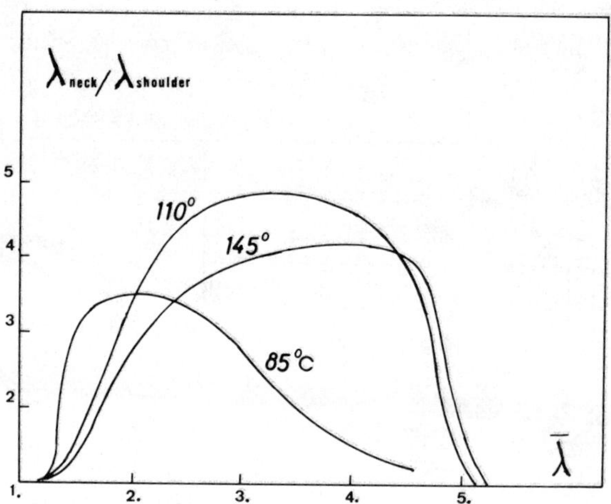

FIG. 10. Degree of "necking" in uniaxially drawn polypropylene as a function of elongation ratio and temperature.

semicrystalline polymers remains highly desirable also from the point of view of the technologist who wants to know how the mechanical behavior can be modified by means of appropriate changes in process conditions.

MOLECULAR ORIENTATION AND FROZEN-IN STRESS

The molecular orientation in amorphous polymer samples submitted to uni- or biaxial extension in the rubbery state may be frozen-in by more or less rapid cooling to temperatures below T_g. In general, molecular mobility in the glassy state is insufficient for disorientation to occur within a measurable span of time, even in the absence of external stress, notwithstanding the inherent tendency of the molecular chains to revert to the unoriented, most probable, random state.

The frozen-in elastic potential energy will, however, be dissipated as soon as the sample is heated to temperatures in or above the glass transition range, as a result from enhanced molecular mobility. In the absence of external constraints the energy dissipation will manifest itself as a dimensional instability or shrinkage. On the other hand, if the sample is restrained from shrinkage by maintaining its length constant during the heating process, an external force has to be applied, named shrinkage force or retractive force [6, 15]. Shrinkage force analysis permits a rapid determination of the maximum shrinkage stress (defined as the maximum shrinkage force per unit of cross-section of the oriented sample exerted during heating) which should be equal in absolute value to the frozen-in tensile stress.

In the case of an oriented specimen of an ideal crosslinked rubber, no energy dissipation would occur if the specimen at constant length is heated through the glass transition range up to a temperature T well above T_g, and the shrinkage stress would rapidly obtain a constant value completely determined by the stored elastic energy at temperature T. In an amorphous uncrosslinked polymer sample, however, the internal stress will relax more or less rapidly depending on the value of $T-T_g$ and the shrinkage stress as a function of temperature will pass through a maximum and, ultimately, vanish. It has been well established experimentally [6, 15] that the maximum shrinkage stress is practically independent of the rate of heating, as shown in Figure 11 for uniaxially oriented specimens of unplasticized polyvinylchloride. The shape of the curve determined by the temperature dependent rate of stress relaxation may be very different from one polymer to another and generally depends on the thermomechanical history of the sample (deformation conditions, cooling rate) as it is obvious from Figures 11 and 12.

It has been often assumed, until now, that the maximum shrinkage stress is identical with the frozen-in internal entropic stress, resulting from the deformation of the rubberlike molecular network and that its value should be directly related, therefore, to orientation birefringence. Rudd and Andrews [21] found a unique, nonlinear relationship independent of thermomechanical history in the case of uniaxially oriented polystyrene filaments. The results of the IUPAC Working Party, based on numerous data obtained in different laboratories, reported by Jones [6], can be represented by a linear relationship for uniaxially oriented polystyrene sheets of variable thermomechanical history and the stress–optical coefficient (SOC) calculated from this relationship, -5200 Brewsters (1 Brewster $= 10^{-12}\ m^2/N$), is in good agreement with other literature data. A similar result was reported by Pinnock and Ward [22] for amorphous polyethylene terephthalate fibers at low elongation ratios.

Further work in our laboratory has shown, however, that a unique linear relationship does not exist, in particular if the polymer samples have been oriented at high strain rates (or at low temperatures), followed by rapid quenching below T_g. In general, the calculated SOC is not constant but decreases (in absolute value) with increasing internal stress in accordance with the results reported in [21]. Rudd and Gurnee [23] had already observed some years ago that the birefringence of polystyrene samples rapidly stretched at temperatures near T_g and, subsequently, maintained at constant strain, may pass through a maximum

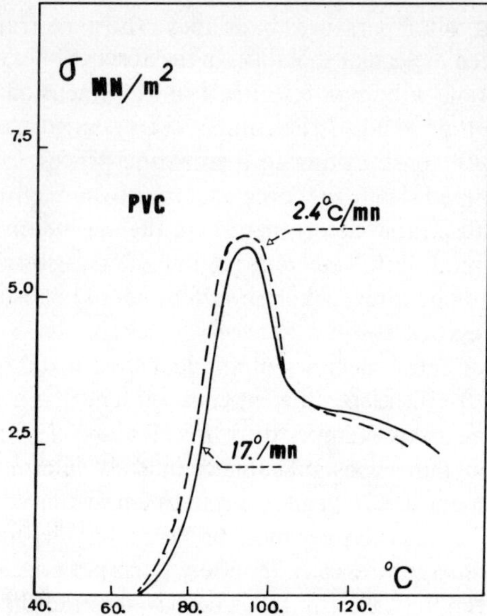

FIG. 11. Effect of heating rate on shrinkage stress diagrams of uniaxially oriented PVC.

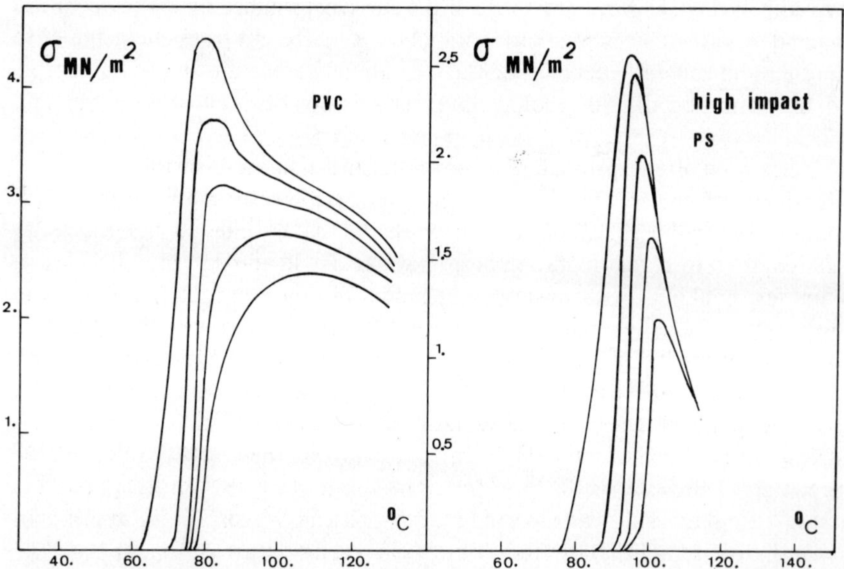

FIG. 12. Typical shrinkage stress diagrams for PVC and polystyrene of different thermomechanical histories.

as a function of time. They explained this peculiar behavior by assuming that the total birefringence is the algebraic sum of two components, a positive one which rapidly decays as an exponential function of time, and a negative component which decays much more slowly in the same temperature region. Only the latter component is associated with molecular chain orientation, whereas the positive birefringence would be related to local distortions, in particular side group twisting which should partially align the phenyl groups in the direction of stress. The presence of both orientation birefringence and distortional birefringence in strained polymers has been frequently discussed in the literature, and experimentally investigated by means of dynamic birefringence methods [24]. The exact nature of the local distortions, giving rise to positive or negative birefringence (depending on the chemical structure of the polymer) is mostly unknown but is generally attributed to segmental motions against rotational energy barriers, bending of valence angles or changes in primary and secondary bond lengths. All these molecular deformation mechanisms will be associated with an increase in internal energy, dU, which has to be added to the change in configurational entropy $-TdS$ (corresponding to the network deformation) in order to find the total amount of stored elastic energy.

Depending on the rate of quenching, a more or less important part of the total elastic energy will be frozen-in below T_g and the internal stress should be considered, therefore, as a sum of two contributions:

$$\sigma_i = \lambda(\partial U/\partial\lambda - T\partial S/\partial\lambda) \tag{7}$$

$$= \sigma_d + \sigma_s \tag{8}$$

where σ_d and σ_s are the frozen-in energy elastic (or distortional) and entropy-elastic stresses, respectively, and $U - TS$ is the stored free energy per unit volume, also designated by the symbol W (work of deformation per unit volume).

The relative contribution of the first term in eqs. (7) and (8) increases with increasing strain rate and decreasing deformation temperature [6], as easily explained by the much smaller relaxation times associated with the distortional stresses compared to the rubber elastic stress. This is particularly obvious in the case of polyvinylchloride, as shown in Figures 11 and 12. The distortional stresses relax and vanish rapidly at temperatures close to T_g giving rise to a relaxation peak quite distinct from the rubber elastic stress plateau which extends to temperatures significantly higher than the deformation temperature before it starts to vanish as a result from relaxation.

The remarkable thermal stability of the rubbery network in polyvinylchloride which must be due to a relatively high cohesive energy of the temporary network junctions is quite probably related to the occurrence of many small crystallites associated with the presence of short syndiotactic sequences in the polymer chains. The effect of this small amount of crystallinity on the viscoelastic behavior of PVC has often been invoked since the early investigations of Tobolsky and co-workers [25]. The crystalline nature of at least part of the network junctions in PVC explains why the temperature region in which rubberlike behavior during stretching is observed, is much broader than for completely amorphous polymers such as atactic polystyrene.

Because of the fast relaxation of distortional stresses at temperatures close to T_g, the amount of frozen-in distortional stress below T_g is strongly dependent on the rate of cooling through the glass transition region. Taking into account the effects of strain rate, deformation temperature, and cooling rate, it can now be understood why certain authors found a linear relationship between maximum shrinkage stress and birefringence in oriented polystyrene whereas others did not. Only the entropic stress will, in general, be directly related to the macroscopic recoverable strain and to other physical effects characteristic of molecular orientation. The presence of internal distortional stresses may, however, seriously affect certain aspects of the mechanical behavior of oriented films and sheets, as will be further discussed below.

The considerable difference in relaxation rate of the distortional stresses in PVC, compared to the decay of entropic stress makes it relatively easy to separate the two kinds of stresses quantitatively by means of an isothermal relaxation experiment on an oriented sample rapidly heated to a temperature a few degrees above T_g, as illustrated in Figure 13. A more accurate method of analysis, which may be applied to an oriented film of any amorphous polymer, consists in measuring the maximum shrinkage stress in specimens previously submitted to increasing amounts of shrinkage of known magnitude at a temperature close to T_g. Indeed, if a sample of oriented polymer is allowed to shrink a small amount, the associated decrease in internal stress will be mainly due to relaxation of distortional stresses and it may be assumed, therefore, that the residual internal stress after a sufficient amount of shrinkage is essentially of entropic nature and, hence, completely determined by the residual amount of recoverable strain. This hypothesis is confirmed by the experimental results in Figure 14 showing that the residual amount of internal stress after partial shrinkage of specimens oriented at the same deformation temperature T_o, is only dependent, after a sufficient degree of shrinkage, on the residual amount of recoverable strain, irrespective of the initial value of the latter parameter.

The observed relationship between entropic internal stress and recoverable strain found for various vinyl polymers, in particular atactic polystyrene and chlorinated PVC [15], is in good agreement with the predictions of the ideal rubber elasticity theory based on Gaussian chain statistics [26], leading to the following expressions for the difference in any pair of principal stresses:

$$\sigma_1 - \sigma_2 = \lambda_1 \partial W / \partial \lambda_1 - \lambda_2 \partial W / \partial \lambda_2 \tag{9}$$

$$= -T(\lambda_1 \partial S / \partial \lambda_1 - \lambda_2 \partial S / \partial \lambda_2) \tag{10}$$

$$= G(\lambda_1^2 - \lambda_2^2) \tag{11}$$

where G is the equilibrium shear modulus at temperature T. The following eqs. (12) and (13) are valid then in the case of uni- and biaxial extension, respectively:

$$\sigma_s^u = G(\lambda_1^2 - 1/\lambda_1) \tag{12}$$

$$\sigma_s^l = G(\lambda_1^2 - 1/\lambda_1^4) \tag{13}$$

The use of eqs. (12) or (13) allows the determination of the entropic part of frozen-in internal stress by means of a simple extrapolation to zero amount of shrinkage, as illustrated in Figure 14 for the case of a biaxially-oriented film of chlorinated PVC.

The effect of deformation temperature on the relationships (12) and (13)

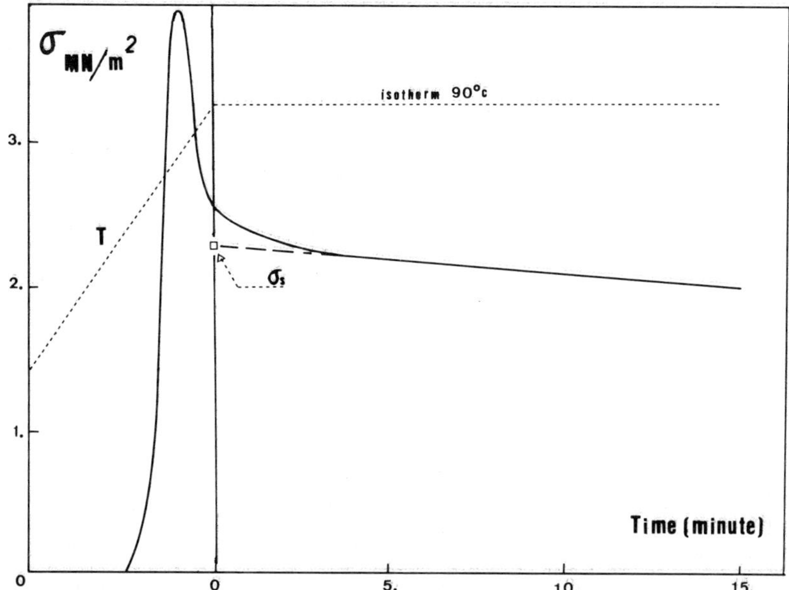

FIG. 13. Shrinkage stress of biaxially oriented PVC specimen heated to 90°C, followed by iso-thermal relaxation. Determination of internal entropic stress by means of graphical extrapolation.

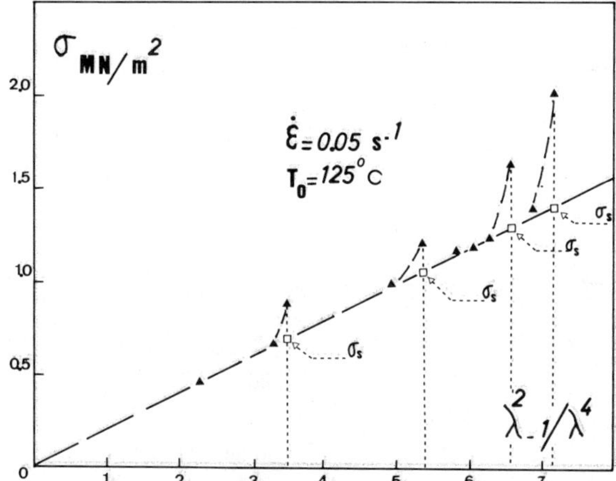

FIG. 14. Residual internal stress after partial shrinkage as a function of residual recoverable strain for biaxially drawn films of CPVC. Determination of internal entropic stress by means of graphical extrapolation.

in the case of a "high-impact" polystyrene is indicated in Figure 15: As expected, the value of G decreases rapidly with increasing temperature, reflecting the gradual disappearance of temporary network junctions as a result from enhanced molecular motions at higher temperatures. The number of network junctions or the corresponding number of effective network chains may, in principle, be calculated from G which according to the statistical theory of rubber elasticity is given by:

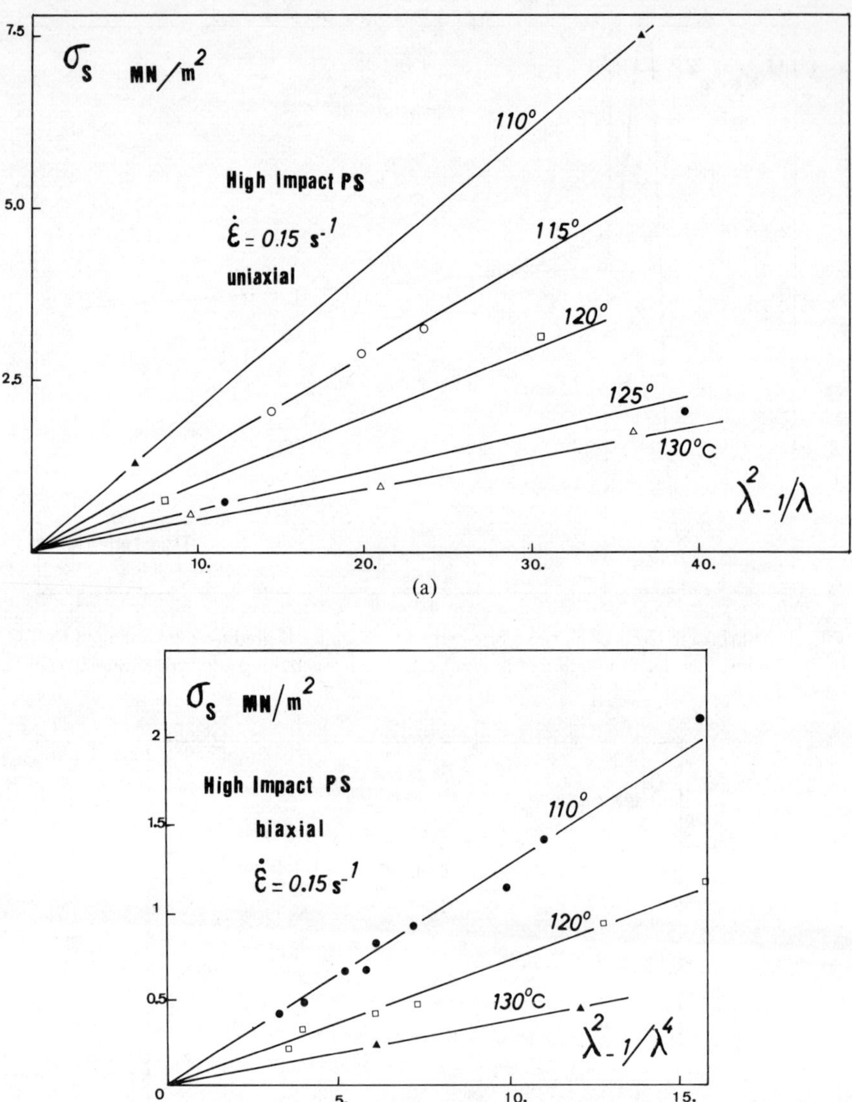

FIG. 15. Internal entropic stress as a function of recoverable strain for different deformation temperatures (high-impact polystyrene): (a) uniaxial extension; (b) biaxial extension.

$$G = gN_e kT \tag{14}$$

where N_e is the number of effective network chains per unit volume, k the Boltzmann's constant and g a "front factor" whose value may vary between 0.5 and 1 according to different theoretical estimates [27, 28]. In recent publications [29, 30] it has been suggested that g increases with increasing functionality of the network junctions. According to Graessley [29], front factor g and functionality f would be simply related as follows:

$$g = (f - 2)/f \tag{15}$$

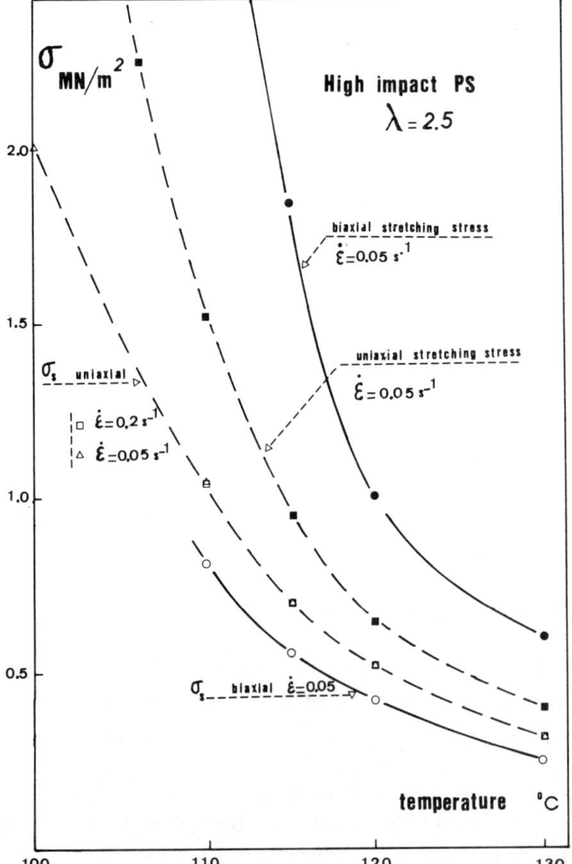

FIG. 16. Comparison between stretching stress and internal entropic stress in uni- and biaxial extension, as a function of deformation temperature (high-impact polystyrene).

Since the network topology in general and the functionality, in particular, are unknown parameters it is not possible to derive any accurate values for N_e with the aid of eq. (14). In spite of this uncertitude, the order of magnitude of N_e calculated in this way is quite acceptable and corresponds to an average chain length of some hundreds of backbone atoms between two successive junction points in the network existing between T_g and $T_g + 30°C$, but the average chain length increases rapidly with increasing temperature.

It is obvious from Figure 15 that the values of G calculated from the biaxial extension experiments are, in general, lower than those corresponding to uniaxial extension; this difference decreases with increasing temperature and could be due to a systematic experimental error. If, however, this observed difference is real, it might be associated with a difference in network topology (number and/or functionality of junction points) resulting in a less rigid network during biaxial extension. Another possible cause will be discussed below.

It is worthwhile to remind here that under identical conditions of temperature, strain rate, and total recoverable deformation, the stretching stresses in biaxial

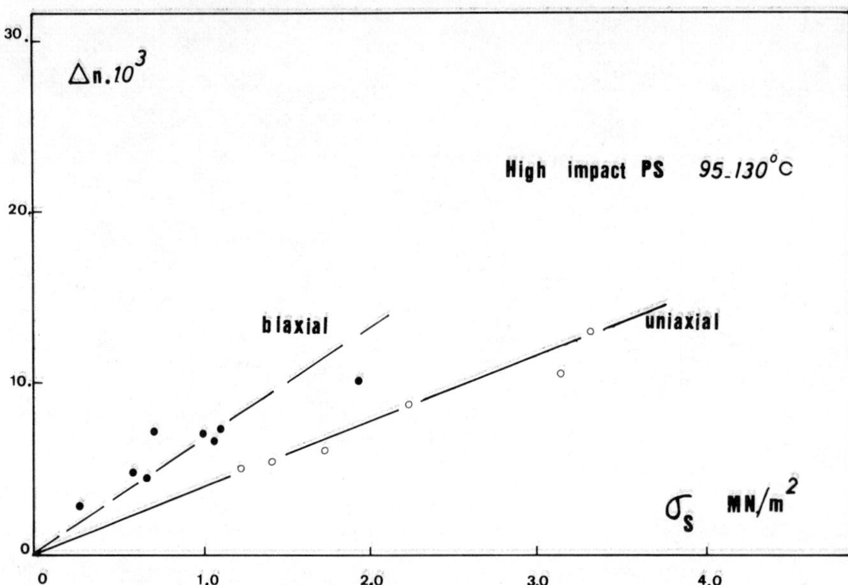

FIG. 17. Birefringence as a function of internal entropic stress for high-impact polystyrene.

extension are about twice as large as in uniaxial extension; to the contrary the internal entropic stress, characteristic of the orientation of the rubberlike network, is actually smaller in biaxial extension, as illustrated for polystyrene in Figure 16.

The results of birefringence measurements, shown in Figure 17, also indicate a systematic difference between uni- and biaxial extension. Birefringence is proportional to the value of internal entropic stress, irrespective of deformation conditions, but the stress–optical coefficient is significantly higher in biaxial compared to uniaxial extension. Whatever the reasons may be for the observed differences between uni- and biaxial extension, the proportionality between birefringence and internal entropic stress in both modes of deformation may be considered as a strong argument in favor of the hypothesis that only the entropic part of the internal stress is directly related to the molecular orientation.

In the case of polyvinylchloride, the simple relationships (12) and (13) are not obeyed, neither in uniaxial nor in biaxial extension (Fig. 18), but as for polystyrene, optical birefringence is again proportional to internal entropic stress with a similar difference in SOC between the two modes of deformation (Fig. 19). The calculated value of the latter coefficient in uniaxial extension (~600 Brewsters) is in good agreement with other experimental data reported in the literature [24, 31]. Kashiwagi and Ward [32] concluded from NMR studies that the behavior of oriented PVC may be interpreted in terms of a pseudo-affine deformation scheme.

The apparently very general character of the linear relationship between birefringence and internal entropic stress, irrespective of the exact shape of the stress–strain curve, lends further support to our thesis that internal entropic

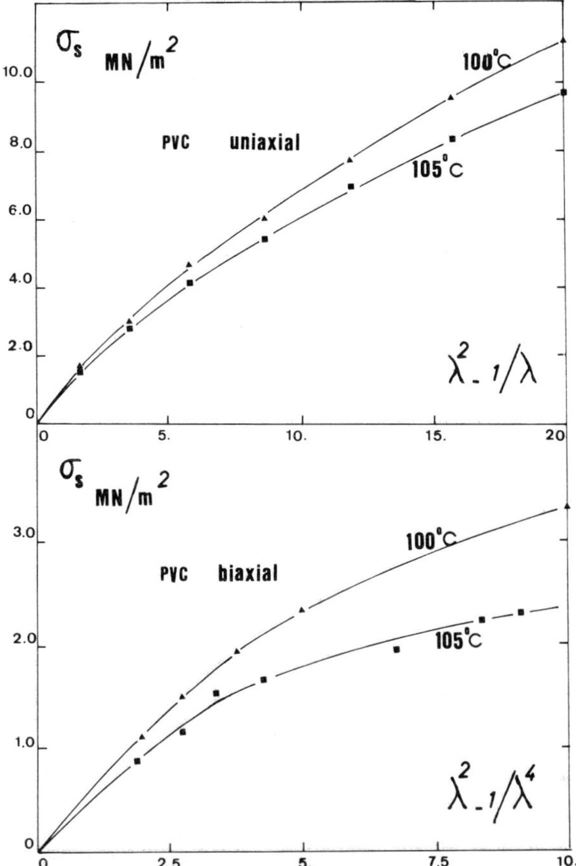

FIG. 18. Relationship between internal entropic stress and recoverable strain for PVC.

stress, determined in the way we have described, is a direct measure of the extent of molecular orientation. It is well known that orientation birefringence can be directly related to the second moment of the distribution of orientation angles θ by means of the Hermans' orientation function [4]:

$$\Delta n = \Delta n_{max}(3\langle\cos^2\theta\rangle - 1)/2 \tag{16}$$

where Δn_{max} is the maximum birefringence at full orientation when $\langle\cos^2\theta\rangle = 1$. The relationship between the orientation function and the internal entropic stress may be written, therefore, as follows:

$$f_a = (3\langle\cos^2\theta\rangle - 1)/2 = \Delta n/\Delta n_{max} = SOC\sigma_S/\Delta n_{max} \tag{17}$$

which shows that the Hermans orientation function for the amorphous rubberlike network is directly proportional to σ_S.

Introducing the Kuhn–Grün model for the stress–optical behavior of rubberlike networks [33] will finally lead to the following expression:

$$f_a = \sigma_S/5N_s kT^g \tag{18}$$

where N_s is the number of statistical segments per unit volume.

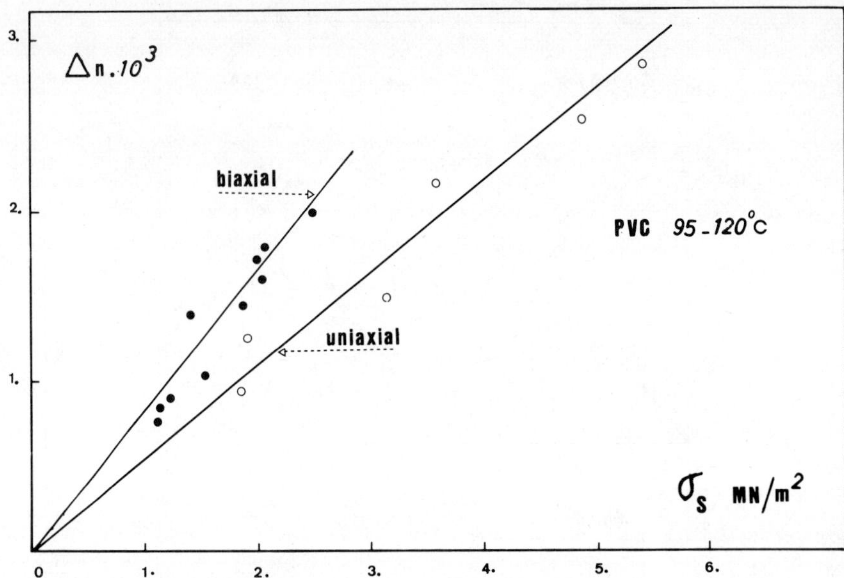

FIG. 19. Birefringence as a function of internal entropic stress for PVC.

If both SOC and Δn_{max} are known for a given polymer, eq. (17) may be used to calculate the amorphous phase orientation f_a from the internal entropic stress. Possible applications of this experimental method to the investigation of amorphous phase orientation in semicrystalline polymers are actually under study in our laboratory.

Summarizing the experimental results discussed in this section, it can be stated that molecular orientation in the amorphous regions of oriented polymer films may be regarded as resulting from the deformation of a rubberlike network: The relationship between internal entropic stress, determined from shrinkage force measurements, and recoverable strain is in good agreement with the predictions of rubber elasticity theory in its simplest form, at least in the case of completely amorphous polymers.

The only polymer for which we have found a more complex behavior, polyvinylchloride, is known to possess a certain degree of local order (presence of "crystallites") giving rise to network junctions remaining persistent at temperatures far above T_g.

In all cases, however, internal entropic stress was found to be proportional to birefringence. In a recent report by Ward and co-workers [34] the birefringence of uniaxially oriented films of amorphous polyethylene terephthalate as a function of recoverable strain was also found to be in accordance with the predictions of rubber elasticity theory, up to relatively high values of elongation ratio ($\lambda \sim 3.5$).

Finally it should be reminded that what we have defined as "internal entropic stress" is not necessarily resulting from entropy changes only. It is generally recognized that the extension of a polymer chain may be accompanied by changes in molecular conformation (relative amounts of rotational isomers), dependent on intramolecular interactions. Theoretical treatments of this effect [35] predict that the "entropic stress" should contain a constant factor, independent of elongation ratio, and associated with the change in conformational internal energy.

Similar to the change in configurational entropy upon straining the rubberlike network, the change in internal intramolecular energy due to the change in average population of various conformers would be determined by the variation of the mean-square vector length of the network chains and be proportional to:

$$\Delta U(\text{intramol}) \sim \Delta S \sim (\lambda_1^2 + \lambda_2^2 + \lambda_3^2 - 3)$$

The relative contribution $(\sigma_u/\sigma_u + \sigma_s)$ of the internal stress $\sigma_u = \lambda \partial U/\partial \lambda$, associated with this change in intramolecular energy, would therefore be neither dependent on the amount of strain, nor on the type of strain.

Published experimental results on the intramolecular energy contribution to the retractive force of crosslinked rubbers show no agreement on the question of its invariance with strain. Most experiments have been performed in uniaxial extension with the notable exception of the work of Boyce and Treloar [36] on the thermoelasticity of natural rubber in torsion. Cunningham and co-workers [34] found that the change in concentration of *trans*-ethylene glycol linkages in uniaxially stretched films of polyethylene terephthalate varied as a function of elongation ratio in at least qualitative agreement with the theory of Abe and Flory [37].

It appears, therefore, well established that the deformation of a rubberlike network, containing either permanent or temporary crosslinks, may be associated in general with a change in intramolecular energy, depending on the chemical structure of the polymer. If its relative contribution were independent on the amount and type of strain, the corresponding energy–elastic internal stress would be indistinguishable from the purely entropic internal stress as determined from shrinkage force measurements.

On the contrary, if its contribution should be different for different types of strain, it might be partly responsible for the systematic differences we found between uni- and biaxial orientation stresses.

In any case, it seems obvious to us that the distortional internal stress σ_d, whose relaxation behavior is very different from the entropic stress, must be mainly due to intra- and intermolecular energy effects which are only indirectly related to the macroscopic recoverable strain, but directly associated with states of local strain or orientation. The probable existence of the latter kind of internal energy contributions to the total free energy of deformation in rubberlike networks had already been emphasized by Treloar [38].

CRYSTALLIZATION AND ORIENTATION IN UNI- AND BIAXIALLY STRETCHED POLYETHYLENE TEREPHTHALATE FILMS

We have already discussed the peculiar shape of the stress–strain relationship

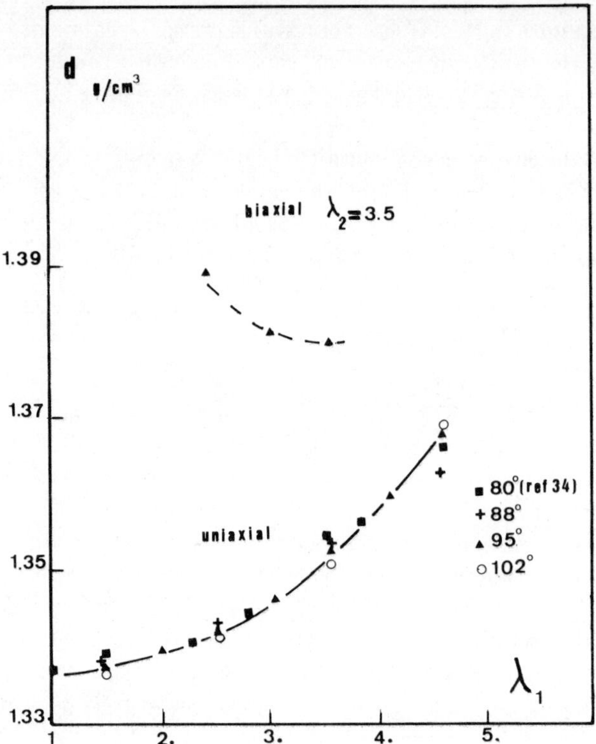

FIG. 20. Density of uni- and biaxially drawn PET films as a function of draw ratio.

in uni- and biaxial extension of initially amorphous PET films, due to strain- or stress-induced crystallization.

In the present section we will report experimental results concerning the structure and properties of uni- and biaxially stretched PET films, obtained under various conditions comparable to those prevailing in industrial practice. The molten polymer (intrinsic viscosity 0.64) was extruded through a slit die (width 45 cm) and subsequently submitted to uni- or biaxial extension under variable conditions of temperature and strain rate. In the case of uniaxial extension, final film thickness was about 40 micrometers; in biaxial stretching the film thickness was reduced to still smaller values. The films were prepared in the laboratories of Rhône-Poulenc–Films Division.

In Figure 20 the density of uniaxially stretched films has been plotted as a function of elongation ratio. The increase of density with elongation ratio is practically independent of temperature in the limited range investigated (88°–102°C), except for the highest value of λ_1 (= 4.5). In the latter case density increases with increasing stretching temperature, undoubtedly due to the enhanced rate of crystallization [39]. It is well established 140, 41] that the onset of crystallization is not the only cause for the density increase in oriented PET: The density of the amorphous phase itself is orientation-dependent, resulting probably from the *gauche-trans* isomerization of the ethylene glycol linkage. Conclusive evidence for this point of view has been recently reported in the al-

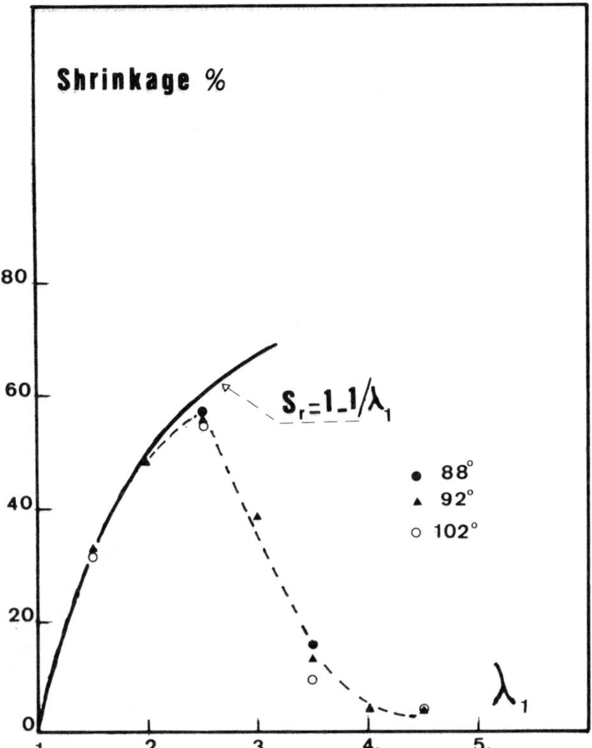

FIG. 21. Shrinkage ratio (after 30 min at 90°C) of one way drawn PET films as a function of draw ratio.

ready cited paper by Cunningham, Ward and others [34]. The density increase with elongation ratio, observed by the latter authors, on PET specimens drawn at 80°C, has also been plotted in Figure 20 and is seen to be in good agreement with our results. The same authors attribute the observed density increase mainly to the increase in *trans*-content of the amorphous phase, at least up to an elongation ratio of 3.5. In the PET films we have examined and which had been drawn at somewhat higher temperatures, part of the density increase must be associated with increased crystallinity when the elongation ratio becomes greater than 2.5.

This is obvious from Figure 21 in which we have plotted the shrinkage ratio of the same films, measured after 30 min heating at 90°C. Up to $\lambda = 2.5$, the shrinkage ratio increases with increasing draw ratio in good agreement with the expected relationship:

$$S_r = 1 - 1/\lambda \tag{19}$$

where the shrinkage ratio S_r is the fractional decrease of length of an oriented sample of draw ratio λ.

For $\lambda \leqslant 2.5$, therefore, the elongational strain is completely recoverable after heating above T_g and may be ascribed to the deformation of an amorphous rubberlike network.

However, for $\lambda > 2.5$, S_r decreases with increasing draw ratio, due to the occurrence of a crystalline phase which confers dimensional stability on the oriented amorphous network. This well known inverse relationship between shrinkage ratio and crystallinity has been used by Smith and Steward [39] in a study of the rate of crystallization in oriented PET fibers.

The formation of an oriented crystalline phase has been further examined by means of wide-angle X-ray diffraction studies. Since it has been well established [42, 43] that the (100) plane, which is nearly parallel to the plane of the PET molecule, tends to become preferentially oriented with respect to the plane of the film, we have expressed our X-ray diffraction results in terms of the orientation angle distribution of the normal to the (100) plane with respect to the direction of stretching, on the one hand, and with respect to the plane of the film surface, on the other hand.

Calling ϕ_{N1}, ϕ_{N2}, and ϕ_{N3} the angles which makes the normal to plane (100) with reference axes 1 (machine direction), 2 (transverse direction), and 3 (direction normal to film surface), respectively, we may write:

$$\langle \cos^2\phi_{N1} \rangle + \langle \cos^2\phi_{N2} \rangle + \langle \cos^2\phi_{N3} \rangle = 1 \tag{20}$$

For random orientation: $\langle \cos^2\phi_{N1} \rangle = \langle \cos^2\phi_{N2} \rangle = \langle \cos^2\phi_{N3} \rangle = 1/3$; for perfect planar orientation: $\langle \cos^2\phi_{N1} \rangle = \langle \cos^2\phi_{N2} \rangle = 0$; $\langle \cos^2\phi_{N3} \rangle = 1$ and finally, for perfect orientation with respect to the stretching direction:

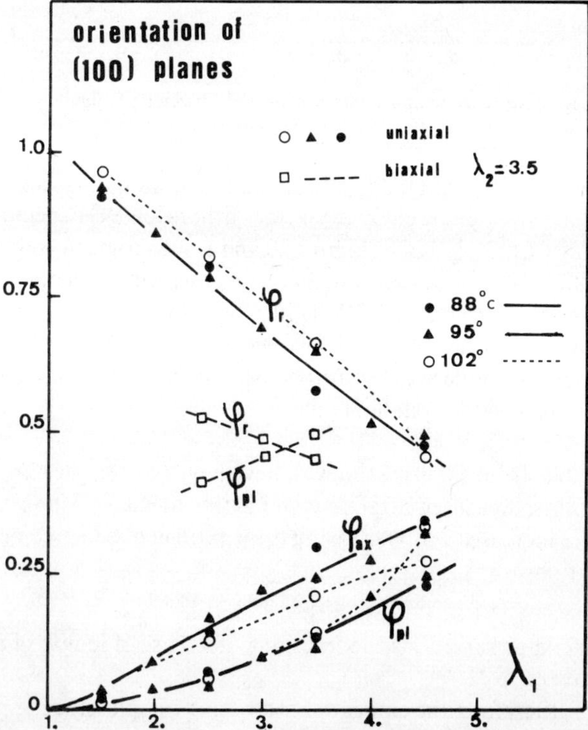

FIG. 22. Orientation of (100) planes in uni- and biaxially drawn PET films as a function of draw ratio.

$$\langle \cos^2 \phi_{N1} \rangle = 0; \quad \langle \cos^2 \phi_{N2} \rangle = \langle \cos^2 \phi_{N3} \rangle = 1/2 \tag{21}$$

In the terminology of Heffelfinger and Burton [43], the second case corresponds to uniplanar and the last case to plan axial orientation. If we admit, now, in a purely formal fashion, that a fraction φ_r of the (100) planes is randomly oriented, a fraction φ_{pl} planar oriented and a fraction φ_{ax} axially oriented with respect to the stretching direction we obtain:

$$\langle \cos^2 \phi_{N1} \rangle = \varphi_r/3$$

$$\langle \cos^2 \phi_{N2} \rangle = \varphi_r/3 + \varphi_{ax}/2$$

$$\langle \cos^2 \phi_{N3} \rangle = \varphi_r/3 + \varphi_{pl} + \varphi_{ax}/2 \tag{22}$$

or finally:

$$\varphi_r = 3\langle \cos^2 \phi_{N1} \rangle$$

$$\varphi_{ax} = 2[\langle \cos^2 \phi_{N2} \rangle - \langle \cos^2 \phi_{N1} \rangle]$$

$$\varphi_{pl} = \langle \cos^2 \phi_{N3} \rangle - \langle \cos^2 \phi_{N2} \rangle \tag{23}$$

A similar analysis has been applied by Heckmann and Spilgies [44] for the characterization of crystalline orientation in injection molded specimens of polyethylene. The average squared cosine values were calculated from the diffracted X-ray intensity distribution by means of well described standard methods [45, 46]. The results obtained on uniaxially drawn films (Fig. 22) show that the fraction of randomly oriented planes decreases rapidly with increasing draw ratio, due to an increase in both axial and planar orientation. For relatively low values of draw ratio, axial orientation increases more rapidly than planar orientation, but for λ-values larger than about 3, planar orientation is considerably enhanced and may, ultimately, become more important than axial orientation if the temperature of stretching is sufficiently high ($>100°C$ in this particular case).

Preferred planar orientation implies absence of transverse isotropy, as obvious from Figure 23, which shows the three principal refractive indices of the same films as a function of draw ratio. It is particularly noteworthy that the refractive index in the transverse direction (n_2) passes through a minimum for $\lambda = 3.5$ corresponding to the rapidly increasing importance of planar orientation, as had been noticed before [43].

Our experimental results concerning biaxially oriented films, obtained by uniaxial drawing, immediately followed by stretching in the transverse direction, are given in Figures 20, 22, and 24. The density of the biaxially drawn films (Fig. 20) is appreciably higher and varies little with draw ratio nor with the temperature between 90° and 110°C. The major part of the density change is probably due to the increase in crystallinity resulting in low shrinkage ratios (between 5 and 10% after 30 min heating at 150°C). The fraction of randomly oriented (100) planes decreases to values of about 0.5 or less, whereas the fraction of axially oriented (100) planes becomes negligibly small, in particular for equal

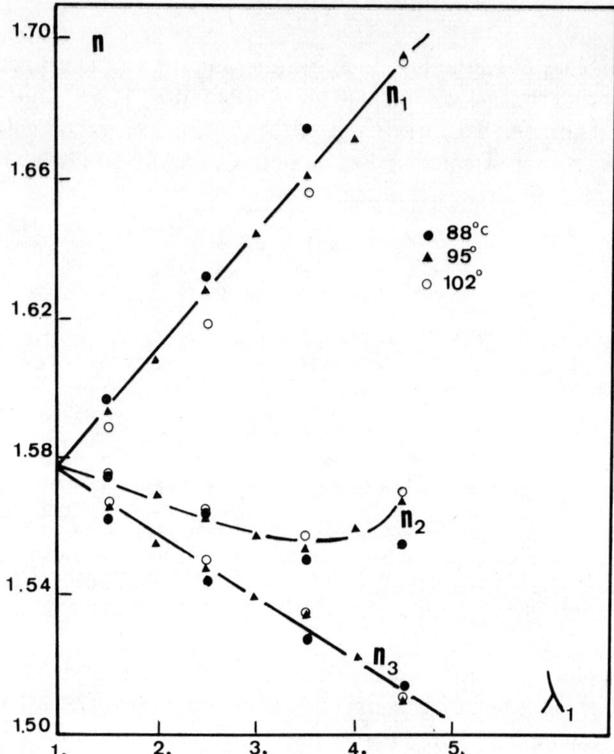

FIG. 23. Principal refractive indices (wavelength = 589 nm) for one way drawn PET films as a function of draw ratio.

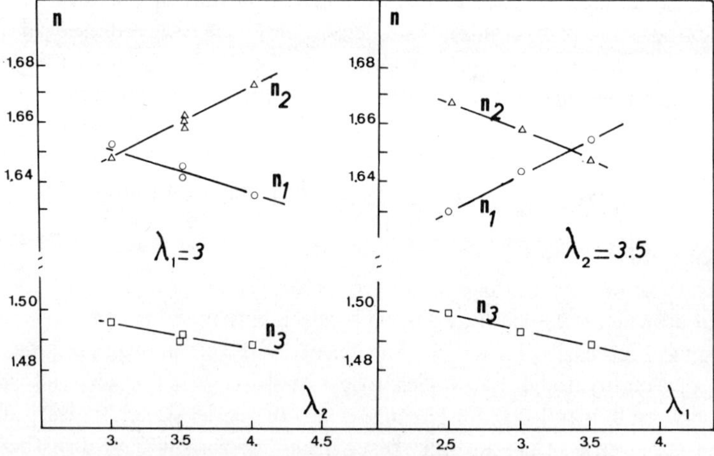

FIG. 24. Same as Figure 23 for biaxially drawn PET films.

values of the draw ratios in both machine and transverse direction (Fig. 22). The difference between the principal refractive indices in the plane of the film increases with increasing difference between the two draw ratios (Fig. 24), whereas the value of the third principal refractive index appears to be mainly determined by the degree of planar orientation.

The relationship between planar orientation and refractive index in the direction perpendicular to the film surface has been examined for a great number of PET films obtained under widely different conditions, uni- or biaxial drawing, followed or not by thermal treatments (heat-setting), or strain relaxation. The results are given in Figure 25 which shows a plot of

$$\frac{\bar{n}^2 - 1}{\bar{n}^2 + 2} \Big/ \frac{n_3^2 - 1}{n_3^2 + 2}$$

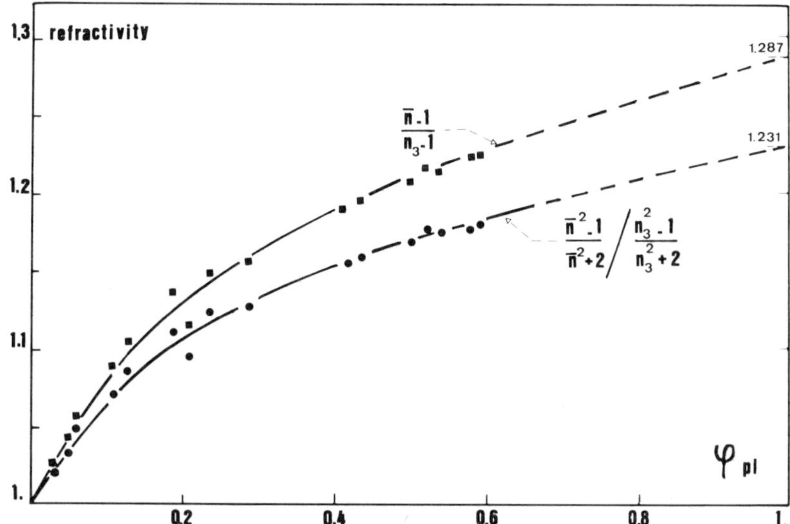

FIG. 25. Refractivity of oriented PET films as a function of planar orientation of (100) planes. Graphical extrapolation to complete planar orientation.

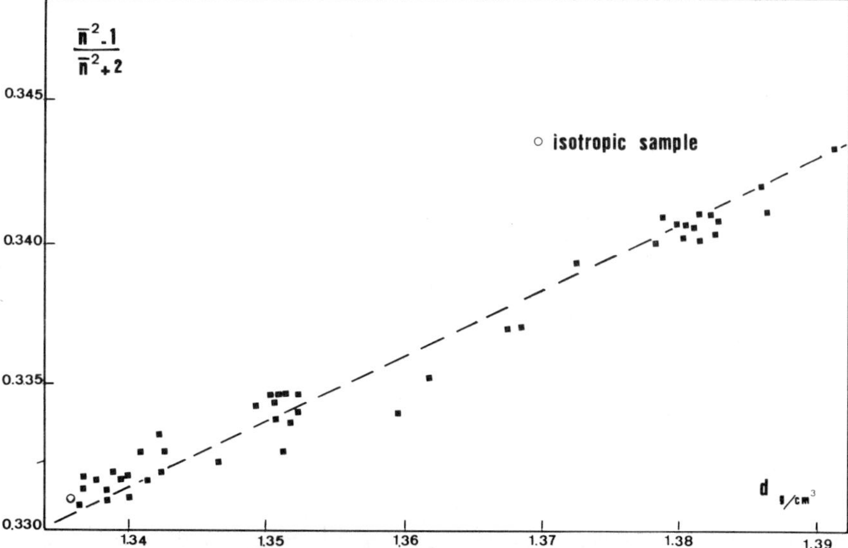

FIG. 26. Variation of refractivity with density for oriented PET films according to Lorentz–Lorenz theory.

as a function of φ_{pl} the fraction of planar oriented (100) planes. The mean refractive index $\bar{n}$ has been calculated from the Lorentz–Lorenz relationship:

$$\frac{\bar{n}^2 - 1}{\bar{n}^2 + 2} \equiv \frac{1}{3} \sum_{j=1,2,3} \frac{n_j^2 - 1}{n_j^2 + 2} = \frac{d}{d_{iso}} \frac{n_{iso}^2 - 1}{n_{iso}^2 + 2} \tag{24}$$

where n_j are the refractive indices and d the density of an oriented sample and n_{iso} and d_{iso} the refractive index and density of an isotropic sample of PET, respectively. Our experimental data plotted as a function of density, according to eq. (24), are shown in Figure 26 and may be represented, irrespective of the degree of crystallinity and state of orientation, by the following relationship:

$$\frac{\bar{n}^2 - 1}{\bar{n}^2 + 2} = 0.2471d \tag{25}$$

if d is expressed in g/cm^3.

According to eq. (25) an isotropic sample of density 1.335 would have a refractive index of 1.574 which corresponds well with our experimental value of 1.577.

Since

$$\frac{n_j^2 - 1}{n_j^2 + 2}$$

is proportional to the density, according to the Lorentz–Lorenz theory, the ratio

$$\frac{\bar{n}^2 - 1}{\bar{n}^2 + 2} \bigg/ \frac{n_3^2 - 1}{n_3^2 + 2}$$

plotted in Figure 25 is assumed to be independent of density and only dependent on the degree of planar orientation.

Extrapolation of the experimental relationship to 100% planar orientation, and making use of eq. (25) leads to an estimated limiting value for $n_3^0 = 1.4958$ for a density of 1.455 g/cm^3, corresponding to 100% crystallinity. An identical value for n_3^0 may be estimated by extrapolation of the ratio $(\bar{n} - 1)/(n_3 - 1)$ as also shown in Figure 25. In the case of equilibrated biaxial extension: $n_1^0 = n_2^0$, and hence:

$$2\left(\frac{n_1^{0^2} - 1}{n_1^{0^2} + 2}\right) + \frac{n_3^{0^2} - 1}{n_3^{0^2} + 2} = 3 \times 0.2471d \tag{26}$$

After substitution of the above mentioned values for d and n_3^0, the following result is obtained from eq. (26):

$$n_1^0 = n_2^0 = 1.716$$

and the maximum birefringence for the completely oriented crystalline film would be

$$\Delta_c^0 = n_1^0 - n_3^0 = 0.220 \tag{27}$$

It is noteworthy, although perhaps fortuitous, that the calculated value for Δ_c^0

is identical with the one calculated by Dumbleton [47] for the intrinsic bire-fringence of crystalline PET obtained by means of an analysis proposed by Samuels [48], applied to the combined results of birefringence, X-ray diffraction, and sonic modulus measurements on PET fibers of various degrees of crystallinity and orientation. In a completely oriented crystalline fiber with transverse isotropy, all (100) planes may be assumed to be parallel to the fiber axis and the agreement between the calculated Δ_c^0 values is, therefore, not unexpected. A maximum birefringence value of 0.23 is given by Kashiwagi et al. [49] who also offer a theoretical explanation of the effect of preferred planar orientation on the characteristic variation of n_2 and n_3 with draw ratio, shown in Figure 23.

The occurrence of preferred planar orientation and the associated absence of transverse isotropy in uniaxially drawn PET films do not allow the customary birefringence $(n_1 - n_2)$ to be used for the characterization of molecular orientation. In studies on the effect of orientation on the properties of drawn PET films, the use of $n_1 - n_3$ as an index of orientation may, however, lead to acceptable results as will be shown in the next section.

EFFECT OF ORIENTATION ON THE MECHANICAL BEHAVIOR OF UNI- AND BIAXIALLY DRAWN POLYMER FILMS AND SHEETS

Amorphous Polymers

The effect of molecular orientation on the mechanical behavior of polymer films and fibers has been extensively discussed in the literature, in particular in [2] and [3]. Recent results of detailed systematic investigations may be found in the report by Jones [6] and in a paper by Retting [50].

The elastic or viscoelastic behavior at small strains, as reflected in properties such as stiffness or creep compliance, is affected to a more or less important degree by the amount of molecular orientation. Young's modulus measured at high speed, in particular by means of sound velocity measurements, has been related to Hermans' orientation function (eq. 16) by Moseley [51] and Ward [52]. If E_u is the Young's modulus of an unoriented specimen and E_{max} the maximum modulus for a completely oriented specimen, the value of the modulus in the stretch direction would be given, in a first approximation, by

$$1 - \frac{E_u}{E} = \frac{\Delta n}{\Delta n_{max}} \cdot \left(1 - \frac{E_u}{E_{max}}\right) \qquad (28)$$

Since, in general, $E_{max} \gg E_u$, the last term in eq. (28) is mostly neglected.

In Figure 27 we have plotted the values of $1 - E_u/E$ for a great number of oriented PVC specimens as a function of Δn; these results are in fair agreement with those obtained by Retting [50] on a different sample of PVC and show that a linear relationship is well obeyed for this polymer at least up to values of about 0.5 for $1 - E_u/E$. Application of eq. (28) would lead to an extrapolated Δn_{max} value of about 13×10^{-3}, in good agreement with the value derived from combined NMR–birefringence data by Kashiwagi and Ward [32].

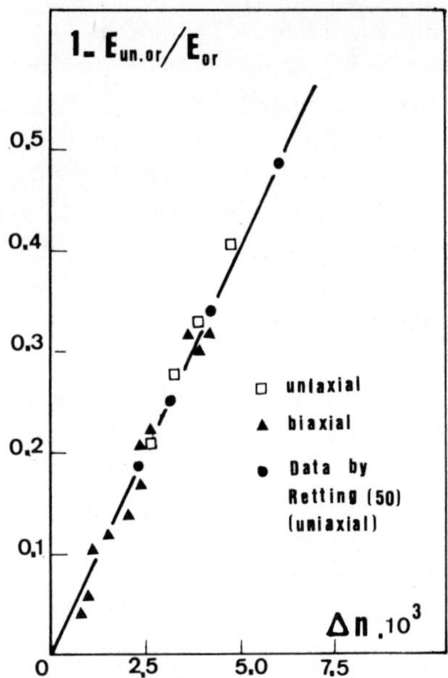

FIG. 27. Variation of Young's modulus as a function of birefringence for uni- and biaxially oriented PVC specimens.

The highest birefringence value we have ever measured for this polymer was obtained on a melt spun filament, after-stretched at 100°C to more than 10 times its initial length, and was equal to 12×10^{-3}. Such a high degree of molecular orientation is, however, seldom obtained in films or sheets of essentially amorphous polymers which do not develop a substantial amount of stress- or strain-induced crystallinity, and an increase of modulus by a factor less than two in the draw direction is, therefore, generally observed in uni- or biaxially oriented polymers of this kind [6, 16, 50].

At larger deformations the tensile behavior of amorphous polymers becomes strongly nonlinear and the exact shape of the stress–strain curve depends on the rate of strain [53]. Characteristic values such as yield strain, rupture stress, and elongation at break, are all affected by molecular orientation. But the quantitative improvement obtained will, in general, depend on the exact conditions (temperature, strain rate) under which the tensile test is performed [6, 16, 50].

As it was noticed before by Samuels [3] the tensile test is in fact a uniaxial extension experiment performed at relatively low temperature and it may be expected, therefore, that if the strain rate during the test is low enough to allow further molecular orientation to take place, the final structure at break may be more or less independent of the initial structure. Examination of published [3, 50] and unpublished data (see below) on amorphous and semicrystalline polymers confirm indeed that often the true tensile strength (rupture stress with respect to the final cross-sectional area at break) varies little with the initial state of uniaxial orientation, unless the strain rate is very high or the temperature during the test very low.

A notable exception is the case of polystyrene [6, 50] which in its unoriented state is so brittle that molecular orientation not only improves the tensile strength but also increases the elongation at break. A similar effect due to a transition from brittle to ductile behavior has been reported by Retting [50] in the case of PVC for which the elongation at break may pass through a maximum with increasing initial orientation if the tensile test is performed at a high strain rate. Prevention of failure at high strain rates is of great technological importance and this property is generally investigated by means of various impact strength tests. Measured values depend on the type of test as well as on the exact conditions under which the test is performed, but whatever test is used, the results obtained are always extremely sensitive to the degree of molecular orientation in the polymer sample. In the IUPAC study reported by Jones [6] a great number of data is given on the effect of uniaxial orientation on the impact strength of polystyrene, obtained in five different laboratories using essentially similar methods based on the Charpy flexural impact test. In spite of an important amount of scattering in the experimental results, all laboratories agree on the spectacular improvement in impact strength for specimens cut at 0° to the direction of orientation, associated with a corresponding decrease in impact strength for specimens cut at 90° to the direction of orientation.

A tenfold increase in impact strength in the direction of orientation may be easily attained for a uniaxially oriented specimen of polystyrene having a moderate degree of orientation ($\Delta n \sim 7 \times 10^{-3}$; internal entropic stress ~ 1.5 MN/m^2) but because of the decrease in strength in the perpendicular directions, uniaxial orientation of films and sheets is of only limited practical use. Biaxial orientation however, results in an improvement of stiffness and strength in all directions parallel to the surface and presents, therefore, a much greater technological interest.

At comparable values of birefringence, the effect of biaxial orientation on the impact strength of polystyrene is somewhat smaller than in uniaxial orientation but is still considerable. A tenfold increase in Izod impact strength for a Δn value of 12×10^{-3} has been reported by Thomas and Cleereman [16] who found a linear increase of impact strength with birefringence. A comparable improvement in the impact strength of biaxially oriented PVC sheets has been found in our laboratory, as shown in Figure 28. In order to account for the difference in stress optical coefficients between polystyrene and PVC, we have plotted the relative increase in impact strength as a function of internal entropic stress which may be considered as a more significant parameter for the comparison of molecular orientation in polymers of different chemical structure [see eq. (18)]. Although different tests have been used, a flexural impact test on a notched sample in the case of polystyrene and a tensile impact test on an unnotched sample in the case of PVC, it is quite obvious that the improvement in impact strength as a result from biaxial orientation is of comparable magnitude for the two polymers.

Many other examples of the considerable effect of biaxial orientation on the mechanical behavior of amorphous vinyl polymers in the nonlinear region and, in particular, on the transition from brittle to ductile behavior, have been reported

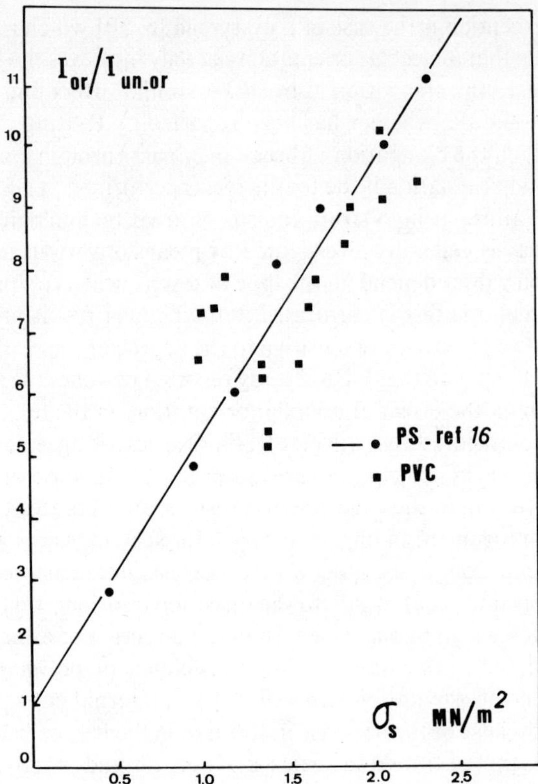

FIG. 28. Relative increase of impact strength as a function of internal entropic stress for biaxially oriented polystyrene and PVC.

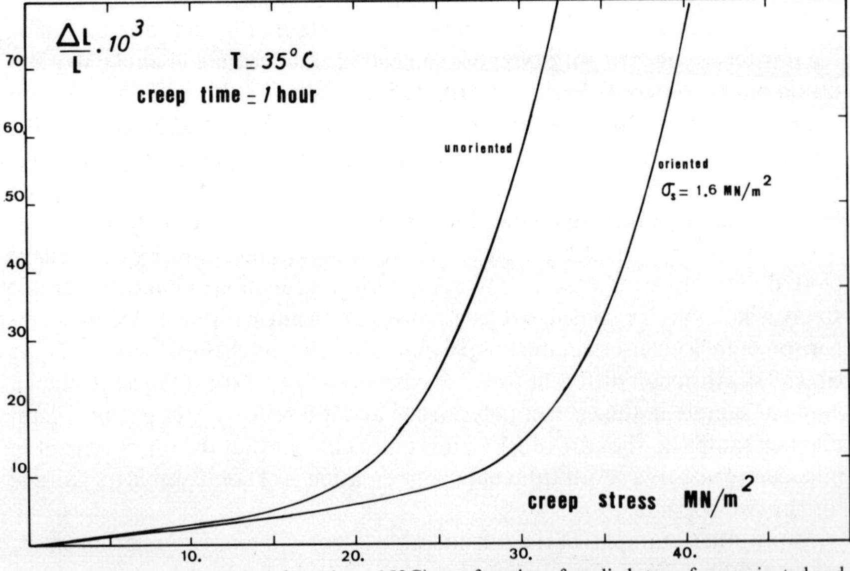

FIG. 29. Tensile creep strain (after 1 hr at 35°C) as a function of applied stress for unoriented and biaxially oriented PVC.

in the literature [16, 54]. Flexural fatigue strength, resistance to stress-cracking, and resistance to weathering are some of the technological properties which have been found to be improved in a quite significant way.

The effect of molecular orientation on the resistance to creep, another property of great practical interest, is also quite significant especially at large values of time (or strain) and becomes particularly important in the region of relatively high stresses where the creep behavior is strongly nonlinear. Figure 29 shows some results, obtained in our laboratory, on the effect of biaxial orientation on the tensile creep of PVC specimens submitted to a constant load at 35°C during one hour. The degree of orientation in the direction of the applied tensile stress is characterized by the value of internal entropic stress. The tensile strain after one hour is a nonlinear function of applied stress and increases quite rapidly if the stress is sufficiently high, in the case of the unoriented specimens. Depending on the degree of molecular orientation, the creep strain in oriented specimens is considerably smaller at a given stress level, and the region of rapidly increasing strains is displaced to much higher values of applied tensile stress.

Because the stress levels at which nonlinear creep behavior becomes important will be lower the higher the temperature, molecular orientation is of particular interest for improving the resistance to creep of objects liable to be submitted to external stresses for long periods of time at relatively elevated temperatures. The maximum temperature at which oriented polymer films, sheets and other objects can be used is determined by the onset of dimensional instability or shrinkage.

Shrinkage may be regarded as a particular kind of creep or recovery process under the influence of the frozen-in internal stress. In oriented amorphous polymers in the glassy state, shrinkage will start at a temperature more or less close to T_g as a result from the rapidly increasing mobility of molecular segments. If the internal stress is high, a strongly nonlinear behavior may be expected in this region of temperatures, and the rate of shrinkage will be dependent on the total amount of internal stress.

In Figure 30 some results are given of nonisothermal shrinkage experiments on biaxially oriented specimens of PVC. Both samples were heated at a rate of 3°C/min and their decrease in length continuously recorded. Although total amount of recoverable strain, internal entropic stress and, consequently, birefringence were all identical in both specimens, their shrinkage behavior is quite different. Sample (1) starts to shrink at a significantly lower temperature, and its rate of shrinkage is much higher at temperatures above T_g. The only difference between the two samples was the amount of internal distortional stress, being of the same order of magnitude as the entropic stress in sample (1) but practically equal to zero in sample (2) as shown by the retractive force diagrams also given in Figure 30.

It may be concluded from these results that a knowledge of the state of orientation (as characterized by the birefringence, e.g.) does not allow a prediction concerning the shrinkage behavior, since the latter is mainly governed by the total amount of frozen-in internal stress (entropy elastic plus energy elastic stress). Moreover, rate of shrinkage in the glass transition region appears to be

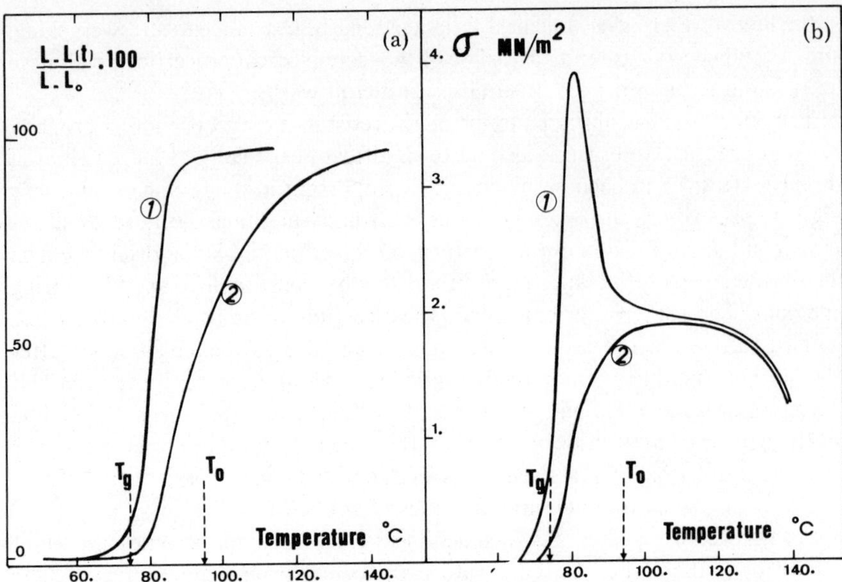

FIG. 30. Nonisothermal shrinkage (a) of biaxially oriented PVC specimens containing the same amount of internal entropic stress, but different amounts of total internal stress, as shown by retractive force diagrams (b); initial extension ratio $L/L_0 = 2.3$.

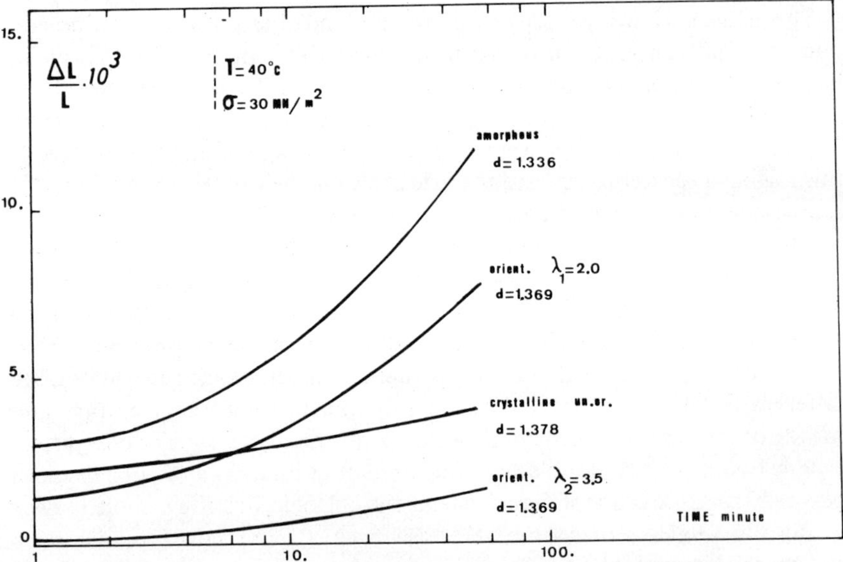

FIG. 31. Tensile creep strain at 40°C as a function of time for various PET specimens (applied tensile stress: 30 MN/m²). Effect of crystallinity and draw ratio.

extremely sensitive to the level of internal stress, due probably to the essential nonlinear character of creep processes in amorphous polymers.

Semicrystalline Polymers

In oriented semicrystalline polymers the rate of shrinkage may remain negligibly small at temperatures far beyond T_g due to the stabilizing effect of the presence of a crystalline phase. Under the influence of external stresses, however, creep may occur even at temperatures below T_g, although the creep compliance is, in general, much smaller than for completely amorphous polymers. The effect of crystallinity on the creep of unoriented PET is shown in Figure 31. The short-time behavior is governed by the modulus of elasticity whose value is only slightly affected by the degree of crystallinity which always remains relatively low in PET ($<50\%$). However, with increasing time the effect of crystallinity becomes more and more important, and the creep curves of the amorphous and the partly crystalline sample diverge rapidly. A low creep compliance may also be obtained by a sufficient amount of orientation, even at low degrees of crystallinity, as also shown in Figure 31. The relatively small influence of the degree of crystallinity on the short-time modulus may be easily explained on the basis of the two-phase series model [55] which assumes linear additivity of the compliances of the amorphous and the crystalline phase:

$$J = XJ_c + (1 - X)J_a \tag{29}$$

where J_c and J_a are the compliances of the crystalline and amorphous phases, respectively, and X the degree of crystallinity. Since $J = 1/E$ the influence of the amorphous phase will become predominant if the value of its modulus is much lower than that of the crystalline phase. By extrapolation of his data on sound velocity in unoriented PET fibers, Dumbleton [47] estimated that the modulus of the randomly oriented crystalline phase would be twice as high as that of the unoriented amorphous phase. In an oriented PET fiber or film in which strain-induced crystallization has occurred, this ratio will be much larger because of the presence of a highly oriented crystalline phase, whose modulus may attain extremely large values.

A maximum value for the modulus of oriented crystalline PET, determined by X-ray diffraction, has been reported by Dulmage and Contois [56] and is equal to $140 \, \mathrm{GN/m^2}$. On the other hand, a value of $18 \, \mathrm{GN/m^2}$ has been reported by Hadley et al. [57] for the modulus of highly drawn PET filaments. Assuming a value of 0.4 for the crystallinity of these filaments, substitution of the above cited values into eq. (29) would lead to an estimated value of about $11 \, \mathrm{GN/m^2}$ for the modulus of the highly oriented amorphous phase, which is an order of magnitude lower than the modulus of the completely oriented crystalline phase. It is reasonable, therefore, to assume that the increase in short-time modulus with draw ratio (see Fig. 32) is, for a large part, due to the rapidly increasing degree of orientation in the amorphous phase.

The predominant effect of amorphous phase orientation, compared to the effect of increasing crystallinity, may also be inferred from the following semiquantitative treatment of the data in Figure 32, based on Samuels' analysis of the modulus of uniaxially oriented semicrystalline filaments with transverse isotropy [48]. According to this analysis, eq. (29) may be written as follows:

$$J = X(1 - f_c)J_c^u + (1 - X)(1 - f_a)J_a^u \tag{30}$$

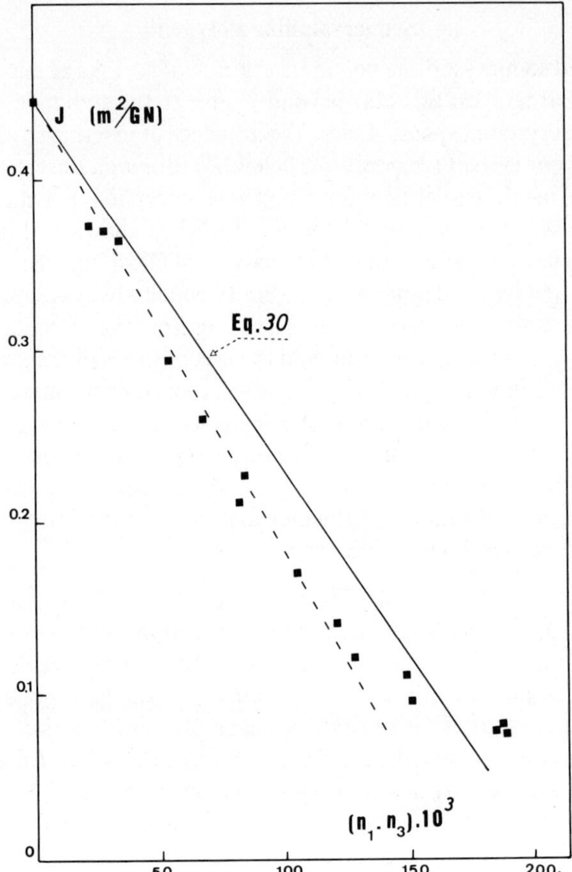

FIG. 32. Inverse of Young's modulus as a function of birefringence for one way drawn PET films. Comparison with eq. (30).

where J_c^u and J_a^u are the short-time compliances of the completely unoriented crystalline and amorphous phases, respectively, and f_c and f_a the values of Hermans' orientation function for the two phases. The measured value of J_a^u is equal to 0.45 m²/GN and on the basis of Dumbleton's results [47] one might estimate that J_c^u would be at least twice as low, i.e., about 0.225 m²/GN.

If we further assume that the birefringence $\Delta n_{13} = n_1 - n_3$ may be expressed by the usual additivity rule:

$$\Delta n_{13} = Xf_c\Delta_c^0 + (1 - X)f_a\Delta_a^0 + \Delta n_f \tag{31}$$

where Δn_f is the (often assumed negligible) contribution of form birefringence, we can estimate the minimum contribution of the amorphous phase to the observed birefringence as follows: At first, a maximum value of the crystallinity X may be calculated from the density with the aid of eq. (32):

$$X = \frac{d - 1.335}{1.455 - 1.335} \tag{32}$$

where 1.455 and 1.335 are the densities (g/cm^3) of the crystalline and amorphous phase, respectively. It is well known that eq. (32) overestimates the degree of crystallinity, if the amorphous phase becomes highly oriented [40].

For Δ_c^0 we adopt the value of 0.22, given by eq. (27); a similar calculation starting from eq. (26) for $d = 1.335$ results in an estimated value for $\Delta_a^0 = 0.20$, corresponding to the birefringence $n_1 - n_3$ of a completely oriented amorphous PET film.

The orientation of the crystalline phase has been calculated from Gaylord's theoretical analysis [58] of stress-induced crystallization in Gaussian networks whose results have been shown to be in good agreement with various experimental data [59, 60]. This analysis results in the following expression for f_c as a function of draw ratio:

$$f_c = \frac{1}{2}\left(\frac{3\lambda^3}{2 + \lambda^3} - 1\right) \tag{33}$$

According to eq. (33), for draw ratios larger than 3, f_c values will be higher than 0.9, in good agreement with the results of Dumbleton on drawn PET filaments [47].

The calculated values of the crystalline contribution to the observed birefringence are given in Table I. It is obvious that the crystalline contribution is much smaller than the amorphous phase contribution in particular for the lower draw ratios. Neglecting the contribution of form birefringence, an approximate value for f_a can be calculated for each draw ratio with the aid of eq. (31); these values are also indicated in Table I. The calculated values for f_a are quite acceptable and somewhat higher than those derived from infrared spectroscopy on comparable PET films of low crystallinity reported by Cunningham [34], or from polarized fluorescence data by Nobbs et al. [61].

Finally, both terms of eq. (30) have been calculated and are also given in Table I, in order to compare the calculated compliance with the measured one (see Fig. 32). The data clearly show that only the amorphous phase gives a substantial contribution to the short-time compliance, the crystalline phase contribution being always very small and varying little in absolute value. Although our analysis confirms the predominant influence of the amorphous phase orientation on the short-time modulus, we found no quantitative agreement between calculated and observed values which is perhaps not unexpected in view of the numerous approximations and assumptions.

For a more refined analysis, independently measured values for the amorphous phase orientation, in particular, would be necessary. Nevertheless it does not appear very probable that a consistent set of experimental parameter values could be found which obey both eqs. (30) and (31) up to high draw ratios, unless the form birefringence Δn_f would have a nonnegligible effect on the observed $n_1 - n_3$ values. Indeed, it is obvious from Figure 32 that at very low degrees of crystallinity ($X < 0.1$) the compliance is a linear function of birefringence, as it was also found in the case of PVC (see Fig. 27). This is in accordance with eqs. (30) and (31) in which the terms referring to the crystalline phase may now be neglected (see Table I). However, linear extrapolation of this relationship to

TABLE I
Birefringence and Modulus of One Way Drawn PET Films

DRAW RATIO	Temperature (°C)	Density (g/cm^3)	X [a]	$n_1 - n_3$	f_c [b]	$f_c X \Delta_c^\circ$	f_a [c]	$X J_c$ (m^2/GN)	$(1-X) J_a$ (m^2/GN)	$X J_c + (1-X) J_a$ (m^2/GN)	$J_{exp} = 1/E_{exp}$ (m^2/GN)
2	95	1.3394	0.036	0.0535	0.70	0.0055	0.25	0.0024	0.328	0.33	0.29
2.5	95	1.3425	0.0625	0.0815	0.83	0.0114	0.37	0.0024	0.268	0.27	0.23
3	95	1.3463	0.094	0.105	0.89	0.0185	0.48	0.0020	0.214	0.21	0.17
3.5	95	1.3517	0.14	0.127	0.93	0.0287	0.57	0.0021	0.168	0.17	0.12
4	95	1.3598	0.206	0.150	0.95	0.0442	0.67	0.0020	0.119	0.12	0.095
4.5	95	1.3678	0.27	0.187	0.96	0.057	0.88	0.0021	0.039	0.04	0.08

[a] Eq. (32).
[b] Eq. (33).
[c] $f_a = \dfrac{\Delta n_{13} - 0.22 f_c X}{0.2(1 - X)}$.

zero compliance leads to an unacceptable low value for Δ_a^0; the data in Figure 32 suggest that at high degrees of orientation the relationship between compliance and birefringence becomes nonlinear.

Dumbleton [47] concluded that his results on drawn PET fibers may be conveniently represented by eqs. (30) and (31), when adopting $\Delta_c^0 = 0.22$ and $\Delta_a^0 = 0.25$. The latter value is significantly higher than ours and was derived by linear extrapolation of the graph of sonic compliance versus birefringence, the highest value measured being 0.138. A deviation from linearity at higher degrees of orientation might, however, result in a lower value for Δ_a^0 and would also lead to a consistent interpretation of Dumbleton's set of experimental data, but the curvature in the sonic compliance versus birefringence plot should then have a sign opposite to the one shown in Figure 32, which is rather unexpected.

Further experimental data are needed in order to obtain more precise information on the intrinsic birefringence of the amorphous and crystalline regions in an oriented polymer. In this respect it may be noted that also in the case of isotactic polypropylene, conflicting data have been reported in the literature [62, 63].

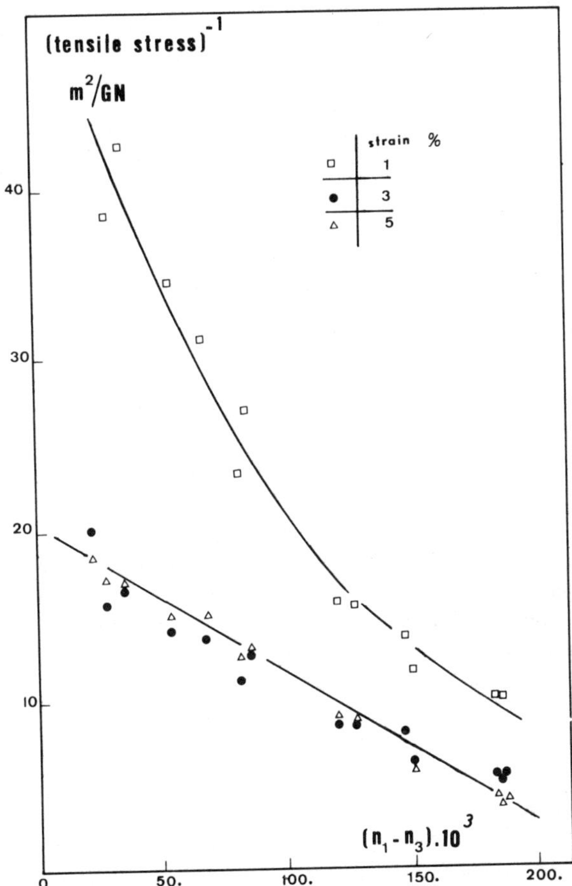

FIG. 33. Inverse tensile stress for different strain values as a function of birefringence for one-way drawn PET films.

TABLE II

Tensile Stress, Elongation at Break, and "True Tensile Strength" of Uni- and Biaxially Drawn PET-Films

DRAW RATIO	Temperature (°C)	Density (g/cm^3)	Post-Treatment	Tensile stress at break (GN/m^2)	Elongation at break %	True tensile strength (GN/m^2)
2 X 1	95	1.3394	–	0.15	212	0.47
2.5 X 1	88	1.3425	–	0.209	168	0.56
2.5 X 1	95	1.3425	–	0.199	166	0.53
3 X 1	95	1.3463	–	0.278	114	0.59
3.5 X 1	95	1.3517	–	0.302	77	0.53
4 X 1	95	1.3598	–	0.366	58	0.58
4.5 X 1	88	1.3618	–	0.410	38	0.56
4.5 X 1	95	1.3678	–	0.384	41	0.54
4.5 X 1	102	1.3685	–	0.407	40	0.57
3 X 3	90 – 110	1.3880	–	0.239	119	0.52
3 X 3	90 – 110	1.3726	heat set 150°C	0.245	128	0.56
3 X 3	90 – 110	1.3860	heat set 200°C	0.220	144	0.54
3 X 3.5	90 – 110	1.3822	–	0.235	117	0.51
3 X 3.5	90 – 110	1.3827	heat set 175°C 4% Strain relax.	0.250	119	0.55
3 X 3.5	90 – 110	1.3930	heat set 175°C 8% strain relax.	0.258	120	0.57
3.5 X 3.5	90 – 110	1.3802	–	0.274	86	0.51
3 X 4	90 – 110	1.3814	–	0.236 0.350	130 61	0.54 0.56

The tensile behavior of uniaxially oriented PET films becomes highly nonlinear for relatively small strains. The data concerning the tensile stress up to strains of 5%, given in Figure 33, show that in the nonlinear region the effect of amorphous phase orientation as characterized by the birefringence Δn_{13}, is similar to the effect on short-time modulus. For a draw ratio equal to 4.5, the tensile stress at a given value of strain is nearly five times as high as for an unoriented film, and in the region of yielding (3%–5% strain) the reciprocal stress values are a linear function of the birefringence.

A similar improvement is found for the tensile stress at rupture, associated with a corresponding decrease in elongation at break. The true tensile strength, i.e., its actual value at the moment of rupture, is fairly constant irrespective of the amount and mode of initial orientation. The values in Table II refer to uniaxially and biaxially oriented PET films, in the latter case both equilibrated and nonequilibrated and having been submitted to various thermomechanical post-treatments. In spite of the different thermomechanical histories of the films, true tensile strength varies very little and has an average value of about 0.55 GN/m^2, which should be characteristic of the "intrinsic" strength of the polymer film.

Samuels [3] has shown that the tenacity (tensile stress with respect to initial cross section) of drawn PET filaments with various thermomechanical histories—data reported by Dumbleton [64, 65]—is essentially governed by the amorphous phase orientation and may be represented by a linear semilog relationship as a function of f_a (see Fig. 34). Our own experimental data on uniaxially oriented films are in good agreement with this relationship up to $f_a = 0.5$, if we adopt the values of Table I.

At higher f_a values, however, the tenacity of the films was found to be lower than predicted by Samuels' relationship. This discrepancy may be, in part, due to a possible overestimation of the amorphous phase orientation in the films of highest draw ratio. As a matter of fact, a lower f_a value at the highest draw ratio would also give better agreement with the calculated short time compliance (Fig. 32); on the other hand, recalculation of Dumbleton's data, adopting a Δ_a^0 value of 0.20 would also result in somewhat better agreement, but still leads to lower tenacity values in the highly oriented films, compared to the filaments. In the case of isotactic polypropylene, Samuels [3] has reported that a unique relationship exists between tenacity and amorphous phase orientation valid for both fibers and films, up to f_a values of about 0.6 in the uniaxially oriented films.

From a linear extrapolation of the semilog plot in Figure 34, Samuels concluded that the maximum obtainable tenacity in a completely oriented PET fiber would be of the order of $1.2\ GN/m^2$, which is twice as high as the above calculated "true tensile strength" of PET films. A comparable twofold ratio between extrapolated maximum tenacity and "true tensile strength" can be derived from Samuels' data on isotactic polypropylene [3].

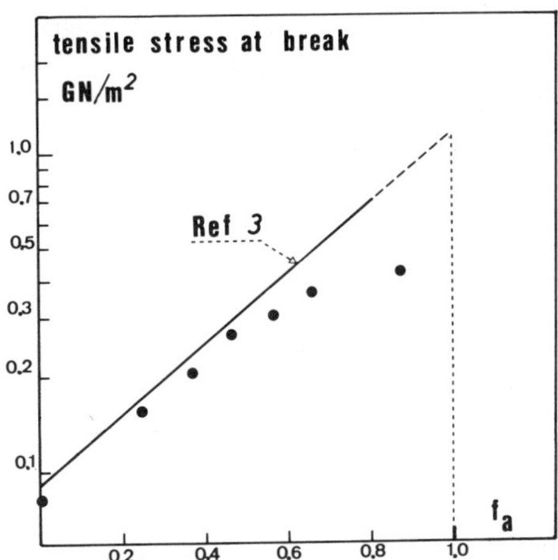

FIG. 34. Tensile strength of one-way drawn PET films as a function of amorphous phase orientation. Comparison with Samuels [3] relationship for drawn PET filaments.

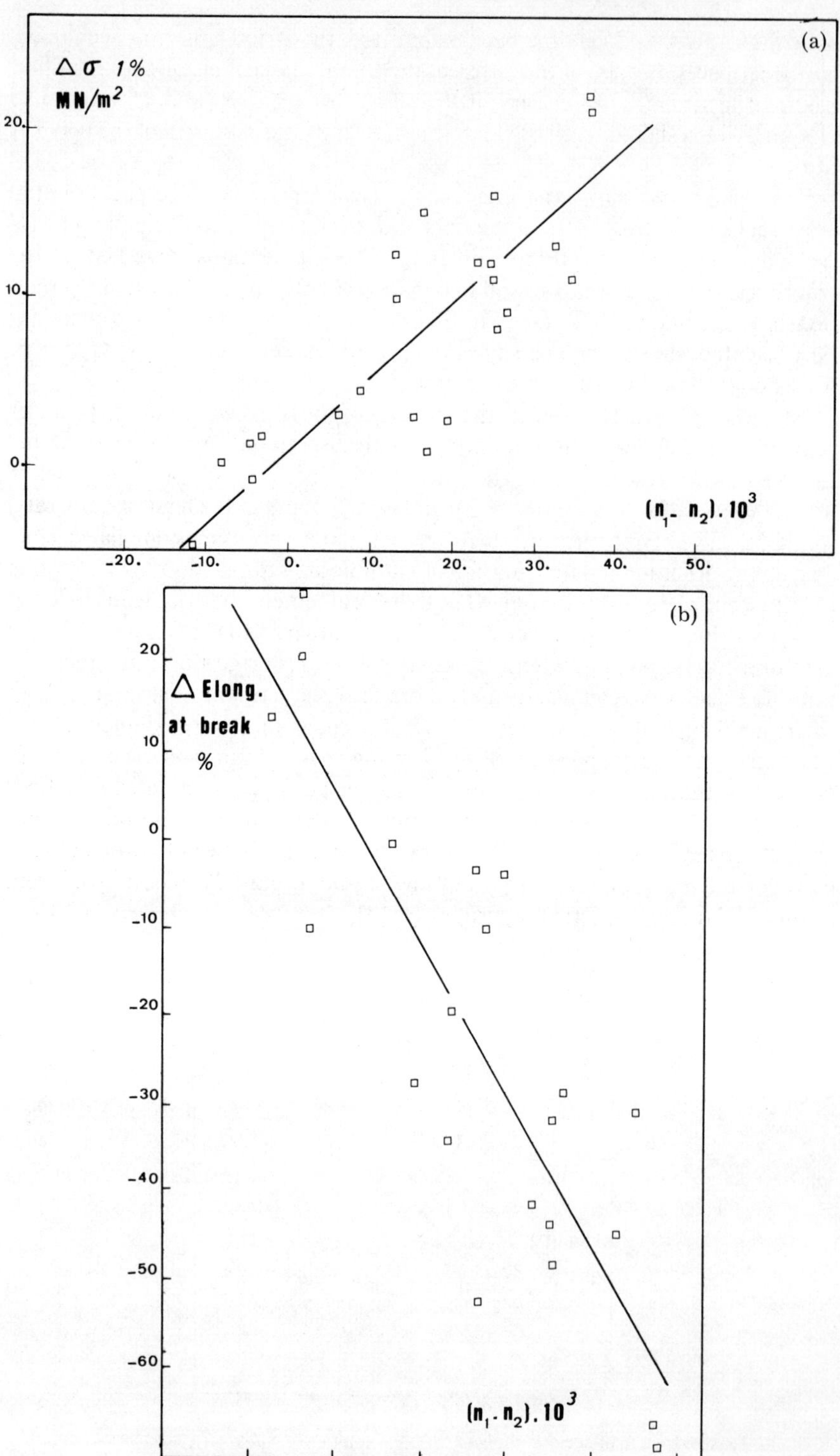

FIG. 35. Anisotropy of mechanical properties in biaxially oriented PET films as a function of birefringence; (a) tensile stress at 1% strain, (b) tensile stress at break, (c) elongation at break.

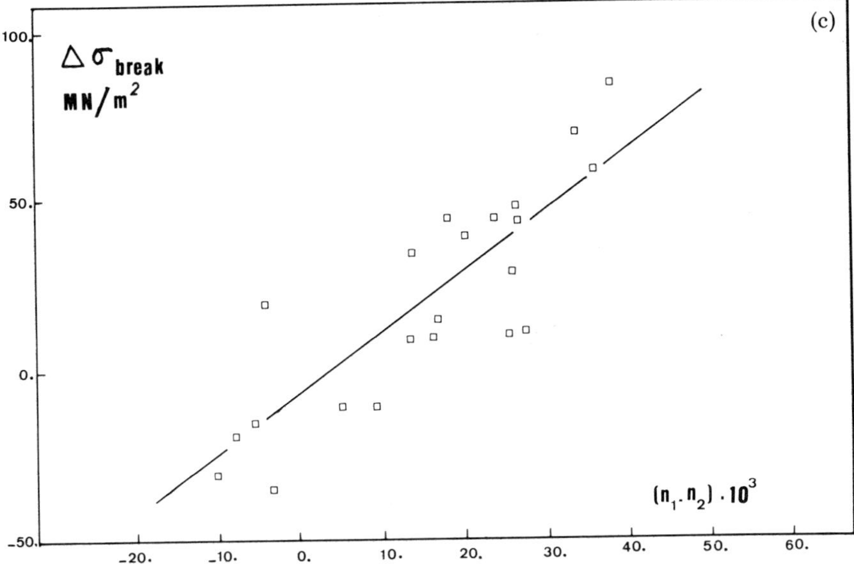

Fig. 35 (Continued from previous page.)

The mechanical properties of biaxially oriented PET films are only slightly lower than those measured in the stretch direction of uniaxially oriented films of the same draw ratio, but much higher than the properties in the transverse direction of one-way drawn films [66]. In nonequilibrated biaxially drawn films, the difference in properties in two perpendicular directions in the plane of the film are, in a first approximation, a linear function of the difference in refractive indices in the same directions, as shown in Figure 35.

CONCLUSION

In a remarkable lecture, presented at the IUPAC International Symposium on Macromolecules, Aberdeen, September 1973 [67] I. M. Ward concluded: "Although our understanding of the mechanical properties of polymers is only as good as our understanding of their structure, this is not a unidirectional process, and there is much to be gained in structural understanding from detailed studies of mechanical behavior."

We would like to add, at least in the particular case of oriented polymers, that a satisfactory understanding of the deformation and ordering processes at the molecular level can only be expected to emerge from detailed studies of macroscopic mechanical behavior if they are combined with the use of various experimental techniques of structural investigation, such as X-ray diffraction, NMR, IR and Raman spectroscopy, polarized fluorescence, of which considerable interest has been shown in numerous studies by Ward and his co-workers, as well as by other groups in many different places [2].

One of the unsolved problems that we have extensively discussed in the present paper is related to the presence of large internal stresses in oriented amorphous polymers, which are not directly associated, apparently, with the configurational entropy changes in the deformed rubberlike network. We hope that neutron scattering studies, presently in progress, combined with an appropriate theoretical analysis [68] may shed new light on the specific molecular deformation mechanisms associated with these internal stresses which become particularly important at high strain rates and/or low deformation temperatures.

The global molecular deformation during orientation of amorphous polymers in the rubbery state is fairly well understood and may be described in a satisfactory manner with the aid of rubber elasticity theory applied to Gaussian networks with temporary junction points. Specific questions concerning the structure and topology of the molecular network require further elucidation, and our work has shown that useful information may be gathered, in this respect, by a systematic comparison of the deformation behavior in both uni- and biaxial extension under identical conditions of strain rate and temperature.

The study of stress–optical properties of oriented polymers is of great interest for the quantitative characterization of molecular orientation in the amorphous phase. In the case of semicrystalline polymers, however, simple birefringence measurements are in general insufficient, and complete knowledge of the three principal refractive indices as a function of density and orientation is required for a comprehensive analysis of the observed birefringence in terms of relative contributions from oriented amorphous and crystalline regions.

As long as the degree of crystallinity remains relatively low, the mechanical behavior of oriented semicrystalline polymers appears to be mainly determined by orientation in the amorphous regions, and may be described, in a first approximation, by means of two-phase models which does not necessarily imply that well separated and clearly recognizable, discontinuous phases really exist. An interesting limiting case, in this respect, is perhaps PVC because of the presence of a rather small maximum content of local order in the unoriented polymer which nevertheless seems to affect its orientation behavior to a certain extent.

Another extreme situation has been shown to occur in the case of highly oriented cold-drawn high-density polyethylene [69] in which the amorphous phase content is so low that its exact value (and not so much its degree of orientation) may become the predominant factor in determining mechanical behavior, in particular, stiffness.

Finally, it must be reminded that not only the mechanical behavior of polymer films and sheets, but other physical properties too may be affected to a more or less important degree by the amount of uni- or biaxial orientation.

Optical properties like transparency and haze, thermal properties such as thermal conductivity and thermal expansion, rate of diffusion, and permeation of gases or low molecular weight compounds, are some of the most important properties from the technologist's point of view, on which the effect of molecular orientation may be highly significant, depending on the chemical structure of the polymer and on the processing conditions under which the molecular orientation has occurred. Discussion of these problems and related phenomena was outside the scope of the present paper.

REFERENCES

[1] G. Schuur and A. K. van der Vegt, see [2], Chap. 12.

[2] *Structure and Properties of Oriented Polymers*, I. M. Ward, Ed., Applied Science Publishers Ltd., London, 1975.

[3] R. J. Samuels, *Structured Polymer Properties*, John Wiley and Sons, New York, 1974.

[4] P. H. Hermans, *Contribution to the Physics of Cellulose Fibers*, Elsevier, Amsterdam, 1946.

[5] R. S. Stein and G. L. Wilkes, see [2], Chap. 3.

[6] T. T. Jones, *Pure* and *Appl. Chem.*, **45**, 39 (1976).

[7] L. R. G. Treloar, *Trans. Inst. Rubber Ind.*, **19**, 201 (1944).

[8] R. A. Dickie and T. L. Smith, *J. Polym. Sci. A2*, **7**, 687 (1969).

[9] D. L. Holt, *Int. J. Mech. Sci.*, **12**, 491 (1970).

[10] C. G. Cornfield and R. H. Johnson, *Int. J. Mech. Sci.*, **12**, 479 (1970).

[11] D. D. Joye, G. W. Pochlein, and C. D. Denson, *Trans. Soc. Rheol.*, **16**, 421 (1972).

[12] L. R. Schmidt and J. F. Carley, *Polym. Eng. Sci.*, **15**, 51 (1975).

[13] J. M. Maerker and W. R. Schowalter, *Rheol. Acta.*, **13**, 627 (1974).

[14] M. O. Lai and D. L. Holt, *J. Appl. Polym. Sci.*, **19**, 1805 (1975).

[15] A. J. de Vries and C. Bonnebat, *Polym. Eng. Sci.*, **16**, 93 (1976).

[16] L. S. Thomas and K. J. Cleereman, *SPE J.*, **28**, 61 (April 1972).

[17] F. N. Cogswell and D. R. Moore, *Polym. Eng. Sci.*, **14**, 573 (1974).

[18] J. E. Adkins and R. S. Rivlin, *Philos. Trans. R. Soc. London, Ser. A*, **244**, 505 (1952).

[19] L. J. Hart-Smith and J. D. Crisp, *Int. J. Eng. Sci.*, **5**, 1 (1967).

[20] A. Peterlin, see [2], Chap. 2.

[21] J. F. Rudd and R. D. Andrews, *J. Appl. Phys.*, **29**, 1421 (1958).

[22] P. R. Pinnock and I. M. Ward, *Trans. Faraday Soc.*, **62**, 1308 (1966).

[23] J. F. Rudd and E. F. Gurnee, *J. Polym. Sci., A*, **1**, 2857 (1963).

[24] B. E. Read, *Proc. Fifth Int. Congr. Rheol.*, **4**, 65, University of Tokyo Press, Tokyo, 1970.

[25] T. Alfrey, N. Weiderhorn, R. S. Stein, and A. V. Tobolsky, *Ind. Eng. Chem.*, **41**, 701 (1949).

[26] L. R. G. Treloar, *The Physics of Rubber Elasticity*, Second Ed., Oxford University Press, London, 1958.

[27] H. M. James and E. Guth, *J. Chem. Phys.*, **15**, 669 (1947).

[28] K. Dusek and W. Prins, *Adv. Polym. Sci.*, **6**, 1 (1969).

[29] W. W. Graessley, *Macromolecules*, **8**, 186 (1975); **8**, 865 (1975).

[30] A. J. Chompff and W. Prins, *J. Chem. Phys.*, **48**, 235 (1968).

[31] R. D. Andrews and V. Chatre, *J. Appl. Phys.*, **40**, 4266 (1969).

[32] M. Kashiwagi and I. M. Ward, *Polymer*, **13**, 145 (1972).

[33] W. Kuhn and F. Grün, *Kolloid-Z.*, **101**, 248 (1942).

[34] A. Cunningham, I. M. Ward, H. A. Willis and V. Zichy, *Polymer*, **15**, 749 (1974).

[35] T. M. Birshtein and O. B. Ptitsyn, *Conformations of Macromolecules*, Interscience Publishers, Inc., New York, 1966.

[36] P. H. Boyce and L. R. G. Treloar, *Polymer*, **11**, 21 (1970); see also: M. Shen, *Pure Appl. Chem.*, **43**, 43 (1975).

[37] Y. Abe and P. J. Flory, *J. Chem. Phys.*, **52**, 2814 (1970).

[38] L. R. G. Treloar, *Polymer*, **10**, 291 (1969).

[39] F. S. Smith and R. D. Steward, *Polymer*, **15**, 283 (1974).

[40] G. Farrow and I. M. Ward, *Polymer*, **1**, 330 (1960).

[41] W. L. Lindner, *Polymer*, **14**, 9 (1973).

[42] W. J. Dulmage and A. L. Geddes, *J. Polym. Sci.*, **31**, 499 (1958).

[43] C. J. Heffelfinger and R. L. Burton, *J. Polym. Sci.*, **47**, 289 (1960).

[44] W. Heckmann and G. Spilgies, *Kolloid Z. Z. Polym.*, **250**, 1150 (1972).

[45] B. D. Cullity and A. Freda, *J. Appl. Phys.*, **29**, 25 (1958).

[46] Z. W. Wilchinsky, *J. Appl. Phys.*, **31**, 1969 (1960).

[47] J. H. Dumbleton, *J. Polym. Sci. A2*, **6**, 795 (1968).

[48] R. J. Samuels, *J. Polym. Sci. A.*, **3**, 1741 (1965).

[49] M. Kashiwagi, A. Cunningham, A. J. Manuel and I. M. Ward, *Polymer*, **14**, 111 (1973).

[50] W. Retting, *Colloid Polym. Sci.,* **253,** 852 (1975).

[51] W. W. Moseley, Jr, *J. Appl. Polym. Sci.,* **3,** 266 (1960).

[52] I. M. Ward, *Text. Res. J.,* **34,** 806 (1964).

[53] L. E. Nielsen, *Mechanical Properties of Polymers and Composites,* Vol. 2, Marcel Dekker, Inc., New York, 1974.

[54] L. S. Thomas and K. J. Cleereman, *SPE J.,* **28,** 39 (June 1972).

[55] M. Takayanagi, K. Imada, and J. Kajiyama, *J. Polym. Sci. C.,* **15,** 263 (1966).

[56] W. J. Dulmage and L. E. Contois, *J. Polym. Sci.,* **28,** 275 (1958).

[57] D. W. Hadley, P. R. Pinnock, and I. M. Ward, *J. Mater. Sci.,* **4,** 152 (1969).

[58] R. J. Gaylord, *Polym. Letters,* **13,** 337 (1975).

[59] W. R. Krigbaum and R. J. Roe, *J. Polym. Sci. A,* **2,** 4391 (1964).

[60] R. Kitamaru, H. Dong, and W. Tsuji, *Int. Symp. Macromol. Chem., Prepr.* **8,** 98, Tokyo-Kyoto 1966.

[61] J. H. Nobbs, D. I. Bower, and I. M. Ward, *Polymer,* **17,** 25 (1976).

[62] R. J. Samuels, *J. Polym. Sci. A2,* **6,** 1101 (1968).

[63] T. Masuko, H. Tanaka and S. Okajima, *J. Polym. Sci. A2,* **8,** 1565 (1970).

[64] J. H. Dumbleton, *J. Polym. Sci. A2,* **7,** 667 (1969).

[65] J. H. Dumbleton, *Polymer,* **10,** 539 (1969).

[66] C. J. Heffelfinger and P. G. Schmidt, *J. Appl. Polym. Sci.,* **9,** 2661 (1965).

[67] I. M. Ward, *Polymer,* **15,** 379 (1974).

[68] H. Benoit, R. Duplessix, R. Ober, M. Daoud, J. P. Cotton, B. Farnoux, and G. Jannink, *Macromolecules,* **8,** 451 (1975).

[69] J. B. Smith, G. R. Davies, G. Capaccio, and I. M. Ward, *J. Polym. Sci. Polym. Phys. Ed.,* **13,** 2331 (1975).

MICROINDENTATION HARDNESS OF ORIENTED CHAIN-EXTENDED POLYETHYLENE

F. J. BALTÁ-CALLEJA and D. C. BASSETT

Instituto de Estructura de la Materia, CSIC, Madrid-6, Spain and J. J. Thomson Laboratory, University of Reading, U.K.

SYNOPSIS

The Knopp and Vickers microindentation hardnesses of oriented chain-extended polyethylene (cold-drawn, then hydrostatically annealed at 5.3_5 kbar) have been measured to establish correlations with some of the morphological features previously investigated (lamellar thickness, orientation). The irreversible deformation increases with time under load and tends towards a characteristic limiting value for each annealing temperature. This limiting value of microhardness is an increasing function of annealing temperature, T_A, and consequently of chain extension, 1. A conspicuous feature of the investigated oriented polyethylene samples is the clear anisotropy exhibited by the indentation pattern. The Vickers microhardness is a maximum when the indenter diagonal lies parallel to the fiber direction and minimum when normal to it. The opposite occurs for the Knoop case. This contrary behavior is due to the shape difference between the two indenters coupled with the high anisotropy of the material. This behavior supports morphological experiments and is a further illustration of the value of this technique in polymer textural studies.

INTRODUCTION

The study of the various mechanisms (microfracture, surface cracking, plastic deformation) which solids suffer beneath point indentations (Vickers, Knoop) and their microstructural implications is a proliferating issue which has been the object of increasing interest for a wide variety of materials (alkali halides, silicates, ionic crystals, metals, alloys, and composites) during the last years [1–9]. Microindentation hardness studies of semicrystalline polymers within the context of a correlation to microstructure are, however, still scarce and limited, as far as we know, to the case of melt crystallized and drawn polyethylene [10] and injection molded poly-4-methyl-pentent-1 [11]. The purpose of the present contribution is to extend the above mentioned investigations to the case of oriented chain-extended polyethylene (CEPE) since this is a powerful model system from both a morphological [12] and a mechanical [13] point of view.

Microindentation hardness (MH) is a potentially useful mechanical property capable of yielding valuable information on anisotropy in highly drawn polymers. The anisotropy of MH, similar to other mechanical properties, seems to be mainly a consequence of high orientation of polymer chains in both crystalline and noncrystalline regions. For anisotropic polymers, for instance, there are different hardness values depending on the direction of test. For uniaxially aligned chains there are at least two values of MH: parallel and perpendicular

Journal of Polymer Science: Polymer Symposium 58, 157–167 (1977)

to the chain direction. From a general viewpoint, MH represents the resistance to deformation at a polymer surface. The use of a sharp indenter offers the advantage of a contact pressure which is independent of indent size and consequently affords a convenient measure of hardness. In this case the stress field essentially leads to irreversible deformation modes such as plastic deformation, viscous flow, and induced densification (phase changes). In anisotropic polymers, at least, an elastic contribution to indentation also occurs. Since the indentation process at the surface is primarily controlled by irreversible deformation under the indenter, the MH value is expected to be directly correlated to the specific modes of plastic deformation in semicrystalline polymers. These are predominantly determined by the structure and packing of the crystals and their connection by tie molecules in the axial direction of the fibrous material and may involve chain tilt and slip, crack formation and chain unfolding [14, 15] as well as other mechanisms.

The ample range of crystal thicknesses generated by annealing drawn PE hydrostatically at high pressure, varying from the common oriented chain-folded to chain extended-structures—where folds and tie molecules tend to disappear [12], coupled with the preservation of the c molecular orientation, offers a very suitable model for establishing a correlation of the indentation mechanism on a highly oriented polymer with its microstructure. Two aspects are the object of the present study:

1. The investigation of the anisotropy of the oriented CEPE material in relation to the specific geometry of the indenter.
2. The microstructural mechanisms at a crystal and molecular level which are responsible for the observed anisotropy.

EXPERIMENTAL

These experiments have used the linear unfractionated polyethylene Rigidex 9 ($\overline{M}_n$ = 11,700; $\overline{M}_w$ = 169,000). A molded sheet of this was cold-drawn at atmospheric pressure to a draw ratio of 8 and then samples of approximately 5 × 1 cm were punched from the drawn sheet and annealed at $5 \cdot 3_5$ kbar as in previous work [12, 16]. The annealing times were 15 min, and maximum temperatures were chosen to lie on either side of the orthorhombic-hexagonal [17] transition. According to high pressure DTA of undrawn polymer [16], this occurs between about 230° and 242°C at this pressure with a peak at 238°C. The effect of annealing is to preserve the molecular orientation while increasing lamellar thickness in the range to ~2000 Å [12, 16]. According to work on Rigidex 2 [12], however, the initial stiffening due to cold drawing is already lost when crystals have thickened to an average of 300 Å; thereafter they behave simply as an oriented array of crystals rather than as a collection of fibers.

The thickness of lamellae has been measured by gel permeation chromatography (GPC) on portions of a specimen following immersion in fuming nitric acid for 3 days at 60°C. This has been shown to be a reliable measure for chain-extended polyethylenes. Values shown in Table I were measured in this way; it should be noted, however, that samples annealed at 235° and 245°C came from one original drawn sheet and those at 238° and 245.5°C from another.

TABLE I

Lamellar Thickness, 1, as Function of Annealing Temperature at 5.3 Kbar

$T_A\,^0C$	$\ell\,(\overset{\circ}{A})$
Original	340
235	510
238	1450
242.5	1440
245	1690

TABLE II

Geometrical Features of the Two Indentation Pyramids Used[a]

	VICKERS	KNOOP
Pyramid proyection	square	rhombic
Edge angle	148º	172.5º.130º
Diagonals' ratio	1:1	7; 1
Penetration depth	L/7	L/30 (+)

[a] L is the indentation diagonal (largest for Knoop) and the two angles given for Knoop are those subtended by the longer and shorter edges.

The microhardness measurements were made with a Leitz microhardness tester. A Vickers (square) and a Knoop (rhombic) pyramidal diamond were used to make indentations. Some of the geometrical features of the two pyramids used are collected in Table II. Indentation lengths were optically measured at a magnification of ×400 and the accuracy reached was ±0.5 μm. The distance between the diamond tip and the specimen surface was 40 μm and the time required for the diamond to reach the specimen surface was 20 sec. Impressions were measured after load removal. Thus, no information about the instant elastic recovery was obtained. The final permanent deformation was, on the other hand, dependent on the length of time of contact under load. The loading cycle was, therefore, controlled at various times ranging between 0.1 and 30 min. In contrast to conventional polymers no delayed recovery was detected.

The MH is defined as the ratio of the force applied, P to the surface of the indentation:

$$MH = K \cdot P/L^2 \; (kg/mm^2) \qquad (1)$$

where L is the length of the indentation diagonal in μm (the largest for Knoop)

FIG. 1. Vickers indentations of oriented CEPE (T_A = 242.5°C) along the fiber axis for various loads ranging between 15 and 300 g showing the typical anisotropic impressions. Time under load: 1 min. (×175).

and K a geometrical factor equal to 1854.4 for Vickers and to 14230 for the Knoop case. The microindentation measurements reported in this work were carried out with a load of 15 g except where specified otherwise.

RESULTS AND DISCUSSION

Anisotropy Effect

The most conspicuous feature in the present hardness measurements is the anisotropy shown by the indentation pattern. This is illustrated in Figure 1 for a series of Vickers impressions made at various loads for the sample hydrostatically annealed at 242.5°C. Anisotropy, which in a previous work [10] was shown to be a function of draw ratio, depends, in addition, on the orientation of the diagonals of indentation relative to the axial direction. This is shown in Figure 2 for the sample annealed at 245°C (oriented chain-extended polymer) in which microhardness is plotted as a function of the angle between the indenter diagonal and the draw direction. The Vickers hardness is a maximum when the

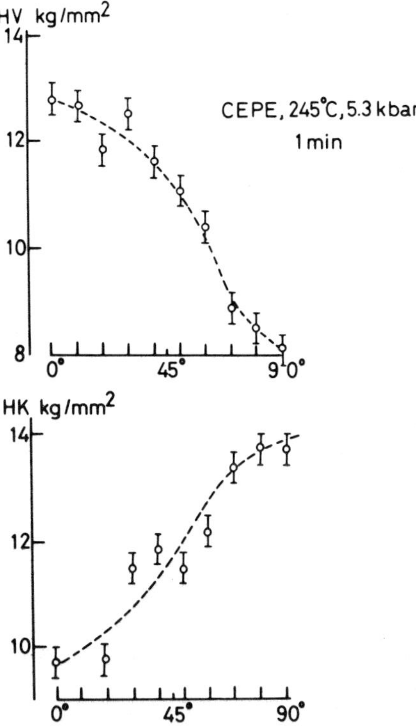

FIG. 2. Vickers and Knoop hardness of oriented CEPE ($T_A = 245°C$) as a function of the angle formed between indenter diagonal and axial direction. Indentation time under load: 1 min. Load applied: 15 g.

indentation diagonal lies parallel to the fiber axis and minimum when normal to it. The opposite occurs for the Knoop case. This contrary behavior which has been previously reported for carbon-fiber-reinforced plastics [4] is essentially due to the shape difference between the two indenters coupled with the high anisotropy of the material. The Vickers indenter imposes a nearly spherically symmetrical strain on the test sample. The stress field developed under it is, however, nonsymmetrical because the material is anisotropic. The stresses are greater along the fiber axis because the material is stiffest and strongest in this direction. Since the shape of the indentation is that of a diamond while the load is applied, and no time-dependent recovery is observed, the altered shape of the final impression must arise instantly on unloading. The observed anisotropy is thus a result of greater elastic recovery of the material in the fiber direction. The strain field under the Knoop indenter is, on the other hand, greater in the direction at right angles to the long axis of the indenter [1]. Thus, in the 0° case the maximum strain is normal to the fiber direction, i.e., parallel to the direction of lowest strength. In the 90° case the maximum strain is parallel to the direction of highest strength. Here the resistance to deformation is high and the indentation consequently small.

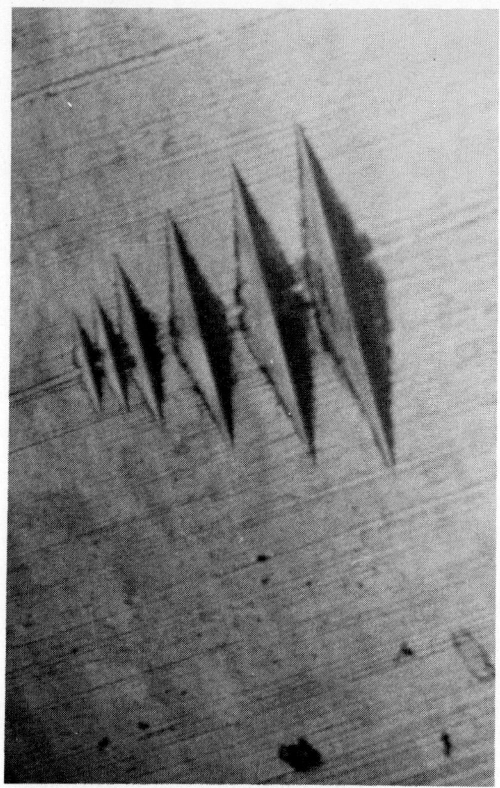

FIG. 3. Typical Knoop impressions normal to the fiber direction on oriented CEPE ($T_A = 242.5°C$) at loads: $P = 15, 25, 30, 50, 100, 145$, and 200 g ($\times 175$).

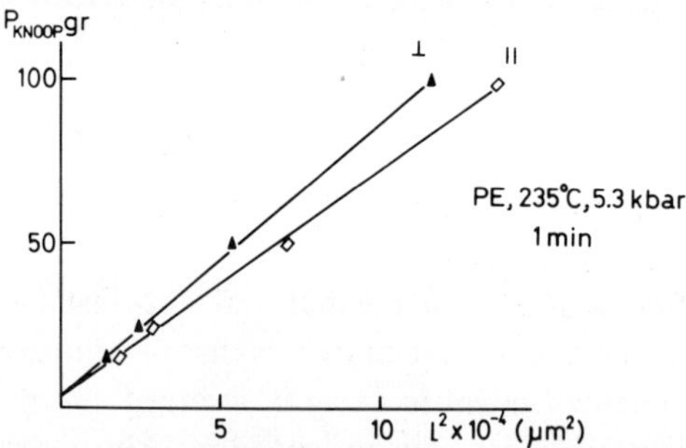

FIG. 4. Load P versus square of Knoop diagonal for oriented hydrostatically annealed PE: $\perp$ perpendicular and $\parallel$ parallel to the axial direction. Indentation time: 1 min.

Load and Loading-Time Dependence

Figure 3 illustrates typical Knoop indentations normal to the fiber direction as a function of load P for the sample hydrostatically annealed at 242.5°C. Figure 4 shows the linear relationship of load P against L^2 of Knoop indentation according to eq. (1) for the sample annealed at 235°C parallel and normal to the fiber direction. The intercept represents the force required to go over from elastic to plastic behavior.

Furthermore in all the samples investigated, the final size of indentation was a function of time underload. This is illustrated in Figure 5 for the sample hydrostatically annealed at 242.5°C by a series of Vickers indentations made in the 0.01–10 min range. Quantitative data for the time dependence of Knoop indentation are shown in Figure 6 for the original and other two hydrostatically annealed PE samples for the 0° and 90° cases respectively. The MH value gradually decreases and finally levels off to an apparently limiting value which depends solely on annealing temperature. In addition, the time dependence as shown in Figure 6 is appearently more conspicuous at short times for the 90° than for the 0° case, suggesting an anisotropic behavior of the noncrystalline regions. The results above clearly show that the deformation of the material under the indenter depends on both the applied load and the time for which it is sustained. The present data are consistent with previous results on drawn PE [10] where it was demonstrated that for lower deformations ($\lambda < 12$) indentation also increased for longer times.

FIG. 5. Typical Vickers indentation along the fiber axis for oriented CEPE ($T_A = 242.5°C$ for loading times: 10^{-2}, 10^{-1}, 10^0, 10^1, 10^2 min. Load: 15 g. ($\times$460).

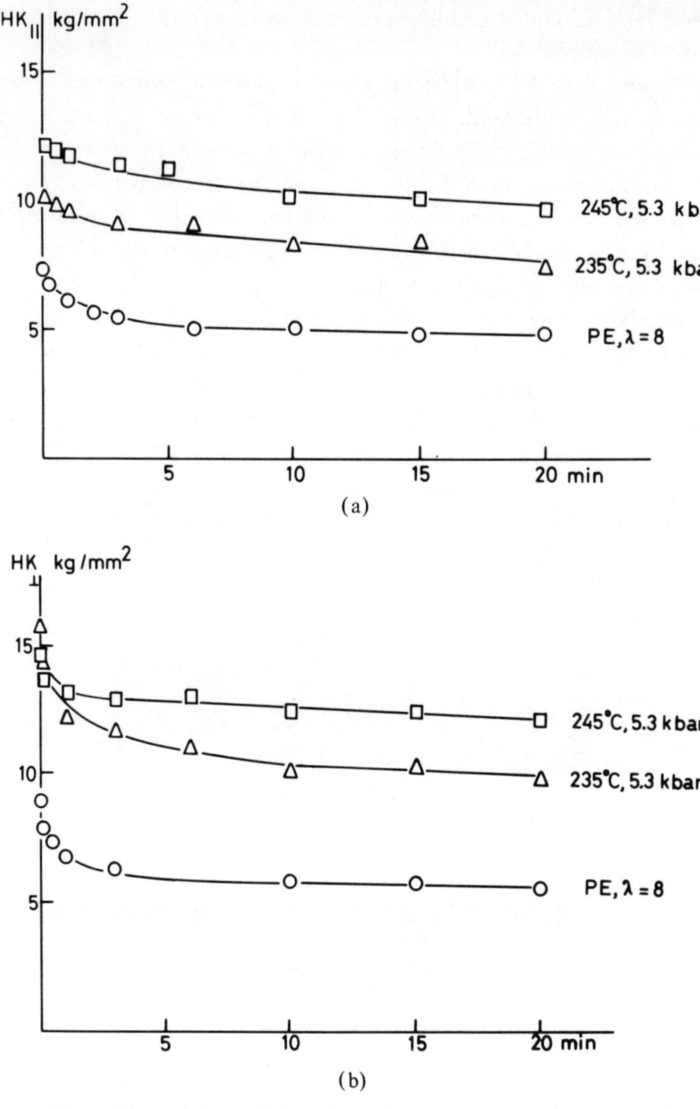

FIG. 6. Loading time dependence of Knoop hardness for original cold-drawn PE and after hydrostatic annealing at 5.3 Kbar at 235° and 245°C respectively. Load: 15 g. Indentation diagonal: (a) parallel and (b) vertical to the axial direction.

Deformation Mechanism and Morphology

The above results show a hardening of PE after annealing at high pressure paralleling the increase in stiffness measured previously [13]. This previous work [12] also showed that once the initial extra stiffening due to cold drawing is lost, the morphology varies only by increasing the chain extension of the oriented lamellae, l. One may attempt, therefore, to find a correlation between the limiting value of MH and l. Figures 7 and 8 indeed illustrate the observed gradual increase of Vickers and Knoop MH as a function of l, finally showing a leveling off tendency for $l > 1500$ Å. The Vickers hardness is larger at 0° than at 90° in accordance with the anisotropy effects discussed above.

Similarly to the case of paraffin crystals [10] the penetration depth of the indenter, d nearly follows an hyperbolic decrease with l, as shown in Figure 9

$$d = Kl^{-1} + d_o \qquad (2)$$

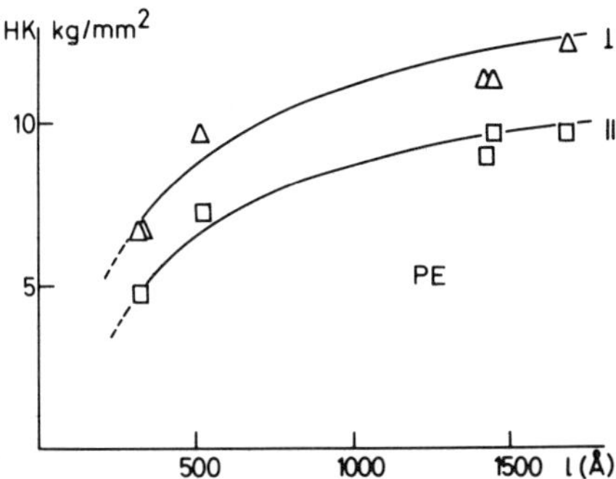

FIG. 7. Knoop microhardness perpendicular and parallel to the axial direction as a function of average molecular extension.

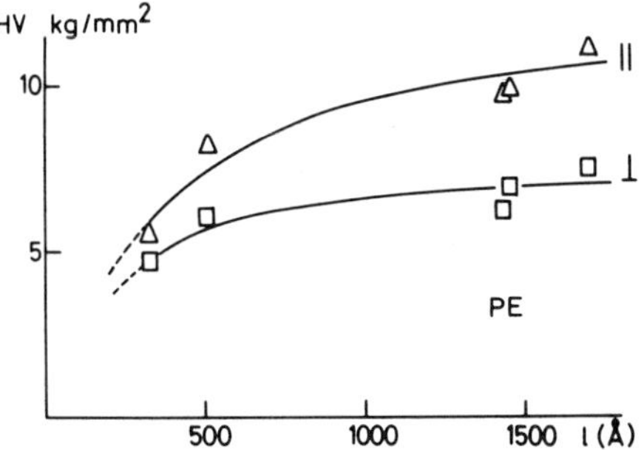

FIG. 8. Vickers microhardness parallel and perpendicular to the axial direction as a function of average molecular extension.

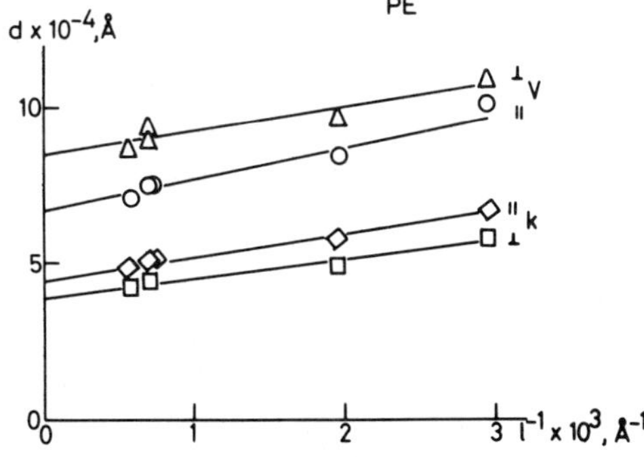

FIG. 9. Relationship between penetration depth, d, of the Vickers and Knoop indenters and inverse of lamellar thickness l^{-1} of oriented original and CEPE. Load: 15 g. Loading time: 20 min.

where K is a constant which is close to 10^7 and d_o is the penetration depth for an infinitely thick crystal. The plastic component of d for CEPE is nearly 9 μm for the Vickers and only 4 μm for the Knoop indenter. These results suggest that for a constant load time, l is the most relevant factor in determining the MH of oriented hydrostatically annealed PE. Oriented PE hydrostatically annealed and paraffins show in fact relatively similar d values as shown in Figure 10. Paraffins which obviously do not exhibit any time dependence of indentation, however, have slightly smaller extrapolated values of d, i.e., they are apparently harder.

This finding is in line with observations on morphology and c-axis modulus [12, 13] which show that notwithstanding the high crystal thicknesses, the interfacial regions in CEPE are well-defined and, e.g., control the c-axis modulus. It is significant that the special morphologies produced in this work are similar to materials of low draw ratio produced at atmospheric pressure in showing time-dependent effects in MH whereas this is not true of higher draw ratios. Evidently high draw ratios are not solely producing greater chain-extension but are also stiffening the interlamellar connections.

For the oriented samples of these experiments, the number of interlamellar connections must be reduced because of the high crystal thicknesses. This reduction is likely to be the cause of the absence of a time-dependent recovery of strain, whereas, of course, it provides no such restriction on a time-dependent deformation. We have, therefore, a further example of how MH measurements can illuminate aspects of polymer texture.

A particular advantage of working with CEPE is that its large crystal sizes offer the prospect of an earlier identification of what happens to crystals when they are deformed. This has already been achieved in part for drawn CEPE specimens [18] and it is hoped that further work will lead to similar progress for MH measurements.

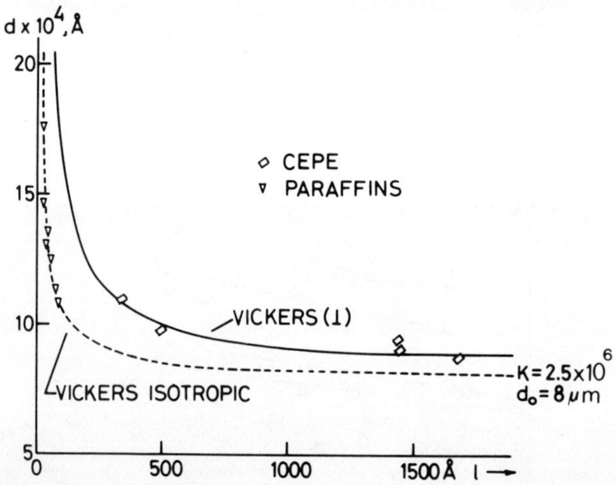

FIG. 10. Comparison of penetration depth of Vickers indentation between melt crystallized paraffins and oriented hydrostatically annealed polyethylene. Direction of test for the latter: perpendicular.

The authors wish to thank Mr. J. Garcia Peña for his help with the mciroindentation experimental work.

REFERENCES

[1] C. A. Brookes, J. B. O'Neill, and B. A. W. Redfern, *Proc. R. Soc. London A,* **322,** 73 (1971).

[2] C. A. Brookes, P. Grenn, P. H. Harrison, and B. Moxley, *J. Phys. D. Appl. Phys.,* **5,** 1284 (1972).

[3] J. B. O'Neill, B. A. W. Redfern, and C. A. Brookes, *J. Mater. Sci.,* **8,** 47 (1973).

[4] A. J. Perry and D. J. Rowcliffe, *J. Mater. Sci.,* **8,** 904 (1973).

[5] B. J. Hockey and B. R. Lawn, *J. Mater. Sci.,* **10,** 1275 (1975).

[6] H. Brown, A. A. Ballman, and G. Y. Chin, *J. Mater. Sci.,* **10,** 1157 (1975).

[7] B. Lawn, R. Wilshaw, *J. Mater. Sci.,* **10,** 1049 (1975).

[8] B. R. Lawn and M. V. Swain, *J. Mater. Sci.,* **10,** 113 (1975).

[9] B. Bethune, *J. Mater. Sci.,* **11,** 199 (1976).

[10] F. J. Baltá Calleja, *Colloid Polym. Sci.,* **254,** 258 (1976).

[11] J. Bowman, N. Harris, and M. Bevis, *J. Mater. Sci.,* **10,** 63 (1975).

[12] D. C. Bassett and D. R. Carder, *Philos. Mag.,* **28,** 513 (1973).

[13] D. C. Bassett and D. R. Carder, *Philos. Mag.,* **28,** 535 (1973).

[14] A. Peterlin, *Macrol. Chem.,* **8,** 277 (1973).

[15] A. Peterlin, *Polym. Eng. Sci.,* **14,** 627 (1974).

[16] D. C. Bassett, B. A. Khalifa, and R. H. Olley, *Polymer,* **17,** 284 (1976).

[17] D. C. Bassett, S. Block, and G. J. Piermarini, *J. Appl. Phys.,* **45,** 4146 (1974).

[18] G. E. Attenburrow and D. C. Bassett, *J. Mater. Sci.* **12,** 192 (1976).

INFLUENCE OF THE ARRANGEMENT OF THE CRYSTALS AND THE STRUCTURE OF THE NONCRYSTALLINE REGIONS ON THE MECHANICAL PROPERTIES OF POLYETHYLENE TEREPHTHALATE

H. J. BIANGARDI and H. G. ZACHMANN

Institute für Physikalische Chemie der Universität Mainz and SFB 41

SYNOPSIS

Amorphous samples of polyethylene terephthalate were stretched with different rates and stretching ratios at different temperatures. Afterwards they were crystallized. The arrangement of the crystals and the conformations of the chains in the noncrystalline regions were investigated by means of X-ray wide-angle scattering, X-ray small-angle scattering, birefringence measurements and nuclear magnetic resonance measurements. The influence of the morphological structure on the mechanical strength was determined. The stretching ratio itself generally does not characterize the orientation state. The structure obtained after crystallization depends mainly on the birefringence of the sample after stretching before crystallization. With increasing birefringence one obtains all intermediate structures starting from the well known spherulites composed of twisted lamellae in the unoriented material up to a paracrystalline lattice of lamellae split up into mosaic blocks in the highly oriented state. Also the conformations of the chains in the noncrystalline regions are influenced by the stretching. With increasing orientation the amount of tie molecules with comparatively large end-to-end distances becomes larger. The improvement of the mechanical properties with stretching is caused not only by the increasing orientation of the molecules but also by the increasing amount of taut tie molecules.

INTRODUCTION

It is well known that the Young modulus and the fracture stress of polyethylene terephthalate as well as of other polymers are increased markedly by stretching and annealing the material [1]. There is the question whether this improvement of the mechanical properties is caused only by the orientation of the chains or if it is induced partly also by structural changes in the noncrystalline regions.

In order to investigate this problem, films of amorphous polyethylene terephthalate which were 200 μ thick, 6 cm long, and 6 cm broad were stretched at different temperatures T_v with different stretching rates w (see Fig. 1). The stretching ratio λ was about 4.4. By this procedure samples with different orientations were obtained which were almost amorphous. These samples were crystallized with their ends being fixed so that no shrinkage could occur. The morphological structure of the samples before and after crystallization was examined by measuring the X-ray small-angle scattering, the X-ray wide-angle

Journal of Polymer Science: Polymer Symposium 58, 169–183 (1977)

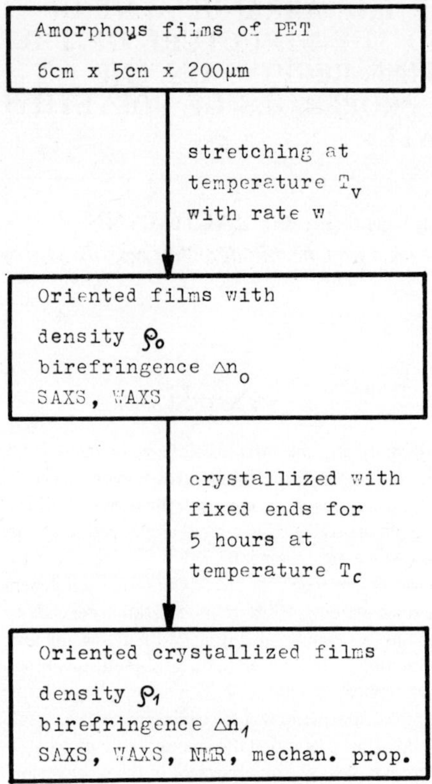

FIG. 1. Treatment of the samples.

scattering, the birefringence, the density, and the nuclear magnetic resonance. In addition the Young modulus, the fracture stress, and the elongation at break were measured and the relationship between the morphological structure and the mechanical properties was determined.

ARRANGEMENT OF THE CRYSTALLITES

Experimental Results and Models

Our investigation shows that the morphological structure of the stretched samples before and after crystallization is determined completely by the birefringence Δn_o of the samples after stretching before crystallization. This is the only reliable quantity to characterize the oriented sample before crystallization. The drawing ratio which is used by many authors characterizes the orientation only as long as the drawing temperature and the drawing rate are kept constant. A birefringence $\Delta n_0 = 40 \times 10^{-3}$ can be obtained, for example, with a drawing ratio $\lambda = 4.4$ at a drawing rate $w = 1\%/\text{min}$ and a drawing temperature $T_v = 80°C$ as well as with $\lambda = 2.2$, $w = 830\%/\text{min}$ and $T_v = 92°C$. On the other hand all values of Δn_0 between 0 and 200×10^{-3} were obtained by us with a constant drawing ratio $\lambda = 4.4$ only by changing the rate and the temperature of drawing.

More details on the influence of the drawing conditions on the orientation are given in a previous publication [2].

With increasing birefringence we obtained all intermediate structures starting from the well known [3, 4] spherulites composed of twisted lamellae in the unoriented material up to a paracrystalline lattice of lamellae split up into mosaic blocks which has been found [5–8] in the highly oriented material. In this paper we will first show the experimental results obtained for each orientation and the models deduced from these results. The proofs that these models are correct will be discussed afterwards.

Figure 2 shows the results for a sample with a very low orientation after stretching. Such a low orientation is obtained, for example, by stretching at T_v = 90°C with a stretching rate w = 0.3%/min to a stretching ratio of λ = 4.4. The birefringence after stretching Δn_0 is 3×10^{-3}, the density ρ_o is 1.339 g/cm^3. The wide-angle X-ray scattering (WAXS) shows an amorphous halo without any visible indication of crystallization or orientation. After crystallization for 5 hr at 240°C the birefringence is changed to $\Delta n_1 = -12 \times 10^{-3}$, the density is increased to ρ_1 = 1.395 g/cm^3. In the wide-angle X-ray diffraction crystal reflexes appear, the 100 and the −110 reflection being concentrated at the meridian. The small-angle X-ray pattern shows a maximum being concentrated also at the meridian.

Some authors [9, 10] have interpreted the X-ray scattering by a model where the chains are lying more or less parallel to the surfaces of the lamellae. We find it more satisfying to assume a structure of twisted lamellae as indicated in the right side of Figure 2. Two possible arrangements of such lamellae can be considered: parallel bundles of twisted lamellae caused by raw nucleation, as discussed by Keller and Machin [11] for polyethylene shown under (a) or flat spherulithes as indicated under (b).

Figure 3 shows the results for a sample with a higher orientation which can be obtained, for example, by stretching at T_v = 90°C with a rate w = 3%/min to an elongation ratio λ = 4.4. The birefringence Δn_0 after stretching is 20×10^{-3}, the density ρ_o is 1.342 g/cm^3. The amorphous halo is no longer isotropic and also some broad crystal reflections can be seen which indicate that the material is not completely amorphous. After crystallization the birefringence and the density are increased and the crystal reflections become more intensive and sharper. The small-angle X-ray scattering which appears after crystallization shows the features of a two-point diagram. All these results can be explained if one assumes that the lamellae are oriented in such a way that the twisting axis is mainly perpendicular to the stretching direction and that the twisting of the lamellae is no longer complete.

Figure 4 shows the results for an orientation which has increased further. The birefringence Δn_0 is now 40×10^{-3}. One has to assume in this case that the twisting of the lamellae has become so weak that one speaks better of bended lamellae. In addition the lamellae are not oriented perpendicular to the stretching direction, they form an angle ω^* with the stretching direction which is smaller than 90°.

Finally, Figure 5 shows the results for the highly oriented samples. While the

Amorphous PET-film after stretching (e.g. $T_V = 90°C$, $w = 0.3 \%/min$)

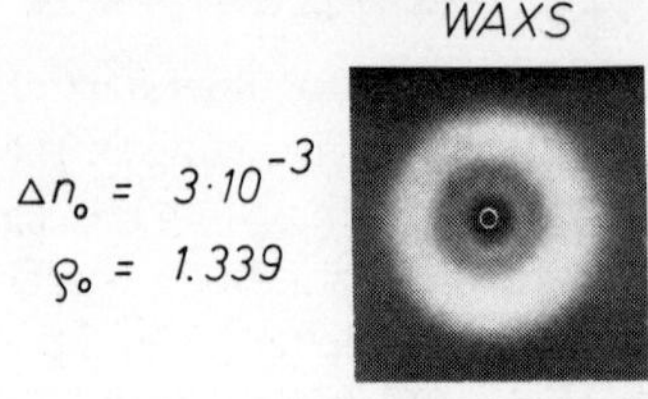

after crystallisation (5 hours, $T_C = 240 °C$)

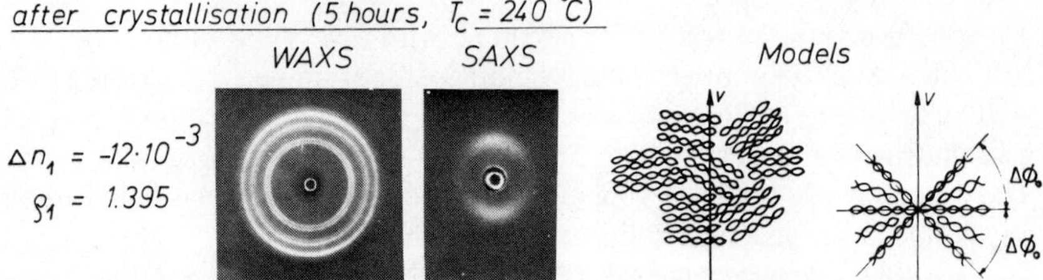

FIG. 2. Wide-angle X-ray scattering (WAXS), small-angle X-ray scattering (SAXS), birefringence $\Delta n_0, \Delta n_1$ and density ρ_o, ρ_1 before and after crystallization. On the right side two possible models for the arrangement of the crystals (twisted lamellae) are given.

Amorphous PET-film after stretching (e.g. $T_V = 90 °C$, $w = 3 \%/min$)

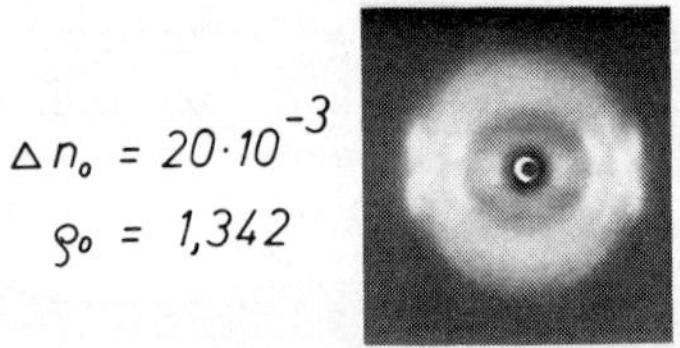

after crystallisation (5 hours, $T_C = 240 °C$)

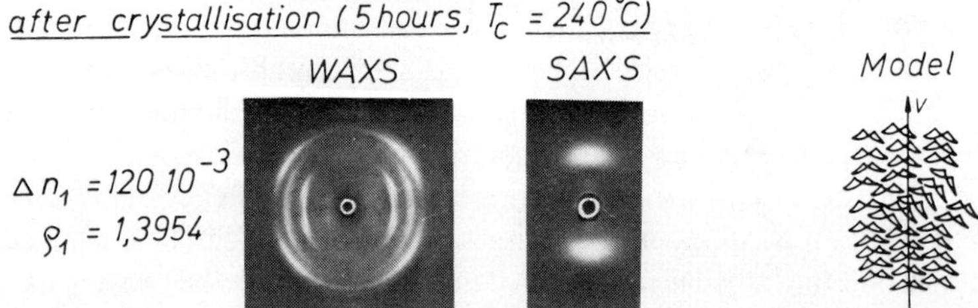

FIG. 3. Wide-angle X-ray scattering (WAXS), small-angle X-ray scattering (SAXS), birefringence $\Delta n_0, \Delta n_1$ and density ρ_o, ρ_1 before and after crystallization. On the right side the model for the arrangement of the crystals is given.

samples with lower orientation show a fiber structure with rotational symmetry about the stretching direction, this highly oriented sample has an uniaxial planar orientation with the 100-planes oriented preferentially parallel to the film surface as described already by Heffelfinger [7]. The angle ω^* has decreased further to 43° similar to the results of Bonart [5] and Berg [12]. The bending of the small

Amorphous   PET-film after stretching (e.g. T_V = 90 °C, w = 16 %/min)

WAXS

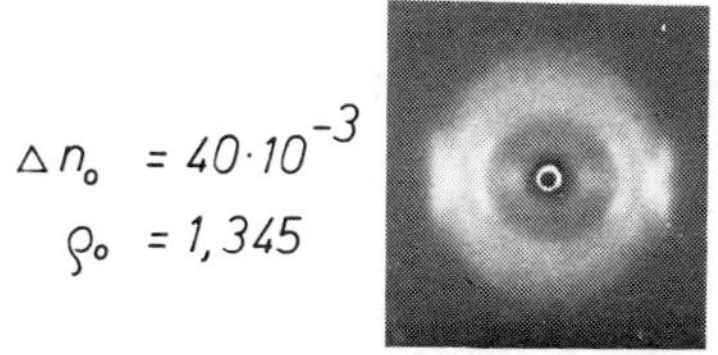

Δn_o = $40 \cdot 10^{-3}$

ρ_o = 1,345

after crystallisation (5 hours, T_c = 240 °C)

WAXS   *SAXS*   *Model*

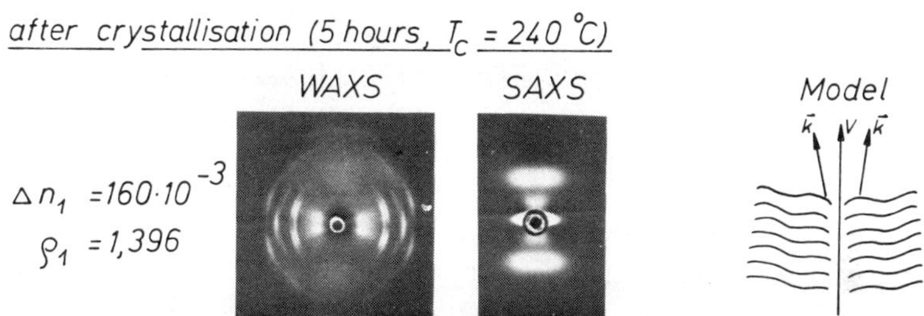

Δn_1 = $160 \cdot 10^{-3}$

ρ_1 = 1,396

FIG. 4. Wide-angle X-ray scattering (WAXS), small-angle X-ray scattering (SAXS), birefringence $\Delta n_0, \Delta n_1$ and density ρ_o, ρ_1 before and after crystallization. On the right side the model for the arrangement of the crystals is given.

WAXS

Δn_o = $190 \cdot 10^{-3}$

ρ_o = 1,359

after crystallisation (5 hours, T_c = 240 °C)

WAXS   *SAXS*   *Model*

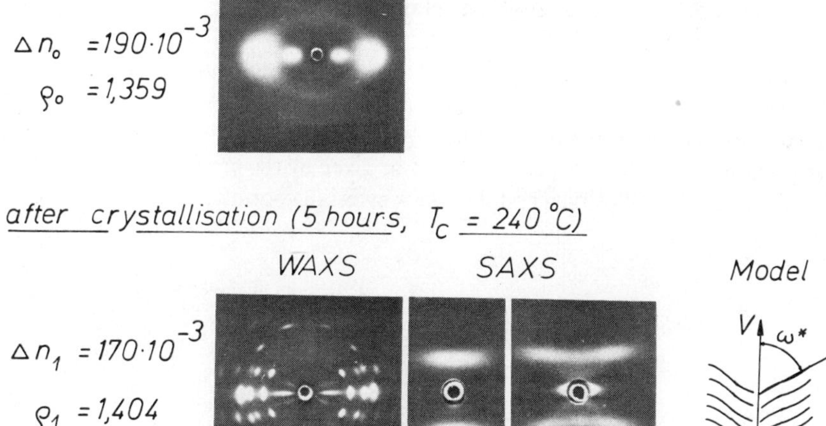

Δn_1 = $170 \cdot 10^{-3}$

ρ_1 = 1,404

FIG. 5. Wide-angle X-ray scattering (WAXS), small-angle X-ray scattering (SAXS), birefringence $\Delta n_0, \Delta n_1$ and density ρ_o, ρ_1 before and after crystallization. On the right side the model for the arrangement of the crystals is given.

angle reflection upwards can be explained by assuming that the lamellae are split into small mosaic blocks, the surface of the lamellae being formed by the 100 plane and the 010 plane. Indications for a mosaic structure have been found also by Fischer and Fakirov [8].

The results discussed up to now are valid for a crystallization temperature $T_k = 240°C$. For lower values of the crystallization temperature, one obtains almost the same arrangements of the crystals. Only the radial and azimuthal half-width of the crystal reflexes increase with decreasing temperature, indicating that the crystals become less perfect and the orientation distribution becomes broader.

The proof that the models proposed are in agreement with the measured X-ray scattering has been given in other publications [2, 13]. There it is also shown that the results are in qualitative agreement with the birefringence measurements. In the following section, we will perform some exact calculations of the birefrigence. The fact that the models are in agreement with the experimental results of several different methods, and that they follow continuously one from the other with increasing birefringence, gives us the feeling that they are, in the main features, the only possible correct models.

Calculations of the Birefringence

According to Pinnock and Ward [14], the index ellipsoid of a polyethylene terephthalate crystal has one main axis lying in chain direction, the second main axis lying perpendicular to the chain direction in the 100 plane, and the third main axis lying perpendicular to the other two. From the polarizabilities given by Pinnock and Ward [14] and by using the relation of Clausius Mosotti with the crystal density [15] 1.515 g/cm^3, one obtains for the refractive indices in the three main directions of the index elliposid the values $\gamma = 1.827$, $\beta = 1.816$, and $\alpha = 1.386$, respectively.

We introduce Cartesian coordinates x_1, x_2, x_3 in such a way that the a-axis of the unit cell lies in the x_1 direction, and the 001 plane lies in the x_1, x_2 plane. The transformation from this coordinate system to one with the axis lying parallel to the main axis of the index ellipsoid may be given by the matrix $\mathbf{S}$. The equation of the index ellipsoid of the crystal in the Cartesian coordinates x_1, x_2, x_3 is then given by

$$(x_1 x_2 x_3)\mathbf{A}_1 \begin{pmatrix} x_1 \\ x_2 \\ x_3 \end{pmatrix} = 1 \tag{1}$$

with

$$\mathbf{A}_1 = \mathbf{S}^{-1}\mathbf{A}_0\mathbf{S} \tag{2}$$

and

$$\mathbf{A}_0 = \begin{pmatrix} \alpha^{-2} & 0 & 0 \\ 0 & \beta^{-2} & 0 \\ 0 & 0 & \gamma^{-2} \end{pmatrix} \tag{3}$$

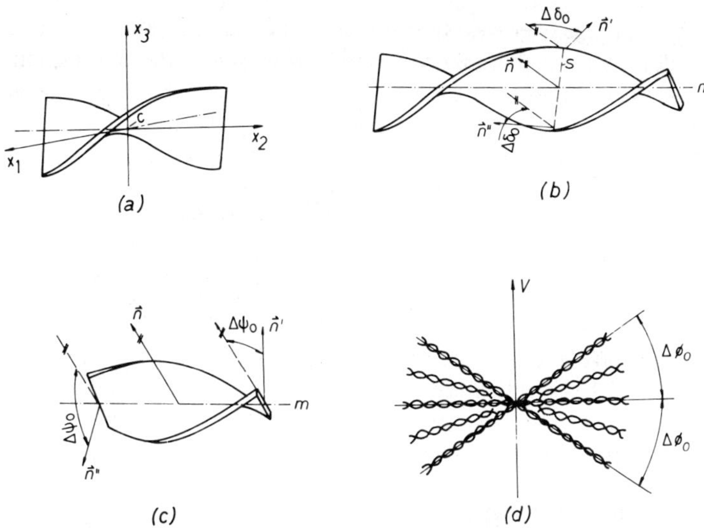

FIG. 6. Definition of the parameters $\Delta\delta_0$, $\Delta\psi_0$, and $\Delta\phi_0$.

In order to obtain the index ellipsoid of the oriented polymer, one has to average the index ellipsoid of the crystal given by $\mathbf{A}_1$ over all orientations occurring in the sample.

We consider first a single twisted lamella situated in a coordinate system in such a way that the crystal in the center of the lamella designated by C has the orientation described above [see Fig. 6(a)]. The shape of the lamella is described by two parameters $\Delta\delta_0$ and $\Delta\psi_0$. $2\Delta\delta_0$ gives the change of orientation of the surface vector of the lamella if one goes along the line s from one border to the other border, that means the angle between the two surface vectors $\mathbf{n}'$ and $\mathbf{n}''$ [see Fig. 6(b)]. $2\Delta\psi_0$ gives the change of the orientation of the surface vector of the lamella if one goes along the line m from one end of the lamella to the other end [see Fig. 6(c)]. If the twisting of the lamella is complete $2\Delta\psi_0 = 360°$.

The index ellipsoid of the crystal at the center of the lamellae C is represented by the matrix $\mathbf{A}_1$ in eq. (2). The surface vector of this crystal lies in the x_3 direction. If one goes along the x_1 axis ($= s$) to the borders of the lamellae, one has a rotation of the surface vector about the x_1 axis with an angle $\Delta\delta$. The values of $\Delta\delta$ lie between $-\Delta\delta_0$ and $+\Delta\delta_0$. Let the matrix $\mathbf{D}_1(\Delta\delta)$ describe a rotation of the coordinate system about x_1 with an angle $\Delta\delta$. The index ellipsoid of a crystal, the surface vector $\mathbf{n}$ of which forms the angle $\Delta\delta$ with the x_3 axis, is then given by

$$\mathbf{A}_2(\Delta\delta) = \mathbf{D}_1(\Delta\delta)^{-1}\mathbf{A}_1\mathbf{D}_1(\Delta\delta) \tag{4}$$

By averaging over all crystals along the x_1 axis, one obtains an index ellipsoid $\overline{\mathbf{A}}_2$ with the matrix elements

$$[\overline{A}_2]_{ij} = \frac{1}{2\Delta\delta_0} \int_{-\Delta\delta_0}^{+\Delta\delta_0} [A_2(\Delta\delta)]_{ij}\, d\Delta\delta \tag{5}$$

In this equation, generally $[\mathbf{X}]_{ij}$ indicates the element of the matrix $\mathbf{X}$ in the row i and column j.

Next we average over the other sheets of the lamellae along the x_2 axis (= m). We designate by $\mathbf{D}_2(\Delta\psi)$ the matrix which describes a rotation of the coordinate system about the x_2 axis with an angle $\Delta\psi$. Therefore, the elements of the $\overline{\mathbf{A}}_3$ for the index ellipsoid averaged over all crystals of the lamellae considered in Figure 6a are given by

$$[\overline{A}_3]_{ij} = \frac{1}{2\Delta\psi_0} \int_{-\Delta\psi_0}^{\Delta\psi_0} [A_3(\Delta\psi)]_{ij}d\Delta\psi \tag{6}$$

with

$$\mathbf{A}_3(\Delta\psi) = \mathbf{D}_2^{-1}(\Delta\psi)\overline{\mathbf{A}}_2\mathbf{D}_2(\Delta\psi). \tag{7}$$

Finally we have to take into account the rotational symmetry about the fiber direction V which is identical with the x_3 axis. We also have to take into account that the lamellae are not lying strictly perpendicular to the x_3 axis; the directions of their axes vary within an angle $2\Delta\phi_0$ [see Fig. 6(d)]. We introduce the angles ϵ and ϑ and designate the matrices which perform the rotations about these angles by $\mathbf{D}_3(\epsilon)$ and $\mathbf{D}_4(\vartheta)$. By performing the averaging indicated above, we obtain as a final result for the elements of the matrix $\overline{\mathbf{A}}_4$ which represents the index ellipsoid for the complete sample

$$[\overline{A}_4]_{ij} = \frac{1}{4\pi^2 \sin\Delta\phi_0} \int_0^{2\pi} d\epsilon \int_{-\Delta\phi_0}^{\Delta\phi_0} [A_4(\epsilon,\vartheta)]_{ij} \cos\vartheta d\vartheta \tag{8}$$

with

$$\mathbf{A}_4(\epsilon,\vartheta) = [\mathbf{D}_4(\vartheta)\mathbf{D}_3(\epsilon)]^{-1} \overline{\mathbf{A}}_3[\mathbf{D}_4(\vartheta)\mathbf{D}_3(\epsilon)] \tag{9}$$

The matrices used here are given in the appendix. The results for some values of the parameters $\Delta\psi_0$ and $\Delta\phi_0$ with a given value $\Delta\delta_0 = 30°$ are shown in Table I. One sees that for $\Delta\psi_0 = 180°$, that means for completely twisted lamellae, one obtains for all values of $\Delta\phi_0$ negative values for the birefringence. In case $\Delta\phi_0 = 90°$ the birefringence becomes zero; this value corresponds for the completely isotropic material consisting of spherulites. With decreasing values of $\Delta\psi_0$ the birefringence turns from negative values to positive values.

The results obtained explain both the negative birefringence obtained for the a-orientation (see Fig. 2) and the change in the sign of the birefringence with increasing orientation (compare Fig. 2 with Fig. 3).

TABLE I

Birefringence Δn for Twisted Lamellae with Various Parameters $\Delta\phi_0$ and $\Delta\psi_0$. $\Delta\delta_0$ is 30°.

$\Delta\psi_0$ \ $\Delta\phi_0$	0°	20°	40°	60°	80°	90°
180°	$-71,4\cdot10^{-3}$	$-63,1\cdot10^{-3}$	$-41,9\cdot10^{-3}$	$-17,9\cdot10^{-3}$	$-2,2\cdot10^{-3}$	0
80°	$-60,9\cdot10^{-3}$	$-52,9\cdot10^{-3}$	$-32,8\cdot10^{-3}$	$-10,0\cdot10^{-3}$	$5,0\cdot10^{-3}$	$7,0\cdot10^{-3}$
30°	$-0,3\cdot10^{-3}$	$5,3\cdot10^{-3}$	$19,4\cdot10^{-3}$	$35,5\cdot10^{-3}$	$46,0\cdot10^{-3}$	$47,4\cdot10^{-3}$

STRUCTURE OF THE NONCRYSTALLINE REGIONS

For the explanation of the mechanical properties, it is of great importance to find out the conformations of the chains in the noncrystalline regions. One has to know which amounts of tie molecules, loops, and chains with one free end occur, and one has also to know something about the length distribution of these noncrystalline parts of the chains. Unfortunately there is no experimental method by which this information can be obtained. One relies completely on indirect methods which give certain quantities to characterize the noncrystalline regions.

One useful method is nuclear magnetic resonance (NMR). The NMR line of a semicrystalline polymer above the glass transition temperature can be separated into a broad and a narrow component. The broad component arises from chains which cannot perform a segmental motion. These are the chains in the crystallites and also some chains in the noncrystalline regions which are so much stretched that their mobility is very much limited. The narrow component arises from chains which can perform segmental motion, that is, from

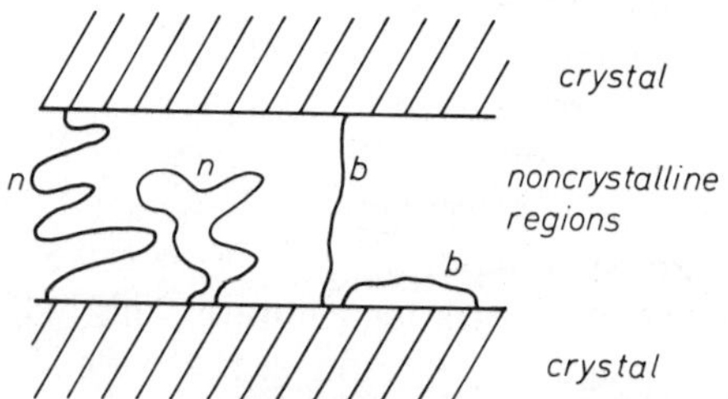

FIG. 7. Different types of chains in the noncrystalline regions.

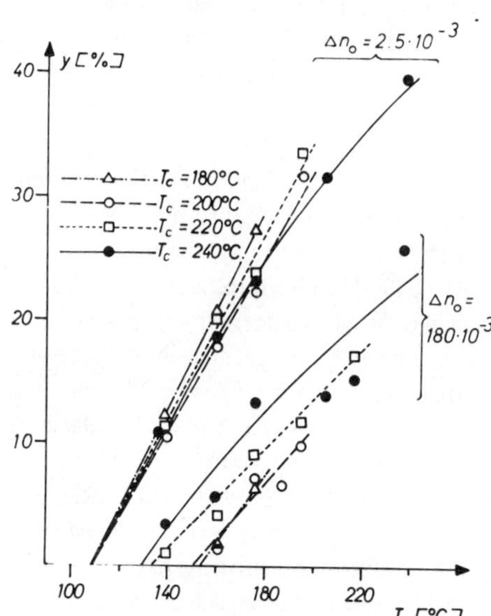

FIG. 8. Mobile fraction as a function of temperature. Δn_0 is the birefringence before crystallization. T_c is the crystallization temperature.

highly mobile chains within the noncrystalline regions. Former investigations [16] have shown that a chain with fixed ends, which means a tie molecule or a loop, must have an end-to-end distance which is less than approximately half of its contour length in order to be sufficiently mobile to contribute to the narrow component. To illustrate this, Figure 7 shows a schematic representation of two crystals with a noncrystalline region between them. The chains designated by n have an end-to-end distance which is smaller than half of the contour length and contribute therefore to the narrow component of the NMR line. The chains designated by b contribute to the broad component.

The fraction of chains which contributes to the narrow component is called the mobile fraction y and is given by the integrated intensity of the narrow component divided by the integrated intensity of the total line. We have determined the mobile fraction of our samples. The NMR line was separated into two components by using a method developed first for unoriented polyethylene terephthalate [17] and improved later for measurements on unoriented and oriented polyamides [18]. In this method the signal is measured with two different modulation amplitudes to avoid any ambiguity in the line separation.

Figure 8 shows the mobile fraction as a function of temperature. The curves on the left were obtained from samples with an initial orientation $\Delta n_0 = 2.5 \times 10^{-3}$. The different curves refer to different crystallization temperatures T_c, as indicated in Figure 8. The curves at the right were obtained from highly oriented samples with an initial birefringence $\Delta n_0 = 180 \times 10^{-3}$. One sees that the mobile fraction increases with temperature. This indicates that, with increasing temperature, more and more chains become mobile as a consequence of the transition from the glassy state to the state with segmental motion. One recognizes clearly that in the highly oriented material the narrow component of the NMR signal appears at a higher temperature than in the sample with low orientation. This agrees with the well known fact that the glass transition temperature increases with orientation. In addition one can see also that the mobile fractions in the highly oriented samples are generally much lower than in the samples with low orientation.

In order to get the total amount of chains with small end-to-end distances which contribute to the narrow component, one has to extrapolate the mobile fraction to a temperature where the mobile fraction has reached its maximum value. In a previous work [17] such an extrapolation using the theory of free volume has been described. We have performed this extrapolation. The results are shown in Figure 9. The extrapolated mobile fraction y_∞ is plotted as a function of crystallization temperature T_c. The parameter is the birefringence Δn_0 before crystallization. One sees that the extrapolated mobile fraction decreases markedly with increasing orientation.

The fraction of noncrystalline chains with large end-to-end distances which do not contribute to the narrow component will be called the rigid noncrystalline fraction. One obtains this fraction by subtracting the mobile fraction from the noncrystalline fraction $1 - \alpha$. α is the degree of crystallinity obtained from the density of the samples. It was assumed that the density of the noncrystalline regions was 1.333 g/cm ^{3} and that of the crystals, according to new measurements

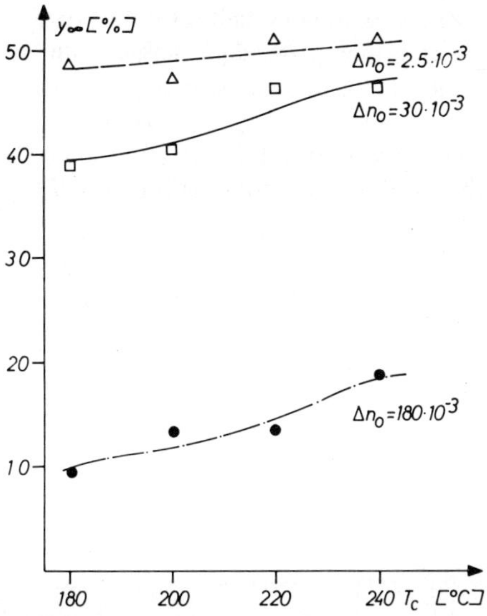

FIG. 9. Extrapolated mobile fraction y_∞ as a function of the crystallization temperature. Δn_0 is the birefringence before crystallization.

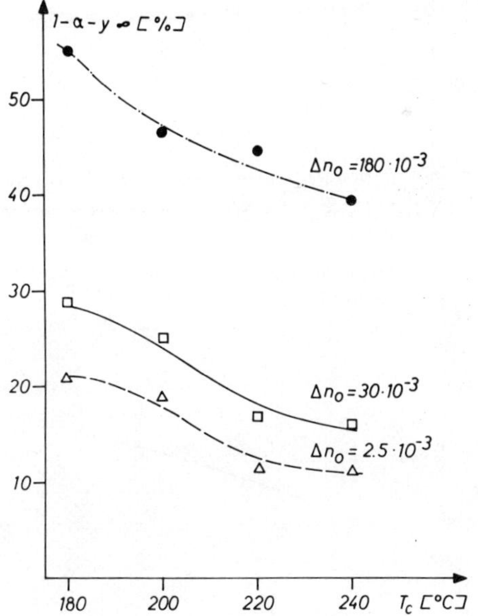

FIG. 10. Rigid noncrystalline fraction as a function of crystallization temperature. Δn_0 is the birefringence before crystallization.

of Fischer and Fakirov [15] 1.515 g/cm^3. Figure 10 shows the results. The rigid noncrystalline fraction is plotted against the crystallization temperature T_c. The parameter is the birefringence before crystallization, Δn_0. One sees that the rigid noncrystalline fraction, that is the amount of noncrystalline chains with comparatively large end-to-end distances, increases markedly with increasing orientation of the sample. In principle the rigid chains can be either stretched loops or stretched tie molecules. We think that if one considers the mechanism of stretching one has to conclude that they are tie molecules.

One must admit that there are some uncertainties in determining the degree of crystallinity and therewith in obtaining the fraction of rigid chains. A critical examination shows however that, by changing the assumptions in a reasonable way, one may shift a little the values obtained but one would not essentially change the main result. For example, if one uses the older values of de Daubeney and Bunn [19] for the density of the crystals, the curves shown in Figure 10 all shift to lower values. If one assumes that the density of the noncrystalline regions in the oriented material is larger than in the unoriented, the difference in the mobile fraction between oriented and nonoriented material would increase.

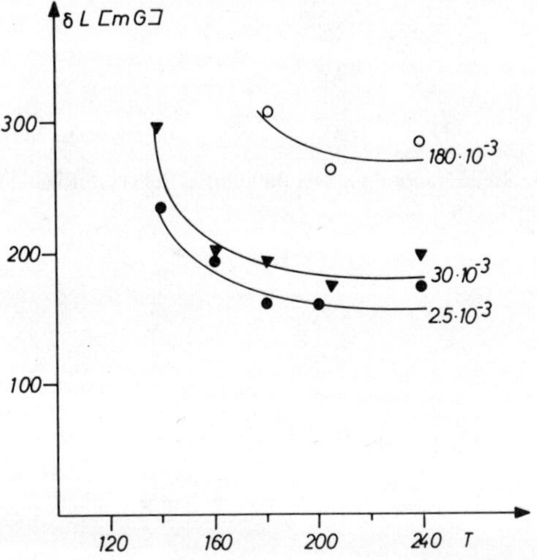

FIG. 11. Line with δL of the narrow component as a function of the temperature T. Parameter is the birefringence Δn_0 before crystallization.

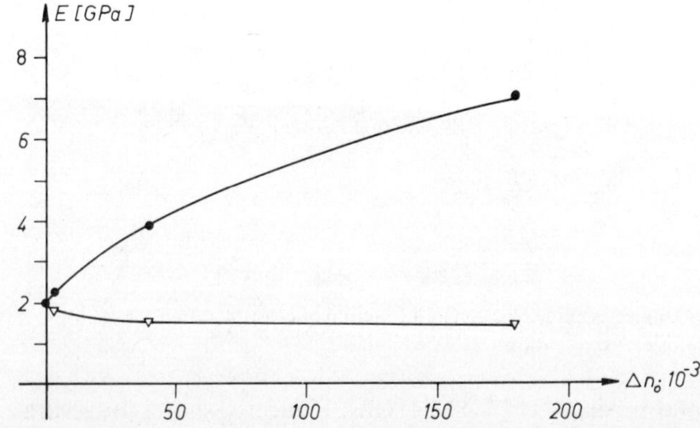

FIG. 12. Young's modulus E for oriented and crystallized (5 hr at $T_c = 240°C$) polyethylene terephthalate as a function of the birefringence Δ_0 before the crystallization. ● = parallel to the stretching direction, ▽ = perpendicular to the stretching direction.

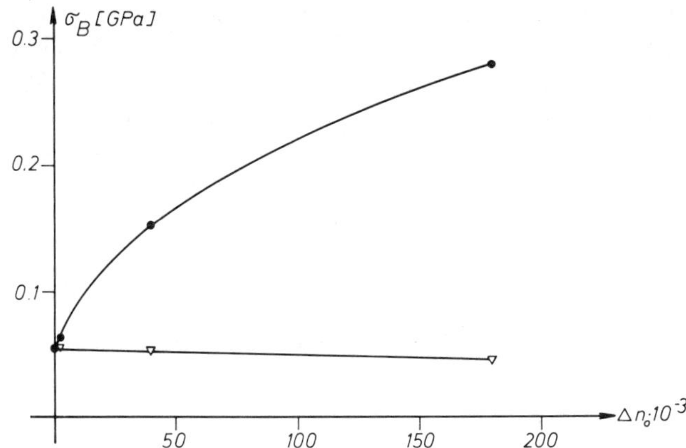

FIG. 13 Fracture stress σ_B oriented and crystallized (5 hr at T_c = 240°C) polyethylene tereph-thalate as a function of the birefringence Δn_0 before the crystallization. ● = parallel to the stretching direction, ▽ = perpendicular to the stretching direction.

Therefore it seems to us proved that the oriented material has a larger amount of almost stretched tie molecules than the unoriented material.

Figure 11 shows the line width of the narrow component as a function of temperature T. Parameter is the birefringence Δn_0 before crystallization. One sees that the line width increases with increasing birefringence. This indicates that, with increasing orientation, not only the fraction but also the mobility of the mobile chains decreases.

MECHANICAL PROPERTIES

Figure 12 shows the Young modulus as a function of the birefringence Δn_0, that is, the birefringence after stretching before crystallization. The Young modulus was measured in the direction of stretching and perpendicular to that direction. One sees that the value of the modulus in the stretching direction increases with increasing orientation, which is in agreement with results reported by Ward [1].

Figure 13 shows the fracture stress which shows a similar behavior as the Young modulus.

The large increase of the modulus and the fracture stress with orientation is explained usually in the following way: In the oriented material one has to overcome valence bond forces in order to stretch and break the material. On the other hand, in the undrawn sample both processes can occur by overcoming only van der Waals forces which are easier than valence bond forces.

The effect of orientation certainly is of essential importance. We think however, that in addition also the differences in the structure of the noncrystalline regions are of essential importance for the mechanical improvement of the mechanical properties with orientations. One has to consider the following: To increase the Young modulus, it seems necessary to increase the number of tie molecules in the noncrystalline regions: otherwise one would be able to deform the sample simply by increasing van der Waals distances. It is also important that the tie molecules are quite taut, so that an increase of their end-to-end distances can be achieved only by deformation of valence bonds and not by changing conformations. To increase the fracture stress, according to De Vries, Statton, and Park [20], the noncrystalline regions should consist of tie molecules of quite uniform lengths. This also is the case if the tie molecules are taut. Therefore, the increase of the rigid noncrystalline fraction with increasing orientation observed by NMR seems to be, in addition to the orientation, an essential reason for the improvement of the mechanical properties.

APPENDIX

The matrices introduced in order to calculate the birefringence are the following ones:

$$\mathbf{S} = \begin{pmatrix} 0.758 & -0.446 & -0.476 \\ 0.294 & 0.885 & -0.362 \\ 0.582 & 0.134 & 0.802 \end{pmatrix} \tag{10}$$

$$\mathbf{D}_1(\Delta\delta) = \begin{pmatrix} 1 & 0 & 0 \\ 0 & \cos\Delta\delta & -\sin\Delta\delta \\ 0 & \sin\Delta\delta & \cos\Delta\delta \end{pmatrix} \tag{11}$$

$$\mathbf{D}_2(\Delta\psi) = \begin{pmatrix} \cos\Delta\psi & 0 & \sin\Delta\psi \\ 0 & 1 & 0 \\ -\sin\Delta\psi & 0 & \cos\Delta\psi \end{pmatrix} \tag{12}$$

$$\mathbf{D}_3(\epsilon) = \begin{pmatrix} \cos\epsilon & -\sin\epsilon & 0 \\ \sin\epsilon & \cos\epsilon & 0 \\ 0 & 0 & 1 \end{pmatrix} \tag{13}$$

$$\mathbf{D}_4(\vartheta) = \begin{pmatrix} 1 & 0 & 0 \\ 0 & \cos\vartheta & -\sin\vartheta \\ 0 & \sin\vartheta & \cos\vartheta \end{pmatrix} \tag{14}$$

REFERENCES

[1] I. M. Ward, *J. Macromol. Sci.*, **B1**, 667 (1967).

[2] H. J. Biangardi and H. G. Zachmann, *Prog. Colloid Polym. Sci.*, **62**, 71 (1977).

[3] A. Keller, *J. Polym. Sci.*, **17**, 291 (1955); *ibid.*, **17**, 351 (1955); *ibid.*, **17**, 447 (1955).

[4] Watkins, Hansen, *Text. Res. J.*, **38**, 388 (1968).

[5] R. Bonart, *Kolloid-Z.*, **199**, 136 (1964); *ibid.*, **213**, 1 (1966); *ibid.*, **231**, 438 (1966).

[6] C. I. Heffelfinger and E. L. Lippert, *J. Appl. Polym. Sci.*, **15**, 2699 (1971).

[7] C. I. Heffelfinger and R. L. Burton, *J. Polym. Sci.*, **46**, 289 (1960).

[8] E. W. Fischer and S. Fakirov, *J. Mater. Sci.*, **11**, 1091 (1976).

[9] E. Liska, *Kolloid-Z.*, **251**, 1028 (1973).

[10] R. Bonart, *Kolloid-Z.*, **211**, 14 (1966).

[11] A. Keller and M. J. Machin, *J. Macromol. Sci.*, **B1**, 41 (1967).

[12] H. Berg, *Chemiefasern Text.*, **3**, 215 (1972).

[13] H. J. Biangardi and H. G. Zachmann, *Makromol. Chem.*, **177**, 1173 (1976).

[14] P. R. Pinnock and I. M. Ward, *Br. J. Appl. Phys.*, **15**, 1559 (1964).

[15] E. W. Fischer and S. Fakirov, *Makromol. Chem.*, **176**, 2459 (1975).

[16] P. Schmedding and H. G. Zachmann, *Kolloid-Z.*, **250**, 1105 (1972); *Colloid Polym. Sci.*, **253**, 441 (1975); *ibid.*, **253**, 527 (1975).

[17] U. Eichhoff and H. G. Zachmann, *Ber. Bunsenges. Phys. Chem.*, **74**, 919 (1970); *Makromol. Chem.*, **147**, 41 (1971).

[18] K. Wangermann and H. G. Zachmann, *Prog. Colloid Polym. Sci.*, **57**, 236 (1975).

[19] P. de Daubeney and W. Bonn, *Proc. R. Soc. London*, **226A**, 531 (1954).

[20] J. B. Park, W. O. Statton, K. L. de Vries, private information.

SUPERMOLECULAR STRUCTURE AND MECHANICAL STRESS DISTRIBUTION AMONG THE CHEMICAL BONDS IN POLYMERS

K. J. FRIEDLAND, V. A. MARIKHIN, L. P. MYASNIKOVA, and V. I. VETTEGREN

A. F. Ioffe Physical-Technical Institute, Academy of Sciences of the USSR, Leningrad, 194021

SYNOPSIS

Stress distribution among the polymer chain segments and parameters of the supermolecular structure of oriented polypropylene, polyethylene terephthalate and polycaproamide films have been compared. The stresses have been determined by infrared spectroscopy, and the structure has been studied by electron microscopy and small-angle and wide-angle X-ray scattering. The polymers studied are shown to have a microfibrillar structure. The sizes of long periods, the lengths of crystallites and amorphous regions are found. Stress distribution among the polymer chain segments is extremely nonuniform. Most of the segments, i.e., 90–95%, are subjected to stresses close to the applied stress, the rest of the segments, i.e., 5–10%, experience stresses which are at least an order of magnitude higher. Highly overstressed segments lie in the amorphous regions of microfibrils. They bear the major part of the applied load (not less than 80–90%). A model for the amorphous regions' structure of microfibrils in oriented polymers is discussed.

INTRODUCTION

Development of an interatomic interaction forces theory for solids has made it possible to calculate the "theoretical strength" of polymeric bodies. Such a calculation was carried out by Kobeco [1], Gubanov and his colleagues [2], *et al.* [3–5]. The results have shown that the "theoretical strength" of an ideal crystalline polymer lattice exceeds the strength of real specimens by factors of tens and hundreds. This difference is usually attributed to the imperfection of the polymeric bodies' structure and the presence of the poorly ordered amorphous regions in them. The structural imperfection leads to a nonuniform distribution of mechanical stresses. Individual chemical bonds are subjected to stresses high enough to rupture these bonds under the action of thermal fluctuations. The rupture of the highly overstressed bonds triggers the fracture process in loaded specimens.

Zhurkov, Vettegren *et al.* [6–8] have observed the overstressed chemical bonds in the stressed specimens of oriented polymers by infrared spectroscopy. In this paper we try to find a relation between the supermolecular polymer structure and the overstressed chemical bonds.

Journal of Polymer Science: Polymer Symposium 58, 185–194 (1977)
© by John Wiley & Sons, Inc.

PREPARATION OF SPECIMENS

Highly oriented films of polycaproamide, polyethylene terephthalate and isotactic polypropylene obtained by uniaxial drawing at temperatures ranging from 20° up to 200°C for polycaproamide, 180–240°C for polyethylene terephthalate and 150°C for polypropylene have been investigated. The drawing ratio was 6.0 for polyethylene terephthalate, 4.5 for polycaproamide, and 9.0 for polypropylene.

X-RAY STUDIES

Small-angle X-ray patterns were recorded using the apparatus constructed on the basis of the industrial device DRON-1 with the point collimation which provided the resolution of 400 Å. The tube BSV-10 with $CuK\alpha$ radiation filtered by Ni was used. The sample-to-film distance was 200 mm. The X-ray patterns were recorded in a IFO-401 microdensitometer.

Wide-angle X-ray scattering reflections were recorded using the device DRON-1.

ELECTRON MICROSCOPIC STUDIES

The surface of the specimens was replicated by platinum and quartz and the films obtained were studied in an JEM-5Y electron microscope. Ultrasonic disintegration of the specimens followed by the negative shadowing of the disintegrated products by phosphorotungstic acid was also used [9].

INFRARED SPECTROSCOPIC STUDIES

A tensile stress directed along the axis of orientation was applied to the oriented specimens using the device described in [10]. The transmission spectra of the stressed specimens were recorded in a DS-403G spectrophotometer in polarized light, the degree of polarization being 97%. The transmission coefficient was measured with the accuracy of 0.3%.

EXPERIMENTAL RESULTS AND DISCUSSION

Supermolecular Structure of the Specimens Investigated

The supermolecular structure of specimens was studied with the help of electron microscopy. The micrographs obtained from the surface replicas of oriented films and the transmission electron micrographs of disintegration products of the same film do not differ from each other qualitatively and thus the experimental data can be demonstrated by Figure 1 where the micrographs for one of the polycaproamide specimens are shown.

The small-angle X-ray patterns have also turned out to be qualitatively the same for all the samples. They are in the form of meridional streaks perpendicular to the axis of orientation (see Fig. 2). According to [11], such X-ray

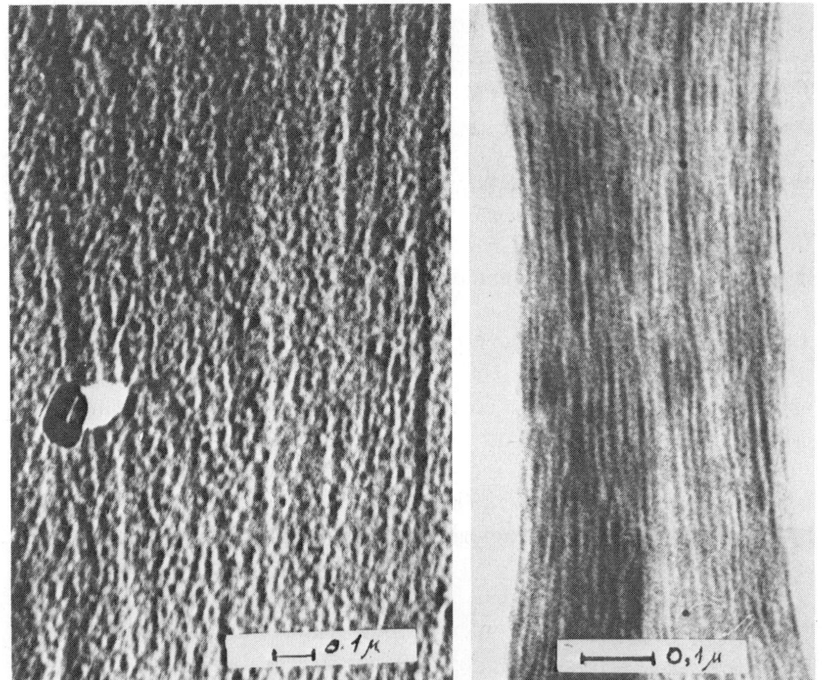

FIG. 1. Surface replica and products of ultrasonic disintegration of drawn nylon-6 film.

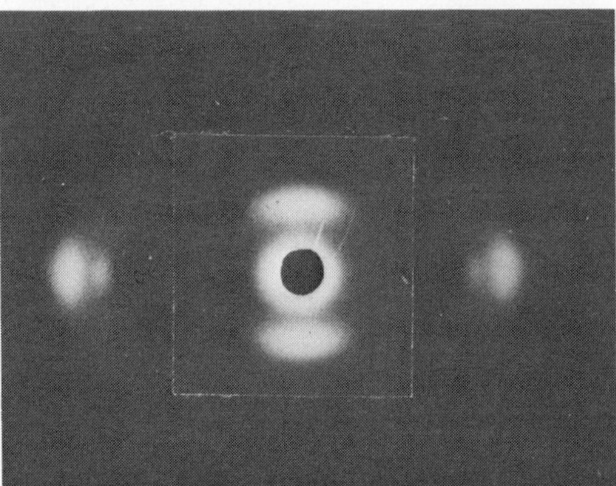

FIG. 2. Wide- and small-angle X-ray patterns of drawn nylon-6 film.

patterns indicate that there is an alternation of regions with different densities, i.e., amorphous and crystalline ones, along the fibril axis, which gives in all the "long period" $\mathcal{L}$. The shape of the small-angle X-ray patterns (the meridional streaks) points to the fact that crystalline regions are in the form of rectangular parallelepipeds [12].

The wide-angle X-ray pattern is shown in Fig. 2. It contains a large number of narrow reflections, which implies that the crystallites of the microfibrils are rather perfect and have a high degree of orientation along the axis of drawing (the misorientation angle is as small as 5–8°).

The data obtained enable us to develop a model regarding the structure of the oriented polymers studied (Fig. 3). Hess [13] was the first to introduce such a model and at present it is in general accepted by many investigators.

Let us consider some features of this model in more detail.

As follows from [14–16], the polymer molecules in the crystalline regions have a regular structure. They are in the form of a flat extended chain consisting of trans isomers for polycaproamide and polyethylene terephthalate and in the form of an isotactic spiral built up of trans and gauche isomers for polypropylene. For convenience of further discussion they are represented by straight lines in Figure 3.

In the amorphous regions the regular structure of the polymer molecules is distorted. Along with the regularly constructed segments consisting of trans isomers, the segments with gauche isomers appear in polycaproamide and polyethylene terephthalate. For polypropylene, the alternation of trans and gauche isomers is distorted. In the sites of the regular structure distortion, the axis of polymer molecules undergo kink. In order to take into account this fact, a part of the molecules in the amorphous regions is shown in Figure 3 by bent lines.

For discussion of results we shall need the data on the structure of amorphous regions. Let us first of all dwell upon determination of their longitudinal sizes. For this purpose the sizes of the long period $\mathcal{L}$ were found from the angular position of small-angle reflections. The longitudinal sizes of the crystal-line regions $\mathcal{L}_{ae}$ were then determined from the half-width of wide-angle reflections. The reflections used in these calculations are listed in Table I. The value of the longitudinal size of the amorphous region $\mathcal{L}_a$ was found from the expression

$$\mathcal{L}_a = \mathcal{L} - \mathcal{L}_{ae} \tag{1}$$

The values of $\mathcal{L}$, $\mathcal{L}_{ae}$ and $\mathcal{L}_a$ are listed in Table II.

We shall also need the data on the number of polymer chains passing from one crystallite into another. Unfortunately, direct methods for the determination of this value have not yet been developed, so we tried to estimate this value in-directly.

The intensity of a small-angle X-ray scattering reflection was estimated. The value obtained was used to determine the difference between the densities of

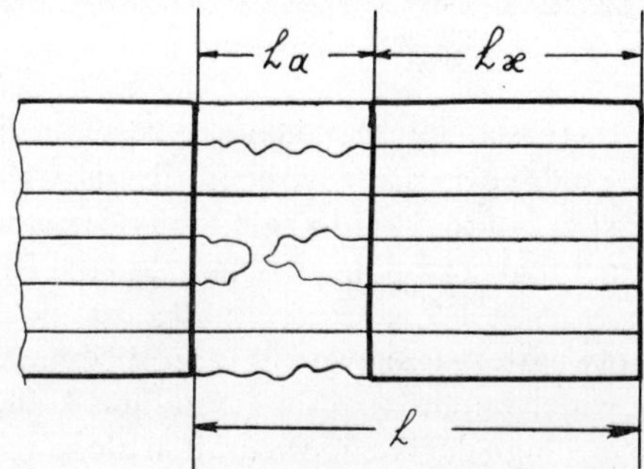

FIG. 3. The model of supermolecular structure of drawn polymer.

TABLE I

Wide-Angle X-Ray Reflections and IR Absorption Bands Used for Measurements[a]

Polymer	X-ray reflections	IR bands, cm^{-1}	Regular segment structure	α, $\dfrac{cm^{-1}mm^2}{kg}$	$\dfrac{K_\Sigma}{K_o}$	References
Polypropylene	[113]	975	$(TC)_4$	0.026	2	17
Polyethyleneterephthalate	[103]	976	T	0.084	3	18
Polycaproamide	[002] [004] [006]	930	$(T)_6$	0.022	2	19

[a] T is trans isomer, C is gauche isomer, the figures show the number of monomer units in the conformation given.

TABLE II

Number of Taut Tie Chains in the Polymers Studied

Polymer	Temp. of drawing °C	L, Å	L_a, Å	L_x, Å	$\dfrac{L}{L_a}$	$\dfrac{N}{N_x}=\dfrac{\delta}{\overline{\Sigma}}$	$\dfrac{N}{N_x}=\dfrac{n\,L}{n_o\,L_a}$
Polypropylene	150	150	70	80	2.1	0.09	0.11
Polyethyleneterephthalate	240	170	70	100	2.4	0.08	0.07
	235	170	80	90	2.1	0.08	0.06
	220	160	80	80	2.0	0.10	0.08
	205	140	90	50	1.6	0.10	0.07
	180	140	90	50	1.6	0.08	0.06
Polycaproamide	200	100	50	50	2.0	0.03	0.03
	150	90	40	50	2.3	0.03	0.04
	100	90	30	60	3.0	0.04	0.04
	20	80	30	50	2.8	0.05	0.05

amorphous and crystalline regions. In accordance with literature [11] these estimates have shown that the density of the amorphous regions differs from that of the crystalline ones by more than 10%. This leads to the conclusion that a major part of the chains are the "tie" chains.

As a rule not all the chains are considered to be the tie chains, since some of them after leaving the crystallites can form folds and then enter the same crystallite again. The folds are obviously due to distortion in the regular structure of the polymer molecules. We have found the concentration of the regularly constructed chain segments in polypropylene and polyethylene terephthalate specimens from the intensity of the infrared absorption bands corresponding to such segments. It turned out that 70–80% of chain segments in the amorphous regions are regularly constructed. Consequently, it can be inferred from these measurements that a major part of chains pass from one crystallite into another without forming any folds.

Stress Distribution among Chemical Bonds

In order to determine mechanical stresses acting on chemical bonds we have
made use of the fact that the natural frequencies of the atomic vibrations shift
under the action of stresses. As shown in [6–8], the value of the shift grows in
direct proportion to the stress on bonds:

$$\Delta\nu = \alpha\Sigma \tag{2}$$

It is known from the theory of IR spectra that the interatomic vibrations are
not localized at individual chemical bonds but embrace the chain segments of
a definite length and structure. Vibrations of the segments produce symmetrical
absorption bands in the IR spectrum, which superimpose and form the experimentally observed band.

Selecting different absorption bands for measurements and using eq. (2), we
can find stresses acting on chain segments with different structure. Let us divide
all the segments into two groups: the first one includes the regularly constructed
segments, the second one includes those segments in which the regular structure
of molecules is distorted and their axis is bent.

Let us consider at first the experimental results for the second type of segments. In order to determine stresses, the bands with the maxima at 502, 900
and 1042 cm^{-1} for polyethylene terephthalate corresponding to segments with
gauche isomers [17] were used. After application of tensile stress to the specimens, the position of the maximum and the shape of such bands remained the
same. This means that the stresses acting on the segments where the regular
structure of the molecular chains is distorted are too small to be registered by
the spectroscopic technique. At the same time the absorption coefficient of the
bands studied decreases under load due to transition of the folded gauche isomers
into more extended trans isomers. When such a transition occurs, the length of
the molecule increases, which is likely to make a quick stress relaxation easier.

Let us turn now to the experimental results for the regularly constructed chain
segments. The length and structure of the segments and the frequencies at which
the maxima of the bands corresponding to them lie are listed in Table I. Figure
4 shows one of the selected bands, i.e., at 975 cm^{-1} for polypropylene. Before
load application to the specimen, the band is symmetrical. This means that the
lines forming this band lie symmetrical with respect to the center. When a stress
is applied, the maximum is shifted in the low-frequency direction, and the band
shape becomes asymmetrical: a wing with a well-expressed edge 25 cm^{-1} away
from the maximum is formed on the long-wave side. Consequently, the absorption lines are shifted in the low-frequency direction by different amounts.
It follows from eq. (2) that the magnitude of the shift grows in direct proportion
to stresses acting on chain segments. Different shifts of the lines then show that
the stress is distributed over the segments of the polypropylene molecules
nonuniformly. Qualitatively similar changes in the absorption bands corresponding to the regularly constructed segments of the molecules were observed
for polyethylene terephthalate and polycaproamide; under the action of a tensile
stress, the maximum of the bands shifted in the low-frequency direction and the

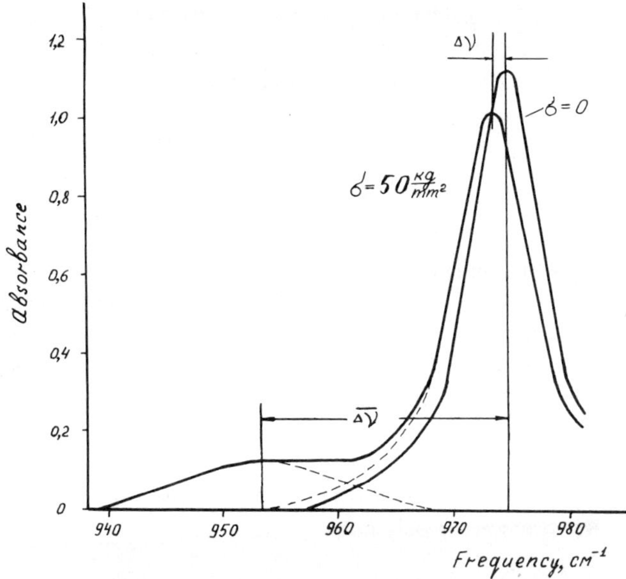

FIG. 4. The shape of absorption band 975 cm^{-1} for polypropylene.

shape became asymmetrical. Therefore, the stress in these polymers is also distributed over the chain segments nonuniformly.

In order to carry out a quantitative evaluation, we divide all the regularly constructed segments into two groups. The first one includes the segments the lines for which are shifted only slightly and form the maximum of the band; the second group contains the segments the lines for which are shifted by considerably greater amounts and form a long-wave wing. Let us separate the maximum from the wing as shown in Figure 4 by a dotted line. Using eq. (2) we can find the average stress on the segments absorbing in the wing:

$$\overline{\Sigma} = \frac{\Delta \nu}{\alpha} \tag{3}$$

where $\Delta \nu = \nu_o - \nu_\Sigma$, ν_Σ and ν_o being the position of the center of gravity of the long-wave wing for a stressed specimen and of the band maximum for an unstressed one, respectively. In order to determine the concentration n of the segments absorbing in the wing we employ the Lambert–Beer law:

$$n/n_o = S_\Sigma K_o / S_o K_\Sigma \tag{4}$$

where S_Σ, S_o and K_Σ, K_o are the areas and molar absorption coefficients for the long-wave wing of a stressed specimen and for the absorption band of an unstressed one, n_o is the total concentration of segments in the specimen. The values of α and K_Σ / K_o are listed in Table I.

It is clear from the spectrum shown in Figure 4 that at a tensile stress of 50 kg/mm^2 the stress on the segments absorbing in the long-wave wing is 900 kg/mm^2 and the concentration of such segments is 5%. The stress on the segments absorbing in the band maximum and constituting the remaining 95%

equals the applied stress, i.e., 50 kg/mm^2. Similar results were obtained for other polymers: several percent of the molecular segments in a stressed specimen are subjected to stresses an order of magnitude higher than the applied stress. Further we shall call such segments the overstressed segments.

In What Regions of the Polymer do the Overstressed Molecular Segments Lie?

As noted above, it is normally assumed that the overstressed segments occur in the amorphous regions of the polymer. This assumption was checked in [20]. The concentration of the overstressed segments was shown to grow in direct proportion to the concentration of amorphous regions in the specimens. Moreover, concentration of such segments depends on the perfection of the amorphous regions' structure: the higher the orientation of polymer molecules in the amorphous regions, the lower is the concentration of the overstressed bonds and stress on them. These facts indicate that the overstressed segments are situated in the amorphous regions of polymers.

Taut Tie Polymer Molecules and Overstressed Segments

As seen from Figure 3, a part of molecules in the amorphous regions can pass from one crystallite into another without folding. Such molecules must be regularly constructed, well oriented and have a high modulus of elasticity. When the specimen is stretched, they are the first to experience high stresses. Further we shall call such molecules taut tie molecules.

As shown above, the overstressed segments are regularly constructed and experience high stresses. It has been shown in [20] that the overstressed bonds have an extremely high degree of orientation. It is therefore natural to suppose that they form a part of taut tie molecules. Let l_o be the length of a regularly constructed segment and N the number of taut tie chains. Then it follows from the model represented in Figure 3 that the number of overstressed segments n equals:

$$n = N \, \mathcal{L}_a/l_o \tag{5}$$

The total number of the regular segments in the specimen n_o is:

$$n_o \approx (N_a\mathcal{L}_a + N_{ae}\mathcal{L}_{ae})/l_o \tag{6}$$

where N_{ae} and N_a are the number of chains in the crystallite and amorphous region, respectively. Since $N_{ae} \approx N_a$, we obtain:

$$n_o \approx N_{ae} \, \mathcal{L}/l_o \tag{7}$$

Combining (5) and (7) we get:

$$N/N_{ae} \approx n\mathcal{L}/n_o\mathcal{L}_a \tag{8}$$

As has been noted above, the stresses on overstressed regions reach high values exceeding at least by an order of magnitude the applied stress. At the same time, the stresses on the bent segments of the molecules are so small that they cannot be measured by the spectroscopic technique. Therefore, the assumption that the major part of the stress applied to the sample is experienced by taut tie molecules is obviously valid. Since there is a series coupling between the crystallites and the amorphous regions in a fibril, the stress acting on the taut tie molecules is inversely proportional to their number:

$$N/N_{ae} = \delta/\overline{\Sigma} \tag{9}$$

If the assumption is valid, the values of N/N_{ae} estimated from eqs. (8) and (9) must coincide. We have compared the values of N/N_{ae} derived from these expressions for different specimens of polypropylene, polyethylene terephthalate, and polycaproamide. The value of the long period and the length of the amorphous regions was changed by varying the temperature of the orientational drawing. The values obtained have been compared for stresses close to the breaking one. The results are listed in Table II. As can be seen, the agreement between the values of N/N_{ae} estimated from eqs. (8) and (9) is good. Therefore, the taut tie molecules bear the major part of the applied stress.

CONCLUSION

The results of IR spectroscopic, electron microscopic, and X-ray studies of oriented semicrystalline polymers lead to the conclusion that the overstressed segments of macromolecules lie in the amorphous regions of microfibrils and are oriented in the direction of the applied stress. In spite of the fact that the number of such regions amounts to only 5–10% of the total number of the chain segments in the specimens, they bear the major part of the applied stress.

REFERENCES

[1] P. P. Kobeco, "Amorphous Substances," *Izd. An SSSR,* M.-L., 1952, p. 217.
[2] A. I. Gubanov and A. D. Chevychelov, *Fiz. Tverd. Tela,* **4,** 928 (1962).
[3] K. E. Perepelkin, *Angew. Makromol. Chem.,* **22,** 181 (1962).
[4] H. H. Kausch and J. Becht, *Kolloid Z. Z. Polym.,* **250,** 1048 (1972).
[5] D. S. Boudreux, *J. Polym. Sci., Polym. Phys. Ed.,* **11,** 1285 (1973).
[6] S. N. Zhurkov, V. I. Vettegren, V. E. Korsukov, and I. I. Novak in *Fracture 1969 (Proc. Second Int. Conf. Fracture,* Brighton, England), P. L. Pratt, Ed., Chapmann and Hall, London, 1969, p. 545.
[7] V. I. Vettegren and I. I. Novak, *J. Polym. Sci., Polym. Phys. Ed.,* **11,** 2135 (1973).
[8] R. P. Wool and W. O. Statton, *J. Polym. Sci., Polym. Phys. Ed.,* **12,** 1575 (1974).
[9] V. A. Marikhin and L. P. Myasnikova, *Mekn. Polim.,* N2, 364 (1971).
[10] E. E. Tomashevskii and A. I. Slutsker, *Zavod. Lab.,* **89,** 934 (1963).
[11] K. Hess, *Chem. Ind.,* **80,** 129 (1958).
[12] D. Ya. Tsvankin, Doct. dissertation, 1970.
[13] K. Kess, H. Mahl, and E. Gutter, *Kolloid-Z.,* **155,** 1 (1957).
[14] G. Natta and P. Corradini, *Nuovo Cimento, Suppl.,* **15, 1,** 9 (1960).
[15] R. de P. Daubeney, C. W. Bunn, and C. J. Broun, *Proc. R. Soc. (London),* **226A,** 531 (1954).

[16] G. Natta and P. Corradini, *Nuovo Cimento,* **15, 1,** 40 (1960).

[17] Yu. V. Kissin, V. I. Tsvetkova, and N. M. Chirkov, *Eur. Polym. J.,* **8,** 529 (1972).

[18] P. G. Schmidt, *J. Polym. Sci.,* **A1,** 1271 (1963).

[19] J. Jakes, P. Schmidt, and B. S. Schneider, *Coll. Czechoslov. Chem. Commun.,* **30,** 996 (1965).

[20] S. N. Zhurkov, V. I. Vettegren, V. E. Korsukhov, and I. I. Novak, *Fiz. Tverd. Tela,* **11,** 290 (1969).

STRUCTURE AND PROPERTIES OF HIGHLY ORIENTED POLYMER FIBERS

S. FRENKEL

Institute of Macromolecular Compounds, Academy of Science of the USSR, 31 Bol'shoy Prospect, Leningrad, 199004 USSR

INTRODUCTION

During the past 10 years a number of methods have been proposed to obtain "superfibers" from organic polymers. The term "superfiber" implies tenacities of the order of 200 kg/mm^2 and elastic moduli not less than 10^4 kg/mm^2. Up to relatively recent times, for reasons which will become obvious later, superfibers were spun from rigid-chain polymers able to form a nematic phase in solution. It was assumed that the nematic domains play part of a specific kind of "stocks" for the structure of the future fiber, and already at low initial draw ratios the boundaries between the domains disappear, and in the whole volume of the system a uniaxial parallel packing of macromolecules is established which is fixed later by elimination of solvent and crystallization. Such a structure in which approximately 100% of chains passes through the cross-section of the fibers and the only form of defects is due to the joints between the ends of subsequent macromolecules, warrants the above mentioned high tenacities and moduli.

Contrary to that, in flexible-chain polymers, crystallization proceeds as a rule with chain folding, which leads immediately to a strength deficit. In 1962 Flory formulated his famous "veto" consisting in the impossibility (irrespective of other details of structural organization) for a chain fraction of more than one-half to pass from one arbitrarily chosen folded crystallite into another [1]. Conformably to fibers these considerations were modified by Hosemann and Bonart [2], and later defined more accurately by Peterlin [3].

It follows from all these studies which obtained the strongest experimental support in a series of investigations of S. N. Jourkov's school (see, for instance, [4]) that responsible for the tenacities and moduli of flexible chain polymer fibers are not the crystalline but the amorphous regions. As to a quantitative estimate of the strength deficit, one must keep in mind that when a fiber is loaded, only a fraction of chains in the amorphous regions (the "tie chains") are strained and immediately resist the external tension. Moreover according to Peterlin, the number of these "strained tie chains" is not higher than 5% as compared to the whole number of chains in a cross-section of the crystallite normal to the crystallographic "c"-axis. (It is assumed that the average orientation of c-axes of individual crystallites coincides with the macroscopic orientation.)

Relatively recently in a series of studies carried out in our laboratory and independently by Porter in the United States [5] and by Keller in Great Britain

Journal of Polymer Science: Polymer Symposium 58, 195–223 (1977)

(see his main lecture in this symposium) it was shown that an intense folding in the course of fiber formation is not a unavoidable harm, and therefore one can decrease essentially the strength deficit of flexible-chain fibers so that their tensile properties may approach those of rigid-chain superfibers.

Though the three above mentioned research groups came to similar conclusions in respect to the resulting structure of flexible-chain superfibers, their fundamentals in resolving the problem differed substantially. In our studies we used the general principles of the kinetic theory of liquids and solids [6]* and the thermokinetic concept following from these principles [7, 8, 9]. Some applications of this concept to the problem under discussion were presented by me [9] and V. G. Baranov [10] on the Twelfth Prague Microsymposium in 1973.

The following represents further developments of these studies during the last 3 years.

FLORY'S CRITERION AND THE TOPOMORPHISM CONCEPT

Since just the chain flexibility is responsible for folding during crystallization and therefore for the deficit of tenacity, one must start with a definition of an absolute criterion of flexibility of macromolecules. Such criterion was first proposed by Flory in his two classical papers in 1956 [11]. Flory introduced a flexibility parameter f given by

$$f = (Z - 2)\exp(-\epsilon/kT)/1 + (Z - 2)\exp(-\epsilon/kT) \tag{1}$$

where Z is the coordination number of the quasi-lattice (which plays part both in the old Flory–Huggins theory and in the new theory of Flory [12]), and ϵ is the difference in energies of a "flexible" and "rigid" skeletal bond. The paper [11] was written before the discovery of folded chain single crystals; assuming that crystallization proceeds with chain extension, Flory meant by "flexible" the gauche and by "rigid" the trans rotamers. Irrespective of this detail, the parameter f has a meaning of the molar fraction of "flexible" bonds in the chain. Flory's strict calculation has shown that at $f \leqslant 0.63$ ($0.63 = 1 - 1/e \cdots$) there must exist a volume fraction φ_2 of the polymer, being the larger the lower is the molecular weight and the higher is f at which a spontaneous formation of the nematic phase occurs. When f equals 0.63, the corresponding concentration φ_2 equals unity which may lead to kinetic hindrances that we shall consider later.

On the contrary, when $f > 0.63$ and in the absence of causes that can provoke a conformational transition (for example, of the coil–helix type), in no case can the nematic phase form itself spontaneously.

The general conclusions of Flory's theory were brilliantly verified approximately 15 years after its publication (cf. [13]). However keeping in mind that one of the premises on which the main calculations were founded turned out to

---

* Not long before his death in 1952 J. Frenkel (see [6a]) proposed to write a new version of his "Kinetic Theory of Liquids" which should be just entitled "Kinetic Theory of Liquids and Solids."

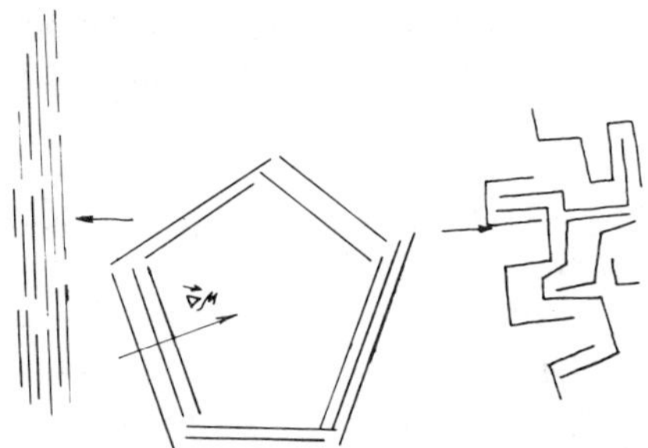

FIG. 1. Schematic representation of "osmotic traps."

be fallacious (crystallization as a rule does not involve the disappearance of gauche rotamers and chain extension) it seems advisable to define the energy ϵ in another way. Most comprehensibly its physical sense can be understood with help of the model of "osmotic traps" (Fig. 1).

Let us imagine that we are filling the solvent initially with rod-like macromolecules (in Fig. 1 the flat model of the situation is presented). Due to orientational disorder the ends of some molecules lean against the sides of other molecules, and as a result some elements of volume appear which are inaccessible for any chain segment. Only these volume elements are the "osmotic traps." On their boundaries arises a very strong gradient of the chemical potential μ_2; if as usual one calculates μ_2 on the base of one monomeric unit, the osmotic force acting per chain of the degree of polymerization P equals

$$\langle \partial\mu_2/\partial r \rangle P \tag{2}$$

This force tends to drive the macromolecule in the trap; if l is the linear dimension of the trap, then the energy corresponding to this force is

$$\Delta E = \langle \partial\mu_2/\partial r \rangle Pl \tag{3}$$

and we must compare this energy with the energy ϵ appearing in eq. (1), attaching to the latter energy any kind of a priori meaning.

The occurrence of the osmotic traps makes the system thermodynamically unstable; to attain equilibrium it must redistribute chain segments to make their concentration equal in the whole volume of the solution. This redistribution can be performed in two ways:

(1) The molecules undergo a series of consecutive kinks and thus turn into usual statistical coils evenly distributed in the solution. To this case corresponds the unequality $\Delta E > \epsilon$.

(2) $\epsilon \gg \Delta E$. Then the only possible way to attain equilibrium will be to "sacrifice" the entropy, to pack the molecules in parallel rows and thus to transform the average value of $\langle \partial\mu/\partial r \rangle^2$ (the averaging extending over the whole volume) into zero, and consequently decreasing the Gibbs free energy ($\Delta G < 0$) in spite of the decrease in entropy ($\Delta S < 0$).

This approach that does not need the introduction of a quasi-lattice defines the part ϵ of the internal energy of the macromolecule as the energy of one effective bend. It is obvious that the lower will be ϵ as compared to ΔE the higher will be the amount of kinks per macromolecule. Since however one bend (or one fold) in any case involves several bonds, the number ϵ for a chain with the rotational-isomeric mechanism of flexibility must exceed the energy difference for gauche and trans isomers.

In the case of semi-rigid (persistent, or "worm-like") chains, one must solve, in general, a similar problem. The only difference is connected with the initial state of chains which are already bent or rather curved; one must find now what will be energetically preferable: to unbend the chain or to coil it further. The energy ϵ which was localized in a short part of the chain in the former case now is "spread" along sufficiently long (of the order of the persistence length) chain portions.

The foregoing discussion was relative to the formation of the liquid crystalline phase. However, one can treat from the same positions the crystallization from the melt. In this case Z in eq. (1) acquires the reality of the coordination number of the future crystalline lattice, and the energies ϵ and ΔE are bound in the same manner as formerly. We shall point however our attention to the fact (which will be discussed in more detail later) that in case of $\epsilon \approx \Delta E$ the coiling or uncoiling may become equally probable. Correspondingly, in these hypothetical conditions the crystallization from the melt may produce either folded-chain or extended-chain crystals (FCC or ECC). The lattice parameters of these two kinds of crystals will be identical, but from the free energy viewpoint these two crystalline forms are not equivalent.

Without further detailization we can indicate two obvious sources of this nonequivalence. The presence of folds (i.e., of gauche rotamers—a factor connected with the internal energy of the macromolecules) in the case of FCC and as a consequence a surplus contribution of the energy of the **ab** crystalline faces. In other words, on the $G-T$ phase diagram, with all other conditions being kept equal, the ECC will be represented by a point with a lower free energy than the corresponding point for FCC. The situation becomes even more obvious if we consider the free energy–pressure $(G-P)$ phase diagram. But what is right for a point describing a state must also keep for the phase lines; therefore the habitual phase lines representing the crystalline and amorphous states on the $G-T$ or $G-P$ diagrams must be blurred and transformed into "phase corridors" (the necessity of such a blurring for the amorphous phase follows from the fundamental condition of the continuity of the free energy as a function of T or P) these corridors being two-dimensional sections of multidimensional phase corridors in the coordinates free energy–intensive parameters. The number of the latter may be high; in addition, due to the fact that each macromolecule represents by itself a small system (in T. Hill's sense [14]), in special cases when the solvent or the diluent actively and selectively interacts with the chains [9] the concentration also may turn into an intensive parameter.

Thus we postulate the existence of a peculiar energetic polymorphism of polymeric crystals which will be noted in the following as crystalline topomorphism. It is a priori obvious that the necessary conditions for the realization of topomorphism are finite values of the parameter f. At $f = 0$ any possibility of folding disappears, and *only in this case* the phase corridor degenerates into a phase line.

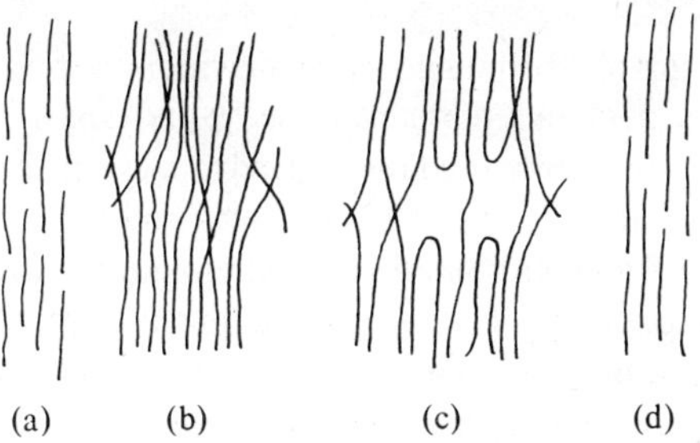

FIG. 2. Models of fiber structures: (a) rigid-chain polymers; point defects on end joints; (b) fringed-fibrillar Hearl's model; (c) HBP model; (d) an idealized ECC model.

The consequence of these (up till now purely formal) considerations consists in an *a priori* probability of four types of structures of oriented fibers as shown in Figure 2.

At very low f the formation of ECC is energetically inevitable; at moderate but subcritical f most probable is the formation of fibrillar structures of Hearl's type [15] (it is equally difficult to fold or uncoil the persistent chains), and with still increasing f the crystalline topomorphism may develop leading either to structures of Hosemann–Bonart–Peterlin (HBP) type or to the ECC structures morphologically quite identical to those for rigid-chain fibers. If we remember now what was the reason for the strength deficit in flexible-chain polymer fibers, it will become quite clear that the problem of regulating the tensile properties is indissolubly connected with the possibility of regulating the growth of one definite topomer during spinning.

The scheme in Figure 2, in addition to its illustrativeness, is also convenient for it allows, still using Flory's criterion, the introduction of such a semitechnological notion as "orientational pliance." Not going into details we shall denote it provisionally as the probability of attaining the structures a or d of Figure 2 at a given f. All recent experience of polymer physics and technology shows that this probability is characterized by a profound minimum at close to critical f values approximately corresponding to cellulose esters (Fig. 3).

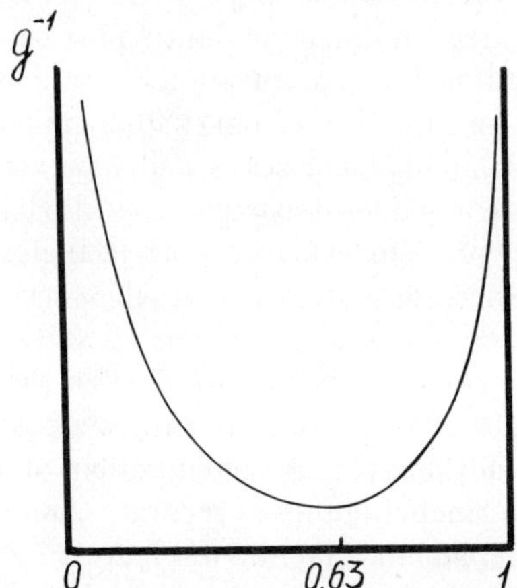

FIG. 3. The orientational pliance diagram. For the designation g^{-1} see Table I.

The sense of this graph at first sight is self-evident. At $f \to 0$ the macromolecules simply cannot pack in a nonparallel way; one needs but an insignificant external influence to overcome the kinetic hindrances. In the case of absolutely flexible chains when ϵ in eq. (1) approaches zero and $f \approx Z - 2/Z - 1$ is close to unity, due to the high kinetic flexibility (which is characterized by the same energy ϵ) there is no difficulty (as it seems) to uncoil wholly the chains with the subsequent formation of ECC. The minimum is tentatively explained by the difficulty of an "independent choice" ($\epsilon \approx \Delta E$) and by the sufficiently high kinetic rigidity (large ϵ) hindering any external intervention. However all this logic applies only for the "first sight."

Indeed, at $f < 0.63$ the chains, following Flory's theory, must at some critical concentration φ_2 form a nematic phase. However it is time now to turn from abstractions to reality and to take into consideration some technological inconveniences, both of thermodynamic and kinetic nature. The thermodynamic inconveniences are most spectacular in the case of absolutely rigid macromolecules of the type, for instance, of the poly(para-benzamide) or its analogue poly(para-phenyleneterephtalamide) (the superfibers "Fiber B" or "Kevlar" in the United States). It should be reminded that the melting and solution temperatures (the dissolution being considered as melting in the presence of the solvent) are given by similar formulae [6, 16]:

$$T^* = \frac{\Delta H_1 + \Delta H_2}{\Delta S_1 + \Delta S_2} \tag{4}$$

ΔH and ΔS being the corresponding enthalpy and entropy changes and indexes "1" and "2" designating conformational and configurational contributions. The value ΔH_1 in a simple and obvious manner is connected with ϵ, being the larger the higher is ϵ. As to ΔS_1 this value at $f = 0$ also equals zero, for the rigid macromolecule does not change its conformation. Thus the creation of disorder in the melting system is possible only in such conditions when it possesses a sufficient volume where the rodlike macromolecules could be disposed at random. This leads to a well known result: absolutely rigid-chain polymers in general do not melt (i.e., their theoretical melting temperature is higher than their decomposition temperature), and their solubility must be very low. This corresponds to the facts, and the low concentration is a serious drawback of all the spinning technologies for superfibers from rigid-chain polymers.

As f increases, the solubility increases as well; however the concentration at which the nematic phase would form spontaneously also increases. In some cases this concentration may prove to be inaccessible, in analogy with the inaccessibility of the melting temperature. In other cases this concentration is accessible, but then a new hindering factor appears: The viscosity which increases proportionally to $\varphi_2{}^a$, the exponent a being 5 for flexible-chain polymers and substantially higher at subcritical fs. But this means a possibility of the kinetic "freezing" of energetically unfavorable configurations of the system as a whole; together with the high kinetic rigidity of separate chains this leads to the occurrence of the minimum on the diagram in Figure 3.

The situation in the region of the minimum of orientational pliance is shown

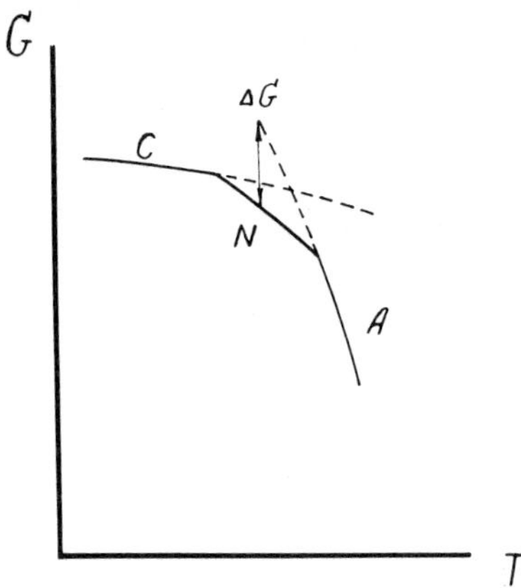

FIG. 4. The G–T diagram for a polymer with a subcritical f value. C, crystalline phase; N, nematic phase; A, amorphous phase.

on the G–T diagram, Figure 4. According to DiMarzio [17] this would just be the shape of the phase diagram for a rigid chain or semirigid polymer able to form spontaneously a nematic or some other liquid crystalline phase. The closer f is to its critical value, the narrower is the range of existence of this phase; as a consequence it becomes more and more probable that the system due to the mentioned kinetic hindrances will "jump over" the point corresponding to the equilibrium phase transition and as a result will crystallize in an energetically unfavorable (as would seem) state shown by the intersection of lines "A" and "C." It is not necessary however to interpret the arising situation as a formation of an overheated crystal. Much more simply, in a system of moderately flexible macromolecules the different topomers can already appear. Of course the system may also undergo the glass transition without any considerable ordering.

The condition for the formation of a nematic phase in the situation under consideration, i.e., at a temperature lower than that corresponding to the intersection of lines "A" and "N," is settled by the comparison of the energy fall ΔG (see Fig. 4) which is nothing else than the product $\Delta E m_2$ (m_2 being the molar fraction of macromolecules) with the activation energy of viscous flow. The latter must be less than ΔG (it should be mentioned that the same logic is adopted when one analyzes the kinetics of "ordinary" crystallization). If this condition is fulfilled, then the further decrease in temperature must lead to the transformation of the nematic phase into ECC.

Kinetic hindrances of quite a different nature arise on the right side of the orientational pliance diagram. In this case the crystallization must proceed at such conditions where the rate of uncoiling of the chains is *higher* than the rate of monomolecular nucleation [18] (which causes the subsequent growth of FCC). Since at high fs the latter rate is very high, the situation seems to be kinetically and thermodynamically helpless, at least in a one-step technology (for the case of recrystallization, see [4]). However in reality just thermodynamic, or more precisely thermokinetic [9], considerations allow to get out relatively easily from the apparent technological impasse avoiding complicated many-step processing. As will be shown, it is not at all necessary to unfold the chains on their whole length; however one must consider for this reason in addition to f the degree of coiling $\beta = h/L$, h being the end-to-end distance, and L the contour length of a chain. Just as f, β also has its critical value after attaining which (during uncoiling) all the events take the same course as for a system with a subcritical f and the same "static" degree of coiling. There is however an essential difference: In our case of partly uncoiled flexible chains the viscosity is substantially lower than in the "equivalent" system of semirigid macromolecules.

Thus we have formulated two quite definite kinetic and thermodynamic problems appearing when the orientational pliance diagram is analyzed more carefully. We shall turn now to the solution of these two problems.

THE OVERCOMING OF THE KINETIC AND THERMODYNAMIC HINDRANCES IN THE REGION OF THE MINIMUM OF ORIENTATIONAL PLIANCE

Self-Organization of the Cellulose Diacetate

As the only scarce confirmation of his theory [11] Flory quoted (it should be remembered that his two papers were written in 1954) a paper published near the beginning of the 1950's [19] where a striking effect of spontaneous elongation of cellulose diacetate was described. The authors [19] used the following *modus operandi*. An amorphous (this is essential!) film was pre-elongated up to 30% and then immersed in a special bath containing 96% (weight) of water, 2% of phenol, and 2% of sodium sulfate, and heated up to 98°C. There followed a rapid selfelongation in the direction of the initial draw, the final elongation ratio attaining 200–300%. Flory had to base his conclusions on the macroscopic effect only, for the authors [19] could not estimate the transformation on the molecular level.

We repeated these experiments using the same *modus operandi*. However we were able to measure the changes in molecular orientation by means of polarized luminescence using a polarizing fluorophotometer Jasco FOM-1 (Japan).* The results are shown in Figures 5 and 6. In the main features they are identical for films and fibers (though the effect on fibers is less pronounced, probably due to the fact that one cannot obtain them in a purely amorphous form; a minute fraction of folded chain topomers—see Fig. 4—may hinder the development of ordering in the whole volume of the sample).

We see that the spontaneous elongation approximately up to 200% is due to an extremely high degree of molecular orientation (with $\langle \cos^2\theta \rangle \approx 0.8$). Therefore this process in reality means the mentioned "fall" along the arrow ΔG in Figure 4 on the "equilibrium line" corresponding to the nematic state. However the transformation is not reduced to the decrease of the activation energy only owing to the bath. The pre-elongation is essentially necessary; without this pre-elongation nothing happens to the film immersed in the same

---

* Cf. Prof. Ward's chapter in this Symposium.

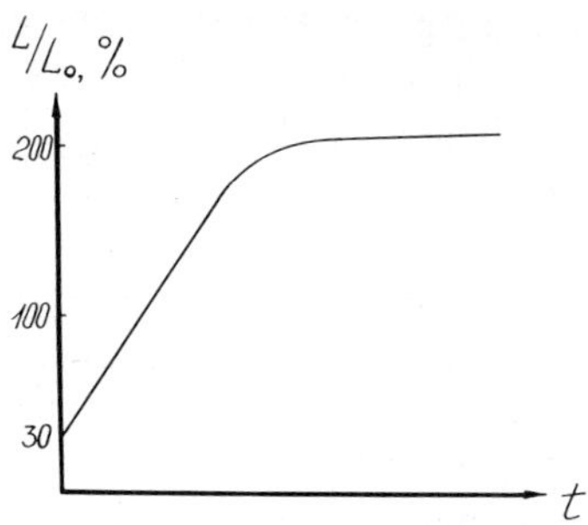

FIG. 5. The self-elongation of cellulose diacetate. At temperatures approximating 98°C the plateau region is attained in 1 or 2 minutes.

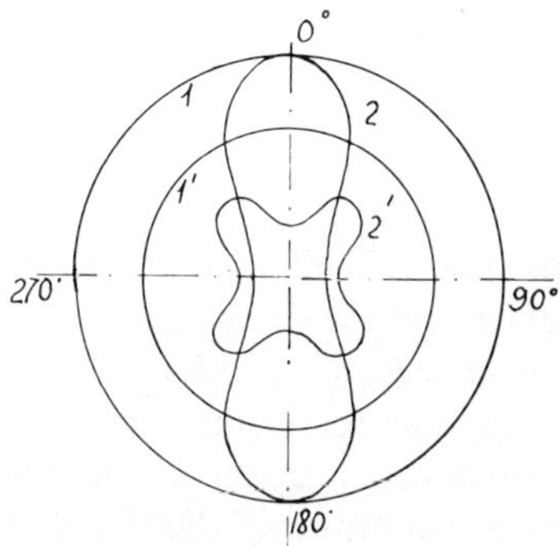

FIG. 6. Angular distribution of polarized luminescence with parallel and crossed (marked by a prime) polaroids before (1) and after (2) the self-elongation of cellulose diacetate.

bath, neither on the macroscopic nor on the microscopic level. Moreover, there does not appear any kind of "polynematic" (which would be a liquid crystalline analogue of "polycrystalline") phase which could be observed if present using a polarizing microscope.

Thus a purely kinetic interpretation of the phenomenon described does not hold. Similarly, a purely thermodynamic interpretation founded on a possible change of the flexibility parameter f due to some selective polymer–solvent interaction (this is just the case when concentration can become an intensive parameter) fails as well. As was shown in [9] due to a change of Flory's interaction parameter χ the flexibility parameter initially being equal to f_0 [given by eq. (1)] can decrease to a value

$$f = \frac{f_0 \exp(-\Delta\chi\varphi_1/kT)}{1 - f_0 + f_0 \exp(-\Delta\chi\varphi_1/kT)} \tag{5}$$

This formula means, evidently, the change of the energy of an effective bend by the value $\Delta\chi\varphi_1$ followed either by the widening of the nematic region on a diagram of the Figure 4 type or more generally by its appearance if in the initial state it was absent [9]. Yet this still does not explain the necessity of the preorienting of the film or of the fiber. One has an impression of some kind of inversion in the time as compared to the situation corresponding to the left extreme of the orientational pliance diagram. If in this latter case one can consider the relatively diluted nematic solution as a number of "stocks" of the future fiber structure, those "stocks" being materialized as the nematic domains, and a low draw rate (and ratio as well) orients these domains in one direction transferring the system into a state of a "nematic single crystal," in the case under consideration, with the diacetate, all proceeds in the reverse succession. It is possible of course starting from the principle of equivalence of internal and external stresses [20] to assume that, due again to the high viscosity, the internal stresses created during the pre-elongation has no time to relax and play the same part as the external draw in the spinning process. However this interpretation also fails, for the pre-elongation is necessary for the initiation of the formation of the nematic phase *per se* which develops itself further spontaneously. One should add that the polarized luminescence does not reveal any observable traces of the molecular orientation due to the pre-elongation by itself.

Therefore there remains but a single interpretation.

(1) The pre-elongation by 30% allows some critical value of the degree of coiling β^* to attain at which the formation of folded topomers becomes energetically unfavorable.

(2) Though on this stage it is impossible to observe the molecular orientation on relatively short parts of the chains a certain orientation of the *vectors* **h** (the coils' end-to-end distances) along the future texture axis arises.

Let us turn now to the right extreme of the orientational pliancy diagram in an attempt to show the reality of the existence of a critical degree of coiling and its correspondence to the critical value of the effective flexibility parameter.

TOPOMORPHISM AND THE ORIENTATIONAL CRYSTALLIZATION OF FLEXIBLE-CHAIN POLYMERS

The occurrence of topomorphism *per se* does not need any special evidence, since the growth of ECC under high pressures is a well and for a long time established fact. After the well known studies of Pennings [21] and the already mentioned works of Porter, it became obvious that a FCC fraction may appear in a peculiar equilibrium with ECC (the "shish-kebab" structures) as result of crystallization from solution or melt in an external mechanic field. In my paper [9], in order to interpret these as well as our own results, I used the thermokinetic equivalence principle which claims that the thermodynamic behavior of a system of stretched flexible-chain macromolecules (in a melt or in a solution) is equivalent in its essentials to the behavior of a nonperturbed system of rigid-chain macromolecules of the same structure.

The most simple presentation of this principle is given by an analogue of eq. (5) for the effective flexibility parameter where the thermodynamic addition to the internal energy $\Delta\chi\varphi_1$ is replaced by a mechanic one $\Delta\epsilon$ being a part of the external field which is "pumped over" in every chain:

$$f = f_0\exp(-\Delta\epsilon/kT)/[1 - f_0 + f_0\exp(-\Delta\epsilon/kT)] \tag{6}$$

In terms of the G–T diagram it means a possibility in principle to "throw over" the system from a state I where only FCC may form in a state II where a nematic phase arises followed by the formation of ECC (Fig. 7; compare it with Fig. 4).

Since in static conditions the free energy of ECC is somewhat lower than the energy of FCC after the accomplishment of the crystallization, the phase line ECC-II must "sink" on a level somewhat lower than the phase line FCC-I, the corresponding energy difference being disengaged as the heat of crystallization. The "transfer" shown in Figure 7 can be treated in the usual terms of the increase of temperatures of melting and crystallization owing to the anisotropy of the thermodynamic properties of oriented polymers. For the analysis of the stream–thread transition it is convenient in addition to the G–T diagram to follow the "deformation" of the T–φ_2 phase diagram in the presence of a longitudinal velocity gradient (Fig. 8). As was shown in [9] in connection with the so-called orientational catastrophe (cf., [9], Fig. 13) the velocity gradient alters the phase diagram in such a manner that some state point A corresponding at rest to a homogeneous solution abruptly "sinks" in the newly-formed heterogeneous region.

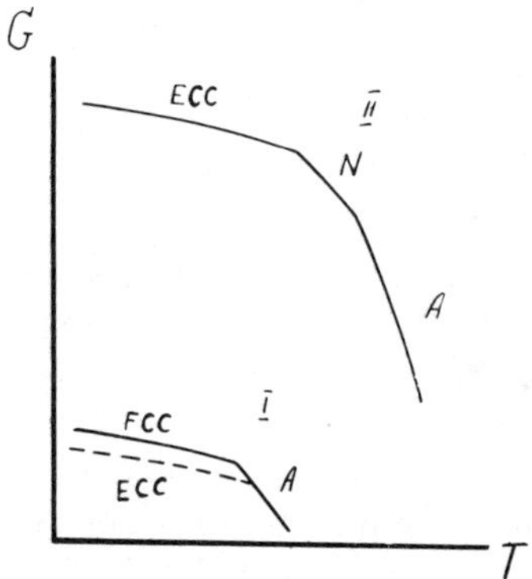

FIG. 7. The transformation of the phase diagram due to an external mechanical field. For details see text.

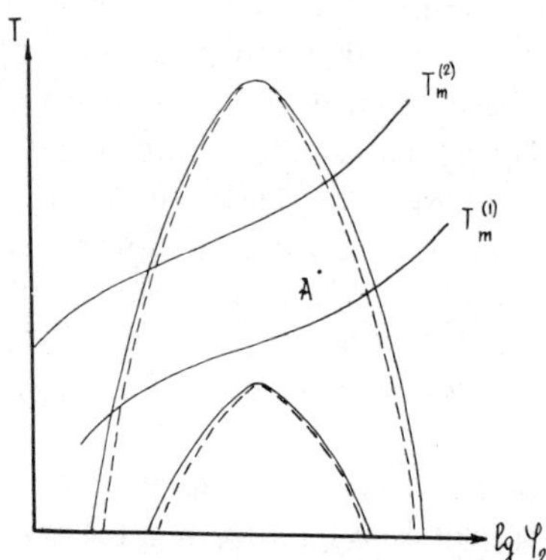

FIG. 8. The transformation of the coexistence diagram due to an external mechanical field. The lower set of curves, the system at rest. The upper set of curves, the same system in dynamic conditions. For other details see text.

From the position of thermokinetic relativity, there is no difference whether the system in reality "sunk" under the binodal and spinodal lines due to an abrupt decrease of temperature, or found itself under these lines due to an abrupt change of its own shape. In both cases a spinodal phase separation begins, however in the second case the compositional fluctuations propagate in a definite direction, namely opposite to the direction of flow. To bring Figure 8 in full accord with Figure 7 one must assume in addition that before the imposition of the external field, the curve corresponding to the concentration dependence of the melting temperature was situated *below* the point A, and after the imposition of the field it also shifted *above* the point A. Thus simultaneously two phase transitions occur, characterized by different kinetic and thermodynamic parameters. Of importance is however the fact (following immediately from the equivalence principle) that both those transitions are anisotropic.

Just here an additional influence of the external field shows up being neglected in the straightforward formulation of the equivalence principle by eq. (6).

First, to characterize the elongation one must take into account the Poisson coefficient, playing the role of a complementary packing parameter and, in the case of the longitudinal flow of a solution (in connection with normal stresses), promoting the ejection of the solvent from the stream. Second, and principal, the external field provides the macroscopic orientation, thus essentially promoting the liquidation of the domains on the "nematic step" *before* these domains had a chance to appear.

From this point of view, the coiling parameter β is more convenient for detailed calculations than the flexibility parameter f because the h value present in the β ratio has a vectorial nature, and it is just the orientation of the vector $\vec{h}$ relative to the draw direction that plays the decisive role in the "choice" of the mode of nucleation [10, 22–24]. As to the parameter f defined by eq. (6), its importance consists in two features: first, it provides the stability of that critical value of β when the $\vec{h}$ vector begins to orient itself along the field (at low β for obvious reasons this is impossible), and second (and in this instance the equivalence principle completely retains its value), it provides the thermodynamic necessity of the formation of the nematic phase.

However after the occurrence of the initial orientation a positive feed-back establishes between β and f, with β tending to unity and f to zero.

Let us turn now to the lower group of curves in Figure 7. The energy difference between FCC and ECC compel us, as was already mentioned, to substitute the phase line for the crystalline state by a phase corridor, and the request of the continuity of G in relation to the intensive parameters compel us to proceed in the same way with the amorphous phase. The corresponding diagrams are shown in Figure 9; one should point that the pressure on the $G-P$ diagram can be substituted by the stretching stress which will better reflect the conventional way of fiber spinning.

The second feature of these diagrams consists in the transformation of the transition *point* into a transition *region* having a shape of a uneven tetrangle and degenerating into a point only at $f = 0$. It could be shown further that the real boundaries of the corridor are somewhat elevated over the boundaries for

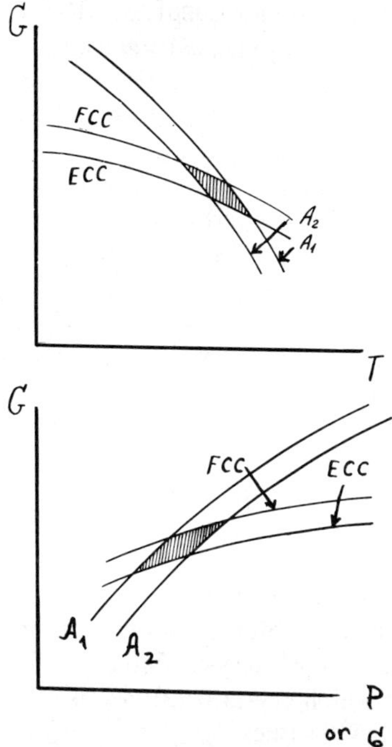

FIG. 9. Phase diagrams representing crystalline topomorphism. For details see text.

pure ECC and FCC because in the flexible-chain polymers at no condition can a whole crystallinity be attained. A direct calculation shows that this is due to the "exhaustion" of the conformational entropy [23] and correspondingly to the straining of uncrystallized parts of the chains. (The role of defects related to end groups is negligible.) Just those strains are responsible for the excess energy in semicrystalline polymers with folded crystallites as compared to systems where ECC are present.

However, in any case, just the situation in the transition region controlled by the external field, concentration and parameters β and/or f define the resulting structure and consequently the mechanical properties of oriented fibers. Of course the higher is f, the more this tangible structure will differ from the idealized schemes presented in Figure 2.

The problem of the manifestations of topomorphism in the transition region is complicated and cannot be resolved in usual terms of the kinetic or thermodynamic theory of phase transitions. The experimental analysis of the situation in the transition region (however, to be precise, just the results of such an analysis led us to the formulation of topomorphism and to the presentation of phase diagrams in the form of Fig. 9) was carried out by two methods.

(1) In isothermal, quasi-static conditions. In these cases a slightly crosslinked polymer—polyethylene or polychloroprene—was melted, i.e., transferred into the elastic state, then the melt was stretched to some amount measured by the macroscopic draw ratio λ, the temperature was lowered, and the crystalli-

208   FRENKEL

zation started. Obviously λ is proportional to β, the latter parameter characterizing now the uncoiling of chain parts between the crosslinks.

(2) In nonisothermal, dynamic conditions, using laboratory devices for melt or solution spinning. In this case the stream undergoing the transformation formally can be divided into several zones differing by temperature (or solvent content), temperature (or composition) gradients, and velocity gradient. It is always possible to distinguish a rubberlike zone which can be characterized by an average temperature, effective λ (and consequently β or f), and an average velocity gradient $\dot{\gamma}$. The latter must be sufficiently high to make the mean life time τ of the nodes of the quasi-network of the melt substantially larger than $1/\dot{\gamma}$. The condition $\tau \gg 1/\dot{\gamma}$ is simply the condition of the stability of high elasticity in the zone under consideration. As to the temperature at which crystallization starts, in accord with comments concerning Figures 7 and 8, the stretch *per se* increases it as compared to static conditions. Thus there is no substantial physical difference between the two types of experiments; however in the second case all events proceed at a much higher rate (regulated by the given $\dot{\gamma}$ and temperature in the crystallization zone).

Figure 10 shows the results of experiments of the first type for polyethylene and polychloroprene. One can observe not only a pronounced nonmonotonity of the dependence of the degree of crystallinity (or of a related value) on β (or a related parameter) at which crystallization was carried out, but also a discontinuity at relatively low β values.

Figure 11 shows—irrespective of the type of experiments—the degrees of crystallinity and free energy changes at nucleation (calculated in a monomolecular approximation [24]) and also the melting temperatures of polyethylene crystallized at different β.

The calculations were made either taking into account or neglecting the superficial energy; necessary numerical data were obtained from [16]. Again, one can observe abrupt discontinuities in the region $0.2 \leqslant \beta \leqslant 0.3$.

It was of interest to estimate the correlation for the effective values of f and β. Assuming for f_0 an initial value 0.7 and transforming eqs. (1) and (6) to the form

$$f = \frac{(Z-2)we^{-a}}{1+(Z-2)we^{-a}} \tag{8}$$

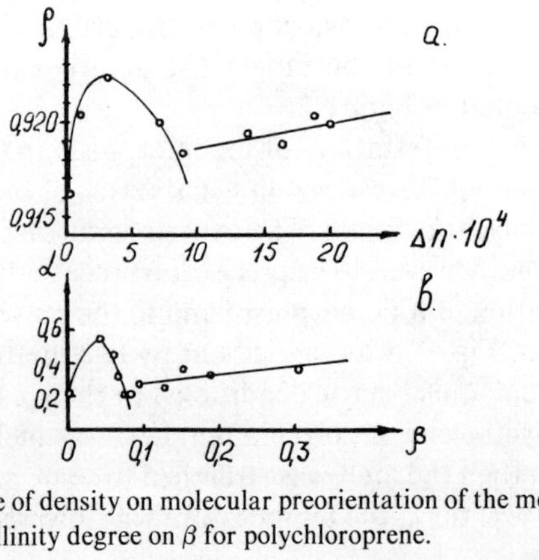

FIG. 10. (a) Dependence of density on molecular preorientation of the melt for polyethylene; (b) dependence of the crystallinity degree on β for polychloroprene.

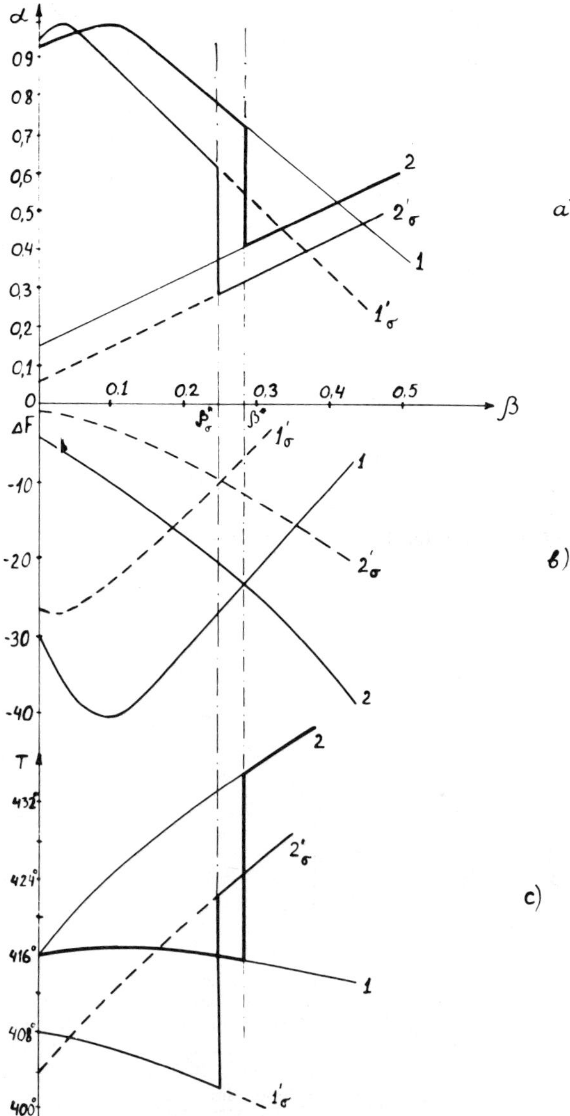

FIG. 11. (a) Dependence of the degree of crystallinity on extension for FCC (curve 1) and ECC (curve 2) without and with taking into account the superficial energy (curves $1'_\sigma$ and $2'_\sigma$ respectively); (b) free energy changes during crystallization as functions of extension for both types of crystallization (same symbols as in Fig. 11a); (c) dependence of the melting temperature on extension for crystals of both types (same symbols as in Fig. 11a).

where $w = \exp(\epsilon/kT)$ characterizes the initial flexibility and $a = \Delta\epsilon/kT$ the additional energy contribution due to stretch we can write for β

$$\beta = \frac{sha}{\sqrt{sh^2a + w^2}} \qquad (9)$$

The correlation for β and f for a given $Z = 6$ is presented in Figure 12. The

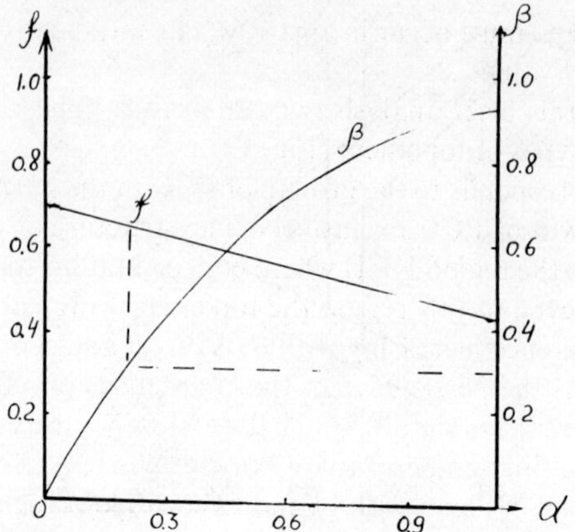

FIG. 12. Dependence of the parameter β and of the effective flexibility parameter f on extension; the initial flexibility parameter $f_0 = 0.7$; $Z = 6$. The dashed lines show that the critical value $\beta \approx 0.3$ (see Fig. 11) corresponds to $f \approx 0.66$ which practically coincides with the critical value $f_0 = 0.63$ necessary, according to Flory, for the formation of a nematic phase.

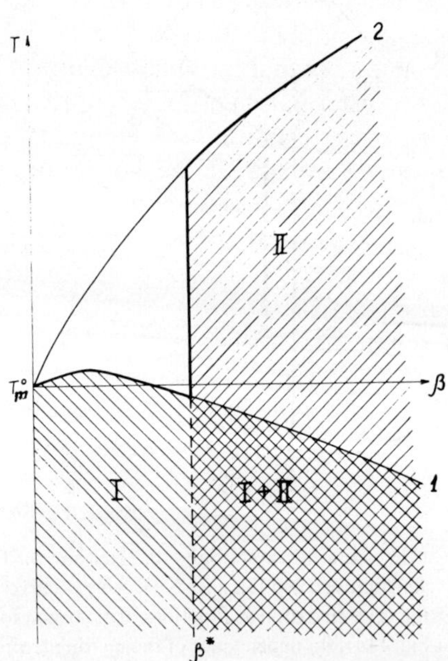

FIG. 13. A phase diagram of state illustrating regions of growth and existence of FCC (I) and ECC (II).

main point of interest consists in the fact that the critical β corresponds closely to Flory's critical value of f, and that practically a complete uncoiling proceeds at f values sufficiently distant from zero. Approximately the same result would be obtained for the maximum (for $Z = 6$) value of $f = 0.8$. Thus the transition from the formation of FCC to the formation of a nematic phase and the subsequent growth of ECC must occur in a narrow and sufficiently distant from zero range of effective f values.

To summarize this brief analysis, we can draw a "phase diagram" for the formation of two types of topomers (Fig. 13).

The region I corresponds to the growth of FCC in the pure form, and the region II to the growth of ECC exclusively. The structure of real fibers is most probably related to the region I + II where both crystalline topomeric forms can coexist. However even in this region the topomers will melt at different temperatures, the difference increasing with β. A direct and conclusive verification of this prediction is the persistence of the characteristic reflections for the c-texture superimposed on the diffuse halo at temperatures higher than the "conventional" melting temperature of polypropylene (see below).

The discontinuities in Figures 10, 11 and 13 correspond to the transition region of the $G-T$ diagram in Figure 9(a). In full accord with this diagram, the higher β was at which the crystallization was carried out, the higher also is the melting temperature (cf. the line 2 on Fig. 13). The existence of topomers in a quite natural way explains the well known difficulties and misunderstandings connected with the search of an "equilibrium" melting temperature. In fact, this search simply means a kind of wandering within the "transition tetrangle." As to the true equilibrium melting temperature, it should be the melting temperature of an infinite (corresponding to Onsager's criterion; in fact a macroscopic crystal would suffice) single crystal of the ECC type.

I would like to note once more that the "phase diagram" of Figure 13 represents not a diagram of transformation of FCC into ECC (the transformation is possible but on the necessary condition of a preliminary fusion of one of the topomers), but a diagram of the change of the mode of crystallization due to the change of the initial state of the melt characterized by the parameter β. Thus, it is a case of true singularity of the thermodynamic potential which is characteristic for a second order transition [25]. Apparently the second order transition becomes more obvious if one remembers that according to the theory of Gibbs and DiMarzio [26] such a low-temperature transition connected with the freezing out of all the gauche rotamers is, in principle, possible (it follows also from eq. (1) when ϵ is treated as the energy difference of gauche and trans rotamers). However, in ordinary conditions this transition is kinetically unaccessible, for it should occur at 50° below the glass transition temperature. Yet $\beta = 1$ just corresponds to a chain containing no more gauche rotamers. Consequently the second order transition can be relatively easily accomplished in a round-about manner [at high pressure or extension—see Fig. 9(b)] and necessarily would forego the first order transition, i.e., crystallization.

These considerations explain the existence of the "amorphous" phase corridor and allow the treatment of the ECC as crystals formed in a subsequence (or coincidence) of a second and first order transition, whereas the FCC result from a first order transition which is preceded or accompanied only by a redistribution of gauche rotamers along the chain (they concentrate on the folds).*

---

* Note added in proof. Later the critical phenomena at $\beta = \beta*$ were interpreted in terms of a "behavioral transition" according to Prigogine. For a brief review see S. Frenkel, *Man-Made Fibers*, (*Russian*), **3,** 11 (1977).

Now we must try to estimate the correspondence of the true structures of flexible-chain superfibers to the idealized structures presented in Figure 2. Of course a statement that these true structures are semicrystalline polymers "stuffed" by a mixture of FCC and ECC, or that they represent some intermediate between ECC and FCC will contain no information at all, though nothing in such a statement contradicts logic. Therefore we shall turn to relatively limited (presently) direct experimental data connected not with the genesis of ECC or FCC, but with immediately observable structures and properties of superfibers, and at first their thermomechanical properties. We shall begin with flexible-chain polymers where this problem is complicated by the pronounced topomorphism.

STRUCTURE AND THERMOMECHANICAL PROPERTIES OF HIGHLY ORIENTED FIBERS MADE FROM FLEXIBLE CHAIN POLYMERS

Using the general principles stated in the preceding section, it is possible to obtain without specific difficulties in a one-step process polyethylene, polypropylene, and nylon 6 fibers with tenacities up to 200 kg/mm^2. The limit of 100 kg/mm^2 also can be relatively easily surpassed in a one-step "wet" spinning process for poly(meta-phenyleneisophtalamide) ("Phenylon" in USSR; "Nomex" in the United States; the DuPont standard for "Nomex" is approximately 50 kg/mm^2). The elastic modulus for polyethylene can be approached in such a process approximately to 6000–7000 kg/mm^2, the data being yet distant from the theoretical value of the order of 25,000 but already close to the limit of 10^4. The problem of the theoretical strength will be discussed later; here we shall present some "secondary" data.

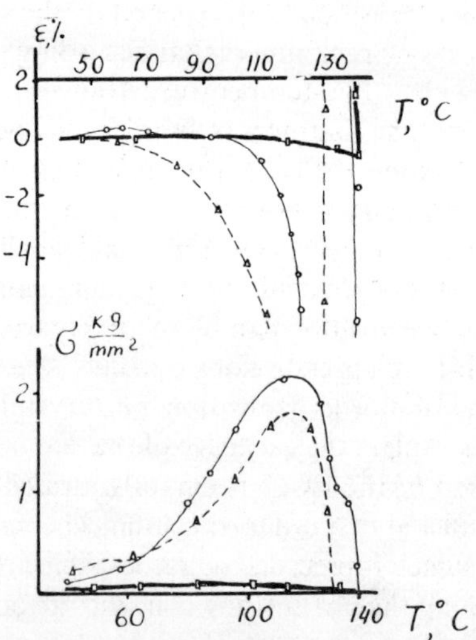

FIG. 14. Thermomechanical curves (upper graph) and curves of isometric heating. Dashed lines, an ordinary (reference) fiber; continuous thin lines, a fiber obtained by the method of orientational crystallization; thick lines, fibrils torn out of a broken fiber. High density polyethylene.

If a specially spun polyethylene fiber with a tenacity of approximately 150 kg/mm^2 is torn, macroscopic fibrils having a length of approximately 15 cm are developed and their tenacity approaches already 400 kg/mm^2. When they are torn in their turn one can observe tertiary structures of a whisker type, i.e., probably pure ECC, but due to their fragility it is at present impossible to investigate them.

The occurrence of fibrils comparable in tenacities with linear single crystals obtained by solid state polymerization of crystalline trioxane [4] allows us to expect some unusual thermomechanic properties of the starting fibers. The experiments were carried out on polyethylene and polypropylene super-fibers; the thermomechanical elongation curves for a constant load are shown in Figure 14. Figure 15 represents the curves of isometric heating and DTA together with the X-ray patterns corresponding to different parts of the curves.

The characteristic feature of the thermomechanical curves for the fibrils is the abruptness of deformation near the melting temperature. Such a behavior would be expected for rigid-chain fibers (if they would melt); since, however, we deal with typical flexible-chain polymers, one must conclude that such a behavior of fibrils correlates most closely with the model "d" of Figure 2.

This conclusion remains same for isometric heating curves for fibrils; however it contradicts to some extent the corresponding data for the initial fibers. Their behavior (a gradual deformation or increase of internal stresses) allows us to conclude that structures of the HBP type [Fig. 2(c)] are also present. On the other hand, one can distinguish a "tail" for polyethylene penetrating in the temperature range above the "conventional" melting temperature, and for polypropylene one can observe a well developed "shoulder." In the regions of the "tail" and "shoulder" the DTA curves reveal distinct endothermal peaks; in the case of polypropylene this peak is shifted to a temperature 15–20° higher than the "conventional" melting temperature. The positions of the peaks are close to the regions where the fibrils abruptly lose their temperature resistance. Such a behavior is in full accord with the "phase diagram" of Figure 13, i.e., it leads to a definite conclusion that the fibers contain ECC. The same conclusion follows from the X-ray patterns, especially for polypropylene. While the long periods disappear, the wide-angle pattern partly persists; in addition, on the low-angle pattern a pronounced equatorial scattering appears which can be ascribed to fibrils or to the above mentioned tertiary whiskerlike structures.

Following Keller's criterion (see his chapter in this Symposium) the occurrence of a "shoulder" or of an endothermal peak are already by themselves unequivocal indications of the presence of ECC in the oriented system.

At last, electron micrographs (Fig. 16) obtained with two different microscopes and by different modes of preparation of samples, reveal in the structure of polyethylene superfibers practically continuous, somewhat tortuous fibrils arranged in the direction of macroscopic orientation and connected laterally by ties, so that the fibrillar system in the whole forms within the fiber a peculiar three-dimensional frame.

Probably these electron-microscopic fibrils represent the above mentioned tertiary structures unaccessible for any mechanical testing.

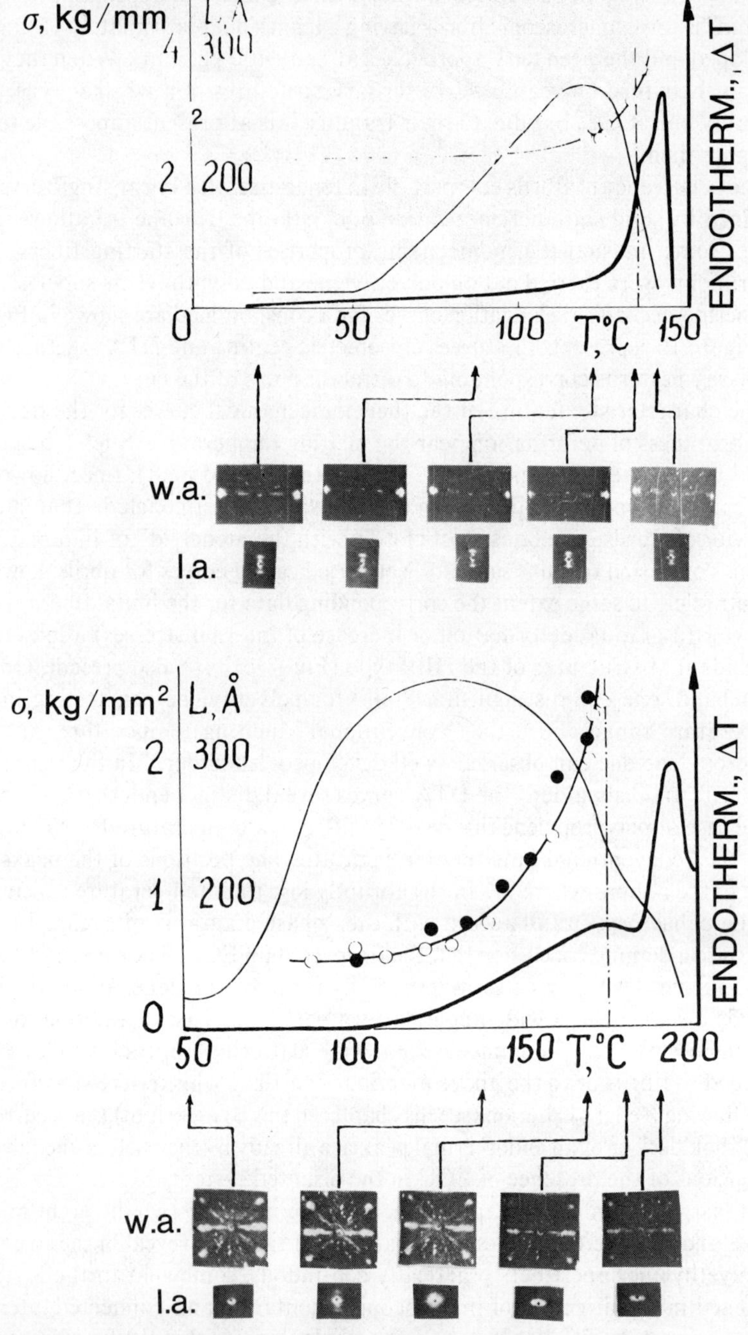

FIG. 15. DTA curves (thick lines) and isometric heating curves (thin lines) for superfibers. The experimental points correspond to X-ray large periods. The wide- and low-angle X-ray patterns correspond to temperatures shown by arrows (the draw direction is vertical) (a) polyethylene; (b) polypropylene.

FIG. 16. Electron micrographs of replicas from split surfaces before (a) and after (b–c) etching of high-density polyethylene samples crystallized in conditions of molecular orientation. The orientation direction and scale are shown on the micrographs.

One more series of rough but in a way illustrative experiments was made on polyethylene superfibers or fibrils torn out of them. After etching with nitric acid, a fragile ephemer remainder was obtained amounting to 10–15% in weight as compared to the initial sample, and mouldering into fine powder at any attempts to manipulate with them.

All data presented together with relatively low (in comparison with theoretical values) tenacities and moduli makes most probable the following model of the structure of superfibers. Approximately 80–90% of the fiber material represents the usual semicrystalline matrix with FCC and a relatively low amount of tie chains. However this matrix practically does not react on the external load. This load is perceived by the ECC grown during the orientational crystallization, which are not a passive filler but form a continuous spacial frame responding to the external forces at low deformations and sufficiently hindering the development of deformations or internal stresses on heating. It is important that already on the structural level of fibrils all relaxational transitions disappear which are readily checked by zonic methods in the initial fibers.

The question why only a ECC *frame* is generated was discussed in some detail earlier [9, 10, 27]. Probably, in regions adjoining the newly grown ECC a local stress relaxation occurs and consequently a local decrease in β. This results in a transition to the growth of FCC and the amorphous "matrix." In Figure 7 this would correspond to a "reverse transfer" of the system from state II into state I. In a similar manner Porter [5] explains the formation of shish-kebab structures during high pressure extrusion.

This discussion does not mean that there exists some fundamental thermodynamic or thermokinetic prohibition for the formation of macroscopic linear crystals which are 100% ECC. However such crystals will be fibers no longer in the usual sense of the word.

FIBERS FROM SEMIRIGID POLYMERS AND THE PROBLEM OF THE SO-CALLED THEORETICAL STRENGTH

The model discussed above explains fairly the relative distance of the really accessible fiber tenacities and moduli from the corresponding theoretical limits. However the problem of theoretical strength needs some reconsideration *per se* because in many publications it was mistreated by use of a vulgar representation of a chain macromolecule as a metallic joint-pin construction or as a continuous thread. For reasons quite obvious (for a physicist) such an approach to the problem is absolutely forbidden because isolated macromolecules obey the laws of quantum but not macroscopic mechanics. Therefore in the following we shall mention only a *mechanical equivalent* of strength of individual chains.

Though the problem of this equivalent was treated with sufficient rigor by Kauzmann and Eyring [28] more than one third of a century ago, up to this date no satisfactory approach was proposed for a direct experimental analysis of the theoretical strength of oriented polymers. Only quite recently such approaches were realized starting from the thermofluctuational (kinetic) concept of strength of solids developed by S. N. Jourkov and his school [4].

The expression for the long-duration strength known as Jourkov's equation and connecting the lifetime τ before the break of a body subjected to a stretching stress σ at a temperature T has the form

$$\tau = \tau_0 \exp\left(\frac{U_0 - \gamma\sigma}{kT}\right) \tag{10}$$

τ_0 being the unit of the molecular time scale, common to all materials and equal (within a decade) to 10^{-13} sec; U_0 for simple bodies coincides with the heat of sublimation and for polymers is approximately 40% lower than the heat of thermal breakdown of chain bonds, and γ is a structure-susceptible coefficient characterizing the structural inhomogeneity of a real body and, consequently, the inhomogeneity of distribution of local stresses on interatomic bonds.

As follows from eq. (10), one can perform measurements in a manner to reduce them to some standard time of testing $\tau = \tau^*$; then one can rewrite eq. (10) in the form

$$\sigma_t = \frac{1}{\gamma}\left(U_0 - kT\,ln\,\frac{\tau^*}{\tau_0}\right) \tag{11}$$

σ_t being the tensile strength (or tenacity for fibers), i.e., a measure of real strength. For different polymers the temperature dependence of σ_t (if γ remains constant during the test) thus must be presented by straight lines of the type shown in Figure 17, holding practically for all fibers. The parameter γ calculated from such graphs is always ten or hundredfold higher than its theoretical limit $\gamma = \gamma_0$, the latter being formally equal [4] to the cube of average interatomic distances, i.e., $\approx 2.1 \times 10^{-23}$ cm^3. In an ideal body (without any defects) the stresses would be distributed absolutely homogeneously, and γ would equal γ_0. Thus we can introduce a "conditional theoretical strength" $\sigma_{th} = \sigma_t g$ with $g = \gamma/\gamma_0$. We use the term "conditional" because the number 2.1×10^{-23} cm^3 is somewhat arbitrary. In addition, the overloading of some bonds resulting from the extreme inhomogenity of the structure of real bodies was not taken into account. At last only the linear term of the Maclaurin series for $U(\sigma)$ was re-

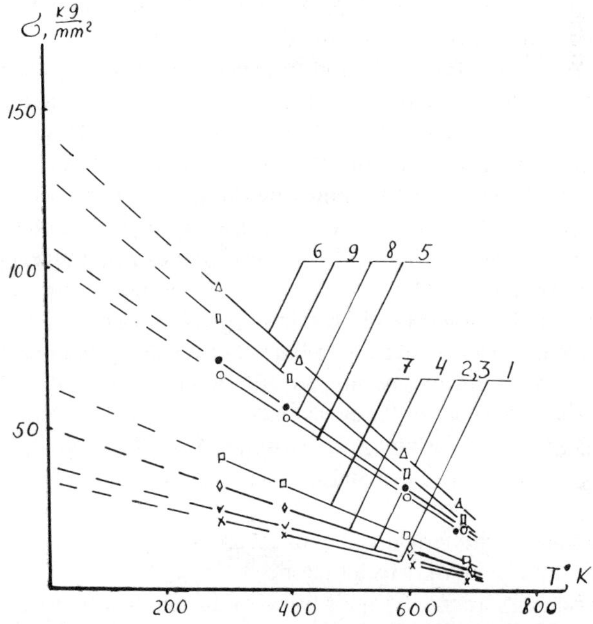

FIG. 17. The graphs for evaluation of the γ coefficient. For details see text. The figures correspond to those of Table I.

tained in eq. (10). Thus the conditional theoretical strength must differ from the "true" one by a factor of an order of unity, being approximately the same for all polymers.

A more direct measure of theoretical strength would be given by the mechanical equivalent of stress on the mentioned overloaded bonds. As was shown by V. I. Vettegren [29], with precision up to 15% this equivalent can be estimated from the shift of some bands (chosen conveniently) of IR spectra of loaded polymers.*

It is preferable to perform such studies on polymer classes of more or less analogous chemical structure. We were happy enough to be in possession of such a class—aromatic polyimides of general structure

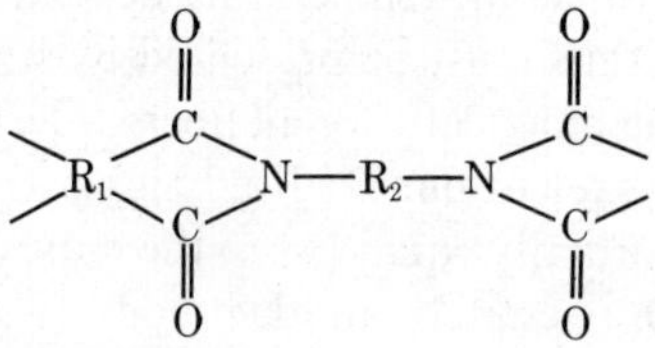

$$
R_1
\begin{array}{c}
\quad \text{O} \\
\quad \| \\
\quad \text{C} \\
\diagup \quad \diagdown \\
\quad \quad \quad \text{N} - R_2 - \text{N} \\
\diagdown \quad \diagup \\
\quad \text{C} \\
\quad \| \\
\quad \text{O}
\end{array}
\begin{array}{c}
\quad \text{O} \\
\quad \| \\
\quad \text{C} \\
\diagup \quad \diagdown \\
\\
\diagdown \quad \diagup \\
\quad \text{C} \\
\quad \| \\
\quad \text{O}
\end{array}
$$

R_1 and R_2 being arylene groups differing in number and configuration of aromatic rings (see Tables I and II). A large quantity of such polyimides was synthesized in the laboratory of the Director of our Institute Professor M. M. Koton. The fibers were obtained in our laboratory by the usual two-step method [30]. The spectral–mechanical studies for the estimation of the mechanical equivalent of stress on overloaded bonds were carried out under the supervision of V. I. Vettegren with courtesy of Prof. S. N. Jourkov in his Department in A. F. Joffe Physical-Technical Institute of the Acad. Sci. USSR.

A remarkable property of polyimides is their "rigid-like behavior", though the individual chains (see Tables I and II) due to "flexible" bonds connected with such atoms or groups as O, CO, CH_2 etc. should belong apparently to polymers with a high thermodynamic flexibility, of the same order as in aromatic vinyl polymers. However as was shown by the late A. P. Roudakov [31] the presence of strongly interacting carbonyl groups results in concordant mobilities of at least pairs of adjacent chains, thus leading to the appearance in polyimides of quasi-ladder properties. Therefore the structure of highly oriented polyimide fibers most probably should be of Hearle's type (Fig. 2b).

The equivalent of sublimation heat for polymers U_c present in Table I was calculated in a purely formal manner as half-sum of the energy of thermal decomposition of simple bonds of the chain and of the interchain interaction energy; the latter was estimated for polyimides also by A. P. Roudakov [31].

The data are given in Tables I and II. The examination of these data reveals the following circumstances:

(1) Though the numerical values of theoretical strength estimated by two methods differ approximately by a factor 2.5–3 (which was to be expected; see the previous remarks about γ_0) their full correlation in two different series of

---

* See also two papers of K.-F. Friedland presented at this Symposium.

TABLE I
Conditional Theoretical and Real Tensile Strength of Oriented Polymers

General formula

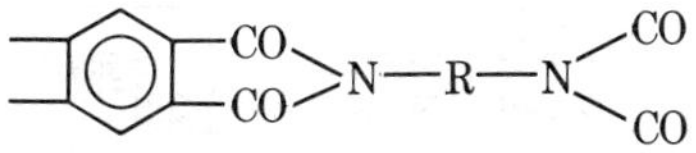

$$\text{(benzene ring)}\underset{\text{CO}}{\overset{\text{CO}}{\Big\rangle}}\!N\!-\!R\!-\!N\!\underset{\text{CO}}{\overset{\text{CO}}{\Big\langle}}$$

R corresponds to numbers 1–9 in the Table.

Polymer class	№	Polymer	σ 25°C kg/mm²	U_0 kcal/mole	γ kcal mm²/mole kg	σ_{theor} 25° kg/mm²	g	ΔH therm	U_c
	1	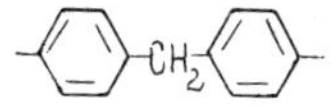 –C₆H₄–CH₂–C₆H₄–	23	45,2	1,248	957	41,6	80	46
	2	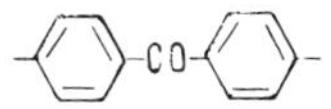 –C₆H₄–CO–C₆H₄–	23	45,0	1,242	950	41,4	80	46
	3	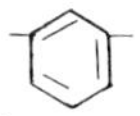 –C₆H₄–	22	45,0	1,296	950	43,2	80	46
X	4	–C₆H₄–S–C₆H₄–	31	40,2	0,765	790	25,5	62	37
	5	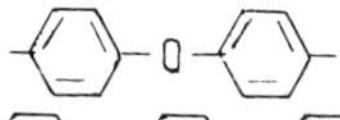 –C₆H₄–O–C₆H₄–	65	44,8	0,435	943	14,5	79	45,5
Polypyromellitimides	6	 –C₆H₄–O–C₆H₄–O–C₆H₄–	84	45,4	0,345	963	11,5	79	45,5
	7	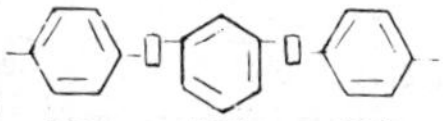 –C₆H₄–O–C₆H₄–O–C₆H₄–	39	45,1	0,732	953	24,4	79	45,5
	8	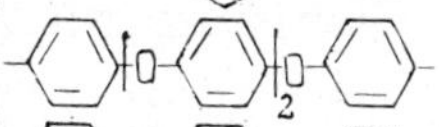 –[C₆H₄–O]₂–C₆H₄–	71	44,6	0,396	935	13,2	79	45,5
	9	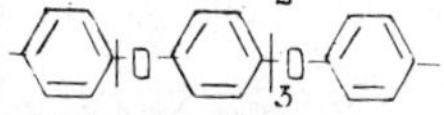 –[C₆H₄–O]₃–C₆H₄–	90	44,5	0,312	933	10,4	79	45,5
	10	Polypropylene	97	30	0,14	450	4,67	62,8	32,4
flexible chain	11	Polyvynilchloride	25	35	0,75	616	25,0	7,4	37,3
	12	Polyacrylonitrile	37	40	0,63	783	21,0	73	37,1
	13	Nylon 6	102	44	0,27	916	9,0	66,4	41,3
	14	Polyethyleneterephtalate	89	37,5	0,236	700	7,86	65	35,5
	15	Nomex	78	36	0,25	650	8,33	66	36,0
RIGID chain	16	Poly(parabenzamide)	115	41,5	0,216	830	7,2	66	38,0
	17	Poly(paraphenylene-terephtalamide)	120	40,0	0,196	783	6,53	66	36,5

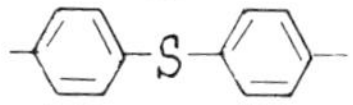

σ_{25}, tenacity at 25°C

U_0, initial activation energy of breakage.

γ, the structure-dependent Jourkov's parameter

σ_{theor}, conditional theoretical tensile strength

g, the supertension coefficient

ΔH(therm), the average heat of thermal splitting of "weak" skeletal bonds

U_c, "the equivalent of sublimation energy for polymers"

tests is quite obvious. It is also obvious that the chemical resources of tenacity (i.e., those connected with the chain structure in itself) for polyimides is higher than for any other fibers, including the rigid-chain superfibers as well.

(2) One can clearly follow the effects of evenness and meta–para isomerism in the oligophenyleneoxide groups of the diamine components of polyimides on

TABLE II

Tensile Strength Equivalents of Macromolecules for Polyheteroarylenes and Some Flexible-Chain Polymers[a]

№	Polymer	Band cm^{-1}	α, $\dfrac{cm^{-1}\cdot mm^2}{kg}$	Σf, $\dfrac{kg}{mm^2}$	$\Sigma f \cdot S \cdot 10^4$ dynes per/chain	$6f$, $\dfrac{kg}{mm^2}$	$g = \dfrac{\Sigma f}{6f}$
1	2	3	4	5	6	7	8
1	$[CH_2-CH(CN)]_n$	1350	−0,08	800	2,5	200	4
2	$[CH_2-CH(CH_3)]_n$	941	−0,05	1000	3,4	130	8
		972	−0,026	1200	4,1		
		1045	−0,04	900	3,1		
		1108	−0,082	900	3,1		
		1168	−0,010	1200	4,1		
3	$HO[CO-C_6H_4-CO-O-CH_2-CH_2-O]_nH$	845	−0,021	1600	3,3	130	12
		976	−0,06	1500	3,1		
4	$[HN-(CH_2)_5-CO-NH-(CH_2)_5-CO]_n$	980	−0,022	1900	3,4	170	11
5	$[OC-C_6H_4-CO-NH-C_6H_4-NH]_n$	782	−0,038	1600	4,6	78	20
6	$[OC-C_6H_4-CO-NH-C_6H_4-NH]_n$	763	−0,025	1900	4,4	210	9
7	polyimide	517	−0,03	1900	4,8	65	30
		605	−0,028	2000	5,0		
8	polyimide	518	−0,028	2000	4,8	84	23
		1015	−0,013	1900	4,6		
9	polyimide	517	−0,029	2000	4,8	70	28
10	polyimide	509	−0,040	1600	4,4	30	53
11	polyimide	519	−0,027	1700	3,4	55	31
12	polyimide	520	−0,017	3500	7,7	160	20
		561	−0,012	3000	6,6		
		1019	−0,018	3300	7,2		
13	polyimide	520	−0,019	3300	7,6	130	24
		1019	−0,014	3000	6,9		
14	polyimide	606	−0,006	3300	9,2	35	94
		647	−0,011	3200	8,9		
		1018	−0,010	3500	9,8		

[a] α, spectral-mechanical proportionality coefficient; Σf, tensile strength equivalent for a single chain; $\Sigma f \cdot S = F$, the breaking load; S, effective cross-section of a chain.

the real strength. This is caused by bending distortions of chain conformations in the "even" polyimides (Nos. 5 and 8 in Table I) and with crystallization hindrances connected with the meta- configuration of the benzene ring (Nos. 3 and 7 in Table I) [32].

(3) The coincidence of energies U_0 and U_c is striking though its physical sense is not clear. Such a systematic coincidence cannot be fortuituous though any attempt of its unequivocal interpretation seems premature.

(4) Striking as well is the practical coincidence of g factors for flexible- and rigid-chain polymers (one should not pay too much attention to the large deviations in cases of poorly crystallizable polyvinylchloride and polyacrylonitrile). Therefore the reverse value of g is a realistic measure of the orientational pliance as it was defined above.

(5) Finally, a careful examination of Table II shows that the "weak" bond in polyimides is the N—C bond, and that it can be strengthened by means of

influencing its electronic configuration when the dianhydride component of the repeating unit is varied.

SOME CONCLUDING REMARKS

After we have shown that a structure of ECC type [Fig. 2(a) or (d)] can be realized in principle for any linear crystallizable polymer, a natural question arises why the tenacities of superfibers made from flexible- or rigid-chain polymers are practically identical. This is true not only for ordinary tenacities presented in Table I but also for "record" tenacities of the order of 200–250 kg/mm^2. Moreover, the theoretical tenacities both for flexible- and rigid-chain polyamides estimated by the more rigorous test of Table II are also practically identical and substantially lower than those of pure and mixed polyimides.

The answer for the flexible-chain polymers consists in the practical impossibility of attaining a "concentration" of ECC frame higher than 15–20% (in accord with g values in Table I) using recent methods. However, for rigid-chain polymers there must exist some other source of the discrepancies between their real structure and the model "a" of Figure 2 which apparently is "obliged" to form itself.

Possibly the reason for the relatively low tenacities of these fibers should be found in technological drawbacks: due to a low solubility it is impossible to spin fibers from a polymer of sufficiently high molecular weight. On first sight, this should not be disastrous if the model "a" of Figure 2 reflects reality. However this model, most probably, deviates from reality. Due to the fact that the chain ends cannot pack in the crystalline lattice, a specific segregation of end-to-end defects must occur immediately leading to a fall in the attainable tenacity.

Another unfavorable property of superfibers situated on both extremes of the orientational pliance diagram is a pronounced tendency for fibrillization which is difficult to overcome without a substantial loss of tenacity. In polyimides with their strong interchain interactions and the most probable structure of the type in Figure 2(b), the fibrillization tendency must be weak and can be, in principle, eliminated completely without damaging their tensile properties.

The nonrealized considerable resource of strength in polyimides is also connected probably with unsufficiently high molecular weights and segregation of defects on the joints of end groups. However the technological routes of improvement of the tensile properties of polyimides are not connected with special difficulties, for in the first step fibers are spun from a readily soluble precursor, the polyamidoacid, and the only remaining problem is not to "lose" the initial degree of polymerization and orientation on the step of secondary thermal treatment. Possessing a unique combination of thermal stability and strength (both attainable and theoretical) the polyimides indisputably are presently the most promising purely organic fiber forming polymers.

There are a number of persons without whose participation this lecture which represents but fractions of a wide research program never could be delivered. The main contribution to the detailed theory of orientational crystallization was made by V. G. Baranov and G. K. Elyashevitch. The main experiments to test this theory were performed by V. I. Gromov and V. V. Krenev. The self-orientation of cellulose diacetate was investigated by N. G. Belnikevitch, L. S. Bolotnikova, E. S. Edilyan, and Yu. V. Brestkin. As was already mentioned the polyimides in a amount necessary to perform systematic studies were placed at our disposal by M. M. Koton and his co-workers. All the studies described in the section on theoretical strength were performed by L. N. Korzhavin and N. R. Procoptchuk, and the data for Table II were obtained (as also was mentioned) in S. N. Jourkov's department of the Physical-Technical Institute of the Acad. Sci. USSR under supervision of V. I. Vettegren. All solution spinnings including the polyimide fibers were performed with the technical assistance of T. P. Poushkina. The X-ray diffraction analysis in our Institute was carried out by Yu. G. Baklagina and B. M. Ginzburg with co-workers; a substantial contribution to the study of the structure of polyethylene and polypropylene fibers was made by V. I. Gerassimov (Polymer Division of the Moscow State University) and A. E. Tchalykh with co-workers (Institute of Physical Chemistry Acad. Sci. USSR). I wish to express to all of them my deep gratitude.

REFERENCES

[1] P. J. Flory, *J. Am. Chem. Soc.,* **84,** 2857 (1962).

[2] R. Bonart and R. Hosemann, *Makromol. Chem.,* **39,** 105 (1960).

[3] A. Peterlin, *Polym. Eng. Sci.,* **14, 9,** 627 (1974).

[4] V. R. Regel, A. I. Slutsker, and E. E. Tomashevsky, *Kinetic Nature of Strength of Solids,* M., "Nauka," 1974.

[5] J. H. Southern, N. Weeks, R. S. Porter, and R. G. Crystal, *Makromol. Chem.,* **162,** 19 (1972).

[6] J. Frenkel, *Kinetic Theory of Liquids.* Dover Publ. Inc., New York, 1955; (a) *ibid.,* Russian Third Edition, "Nauka," Leningrad, 1975 (supplementary notes).

[7] S. Frenkel and G. K. Elyashevitch, *Vysokomol. Soedin.,* **A13,** 493 (1971).

[8] S. Frenkel and G. K. Elyashevitch, in *Relaxation Phenomena in Polymers;* "Khimia," Leningrad, 1972, p. 229, 234, 240.

[9] S. Frenkel, *Pure Appl. Chem.,* **38, N 1–2,** 117 (1974).

[10] V. G. Baranov, R. Zurabyan, and S. Frenkel, *J. Polym. Sci., Symp.,* **44,** 163 (1974).

[11] P. J. Flory, *Proc. R. Soc. London Ser.,* **A234,** 73 (1956).

[12] P. J. Flory, *Trans. Faraday Soc.,* N584, **67,** 8, 2252–2282 (1971).

[13] W. G. Miller, C. C. Wu, E. L. Wee, G. L. Santee, Y. H. Rai, and K. G. Goebel, *Pure Appl. Chem.,* **38, N 1–2,** 37 (1974).

[14] T. Hill, *Thermodynamics of small systems,* New York-Amsterdam, 1963.

[15] J. W. S. Hearle, *J. Appl. Polym. Sci.,* **7,** 1175, 1193, 1207 (1963).

[16] E. DiMarzio, *J. Chem. Phys.,* **35,** 658 (1961).

[17] L. Mandelkern, *Crystallization of Polymers,* McGraw Hill, New York, 1964.

[18] H. G. Zachmann, *Pure Appl. Chem.,* **38, N 1–2,** 79 (1974).

[19] T. G. Majury and N. J. Wellard, *Br. Rayon Research Association* Wythenslave-Manchester, 334, 1949.

[20] L. V. Kuchareva, B. M. Ginzburg, S. Ya. Frenkel, and V. I. Vorob'ev, *Biorheology,* **7,** 37 (1970).

[21] A. Pennings, C. J. H. Scholten, and A. M. Keil, *J. Polym. Sci.,* **C38,** 167 (1972).

[22] V. G. Baranov and G. K. Elyashevitch, *Vysokomol. Soedin.,* **A16,** 611 (1974).

[23] G. K. Elyashevitch, Thesis, Institute of Macromolecular Compounds, Leningrad, 1975.

[24] G. K. Elyashevitch, V. G. Baranov, and S. Frenkel, *Solid State Physics* (Russian), **16,** 2071 (1974).

[25] R. Brout, *Phase Transitions,* Univ. of Brussels, New York-Amsterdam, 1965.

[26] J. H. Gibbs and E. DiMarzio, *J. Chem. Phys.,* **28,** 373, 807 (1958).

[27] V. G. Baranov, V. V. Krenev, and S. Frenkel, *Solid State Physics* (Russian), **17,** 1550 (1975).

[28] W. Kauzmann and H. Eyring, *J. Am. Chem. Soc.,* **62,** 3113 (1940).

[29] V. I. Vettegren and I. I. Novak, *Solid State Physics* (Russian), **17,** 1550 (1975).

[30] Z. G. Opritz, G. I. Kudryavtzev, L. N. Korzhavin, B. M. Ginzburg, and S. Frenkel, *Man-Made Fibers* (Russian), **No. 3,** 61 (1970).

[31] A. P. Rudakov, M. I. Bessonov, Sh. Tuytchiev, M. M. Koton, F. S. Florinsky, B. M. Ginzburg, and S. Frenkel, *Vysokomol. Soedin.,* **12A,** 641 (1970).
[32] Yu. G. Baklagina, N. V. Efanova, and N. N. Prokoptchuk, *Dokl. Acad. Nauk USSR,* **221,** 3, 609 (1975).

DRAWING BEHAVIOR AND MECHANICAL PROPERTIES OF HIGHLY ORIENTED POLYCARBONATE FIBERS

B. FALKAI and G. HINRICHSEN

Bayer AG, 4047 Dormagen, FRG

SYNOPSIS

Fibers from bisphenol-A polycarbonate are typical representatives of the less crystallizing synthetic fibers. Because of the high glass transition temperature, textiles produced from these fibers have outstanding laundry-resistance and noncreasing properties. Polycarbonate fibers can be produced both by the dry or wet-spin and by the melt-spin process. The stretching of the fibers is carried out in accordance with the proposed textile characteristics below or above the glass transition temperature or by a special swelling–drawing process, in the course of which fibers of particularly high tenacity and elastic modulus can be obtained. The fibers produced by the latter process were investigated by a number of physical measurements of their morphological structure (density, birefringence, X-ray wide- and small-angle scattering, differential calorimetry, thermo-mechanical analysis). These measured values of density, birefringence, and heat of fusion show that these fibers have a higher degree of crystallinity than the fibers produced by a melt or dry spinning process. Whereas with dry-spun fibers an increase in tenacity can be obtained by raising the molecular orientation up to a limited tenacity value, it is possible with fibers produced by the swelling–drawing process to virtually double the tenacity due to the higher crystallinity and improved crystal structure (creation of a large number of physical linkages). By means of X-ray small-angle scattering on annealed swollen-drawn polycarbonate fibers, we were able to detect, for the first time, a marked two-phase structure consisting of crystalline and noncrystalline regions with a long-period of about 120 Å. With other crystalline polymers, a period of this kind has been known for many years.

INTRODUCTION

The polycarbonate, which is both easy and economical to obtain from 4,4′-dihydroxydiphenyl-2,2-propane (bisphenol-A), has already made its way as a plastic on account of its good mechanical, thermal, and dielectric properties. The polymer is a typical representative of the low-crystallizing engineering plastics [1]. The fibers produced from bisphenol-A polycarbonate are also moderately crystalline.

The maximum melting temperature for the infinitely extended crystal of the polymer is 255–265°C, and the glass transition temperature, at 148–158°C, is extremely high [2]. Because of this high glass transition temperature, textiles made of these fibers have excellent laundry-resistance and noncreasing properties [3].

From the Stuart molecular model (Fig. 1) it can be seen that the essential cause of the high glass transition temperature and the low tendency to crystal-

Journal of Polymer Science: Polymer Symposium 58, 225–235 (1977)

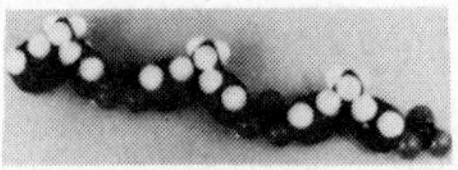

FIG. 1. Structure of polycarbonate shown by chemical formula and molecular model (according to Stuart[12]).

lization is the rigid structure of the molecules. The low rotation of the aromatic rings around the central carbon atom of the bisphenol-A is due less to the dispersion forces of the aromatics than to the rigid arrangement of the molecules.

Polycarbonate fibers can be produced both by the dry-spin and wet-spin processes and also by the melt-spin method. Stretching of the filament yarn is generally carried out at 180–200°C so that the amorphously fixed structures which block the stretching and crystallization have to be loosened in the stretching zone. The fibers drawn by this process possess the textile properties of the conventional type of synthetic fibers, such as a tensile strength of 3–3.5 cN/dtex, elongation at break of 20–30%, and an elastic modulus of 3500–4000 N/mm². With the aid of a coupled swelling–drawing process which is best performed on a wet-spun filament yarn and can be carried out at room temperature because of the marked lowering of the glass transition temperature due to the swelling agent, it is possible to produce fibers of particularly high strength, with 6.0 cN/dtex, 25% elongation, and a modulus of elasticity of approximately 6000 N/mm².

EXPERIMENTAL

Test Material

The wet-spun filament yarns ($\overline{M}_w$ = 110,000) were subjected continuously to a swelling–stretching process with various draw ratios (1:2 to the limit 1:4.8) at room temperature. They were then dried and thermofixed at 225°C under tension and a contact time of 6 sec. The dry-spun yarns ($\overline{M}_w$ = 110,000) could be stretched on a roller at a temperature of 198°C up to a draw ratio of 1:7.5.

Measuring Methods

Measurement of Birefringence. Determination of the birefringence Δn was made by measuring the path difference using a Berek compensator.

$$\Delta n = \frac{\Gamma}{d}$$

Γ = path difference

d = diameter of the fiber

Measurement of Density. The density of the fiber specimens was measured by the buoyancy method in water.

X-Ray Wide-Angle and Small-Angle Scattering. For determining the crystalline part and the molecular orientation, X-ray wide-angle tests were carried out on an interference goniometer. The crystalline part was determined by a process developed by Hermans and Weidinger [4] (comparison of ground intensities). As a measure of the molecular orientation, the intensity quotient on equator and meridian can be selected at an angle of scatter of $2\theta = 17.3°$.

The x-ray small-angle tests were made with a Kiessig chamber; evaluation of the photographic scatter diagrams was made photometrically.

Differential Calorimetry. Differential thermoanalysis was carried out with the differential calorimeter DSC 1 from Perkin Elmer. The heating rate was 16°C/min.

Thermomechanical Analysis. Thermomechanical analysis was carried out with the TMS 1 from Perkin Elmer. The pretension was 0.0015 cN/dtex. The heating rate was 10°C/min.

Mechanical Analysis. The stress–strain diagrams were drawn up on a Wolpert stress–strain tester. The gripped length of the fibers was 500 mm. The test was carried out at a constant elongation rate of 1%/sec. The modulus of elasticity was obtained by differentiation of the stress–strain lines $d\sigma/d\epsilon$.

Microscopic Investigation. The fiber cross-section was made with the Ortholux light microscope from the firm Leitz, and the surface photographs with the transmission electron microscope EM 200 from the firm Phillips.

RESULTS

Whereas heat-drawing at 198°C can be carried out up to a draw ratio of approximately 1:8, swell-drawing at room temperature can only be performed up to a ratio of approximately 1:4.8. The increase in birefringence (Fig. 2) is far more rapid for the swell-drawn material and appreciably higher in its end value. The increase in density (Fig. 3) is also much greater with the swell-drawn filament yarn. A big difference can be observed particularly when the draw ratio is increased from 1:2 to 1:3.

The results of the birefringence and density measurements are fully confirmed

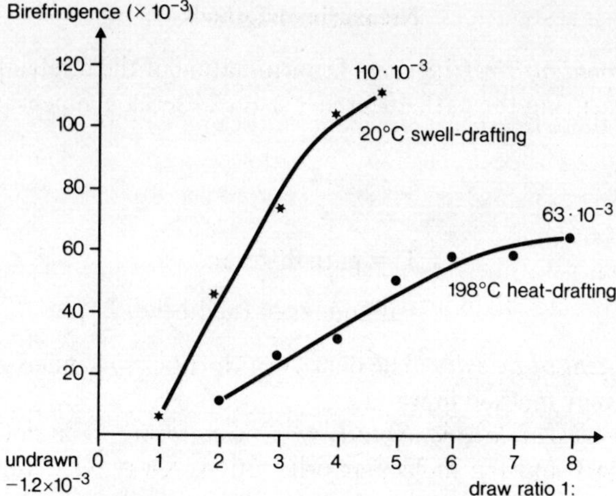

FIG. 2. Change of birefringence with draw ratio.

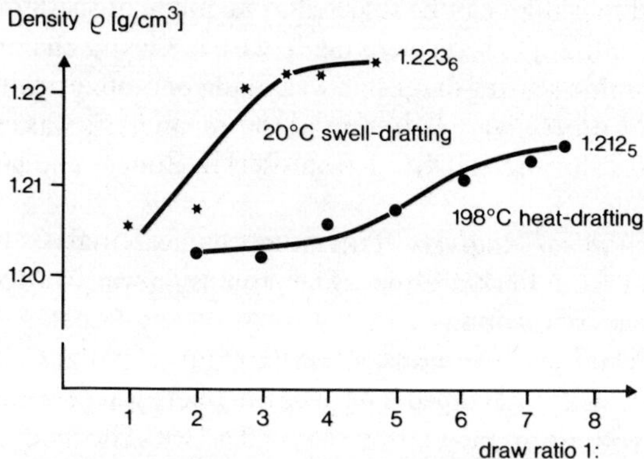

FIG. 3. Change of density with draw ratio.

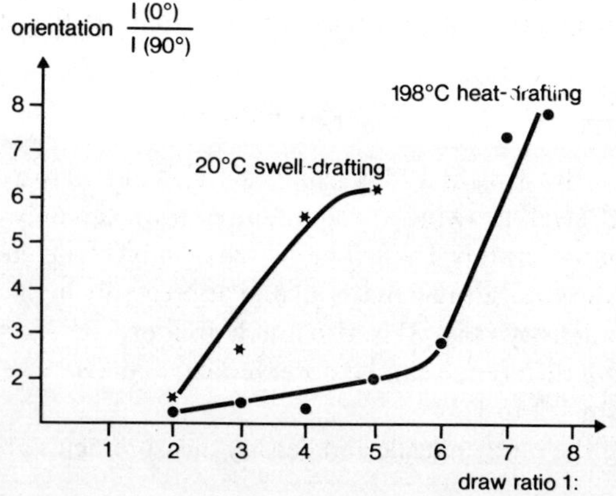

FIG. 4. Change of orientation with draw ratio.

by the X-ray investigations. The molecular chain orientation in fiber direction (Fig. 4) follows a similar course to the birefringence; here too, there is a more pronounced orientation with the swell-stretched specimens, even at low draw ratios. From these findings it is evident that the maximum numerical draw ratio of the swell-drawn specimens must be lower than that of the heat-drawn. The crystalline part as measured by X-ray (Fig. 5) corresponds to the results of the density measurement.

Differential calorimetric measurements (Fig. 6) show a somewhat higher melt enthalpy and a slight increase in the melt maximum from 228°C to 231°C for the swell-drawn specimen. The path of the thermomechanical curve (Fig. 7), with its lower thermoshrinkage, also comes out in favor of the swell-drawn filament yarn; here the thermofixation conditions naturally play a decisive role.

From the stress–strain curves (Fig. 8) of the two filament yarns under comparison at maximum stretching, it can be seen that the swell-stretched specimen has a tensile strength almost twice as high as the heat-stretched one. The differential elasticity modulus curves show characteristic differences and these will be dealt with later.

A periodic super molecular structure of crystallites and noncrystalline interlayers, which it has not been possible to determine for polycarbonate fibers until now, can be seen from the X-ray small-angle scattering graphs of the heat-treated, swell-drawn specimens (Fig. 9). A distinct meridional long-period maximum is clearly recognizable on the diagrams. The distance between this maximum from the primary beam can be used to calculate the mean center of mass of the crystallites at 120–130 Å. From the meridional and azimuthal dimensions of the long-period reflexes, elementary fibril diameters of 60 Å can, according to a theory of Tsvankin [5], be concluded, whereby the crystalline particles are about 75 Å and the amorphous interlayers about 50 Å thick.

On the small-angle diagram of the unannealed specimen, it is only possible to make out an intensive equator scattering, originating from submicroscopic cavities and paracrystalline disturbances in the lattice; long-period reflexes are missing completely. This means, in the case of polycarbonate, that only through an intensive thermal aftertreatment is a two-phase periodic super molecular structure formed.

The cross-section photographs taken with the light microscope (Fig. 10) show a round cross-section in the case of swell-drawn fibers, and a dumbbell-shaped cross-section for heat-drawn fibers. In the surface reproduction taken with the electron microscope (Fig. 11), both the swell-drawn and the heat-drawn fibers show a longitudinally profiled surface on which the elementary fibrils, massed together to form bundles, predominate with a lateral elongation of a few hundred Å. A coarse fibrillation in the direction of the fiber axis, which increases with the draw ratio but is not so pronounced, is also evident on the heat-drawn specimen. This is a morphological consequence of bundles of fibrils sliding past one another in the body of the fiber.

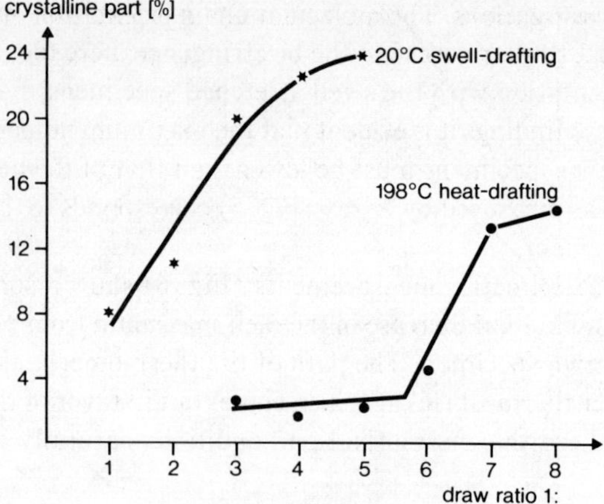

FIG. 5. Change of crystalline part (according to Hermans and Weidinger [4]) with draw ratio.

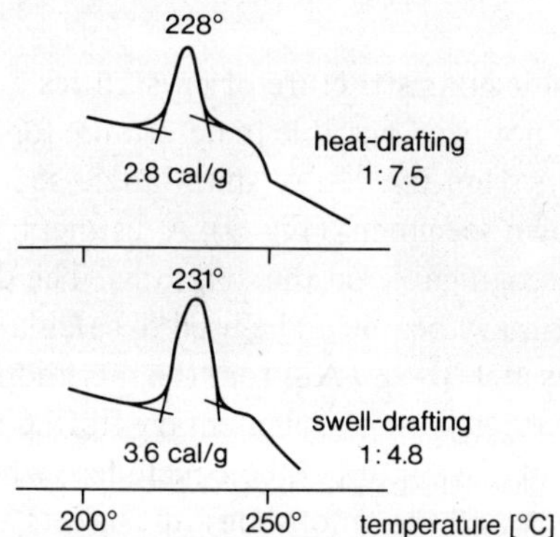

FIG. 6. Differential thermal analysis of heat- and swell-drafted filament yarns.

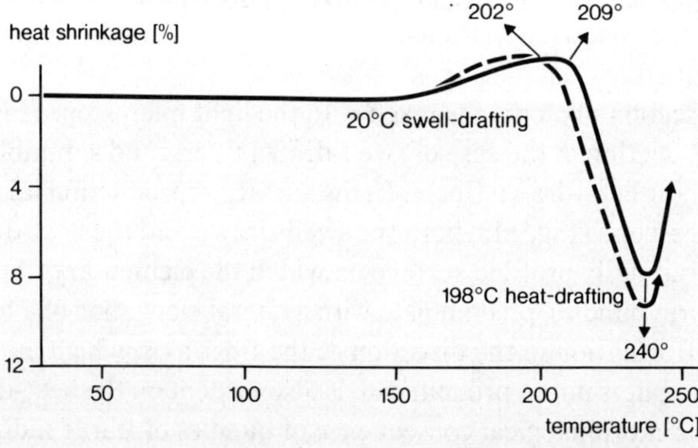

FIG. 7. Thermomechanical analysis of swell- and heat-drafted filament yarn.

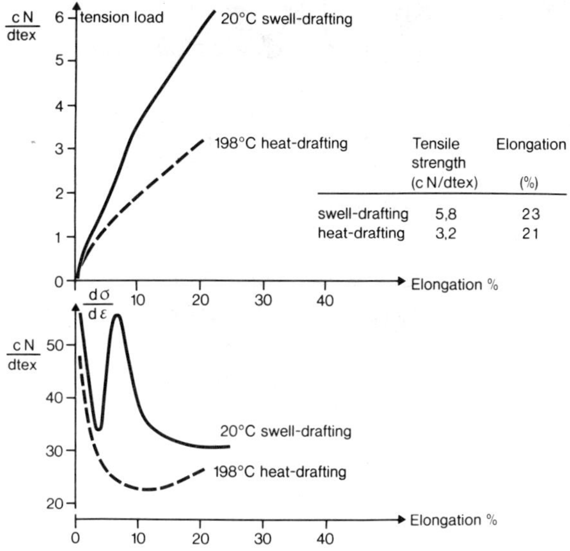

FIG. 8. Stress-strain diagram and differential modulus of elasticity of swell- and heat-drafted filament yarns.

	Tensile strength (c N/dtex)	Elongation (%)
swell-drafting	5,8	23
heat-drafting	3,2	21

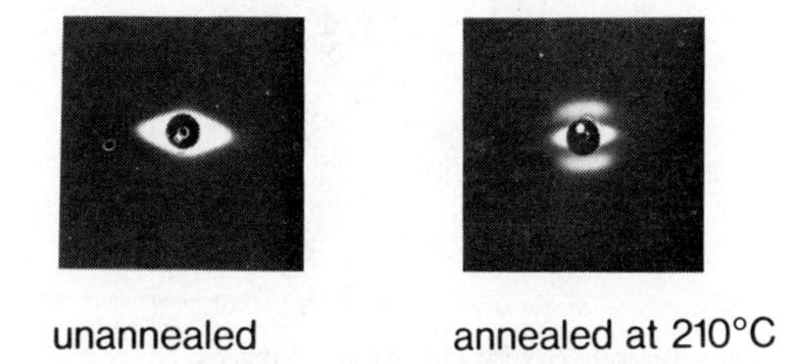

FIG. 9. Small-angle X-ray photographs of polycarbonate filament yarns.

DISCUSSION OF THE RESULTS

When discussing the results, we would like to stress three points.

Drawing Mechanism

The course taken by a curve showing the birefringence and the molecular orientation (see Figs. 2 and 4) as a function of the degree of drawing is extremely dependent on the drawing temperature. The curve for heat-drawn fibers is like the one for swell-drawn fibers when the drawing is carried out below the glass transition temperature of 150°C. However, in such a case, nothing like the high tensile strength is reached and the threads remain amorphous and have a thermoshrinkage of approximately 70%.

When drawing the textile fibers, the amorphously rigid structures already present in the undrawn material must be destroyed. For this, there are two ways of arriving at useful textile fibers: The first is to feed in enough thermal energy to loosen up the structures far enough to permit disentanglement and orientation of the molecules. A more elegant method which gives even better fibers is the

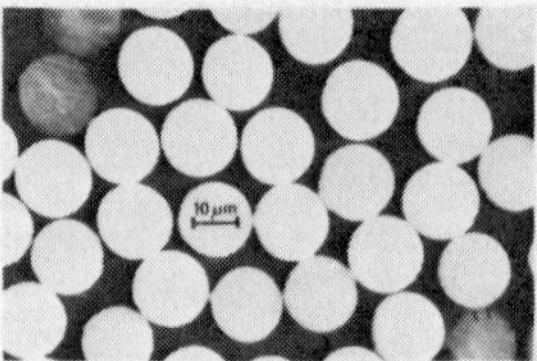

swell-drafted filament yarns

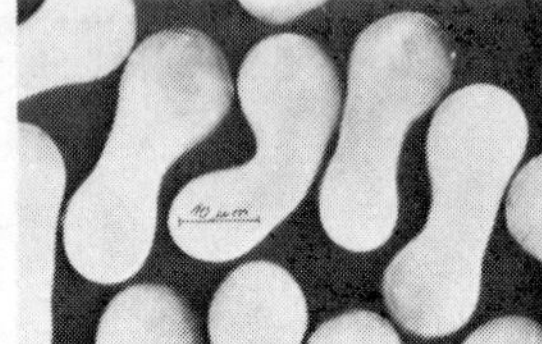

heat-drafted filament yarns

FIG. 10. Light microscopical cross sections.

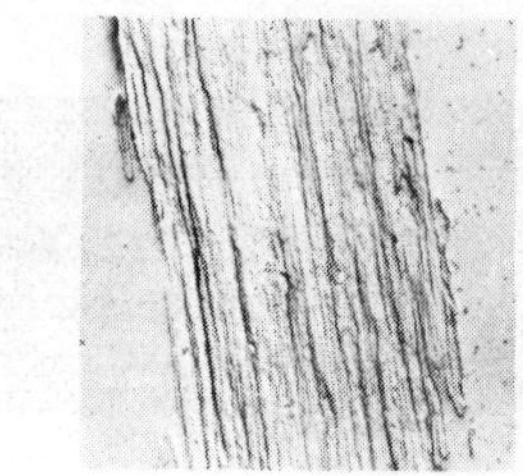

swell-drafted filament yarns

heat-drafted filament yarns

FIG. 11. Electron micrographs of surface replica.

second way, that of bringing about the necessary loosening of the structures by diffusing in a swelling agent, which then acts so to speak as a "molecular lubricant." The higher degree of drawing possible with heat-drawing is explained by the fact that, at the increased temperature, sliding motions occur to a certain extent without any increase in the molecular orientation. It should also be mentioned that the swelling agent evidently has the effect of promoting crystallization—an observation which has been investigated in detail, e.g., on polyethylene terephthalate [6].

Mechanical Data

Compared with heat-drawn fibers, the swell-drawn polycarbonate filament yarns have a far higher tensile strength with the same elongation at break and a somewhat larger elasticity modulus (at 1–2% elongation), i.e., the swell-drawn fibers are of much higher textile quality. When explaining the greater tensile strength of the swell-crystallized fibers, the discussions must cover three particular effects: the molecular orientation, the number of intercrystalline tie molecules, and the size and number of crystallites. As has been known for many years and is standard practice in the production of synthetic fibers, the tensile strength of fibers increases with the degree of drawing, i.e., with the alignment of the molecules in the direction of the fiber. Present scientific discussion is concerned with the question of whether it is the molecular orientation in the crystal ranges or the noncrystalline interlayers which is the important factor [7]. Numerous investigations on a variety of polymers have, however, shown that both orientations increase as the draw ratio rises unless a marked thermoshrinkage is induced [8]. We can therefore proceed on the assumption that our swell-drawn fibers have a higher molecular orientation than the comparable heat-drawn fibers, both in the crystalline and in the noncrystalline ranges, and that this fact leads to the higher tensile strength. From our measurements, we are unable to say anything directly about the number of tie molecules between the crystallites which act as force transmitters. We would nevertheless like to point out that the birefringence of the wet-spun—later swell-drawn—textile fibers is zero or even negative, while the birefringence of the dry-spun8later heat-drawn—fibers is low because of the far higher take-off speed, but is nevertheless much higher than zero. The drawing of virtually nonpreoriented spinning material takes place more evenly and less problematically and ought to lead to a greater number of tie chain molecules.

A high molecular orientation is of no advantage for the tensile strength of the fibers if the chains are able to slide easily past each other on stretching. If this slipping is prevented by some kind of crosslinking, then the strength of the fibers rises considerably. Crystallites in which the molecule chains are fixed in their mutual positions through forces of various kinds constitute physical linkages of this type. Our density and X-ray wide-angle tests now prove that the swell-drawn fibers have a higher proportion of crystallized, i.e., physically crosslinked molecule segments. This property certainly also has a positive effect on the magnitude of the elasticity modulus.

For the interpretation of the differential elasticity modulus curve, reference is made to a detailed paper by Bonart and Schultze-Gebhardt [9]. An interesting point is certainly the distinct elasticity modulus maximum at an elongation of approximately 7% with the swell-drawn fibers. We are of the opinion that this maximum is due to a more uniform distribution of chain lengths in the tie molecules. If we assume, for example, that segments of this kind have a wide distribution of lengths, then, as the drawing progresses, it is always only a few segments which are actually tensioned, while the shorter ones have already been pulled out of the crystallites or have been torn and the longer segments have not

TABLE I

Measurement of Degree of Crystallinity of Drawn Polycarbonate Fibers

Type	Specific volume[1] $[cm^3/g]$	Heat of fusion[2] $[cal/g]$	Specific volume[1] %	Crystallinity X-ray[3] %	Heat of fusion[2] %
swell-drafting 1:4.8	0.8173	3.6	25	24	[14]
heat-drafting 1:7.5	0.8247	2.8	15	14	[11]

1) Measured at 20°C; $\alpha_v = (v_a - v_{mea})/v_a - v_{cr})$;

$v_a = 0{,}8365\ cm^3/g$; $v_{cr} = 0{,}7600\ cm^3/g$ [10].

2) Without consideration of interfacial-energy of crystallites

$a_m = \Delta H_{mea}/\Delta H_{id}$; $\Delta H_{id} = 26.2\ cal/g$. [10].

3) According to Hermans and Weidinger [4].

yet been stressed. Where the distribution is narrow, however, the majority of the tie molecules are, at a given elongation, stressed simultaneously to give a modulus maximum.

Crystallinity and Super Molecular Structure

As mentioned in the Introduction, polycarbonate is among the low-crystallizing high polymers. Our X-ray measurements and measurements of density and heat of fusion make it possible to estimate the degree of crystallinity of the investigated fiber specimens. For the maximum drawn fibers (1:4.8 and 1:8.0) we obtained the values shown in Table I, whereby values given in literature were used for the density of the amorphous and crystalline ranges and the heat of fusion of the ideal crystal [10]. It is evident that the swell-drawn fibers are significantly better crystallized than the heat-drawn ones; there is good agreement for the crystallinities, calculated from the measured density values and the X-ray wide-angle scatter curves. The crystallinities from the heat of fusion therefore turn out too low because the interfacial energy of the crystals in the differential calorimetric measurement lowers the effective heat of fusion of the crystallites. The present measurement results therefore reconfirm that the proportion of crystallized chain segments in the case of polycarbonate is lower than 30% when based on the simplest model of a two-phase system.

Further conclusions on the super molecular structure can be drawn from the X-ray small-angle diagrams. Whereas the diagrams of the drawn unannealed fibers provide no evidence of a periodic super molecular structure, there is clear evidence in the scatter diagrams of the annealed fibers of a meridional long-period reflex which suggests a periodic sequence of crystalline and noncrystalline areas in the direction of the fiber axis at a distance part of about 120 Å. As far as we know, this X-ray picture is the very first to provide evidence that polycarbonate, like most of the other partially crystalline polymers, can form a periodically lamellar super molecular structure when certain production and aftertreatment conditions are applied. We picture the transition from the unannealed to the annealed state as being in line with the suggested model from Fischer *et al.* [1]: In the unannealed material, there are crystalline and amorphous areas distributed statistically, while, as a result of the heat treatment, a disproportionation of the two phases takes place just below the melting point, resulting in a difference in electron density which is responsible for the X-ray small-angle scattering.

Our thanks are due to Dr. Werner, Dr. Druschke, Dr. E. Müller, Dr. Schultze-Gebhardt, Dr. Spilgies and Dr. Wilsing for the production of the filament yarns and for providing us with the measurement results.

REFERENCES

[1] R. Bonart, *Makromol. Chem.,* **92,** 149 (1966).

[2] B. v. Falkai and W. Rellensmann, *Makromol. Chem.,* **75,** 112 (1962).

[3] B. v. Falkai, *Lenzinger Ber.,* **32,** 40 (1971).

[4] P. H. Hermans and A. Weidinger, *Makromol. Chem.,* **64,** 135 (1963).

[5] Tswankin, *Polymer Sci. USSR,* **6,** 2304 (1964).

[6] H. G. Zachmann, *Kolloid Z. Z. Polym.,* **189,** 67 (1963).

[7] R. J. Samuels, *Structured Polymer Properties,* J. Wiley and Sons, New York, 1974.

[8] I. Mitterpach, I. Diacik and M. Jambrich, *Faserforsch. Textiltechk.,* **23,** 115 (1972).

[9] R. Bonart, and F. Schultze-Gebhardt, *Angew. Makromol. Chem.,* **22,** 41 (1972).

[10] J. P. Mercier and R. Legras, *J. Polym. Sci.,* **B 8,** 645 (1970).

[11] E. W. Fischer, H. Goddar and G. F. Schmidt, *Makromol. Chem.,* **118,** 114 (1968).

DIRECT OBSERVATION OF STRUCTURE IN HIGH-MODULUS AROMATIC FIBERS

M. G. DOBB, D. J. JOHNSON, and B. P. SAVILLE

*Textile Physics Laboratory, Department of Textile Industries,
University of Leeds, Leeds, LS2 9JT, U.K.*

SYNOPSIS

A detailed structural characterization of three aromatic polyamide (PPT) fibers has been carried out by means of direct electron microscope observations and X-ray and electron diffraction analysis. Data obtained by the diffraction techniques are supplemented by high-resolution images recorded from fragmented specimens, which provide both a qualitative assessment and quantitative information in terms of crystallite perfection and the distribution of crystallite size. Dark-field images suggest major differences between the macrostructures of various PPT fibers; indeed some specimens are composed of long-range axially periodic assemblies, while others appear to be homogeneous. Such differences in organization may well account for variations in tensile behavior.

INTRODUCTION

Our understanding of the mechanical properties of a material is considerably enhanced when a comprehensive structural characterization at all levels of organization is available. For example, there is now reasonable agreement about the structure of poly acrylonitrile (PAN)-based carbon fibers primarily because the indirect evaluations of X-ray diffraction and other techniques have been supplemented by extensive electron microscope studies which have provided a wealth of information concerning the structure of the crystallites and their interlinking [1]. In contrast, a lack of definitive evidence concerning molecular organization in organic fibers has resulted, in the case of polyethylene terephthalate (PET), in a proliferation of diverse structural models purporting to explain their physical properties [2, 3]. Although direct observation of structure in any polymer is difficult to achieve, because of an inherent sensitivity to the electron beam during electron microscope examination [4, 5], the high-modulus aromatic fibers of poly p-phenylene terephthalate (PPT) developed by DuPont [24] are relatively less sensitive to electrons. Indeed we have reported a preliminary examination of Kevlar 49 in which, for the first time, lattice fringe images reveal the extent and perfection of the crystalline regions [6]. Model studies on beam-stable carbon fibers carried out in conjunction with studies on the beam-sensitive PPT fibers, determined appropriate modes of lattice imaging and showed that a comprehensive analysis of bright and dark-field images can provide better estimates of crystallite size than diffraction analysis alone, to-

Journal of Polymer Science: Polymer Symposium 58, 237–251 (1977)

gether with additional important information concerning crystallite size distribution [7]. The purpose of this paper is to report a detailed characterization of structure in three types of PPT fibers by means of direct electron microscope observations supplemented by X-ray and electron diffraction analysis, and to relate these findings to some basic mechanical properties of the specimens.

EXPERIMENTAL

Materials

Three specimens of high-modulus fiber produced by DuPont have been examined, namely two samples of the commercially available materials Kevlar and Kevlar 49, together with one sample of the experimental version PRD 49. Although the forerunner of Kevlar was Fiber B, a poly p-benzamide (PPB) fiber, it is generally considered that Kevlar, Kevlar 49, and PRD 49, are comprised of PPT [8, 9]. The best evidence on the chemical constitution of these fibers is provided by an analysis which shows that Kevlar 49 is composed of equimolar amounts of terephthalic acid and p-phenylene diamine, thus confirming the fiber as basically PPT [10].

Mechanical Properties

The tensile properties of the three materials were determined, after conditioning at 65% RH and 21°C, by extending 2 cm lengths of fiber to break (30 samples each) in an Instron tensile tester. Average stress–strain curves were plotted from the data, taking into account the diameters measured optically for each type of fiber.

X-Ray Diffraction

Wide-Angle Scattering (WAXS)

X-ray diffraction photographs of fiber bundles were recorded in a flat-plate camera incorporating a lead glass collimator with a 0.3 mm diameter capillary. Both Ni-filtered CuKα and Zr-filtered MoKα radiation have been used. A measure of orientation, Z, the half-height width in an azimuthal direction, was obtained by measurement on the 002 meridional reflections with the polar table of a Joyce-Loebl microdensitometer.

Equatorial and meridonal diffraction traces were recorded in the step-scan mode on punched-paper tape via a scintillation counter, single-channel pulse-height analyzer, and a scaler/timer. Corrections to the diffraction traces were made for polarization, Lorentz factor, and Compton scatter; intensity readings were converted to electron units by equating the areas under each trace with the area under the calculated values of the mean square atomic scattering factor for PPT. The normalized traces were then resolved into individual peaks of mixed

Gaussian/Cauchy shape, together with a polynomial background, by means of a profile-resolution program established earlier with cellulose fibers [11, 12]. Finally, the separated peaks were corrected for instrumental broadening by a Stokes deconvolution procedure using hexamethylene tetramine as a standard. Since the absence of intense higher orders precludes adequate corrections for distortion broadening, an apparent crystallite size was determined from the Scherrer formula, $L_{hkl} = K/\Delta s$, where $s = 2 \sin \theta / \lambda$; K was taken as unity.

Small-Angle Scattering (SAXS)

Small-angle photographs were recorded in a Searle camera with Franks-type optics utilizing the microfocus beam from a Hilger and Watts Y 25 generator.

Electron Microscopy

Specimen Preparation

Specimens for examination in the electron microscope were prepared by either (a) ultramicrotomy, where fibers embedded in Spurr resin were cut longitudinally with a diamond knife in a direction perpendicular to the fiber axis, or (b) fragmentation, where fibers were dispersed by ultrasonic irradiation in water. Sections and fragments were deposited on carbon-coated grids in the usual way.

Instrumentation

A Philips EM300 electron microscope was used throughout the work, operated at 100 kV with the double-condenser mode of illumination. Astigmatism was corrected by observing the phase-structure in a carbon film displayed on a TV monitor via a Plumbicon camera. Time-lapse series of electron diffraction patterns were recorded at various electron densities in order to ensure that electron images were recorded during the lifetime of the specimens. Illumination levels were determined from the exposure meter calibrated in terms of screen current and converted via the known magnification factor and area of the final screen into beam density at the specimen level. In this way suitable levels of illumination could be selected by predetermined adjustment of instrumental settings, particularly condenser-lens current values. In order to shield specimens from the electron beam during instrumental adjustment, a solid-disc beam stop was included with the normal apertures in the second-condenser aperture holder. High-resolution electron microscope images were recorded at magnifications calibrated by measurement on the 1.194 nm lattice fringes from the $(20\bar{1})$ planes of platinum phthalocyanine. All electron images were recorded on Ilford EM4 plates and processed for optimum electron density.

Electron Diffraction (WAED)

Wide-angle electron diffraction patterns were recorded from selected areas of diameter approximately 1.2 and 2.0 μm for fragments and sections, respectively. Calibration of the effective camera length was established by deposition of thallium chloride (a = 0.384 nm) on to the specimens. Equatorial reflections from fragments in all three samples were analyzed from corrected microdensitometer traces using the peak-profile resolution program adapted for use with WAED rather than WAXS. No corrections were made for instrumental broadening or distortion broadening, and apparent crystallite sizes were again evaluated from the Scherrer formula with $K = 1$.

Dark-field Images

Dark-field images were produced by the tilted-beam method; the appropriate equatorial and meridional reflections were defined by a 5 μm copper-foil aperture.

Lattice Images

All lattice images were recorded using the axial mode of illumination and with the objective aperture removed. Because of the inherent beam sensitivity of the specimens, a somewhat complex procedure was developed in order to minimize beam damage and produce images which exhibit meaningful lattice information. The following steps detail this procedure.

(i)   Specimen grids were scanned at low magnification and low beam intensity (condenser 2 fully overfocused) in order to select areas of interest.

(ii)   The solid-disc beam stop was inserted in condenser 2 to protect the specimen from electron damage while the beam was focused at a predetermined setting of condenser 2 and the electron optical magnification adjusted to 100 K.

(iii)   The solid-disc stop was removed and the image of an area adjacent to the area of interest set as close to optimum focus as possible [13]. The disc was then reinserted, condenser 2 fully overfocused, and the magnification reduced to about 3 K.

(iv)   The beam stop was again removed, the precise area of interest selected, and the disc reinserted.

(v)   The magnification was reset to 100 K and the illumination set to an appropriate preselected level by means of condenser 2.

(vi)   A photographic plate was transported ready for exposure.

(vii)   The beam stop in condenser 2 was removed and a 200 μm aperture introduced; simultaneously the camera shutter was opened to expose the photographic emulsion.

TABLE I
Tensile Properties

	MODULUS	BREAKING STRESS	EXTENSION AT BREAK
	GNm^{-2}	GNm^{-2}	%
PRD 49	72.7	2.11	3.42
KEVLAR 49	53.6	2.53	5.26
KEVLAR	45.8	2.83	7.31

TABLE II
Lattice Spacing [16] and Apparent Crystallite Sizes Estimated from X-Ray Diffraction Data

REFLECTION	SPACING	CRYSTALLITE SIZE nm		
hkl	nm	PRD 49	KEVLAR 49	KEVLAR
110	0.433	8.53	6.54	4.81
200	0.388	3.63	5.28	3.95
004	0.322	18.38	17.68	10.98
006	0.215	13.42	10.87	6.28

RESULTS

Mechanical Properties

Values of the initial modulus, breaking stress, and elongation at break, for PRD 49, Kevlar 49, and Kevlar, are given in Table I. It is apparent from this data that the Kevlar 49 sample examined here is very different from the one investigated by Bunsell [14] which had an initial modulus approximately double that of the PRD 49 measured here.

X-Ray Diffraction

Wide Angle

WAXS patterns of PRD 49, Kevlar 49, and Kevlar, recorded with MoKα radiation, are shown in Figure 1A, B, C, respectively. In all patterns, at least ten layer lines can be seen, but the number of reflections on each layer line decreases with the ranking PRD 49, Kevlar 49, and Kevlar. The overall appearance of the patterns, particularly the increasing diffuse scatter beyond the first reflection on each layer line for Kevlar 49 and Kevlar, is remarkably similar to the theoretically predicted patterns from curvilinear crystals illustrated schematically by Vainshtein [15].

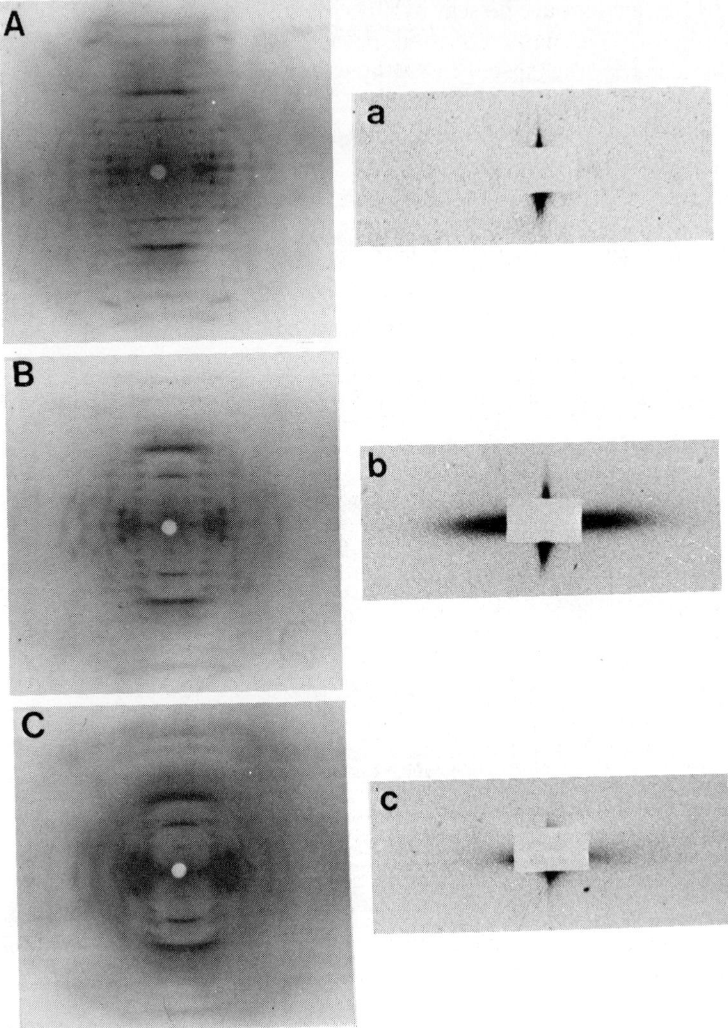

FIG. 1. WAXD patterns from (A) PRD 49, (B) Kevlar 49, (C) Kevlar fibers. (a), (b), and (c) are the corresponding SAXD patterns.

The unit cell of PPT has been evaluated by Northolt [16], Tadokoro *et al.,* [17], and Yabuki *et al.,* [18]. All find that the packing of hydrogen-bonded layers is as shown in Figure 6A; here we will use Northolt's monoclinic (pseudo-orthorhombic) unit cell (a = 0.719nm, b = 0.518 nm, c (fiber axis) = 1.29 nm, and $\gamma = 90°$) to index the reflections. The principal equatorial and meridional reflections and their spacings are given in Table II. Northolt suggests that the most favorable symmetry for the PPT crystal should be $P2_1/n$, but because of a lack of perfect twofold screw symmetry, this is lowered to Pn. Although the majority of reflections from Kevlar 49 and Kevlar (Fig. 1B and C) conform to the Pn space group, there are a few which are anomalous. For example, 010 reflections can be detected in resolved traces of Kevlar, Kevlar 49, and PRD 49, and can be seen in the patterns of PRD 49; 011 is prominent in all patterns, and other 011 reflections are present in PRD 49.

The intensities of many reflections in the PRD 49 photograph are different from those in the other specimens, although the PRD 49 pattern appears to be similar to the 400°C heat-treated sample described by Northolt [16]. The Kevlar and Kevlar 49 patterns appear to be equivalent to the 500°C heat-treated PPT specimen prepared by Northolt; however, there are a number of discrepancies, particularly with respect to the 010 and 011 reflections. These results suggest the presence of alternative packing arrangements for the hydrogen-bonded layer planes.

Crystallite Size

The apparent crystallite sizes found from the resolved profiles for the 110 and 200 equatorial plus the 004 and 006 meridional diffraction peaks, are given in Table II. Apart from the 200 reflection, the ranking is again PRD 49, Kevlar 49, Kevlar. The 200 reflection in PRD 49 is exceptionally broad even allowing for overlap with a number of weak reflections. Confirmation was obtained from low intensity patterns taken with $CuK\alpha$ radiation. In this case we must assume either that the true crystallite size normal to 200 (the number of hydrogen-bonded layers) is low, or that there is considerable distortion in this direction.

Small Angle

SAXS patterns of PRD 49, Kevlar 49, and Kevlar are reproduced in Figure (a), 1(b), and 1(c). The meridional streak is an instrumental artefact; no sample shows any meridional or off-meridional scatter of the type normally associated with chain folding. PRD 49 has no equatorial scatter either, but Kevlar 49 and Kevlar both exhibit an equatorial streak without any discrete interferences. The scatter from Kevlar 49 is much stronger than from Kevlar, and in both cases is similar to that from the needle-shaped voids in high modulus carbon fibers [19,

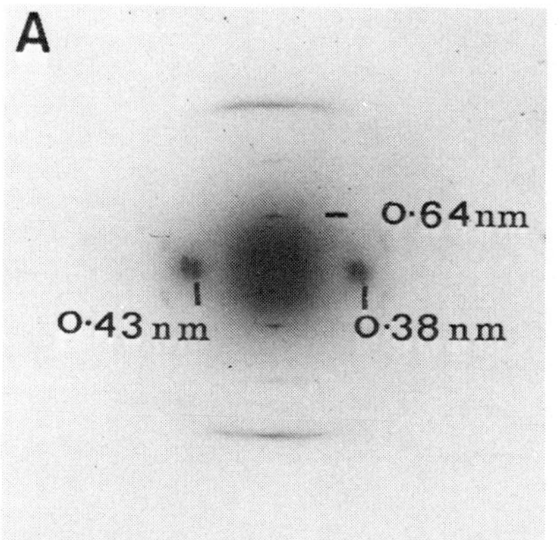

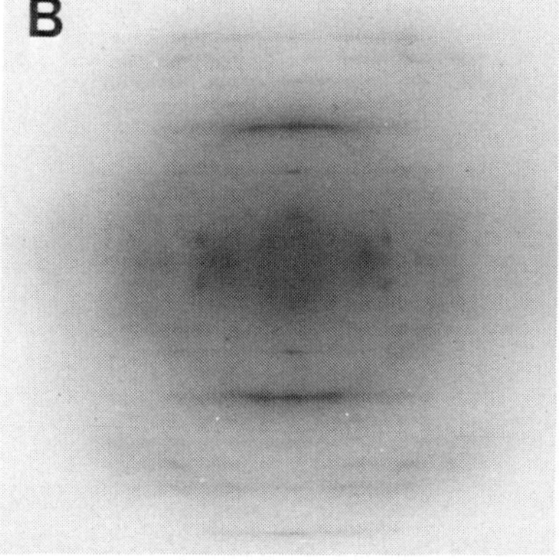

FIG. 2. WAED patterns from (a) a fragment, and (b) a longitudinal section of PRD 49.

20]. The Kevlar SAXS pattern indicates an azimuthal spread of the density discontinuities, but this orientation effect is less pronounced in Kevlar 49. More detailed studies will be reported later, but it is important to note here that there is no evidence for major density fluctuations in PRD 49.

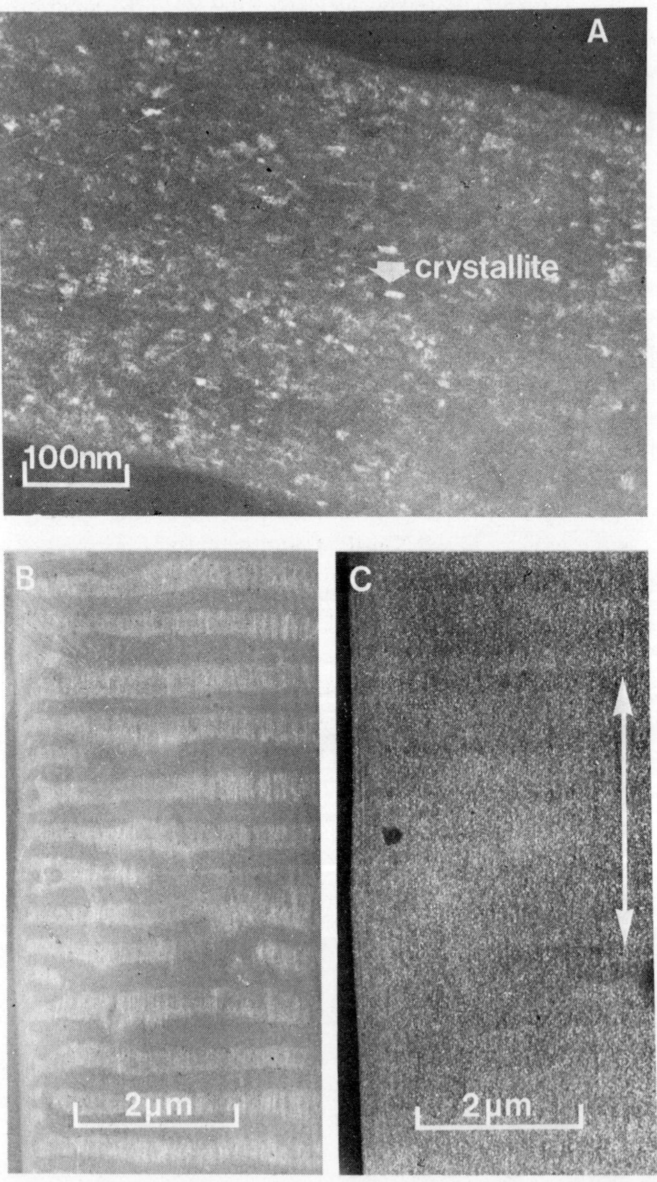

FIG. 3. (a) Dark-field image arising from the 110 and 200 reflections of a PRD 49 fragment. Dark-field images from (b) 006, and (c) 110 and 200 reflections of the same area of a longitudinal section of Kevlar 49.

ELECTRON MICROSCOPY

Electron Diffraction

WAED patterns from a typical section and a typical fragment of PRD 49 are illustrated in Figure 2A and B, respectively. The pattern from the section shows eighteen layer lines, a convincing demonstration of the perfection of order in the material. The principal equatorial and meridional reflections 110, 200, 002, 004, and 006 are visible in both patterns, their spacings are given in Table II and indicated on Figure 2B. Different reflections decay at different rates during exposure to the electron beam; the 110 reflection (0.433 nm) is most stable but does not persist longer than 120 s at a beam density of 1.64×10^{-5} Amm^{-2}. The structural changes in PRD 49 fibers have been studied by means of time-lapse electron diffraction, dark-field, and lattice-imaging techniques; measurements indicate that the decay in molecular order does not fit into either a dose or a dose-rate pattern [7].

The apparent crystallite size measured from resolved profiles in the WAED patterns are given in Table III. These estimates are necessarily limited to one selected area of the material and can only be expected to give an "order of magnitude" result when compared with the average obtained from X-ray diffraction.

Because an electron diffraction pattern represents a plane section of comparatively large area in reciprocal space, it can sometimes reveal additional information not obvious in the X-ray diffraction pattern. Thus it is significant that the 004 and 006 meridional reflections have associated tangential streaks together with weak interference rings. In carbons this indicates the presence of two-dimensional lattices [21, 22], the so-called "turbostratic" layer planes of carbon fibers are a good example; in PPT fibers, we therefore have evidence to suggest the presence of two-dimensional lattices, although a large proportion of the material is in three-dimensional register. Wherever the hydrogen-bonded sheets are not in perfect alignment they may be considered as two-dimensional lattices. Perferred orientation and lattice distortions will tend to limit the extent of the two-dimensional patterns which are essentially due to parallel rodlike intensity distributions in reciprocal space.

Dark-Field Images

Medium-resolution electron micrographs of fiber fragments show breakdown into extensive sheets which contain occasional kink bands. Dark-field images

TABLE III
Apparent Crystallite Sizes Normal to (110) nm

	WAXS	WAED	EM
PRD 49	8.5	7.2	5.4
KEVLAR 49	6.5	4.9	4.9
KEVLAR	4.8	4.8	3.2

FIG. 4. High resolution micrograph of a fragment of PRD 49 showing meridional (M) and equatorial (E) lattice fringes.

derived from the 110, 200 equatorial reflections [Fig. 3(a) is a typical example], reveal a dense population of crystallites evenly distributed throughout the fibrillar ribbons of PRD 49. When meridional dark-field images from the 002 or 006 reflections were recorded, a most unexpected difference was found between the three samples. Kevlar and Kevlar 49 show a banded structure of major periodicity 560 nm with a pseudo half spacing particularly evident near the edges of

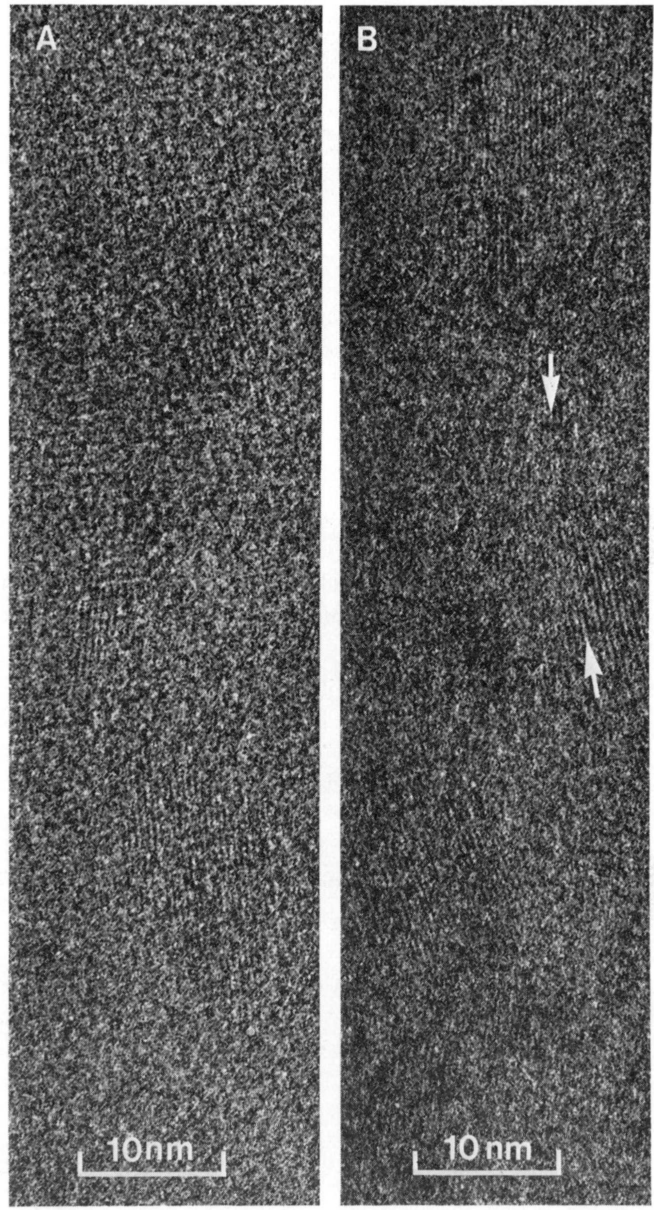

FIG. 5. Micrographs of fragments of (a) Kevlar 49, and (b) Kevlar showing disorientation and curvature of lattice fringes (arrowed).

the section (Fig. 3B), whereas PRD 49 has an even unbanded dark-field meridional pattern. The equatorial 110, 200 dark-field image (Fig. 3C), of the same area as Figure 3B also reveals banding but it is evident that crystallites are present in both dense and less dense bands.

Lattice Images

A most significant result of this work has been the successful imaging of lattice fringes from crystallites in fragments of all three PPT samples discussed here. We believe that this is the first time such high resolution has been achieved in electron beam-sensitive polymers. Typical lattice fringe images are depicted in Figure 4 and Figures 5(a) and 5(b), which are from PRD 49, Kevlar 49, and Kevlar, respectively. Lattice fringes derived from the 0.645 nm [002] meridional, and the 0.43 nm [110] equatorial reflections are indicated by M and E in Figure 4. The origin of the layer planes giving rise to these fringes is indicated diagrammatically in Figure 6(a). In larger areas of micrographs than it is possible to depict here, the extensive meridional fringes M can be seen to form an almost continuous system, however, lateral mismatch between blocks of these fringes indicates the lateral extent of molecular order.

Invariably, the individual arrays of equatorial lattice fringes E are significantly less extensive in the direction of the fiber axis than the meridional arrays M. There are two possible explanations: (i) changes in azimuthal orientation of the (110) lattice planes cause tilting out of the Bragg diffraction condition, (ii) loss of lateral order while the chains maintain axial regularity; of these (i) would seem to be most likely.

Visual assessment of the mutual alignment of arrays of 110 lattice fringes in extensive areas is compatible with the preferred orientation ranking from WAXS evidence, that is PRD 49, Kevlar 49, Kevlar. The necessarily small areas illustrated in Figure 4 and Figures 5(a) and 5(b) are typical of this overall ranking.

Compared to earlier observations on many types of carbon fiber [1], there is very little evidence of lattice distortion in PPT fibers except for the occasional appearance of curved layer planes; a good example of such a set is arrowed in Figure 5(b).

Although it must be emphasized that there is no exact one-to-one correspondence between layer planes and lattice fringes [13], estimates of crystallite size can be obtained directly from the lattice fringe micrographs. Width measurements on individual arrays of the 110 (0.43 nm) lattice fringes, corresponding to the apparent size of crystallites normal to the (110) planes, were made on high-magnification micrographs over 170×90 nm^2 areas in 1.7 nm steps parallel to the length of the arrays. This method, which allows for the varying lengths of lattice perfection and changes in the lateral packing of individual crystallites, gave 300 to 500 readings for each sample.

The three distributions of apparent crystallite size normal to (110) are given in Figure 6(b) together with the mean values. It is important to note that measurements of decay in width of the 110 equatorial electron diffraction peak indicate no decrease in apparent crystallite size for the exposure dose of electrons used to record these images.

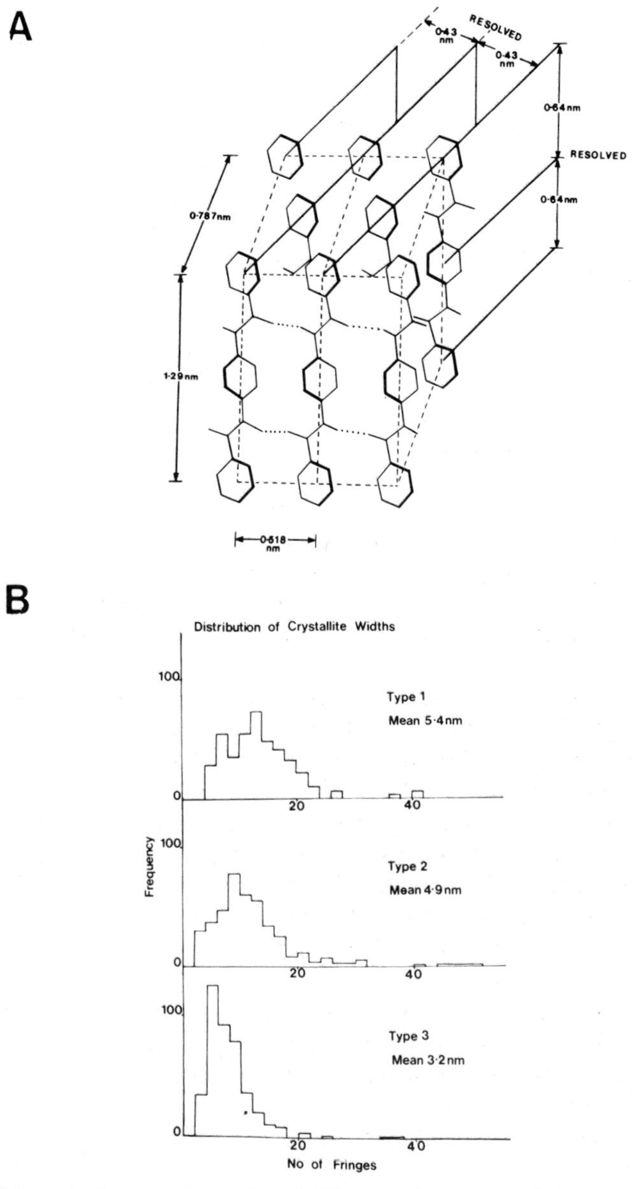

FIG. 6. (a) Schematic diagram of layer planes in PPT crystal structure. (b) Histograms showing distribution of apparent crystallite widths estimated from arrays of 110 fringes. Type 1, PRD 49; Type 2, Kevlar 49; Type 3, Kevlar.

DISCUSSION

Apparent crystallite widths normal to 110 as estimated by X-ray diffraction, electron diffraction, and direct measurement on lattice fringe images, are given in Table II. The discrepancies between the average sizes of crystallite found by

X-ray diffraction, the more selective sizes found by electron diffraction, and direct measurement, can be mainly attributed to the fact that no allowance was made for size distribution in the diffraction measurements, where the larger diffracting units have a disproportionate effect on the intensities and widths of the diffraction peaks leading to higher values of crystallite size. Furthermore, diffraction measurements from equatorial traces neglect crystallites misoriented within a degree or two from the fiber axis, and take no account of the possibility that crystallite width may be a function of orientation angle. Use of the Scherrer formula with $K \simeq 0.6$–0.7 will give reasonable values of mean apparent crystallite width for equatorial reflections.

We have no evidence for chain folding or extensive chain disorder; essentially the extended-chain molecules of PPT from hydrogen-bonded sheets which pack in the three-dimensional crystallographic structure described by Northolt [16]. However, shearing of the sheets will lead to regions of essentially two-dimensional order, as indicated by the presence of normally forbidden reflections such as 010 and 011. Further evidence for two-dimensional order comes from the-layer-like streaking in electron diffraction patterns. There is no reason to invoke the concept of disorder (amorphous zones) in these highly crystalline, high-modulus, high-strength fibers, and we consider that they are best represented by a system of hydrogen-bonded sheets packed predominantly in perfect register but occasionally distorted from perfect packing to give only two-dimensional regularity.

Since SAXS along the equator is absent from PRD 49, but is prominent in Kevlar and Kevlar 49, and because PRD 49 shows no meridional or equatorial dark-field bands, it is logical to suggest that a common structural feature is responsible for both SAXS and the banding observed. It is important to note that lattice fringe images from PPT specimens, unlike those from carbon fibers [1], exhibit no obvious voids or density discontinuities; indeed mismatch of the 0.645 nm meridional repeat is often the only indication of the lateral extent of crystallization. However, as the lattice fringe images were obtained only from fragmented specimens, it is possible that, during preparation, cleavage occurs preferentially along discontinuities.

An unusual feature of some PPT fibers is the occurrence in dark-field images of axial banding derived from the meridional reflections. There are two possible explanations of this phenomenon, namely periodic variations in either crystallinity or crystallite orientation. Preliminary specimen-tilting experiments tend to support the latter proposal and further work is in progress to elucidate the macrostructure of these fibers.

A detailed discussion of mechanical properties will depend not only on molecular aspects (for example, there appears to be a correlation between increasing modulus and preferred orientation found here and summarized for many similar aromatic polyamide fibers by Carter and Schenck [8]) but also on the general hierarchies of organization. Indeed variation of production parameters [23] such as spin–stretch factor or heat treatments, may well determine the packing modes of hydrogen-bonded sheets and consequently the differences in fundamental structure and mechanical performance found between the banded commercial fibers Kevlar and Kevlar 49 and the nonbanded experimental fiber PRD 49.

In conclusion two important points must be emphasized: (1) this work has shown that direct observation and measurement on lattice fringe images can be used to reinforce diffraction data and give an improved characterization of structure, particularly in terms of crystallite size and crystallite size distributions; (2) PPT fibers are considered to be highly crystalline, any disorder being generated by chain termination or defects in sheet packing. From preliminary studies on sections of the bulk fibers, two main types of macrostructure are apparent: that exhibited by PRD 49 can be considered homogeneous, the other, found in Kevlar 49 and Kevlar, takes the form of an axially periodic assembly.

We would like to thank the Science Research Council for financial support, E.I. du Pont de Nemours and Co. Inc. for samples, and Dr. A. M. Hindeleh and Mr. A. Majeed for assistance with the X-ray diffraction studies.

REFERENCES

[1] W. N. Reynolds, *Chem. Phys. Carbon,* **11,** 1 (1973).

[2] D. C. Preversek, G. A. Tirpak, P. J. Marget and A. C. Reimschussel, *J. Macromol. Sci. Phys.,* **B9 (4),** 733 (1974).

[3] E. W. Fischer and S. Fakirov, *J. Mater. Sci.,* **11,** 1041 (1976).

[4] M. G. Dobb and R. Murray, *J. Micros.,* **101,** 299 (1974).

[5] D. T. Grubb and A. Keller, *Proc. 5th Eur. Cong. Electron Micros.,* Institute of Physics, London, 554 (1972).

[6] M. G. Dobb, A. M. Hindeleh, D. J. Johnson and B. P. Saville, *Nature (London)* **253,** 189 (1975).

[7] S. C. Bennett, M. G. Dobb, D. J. Johnson, R. Murray and B. P. Saville, *Proc. EMAG 75,* Bristol, Academic Press, London, 329 (1976).

[8] G. B. Carter and V. T. J. Schenk, *The Structure and Properties of Oriented Polymers,* Ed. I. M. Ward, Applied Science, London, Chapter 13, 1975.

[9] J. Preston, *Polym. Eng. Sci.,* **15,** 199 (1975).

[10] L. Penn, H. A. Nevey and T. T. Chiao, *J. Mater. Sci.,* **11,** 190 (1976).

[11] A. M. Hindeleh and D. J. Johnson, *Polymer,* **13,** 423 (1972).

[12] A. M. Hindeleh and D. J. Johnson, *Polymer,* **15,** 697 (1974).

[13] D. J. Johnson and D. Crawford, *J. Microsc.,* **98,** 313 (1973).

[14] A. R. Bunsell, *J. Mater. Sci.,* **10,** 1300 (1975).

[15] B. K. Vainshtein, *Diffraction of X-rays by Chain Molecules, Elsevier,* London, 294 (1966).

[16] M. G. Northolt, *Eur. Polym. J.,* **10,** 799 (1974).

[17] R. K. Hasegawa, Y. Chatani and H. Tadokoro, *Meet. Soc. Cryst., Jpn.,* Abstr. p 21, Osaka, Japan, 1973.

[18] K. Yabuki, H. Ito and T. Ota, *Sen'i Gakkaishi,* **31,** T 524 (1975).

[19] D. J. Johnson and C. N. Tyson, *J. Phys. D: Appl. Phys.,* **2,** 787 (1969); *ibid.,* **3,** 526 (1970).

[20] R. Perret and W. Ruland, *J. Appl. Crystallogr.,* **2,** 209 (1969); *ibid.,* **3,** 525 (1970).

[21] A. Fourdeux, R. Perret and W. Ruland, *J. Appl. Crystallogr.,* **1,** 252 (1968).

[22] W. Ruland and H. Tompa, *Acta Crystallogr.,* **A24,** 93, (1968).

[23] E.I. du Pont de Nemours and Co.: U.S. Patent, 3,869,430; Br. Patent, 1,391,501.

PROPERTIES OF POLYPROPYLENE FIBERS MADE FROM POLYMER HOMOLOGUE MIXTURES

FRIGYES GELEJI
*Hungarian Viscosa Factory, Central Research Laboratory,
Nyergesujfalu, Hungary*

LÁSZLÓ KÓCZY
Technical University, Budapest, Hungary

ISTVÁN FÜLÖP
Flax Spinning and Weaving Factory, Budakalasz, Hungary

GÉZA BODOR
Polymer Research Institute, Budapest, Hungary

SYNOPSIS

Fibers from mixtures of two polypropylene types were produced with the same conditions. The polymer base was a normal fiber grade polymer to which a high molecular weight polypropylene was added in amounts of 25, 50, and 75%. The extensibility of the fibers changed considerably by increasing the high polymer amount. The tensile properties were measured and compared with measured structural properties, such as crystallinity, crystalline particle size, amorphous orientation, amorphous region derived from small and wide-angle X-ray scattering. New terms, the specific amorphous orientation and the structural deformability index, were proposed, and it was shown that the tensile properties, i.e., strength, elongation at break, initial modulus and sonic modulus, are in very good correlation with them.

INTRODUCTION

It is well known that increasing the molecular weight of the polymer enhances the mechanical properties of the fibers made from these polymers. From the point of view of spinning technology a polymer with low molecular weight, i.e., with high melt index (about 20), is the most suitable. Polymers for normal extrusion processes are inconvenient for spinning. In the course of our experiments we observed that the mixture of a polymer of high molecular weight with a spinning-convenient low molecular weight polymer does not disturb the spinning conditions within limits. In this paper we shall discuss the properties of fibers produced this way.

Journal of Polymer Science: Polymer Symposium 58, 253–273 (1977)

TABLE I
Polymer Composition of Experimental Fibers

Sample	D%	H%	Melt index of the mixture	Melt index of the fiber
A	100	0	16,5	37,6
B	75	25	8,8	15,3
C	50	50	5,3	8,3
D	25	75	2,6	5,7

TABLE II
Maximal Drawability of Experimental Polypropylene Fibers

Sample	Maximum drawing ratio
A	1 : 8,0
B	1 : 6,0
C	1 : 5,0
D	1 : 4,0

EXPERIMENTAL

For the investigations we used a laboratory spinning device containing a Davo Viskosystem extruder. The spinnaret had 40 capillaries with 0.8 mm diameter. The spinning conditions were kept in all cases on the same parameters, i.e.,

Pressure of the polymer melt	50 kp/cm^2
Spinning velocity	540 m/min
Drawing velocity	150 m/min
Drawing temperature	130°C
Fineness of undrawn elementary fiber	64 dtex

In the experiments we used two types of polymers: (1) the polymer suitable for spinning, with melt index 16.5 (type D); (2) the high molecular weight polymer, with melt index 1.04 (type H) (measured at 230°C with 2.16 kp weight).

From these polymers we produced four different fibers with polymer compositions as shown in Table I.

The addition of high molecular weight polymer could reach 75%; above this amount spinning under mentioned conditions was impossible. The spinning was carried out from the polymer granules at 300°C spinning head temperature. The obtained fibers were not drawable to the same extent, the higher was the amount of high molecular weight polymer, the lower was the maximal drawability (Table II).

To be able to make comparisons between the several fibers, we have chosen the region 1:3 to 1:5. The fibers were drawn in the following sense—1:3; 1:3.5; 1:4; 1:4.5; 1:5 except sample D which was drawn till 1:4 maximum.

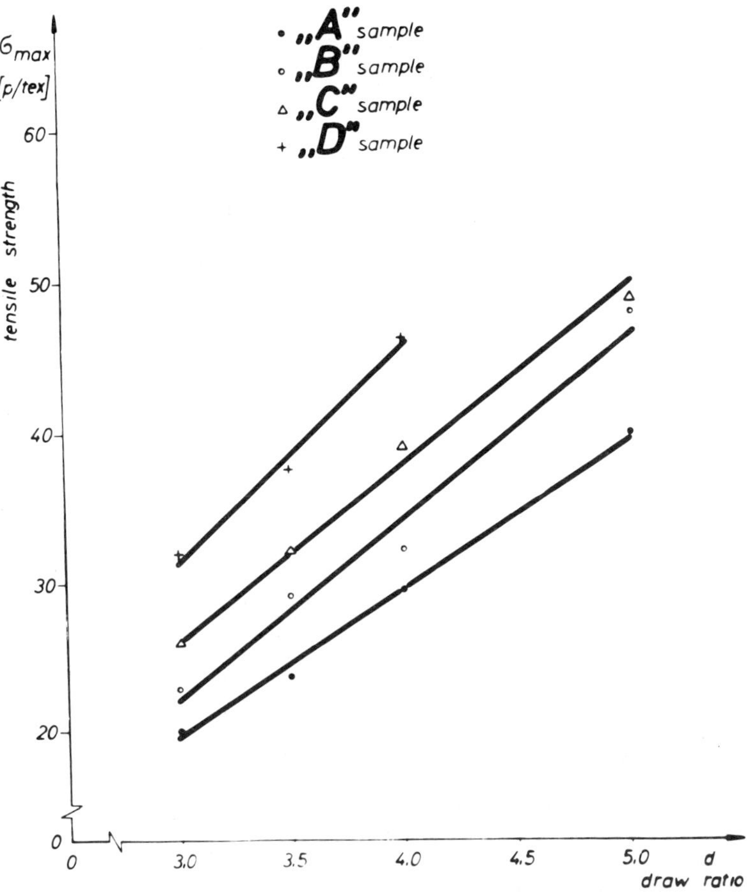

FIG. 1. Dependence of tensile strength in function of draw ratio and composition.

RESULTS AND DISCUSSION

Mechanical Properties as Function of Draw Ratio and H Content

Tensile Strength (σ_{max})

The tensile strength of the fibers is shown in Figure 1. It is clearly visible that in all cases the tenacity improves with growing drawing ratio. The improvement is growing linear in the investigated samples. On the other hand the addition of type H improves further the tenacity; by the highest H content there is about 60% improvement compared with the lower draw ratio.

GELEJI ET AL.

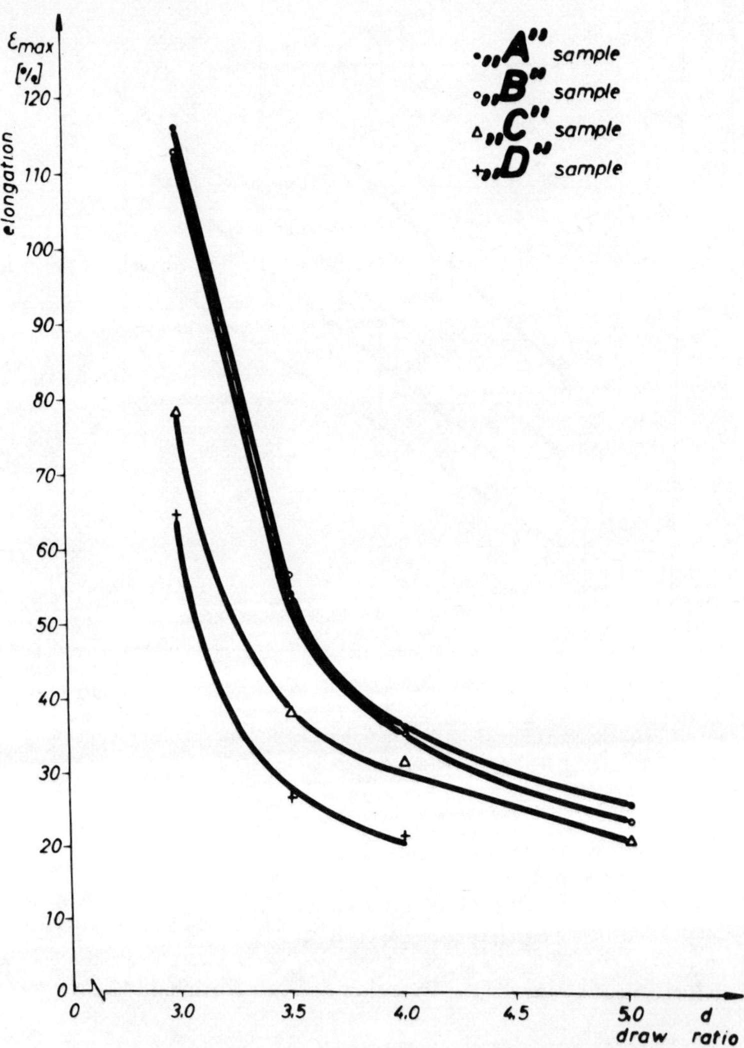

FIG. 2. Change of elongation to break with draw ratio and composition.

TABLE III

The Variation of the Young Modulus in Dependence of H Polymer Content and Draw Ratio

Sample	Draw ratio	E_o /initial/ / p/tex /	E_d /dynamic/ / p/tex /
A	1 : 3,0	260	563
	1 : 3,5	270	707
	1 : 4,0	347	805
	1 : 5,0	476	1097
B	1 : 3,0	254	554
	1 : 3,5	291	765
	1 : 4,0	359	854
	1 : 5,0	527	1098
C	1 : 3,0	288	648
	1 : 3,5	333	793
	1 : 4,0	465	875
	1 : 5,0	561	1146
D	1 : 3,0	298	647
	1 : 3,5	376	847
	1 : 4,0	555	1027

Elongation (ϵ_{max})

The values of the elongation to break correspond completely to that which was expected. The elongation decreases with the draw ratio very intensively and a further lowering may be achieved by the addition of H. The results are shown in Figure 2.

Dynamic and Initial Modulus Measurements

The dynamic modulus was determined by means of ultrasound propagation measurements (Dynamic Modulus Tester PPM-5) (Table III). The dynamic and initial modulus of the fibers varied in the same sense in all samples (Figs. 3 and 4).

Deformation Characteristics

The deformability, i.e., the values of plastic deformation after 5, 10, and 15 p/tex strain, were measured. The results are shown in Figures 5, 6, 7 and 8. We could observe that the deformability of the fibers is more influenced by the correct draw ratio than by the high molecular part of the mixture. At smaller drawing ratios however, the influence of the high molecular part is greater.

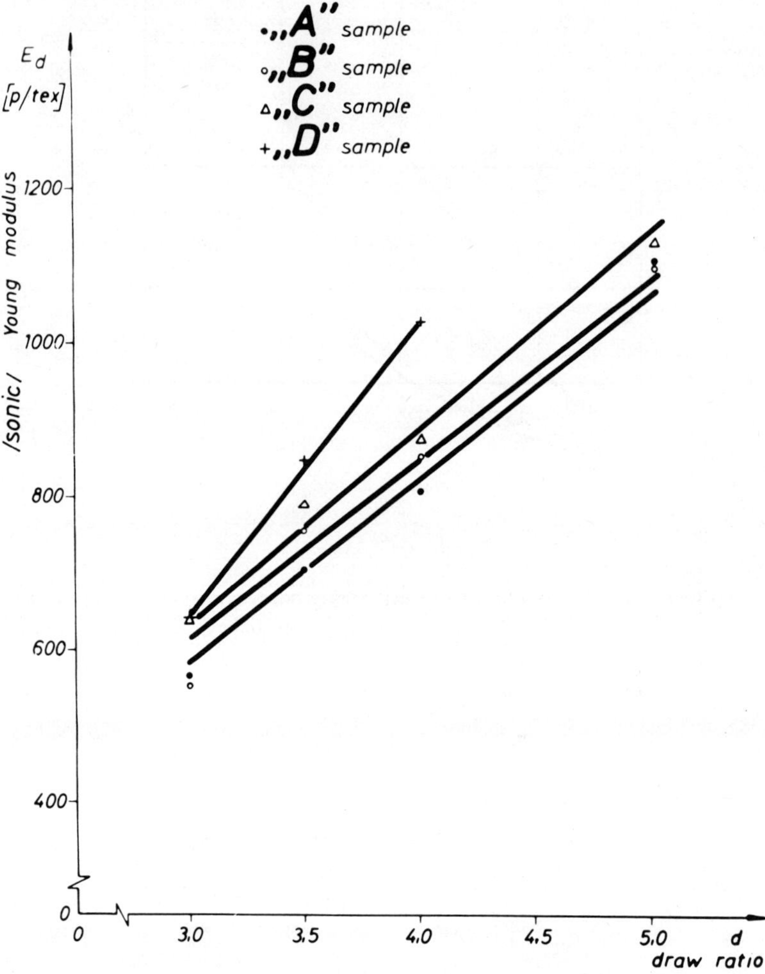

FIG. 3. Change of the dynamic modulus with draw ratio and composition.

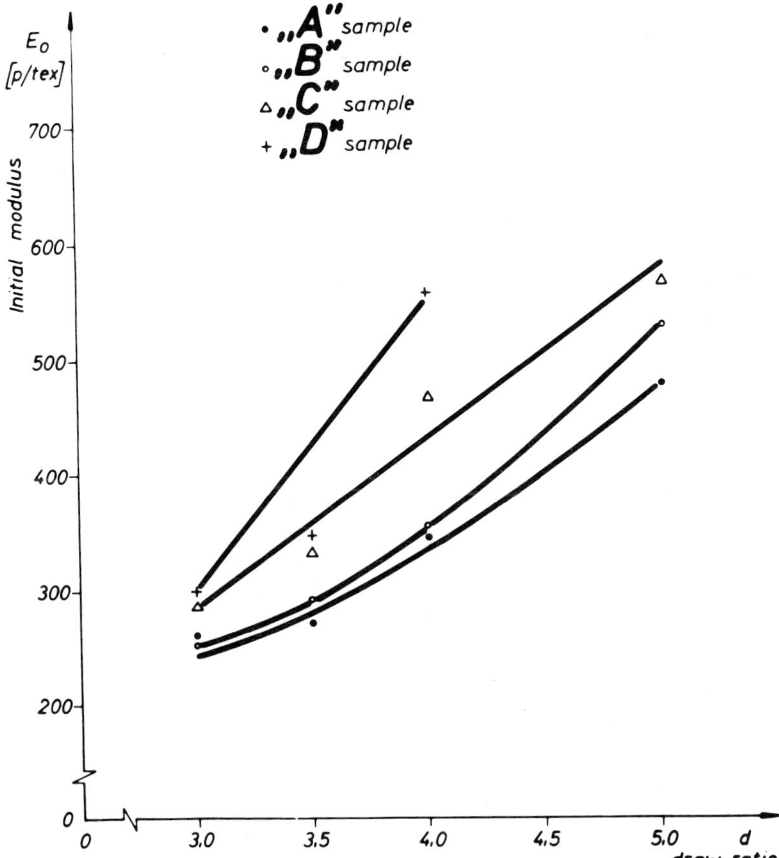

FIG. 4. Change of the initial modulus with draw ratio and composition.

Fine Structure Investigations

Birefringence

First of all the change in birefringence was investigated (Fig. 9). As the results show, the birefringence of fibers with high polymer content follow a logical sequence in the low draw ratio region, the difference in the values getting smaller at the higher draw ratios, all of them approaching the value $\Delta n = 32 \times 10^{-3}$.

Crystallinity

The crystallinity was measured from the density, in a density gradient tube (Table IV).

X-Ray Investigations

The fibers were investigated by small and wide-angle X-ray scattering. The results are listed in Table IV.

The long period, obtained from small-angle X-ray measurements, shows a decreasing tendency in all samples with increasing draw ratio.

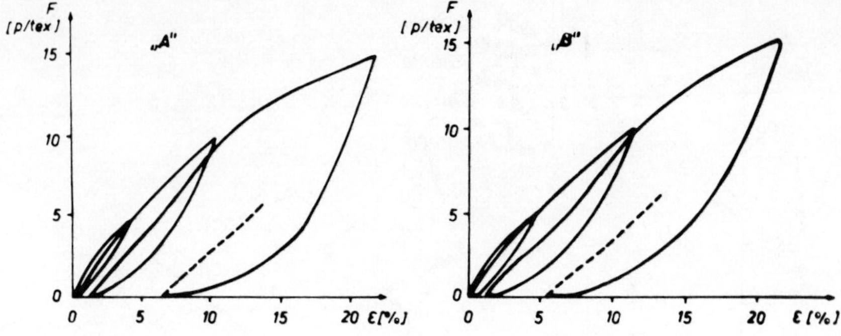

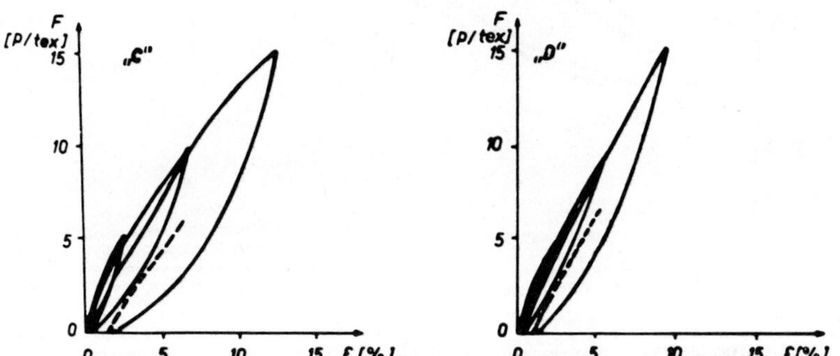

FIG. 5. Change the shape of deformability curves at 5, 10 and 15 p/tex in function of the polymer composition.

The crystalline particle size was determined from the line broadening of the wide-angle reflections.

The values of the amorphous region (T_{am}) were calculated by subtracting the crystalline particle size from the small-angle long period values.

The crystalline orientation was measured at $2\Theta = 14°$ and $17°$, the crystalline orientation data (f_c) were calculated by the method of Krimm and Tobolsky [1]. We could observe that the crystalline orientation in every case improves with the drawing ratio, to reach at the end in all samples about the same value. The starting values, however, show slight differences in dependence of H content in both f_c values ($14°$ and $17°$).

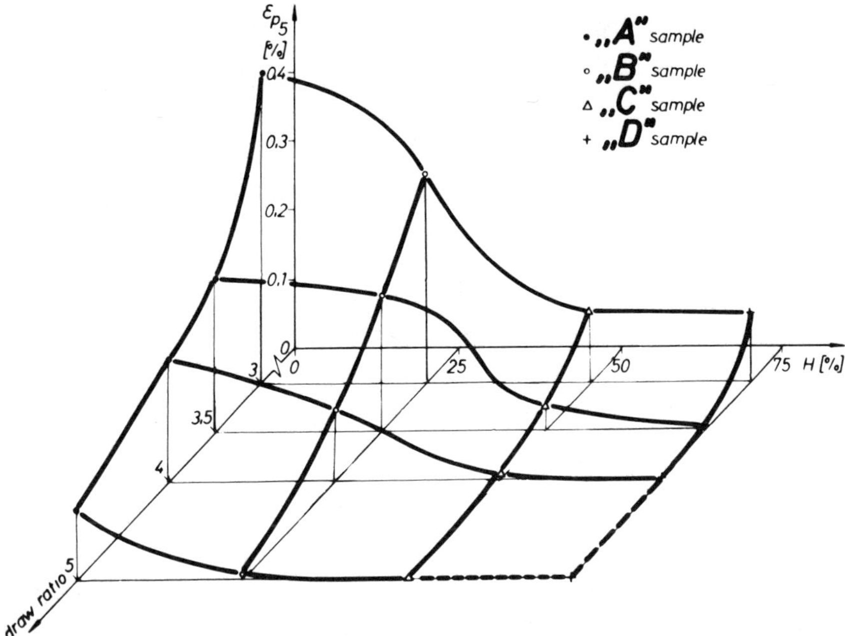

FIG. 6. Change of plastic deformation at 5 p/tex strain in function of draw ratio and composition.

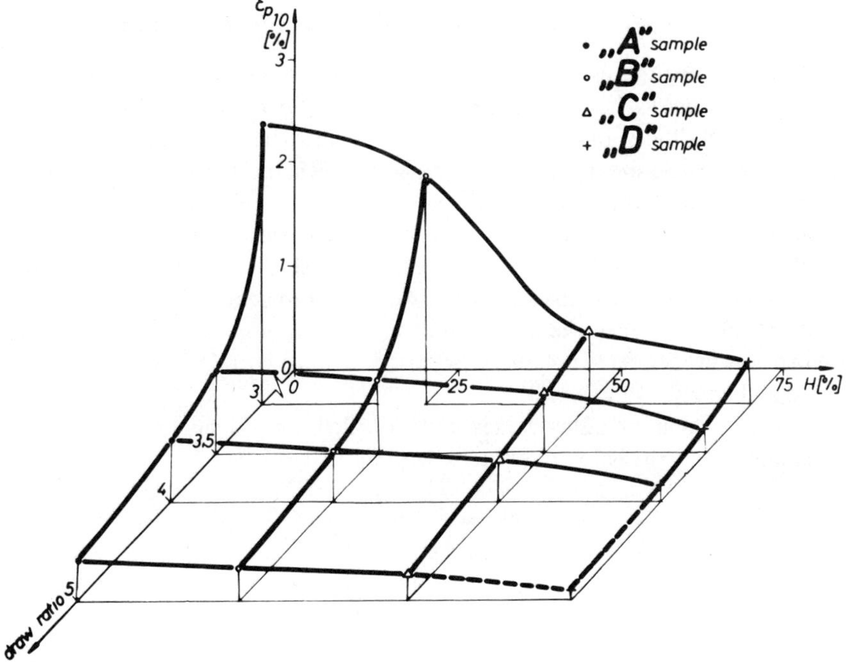

FIG. 7. Change of plastic deformation at 10 p/tex strain in function of draw ratio and composition.

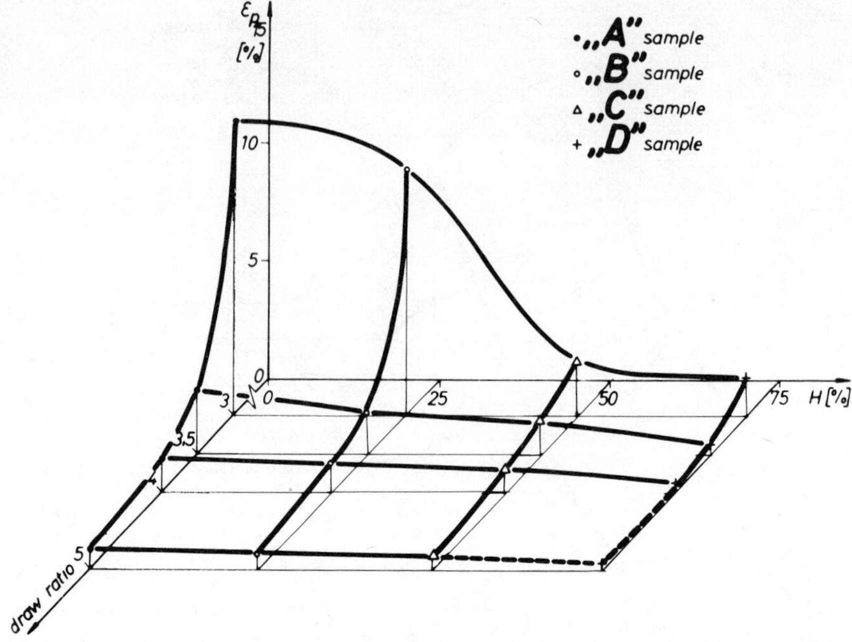

FIG. 8. Change of plastic deformation at 15 p/tex strain in function of draw ratio and composition.

Correlation of the Mechanical and Structural Properties

Besides a practical aspect of our work, our intention was to try to find structural correlations to these mechanical properties of fibers. Although it is obvious that the crystallinity and orientation of crystalline particles, together with the crystalline particle size, have basic influence on mechanical properties in general, we see that these data cannot explain such significant differences in properties which we and all other researchers working in these fields normally obtain. From Table IV we can establish that the values for crystallinity and for crystalline orientation lie very near to each other. According to general opinion, the molecular mobility, hence the deformability, is much more connected with the amorphous properties of polymers. We tried to find a correlation to these measured and calculated values.

From the sonic modulus values, the so-called amorphous orientation was calculated, with the help of Samuel's formula [2]:

$$\frac{3}{2}(\Delta E^{-1}) = \frac{\beta f_c}{E_{t,c}} + \frac{(1-\beta)f_{am}}{E_{t,am}}$$

$$\Delta E^{-1} = \left(\frac{1}{E_u} - \frac{1}{E_{or}}\right)$$

where

E_u = 2.5 × 10^{10} dyn/cm², measured by Bodor,

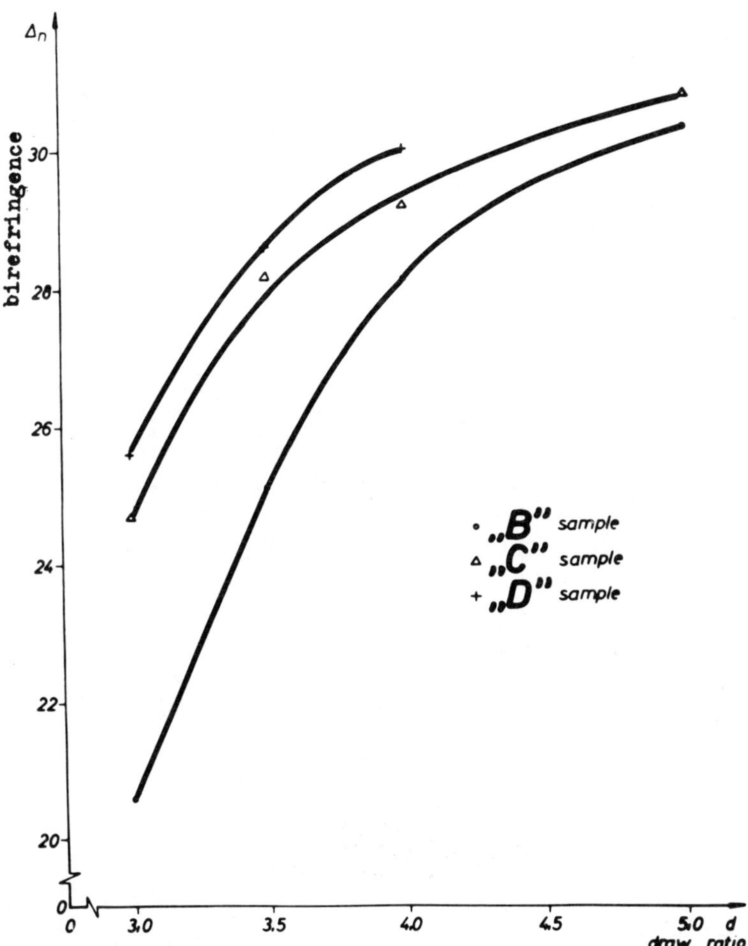

FIG. 9. The change of birefringence in function of draw ratio and composition.

TABLE IV
Values of Fine Structure Measurements

Sample and draw ratio		Crystal-linity %	Small angle X-ray periodicity L /Å/	Crystalline particle size P /Å/	Amorphous region T_{am} /Å/	Crystalline orientation f_{cr}		Amorphous orien-tation f_{am}	$f_{spec,am}$. 10^{-2}	S.D.I.
						$2\theta=14°$	$2\theta=17°$			
A	1:3,0	0,652	158	71,0	87,0	0,850	0,875	0,476	0,55	182,80
	1:3,5	0,649	140	74,0	66,0	0,885	0,900	0,641	0,97	102,98
	1:4,0	0,642	148	74,0	74,0	0,895	0,900	0,723	0,98	102,35
	1:5,0	0,636	156	74,0	82,0	0,900	0,920	0,861	1,05	95,23
B	1:3,0	0,657	170	89,0	81,0	0,850	0,870	0,464	0,57	174,52
	1:3,5	0,683	154	74,0	78,0	0,880	0,894	0,755	0,94	103,30
	1:4,0	0,656	144	68,5	75,5	0,90	0,900	0,778	1,03	97,08
	1:5,0	0,636	120	68,5	51,5	0,90	0,915	0,869	1,68	59,52
C	1:3,0	0,666	160	81,0	79,0	0,875	0,89	0,603	0,76	131,06
	1:3,5	0,660	150	74,0	76,0	0,88	0,895	0,743	0,98	102,35
	1:4,0	0,650	150	74,0	76,0	0,895	0,91	0,798	1,05	95,23
	1:5,0	0,610	140	70,0	70,0	0,900	0,91	0,849	1,10	82,64
D	1:3,0	0,660	154	81,0	73,0	0,895	0,895	0,558	0,76	130,89
	1:3,5	0,645	150	81,0	69,0	0,895	0,905	0,558	0,81	90,75
	1:4,0	0,628	142	81,0	61,0	0,895	0,915	0,828	1,35	73,69

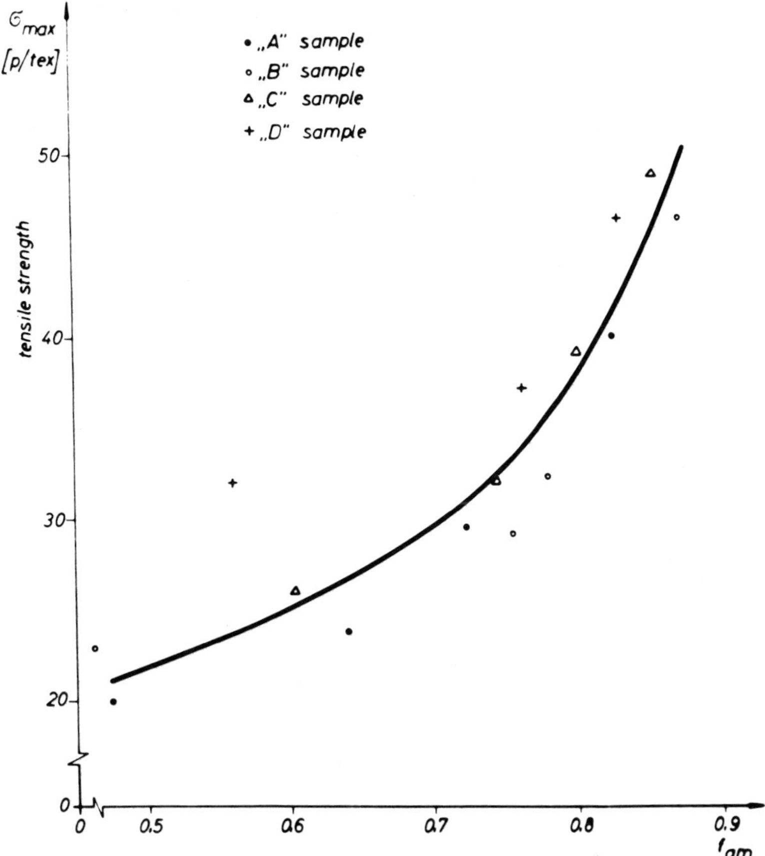

FIG. 10. The characteristic change of tensile strength in function of the amorphous orientation.

$$E_{t,c} = 3.96 \times 10^{10} \text{ dyn/cm}^2,$$
$$E_{t,am} = 1.06 \times 10^{10} \text{ dyn/cm}^2,$$
$$\beta = \text{crystallinity},$$
$$E_{or} = \text{sonic modulus of the oriented sample.}$$

The values of amorphous orientation (f_{am}) increased with the growing drawing ratio in all fiber types and with increasing amount of H added as well.

First we compared the mechanical properties with the values of amorphous orientation (f_{am}).

The tensile strength in function of f_{am} shows characteristic dependence, although the values deviate in some cases (Fig. 10).

The elongation as function of f_{am} has a linear character, however logically in reciprocal manner (Fig. 11).

There is very good functional agreement between f_{am} and the other moduli (E_o and E_d) (Figs. 12 and 13); the character is the same as by the tensile strength.

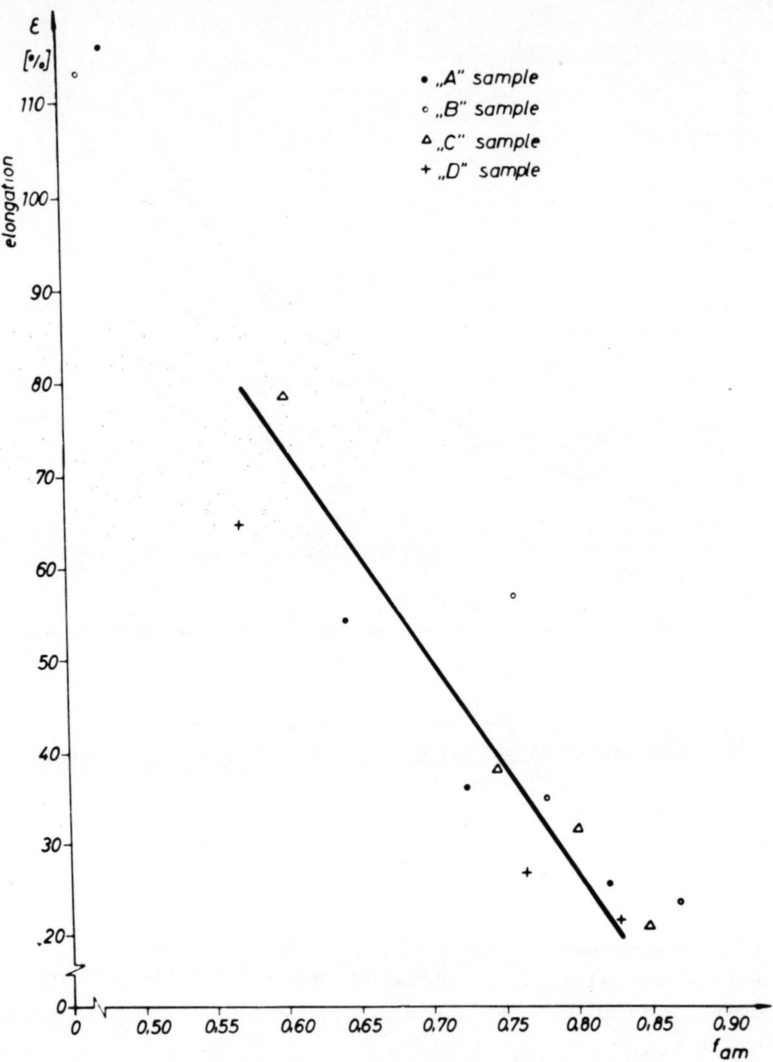

FIG. 11. The characteristic change of the elongation in function of the amorphous orientation.

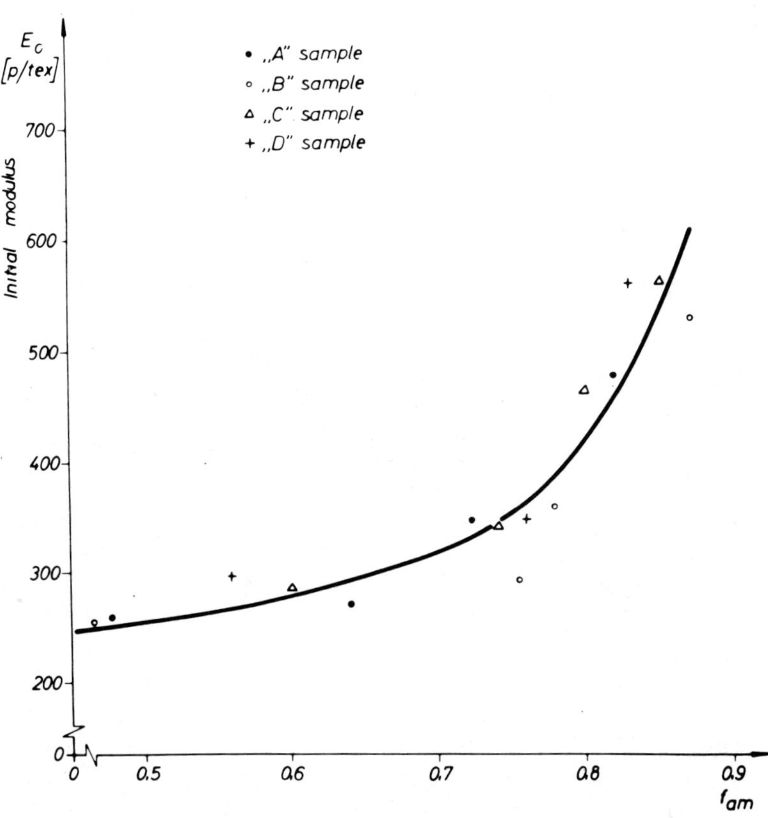

FIG. 12. Variation of the initial modulus with f_{am}.

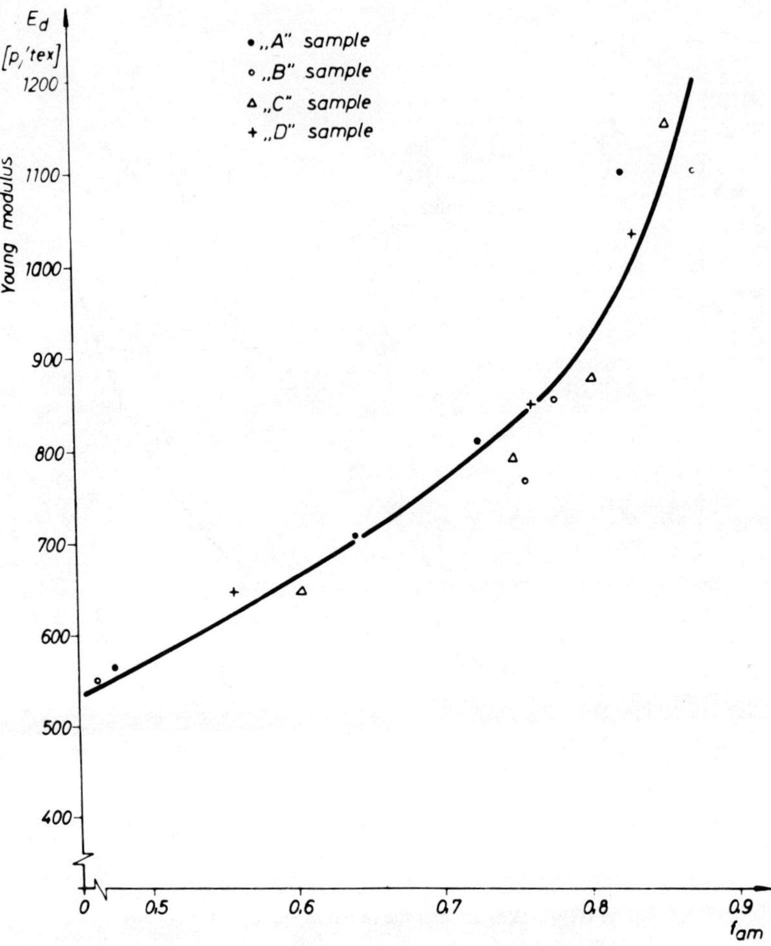

FIG. 13. Variation of the dynamic modulase with f_{am}.

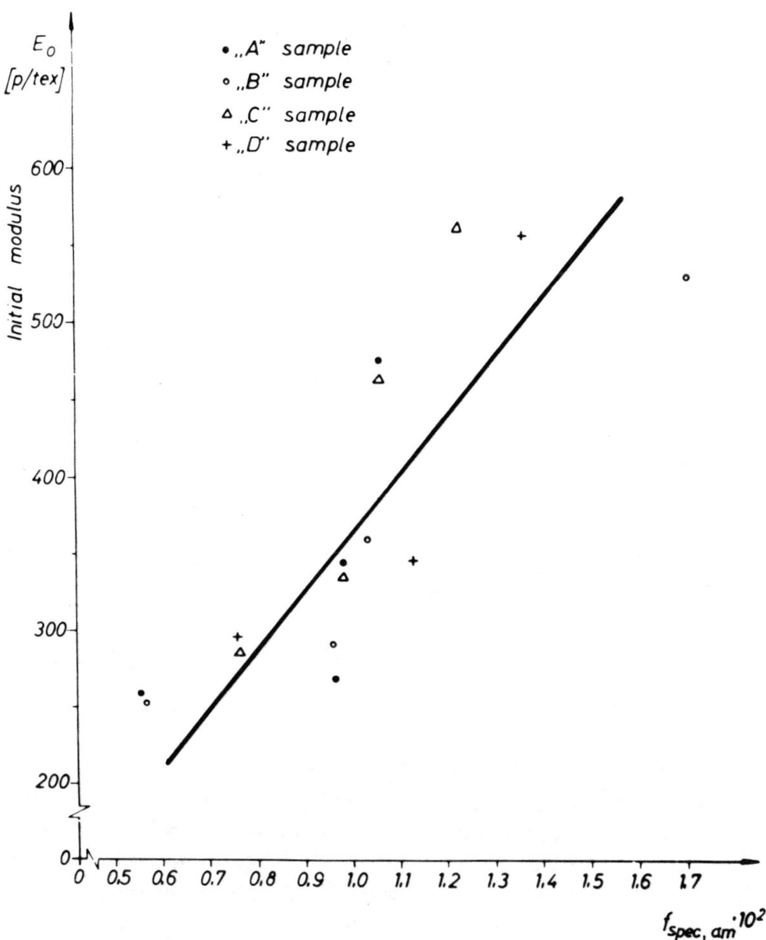

FIG. 14. Variation of the initial modulus with the specific amorphous orientation.

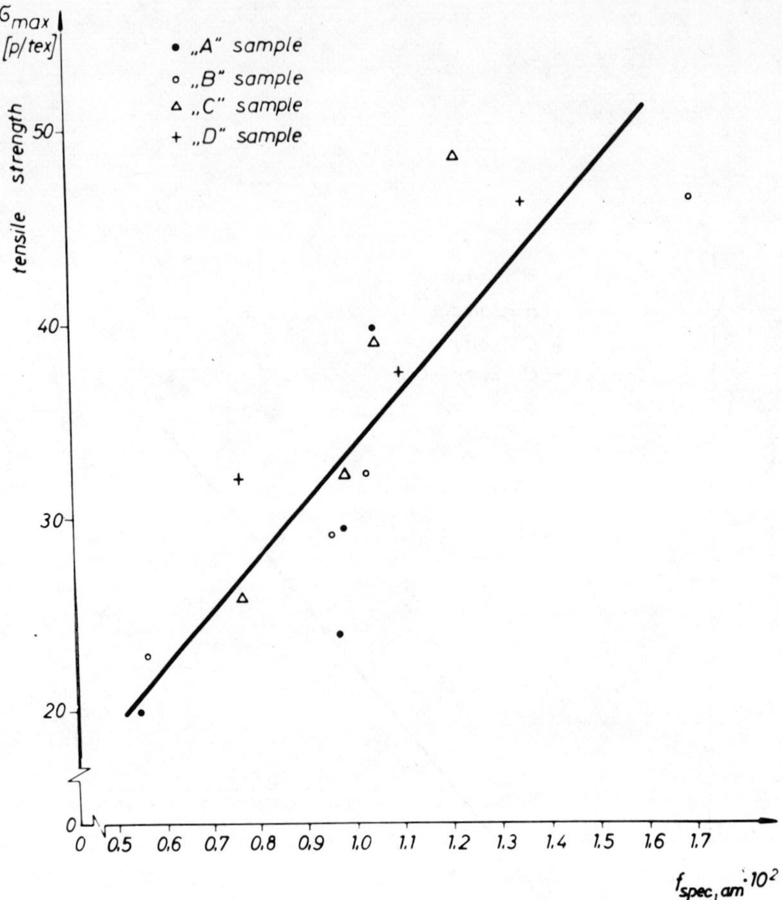

FIG. 15. Variation of tensile strength with the specific amorphous orientation.

Specific Amorphous Orientation

We think that it is interesting to try to combine the values of the amorphous region with those of amorphous orientation. As we already pointed out, the deformability must be very thoroughly combined with the amorphous properties. The deformability has to grow with growing amorphous regions, hence with growing draw ratio—when the amorphous region decreases—it has to decrease.

On the other hand, the orientation of amorphous regions acts inversely on the deformability; it decreases with growing orientation.

The two values could be combined in one index; the amorphous orientation divided by the amorphous region f_{am}/T_{am} is proposed to be called "Specific Amorphous Orientation," $f_{spec.\,am}$. We found that this value is proportional to important properties such as initial modulus, tensile strength and sonic Young modulus (Figs. 14, 15, 16). The reciprocal of this value we propose to call "Structural Deformability Index" (S.D.I.):

$$f_{spec.\,am} = \frac{1}{\text{S.D.I.}} = \frac{f_{am}}{T_{am}}$$

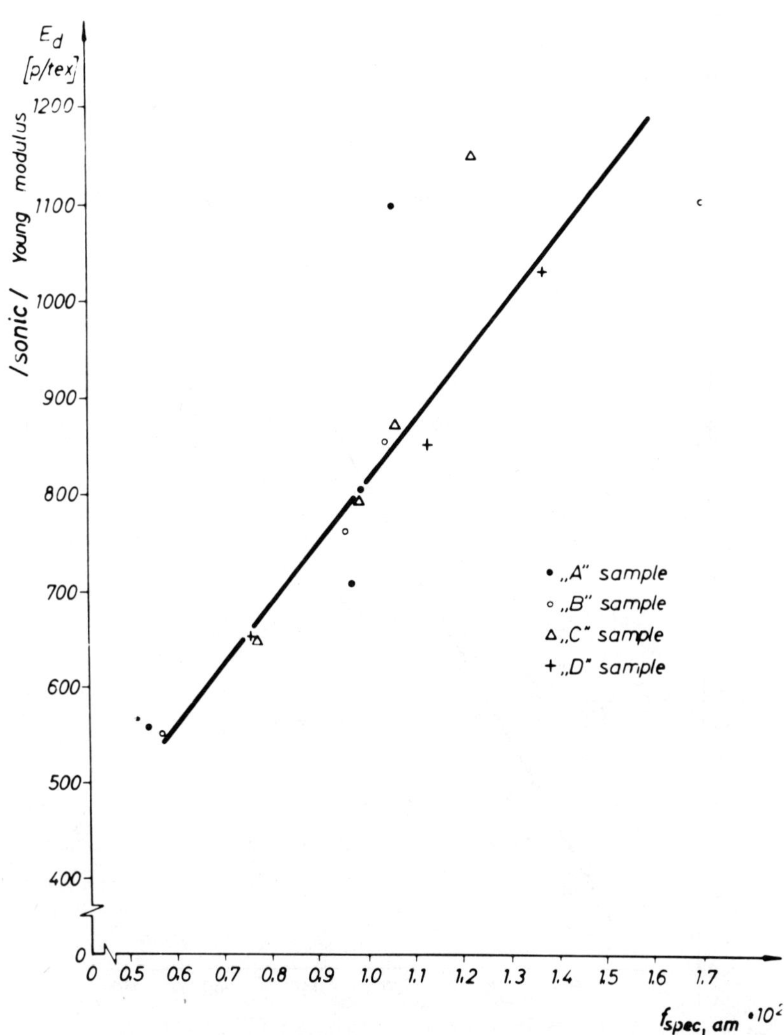

FIG. 16. Variation of the dynamic modulus with the specific amorphous orientation.

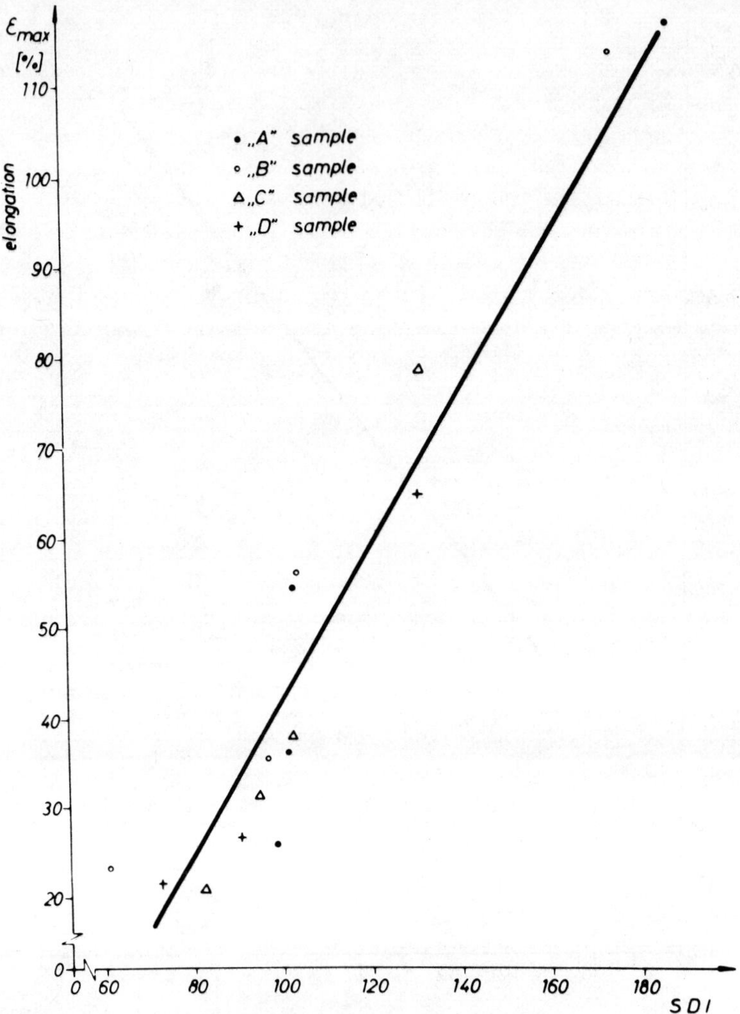

FIG. 17. Variation of the elongation with the S.D.I.

This S.D.I. value is proportional to the elongation of the fibers (Fig. 17). Probably the concept of the specific amorphous orientation will be accepted by the morphologists; the S.D.I. index is nearer to the train of thought of the people working on the field of mechanical properties.

The values for $f_{spec.\ am}$ and for S.D.I. are listed also in Table IV.

CONCLUSION

From the results obtained we came to the following considerations: The basic mechanical properties of the fibers may correlate with the structural characteristics. Among these, the parameters connected with the crystallinity have less exciting influence on the differences in mechanical properties which on one hand may be influenced with the production circumstances or with the molecular weight of the polymer, although we do not want to question that the crystallinity and particle size have importance on influencing these properties.

The amorphous regions as weak points in a chain have greater influence on properties which are tightly connected with tensile properties. All investigated mechanical properties were in some way tensile properties.

The size of the amorphous region decreases with growing draw ratio and with growing amount of high polymer added to the starting polymer. The amorphous orientation increases by the same circumstances.

Taking into consideration that the orientation of the amorphous part may take place only in the "amorphous region" we think it is permitted to correlate the mechanical properties with the specific amorphous orientation and with the structural deformability index, which terms were defined above. The linear results convinced us that these specific values show a very good correlation with the fiber properties.

REFERENCES

[1] S. Krimm, and A. V. Tobolsky, *J. Polym. Sci.,* **17,** 57 (1951).

[2] R. J. Samuels, *Structured Polymer Properties.,* John Wiley Intersciences Publishers, New York, London, 1974.

METHOD FOR THE QUANTITATIVE CHARACTERIZATION OF THE SUPERMOLECULAR STRUCTURE OF HIGHLY ORIENTED POLYMERS

B. J. JUNGNICKEL

*Institut für Polymerenchemie der Akademie der Wissenschaften
der DDR, 153 Teltow-Seehof, DDR*

SYNOPSIS

The morphology of polymers having fiber texture is commonly considered to consist of a fibrillar structure. The author proposes a method for the quantitative description of the internal structure of the fibrils by analysis of X-ray small-angle scattering patterns. A theoretically relevant quantity, "morphology index" (μ) characterizing the fibril morphology can be calculated from structure parameters. Microfibrillar ($\mu \gg 1$) and lamellar ($\mu \ll 1$) structures represent limiting cases of the actual structural order. The dependence of μ on the temperature was determined for a series of fixed ends annealed polyamide-6 filaments. A distinct change from the microfibrillar to the lamellar structure was verified, with slight increase of the microfibrillar character up to 140°C. With drawn polyamide-6 samples the morphology index remains unchanged, in spite of slight changes of individual parameters, such as crystallite size, and weak shearing of fibrils. This suggests that drawing does not lead to essential changes of the internal structure of fibrils.

INTRODUCTION

Monoaxial drawing of fibers and films to a sufficiently high degree results in an axial texture in which highly oriented polymer chains are arranged parallel to the texture axis turned in drawing direction. The supermolecular structure is fibrillar (Fig. 1): As shown in a scheme published by Peterlin [1], extended structural units are arranged more or less parallel to the texture axis. Structural changes caused by thermal and mechanical treatment can be classified as

(1) changes of the internal structure of fibrils,
(2) changes of the mutual arrangement of fibrils,
(3) changes of the interface of fibrils.

The structural characteristics outlined above are highly important with respect to the description and theoretical interpretation of deformation and relaxation processes. This paper is concerned with the first topic.

The individual fibril can be composed of smaller units called microfibrils (Fig. 2). The microfibrils consist of ordered "crystalline" and unordered "amorphous" regions which are arranged in regular succession, leading to a substructure oriented predominantly in fiber direction. For this arrangement Bonart [5] in-

Journal of Polymer Science: Polymer Symposium 58, 275–281 (1977)

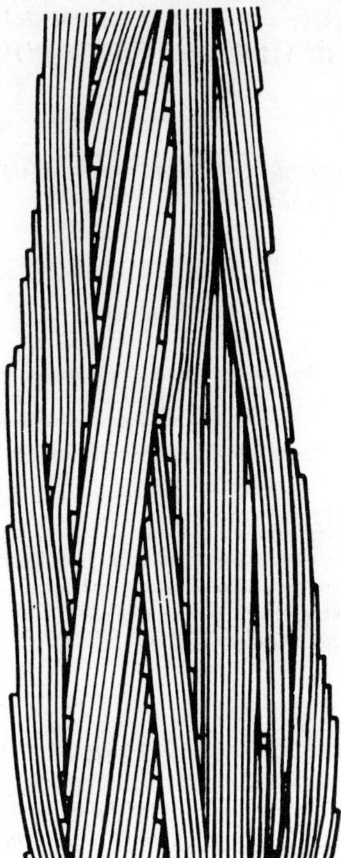

FIG. 1. Fibrillar model of fibrous structure (after ref. [1]).

troduced the term "longitudinal structure." Frequently, the ordered regions are laterally connected in a way that causes the visual impression of crystal lamellae positioned oblique or perpendicular to the texture axis [6] (Fig. 3). The structure along the lamellae is called "lateral structure." These lamellae are composed of individual "mosaic blocks" [7], that is, of crystallites which on appropriate treatment of the sample turn out to be individual morphological units and result in the formation of the individual microfibrils.

THEORETICAL CONSIDERATIONS

Microfibrillar and lamellar structures are visually perceptible limiting cases of the internal morphology of fibrils. Based on this statement, an approach to an analytical description of the real structure will be proposed. To this aim, the mutual three-dimensional arrangement of crystallites in a fibril is considered as a distorted lattice. We suppose, for the sake of later calculations, that there are paracrystalline distortions, which can be treated according to the algorithm of Hosemann [8]. Thus, the scheme described above can be explained as follows:

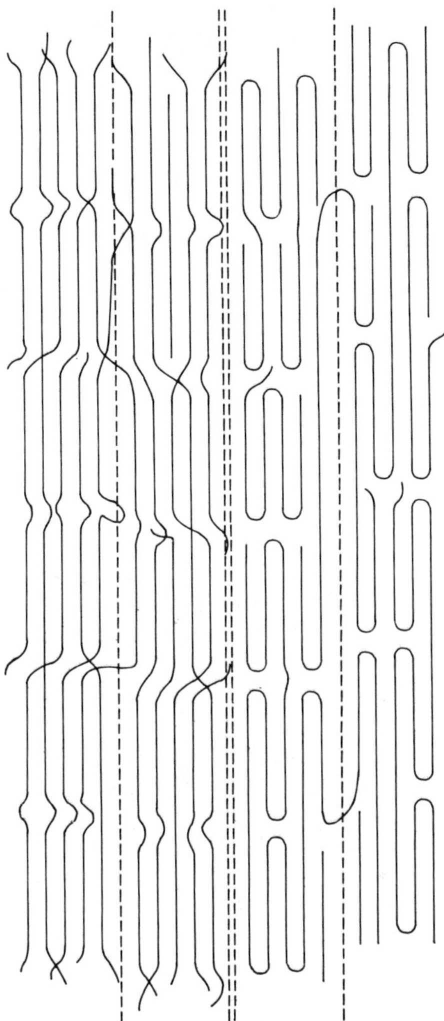

FIG. 2. Microfibrillar substructure of fibrils; left, micellar model after Hess and Kiessig [2]; right, with chain folding after Bonart, Hosemann [3] and Peterlin [4].

If the crystalline order in fiber direction is considerably higher than that in lateral direction, that is, if there is no correlation between the position of laterally neighboring crystallites, then the substructure is microfibrillar. This means that the relative paracrystalline distortions of the longitudinal structure are considerably smaller than those of the lateral structure. In the opposite case the superlattice is particularly complete in the lateral direction. If the relative paracrystalline distortions of the lateral structure are considerably smaller than those of the longitudinal structure, the crystallites are arranged very regularly in lateral direction and the structure is lamellar. This distinction was already suggested some years ago [9].

By mathematical expressions this can be described as follows (Fig. 4): **P** may be the connecting vector of crystallites arranged one upon another (the so-called

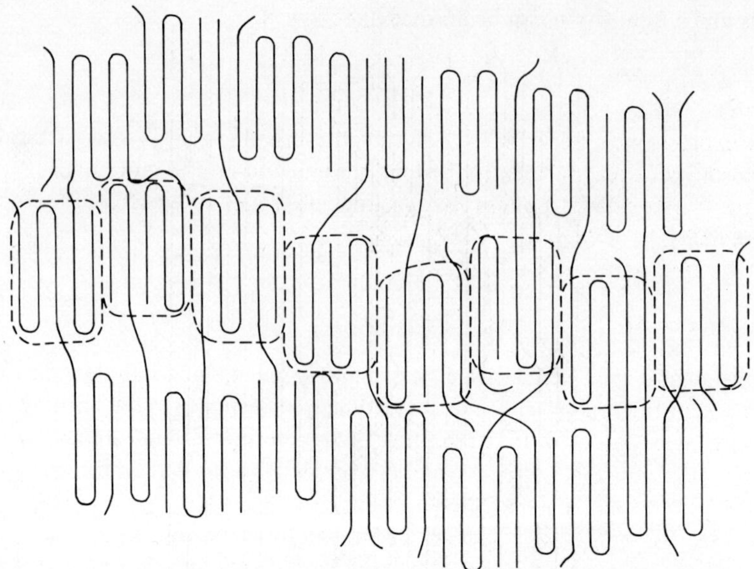

FIG. 3. Lamellar substructure of fibrils. The mean lamella is graphically subdivided in mosaic blocks.

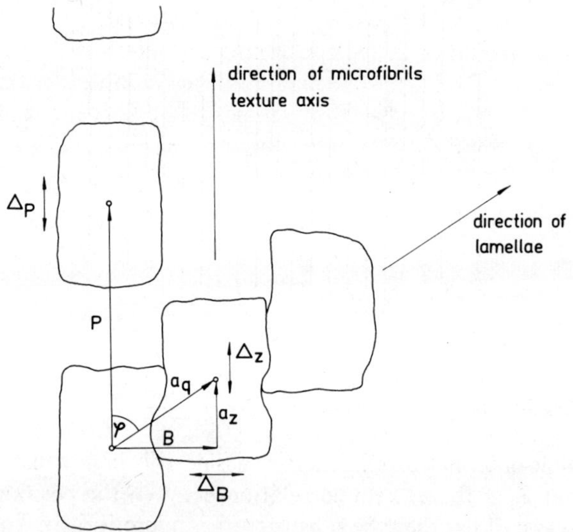

FIG. 4. Scheme explaining the quantities of eq. (2).

long period); $\mathbf{a_q}$ may be the connecting vector of crystallites laying side by side with the components $\mathbf{a_z}$ for the longitudinal shift and $\mathbf{B}$ for the broadness of the crystallites.

Further, Δ_P, Δ_z and Δ_B may be the corresponding parameters of paracrystalline fluctuations. Then the expression

$$\cot \varphi = \mathbf{a_z}/\mathbf{B} \qquad (1)$$

holds and a quantity μ can be defined as follows:

$$\mu = \frac{\mathbf{P}(\cot^2 \varphi \Delta_B^2 + \Delta_z^2)}{\Delta_P^2(\mathbf{B}^2 + \mathbf{a}_z^2)^{1/2}} \tag{2}$$

This expression is obtained by relating the broadness of X-ray small-angle scattering reflexes ascribed to the longitudinal and lateral structure [10].

Considering the introducing suggestions of this chapter, the following condition holds:

$$\mu \ll 1 \frown \text{lamellar order}$$

$$\mu \gg 1 \frown \text{microfibrillar order} \tag{3}$$

The value of μ, here being called "morphology index," of a sample indicates to what extent the real structure differs from or corresponds to the limiting cases of lamellar or microfibrillar structure. The determination of μ of a series of samples reveals in what direction and in which way the structural changes within the fibrils proceed.

As already mentioned, μ can be determined by evaluating X-ray small-angle scattering patterns. The determination of $\mathbf{P}$ and Δ_P can be performed by several methods, e.g., one proposed by Vonk and Kortleve [11]. For determining $\mathbf{B}$, also various methods have already been described in the literature [12]. The values of Δ_z, Δ_B and $\mathbf{a}_z$ can be determined by a method recently published by the author [9, 10, 13].

Two remarks should be added concerning the determination of μ:

1. Even if so-called layer line diagrams are available, the angle φ between the direction of lamellae and microfibrils can, contrarily to the usual interpretation, deviate from 90°;

2. The interpretation of the importance and the definition of μ do not only relate to samples having axial texture but also to double-oriented samples with uniplanar-axial texture.

EXPERIMENTAL RESULTS

The morphology index was determined with two series of polyamide-6 samples subjected to different treatment. With the first series the samples were annealed with fixed ends at different temperatures in vacuum. The results are shown in Figure 5. The untreated sample presents a microfibrillar structure ($\mu = 3$) which becomes more distinct up to 150°C. This was expected because annealing was performed under the condition of fixed ends. At higher temperatures, recrystallization to a higher extent may proceed and results in transformation to the lamellar structure. The value of μ decreases below unity. It should be mentioned that the function $\mu(T)$ depends essentially on the course of the function $\Delta_z(T)$. The parameter Δ_z, describing the correlation of the position of laterally neighboring crystallites, shows in dependence on the annealing temperature a similar course.

Summarily, this method allows to show *numerically*—that is, the essence of the proposed method—the intensive changes occurring within the fibrils during

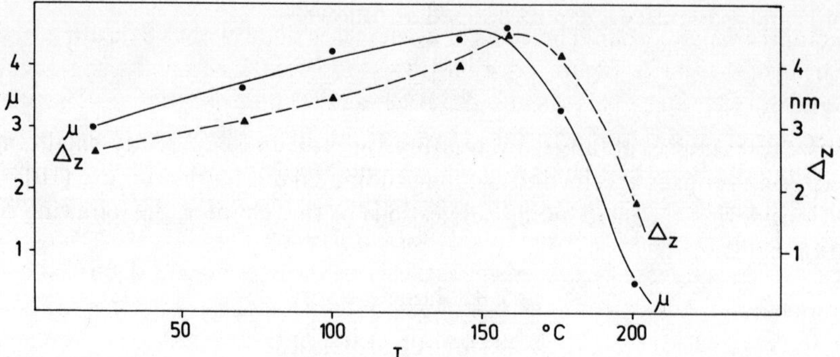

FIG. 5. Dependence of μ and Δ_Z of polyamide-6 fibers on annealing temperature T.

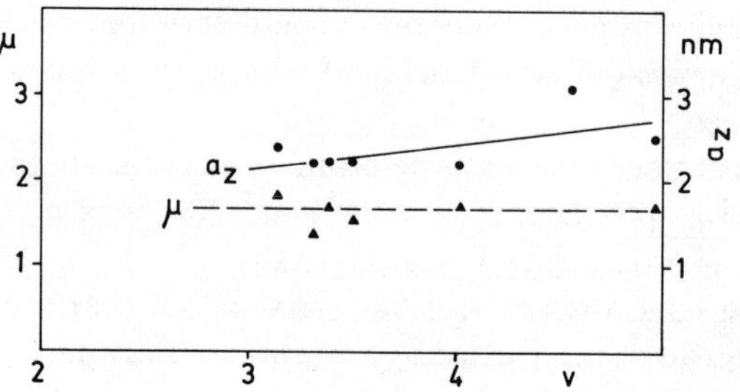

FIG. 6. Dependence of μ and a_Z of polyamide-6 fibers on draw ratio.

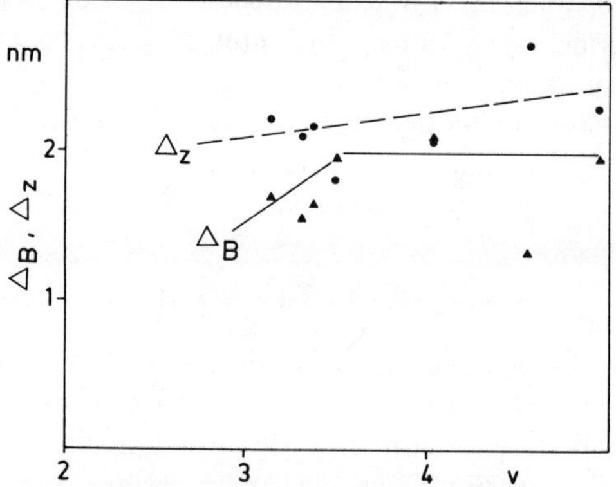

FIG. 7. Dependence of Δ_Z and Δ_B of polyamide-6 fibers on draw ratio.

annealing. This agrees with the established concept of structural changes on annealing [14].

Quite another experimental result was obtained with a series of polyamide-6 fibers which were drawn to different degree. As shown in Figure 6 the value of μ is constant within the drawing range investigated. At low values of the drawing degree the morphology index could not be determined because the small-angle scattering patterns do not present the layer line diagram, which is a necessary basis of our evaluation. According to our results, the character of fibrillar inner morphology is not changed during drawing, although within the fibrils slight structural changes occur. The value of a_z increases slightly, thus indicating slight shearing of the fibrils. Figure 7 demonstrates that also the correlation parameters Δ_z and Δ_B increase. This is caused by the general deterioration of the supermolecular crystallographic order. However, all these changes are only weak. This suggests that the structural changes induced by drawing result predominantly in changes of the state of aggregation of the fibrils, e.g., by slipping processes. Also changes at the interface of fibrils are possible. However, both these processes cannot be proved conclusively by the experiments described here. Summarily, in our opinion the morphology index can be regarded as a useful tool for the quantitative characterization of the fibrillar structure.

Finally I should like to acknowledge the cooperation of the X-ray investigation team of the Institute of Macromolecular Chemistry of the Academy of Science of the CSSR, particularly the cooperation of Dr. Baldrian in evaluating the X-ray small-angle scattering patterns.

REFERENCES

[1] A. Peterlin, *Int. J. Fract.*, **11**, 761 (1975).

[2] K. Hess and H. Kiessig, *Z. Phys. Chem.*, **193**, 196 (1944).

[3] R. Bonart and R. Hosemann, *Kolloid-Z.*, **186**, 16 (1962).

[4] A. Peterlin, *Kolloid Z. Z. Polym.*, **216/217**, 129 (1967).

[5] R. Bonart, *Kolloid-Z.*, **194**, 97 (1964).

[6] K. Hess, H. Mahl, and E. Guter, *Kolloid-Z.*, **155**, 1 (1957).

[7] K. Sakaoku and A. Peterlin, *Makromol. Chem.*, **108**, 234 (1967).

[8] R. Hosemann, and S. N. Bagchi, *Direct Analysis of Diffraction by Matter*, North Holland Publ. Comp., Amsterdam, 1962.

[9] B. J. Jungnickel, M. Teichgräber, and Ch. Ruscher, *Faserforsch. Textiltech.*, **24**, 423 (1973).

[10] B. J. Jungnickel, Dissertation, Berlin, 1973.

[11] C. G. Vonk and G. Kortleve, *Kolloid Z. Z. Polym.*, **220**, 19 (1967).

[12] M. A. Gezalov, V. S. Kuksenko, and A. I. Slucker, *Vysokomol. Soedin.*, **A12**, 1787 (1970); W. O. Statton, *J. Appl. Phys.*, **28**, 1111 (1957).

[13] B. J. Jungnickel, M. Teichgräber, and Ch. Ruscher, *Faserforsch. Textiltech.*, **25**, 95 (1974).

[14] H. G. Zachmann, *Fortschr. Hochpolym. Forsch.*, **3**, 581 (1964); T. Arakawa, F. Nagatoshi, and N. Arai, *J. Polym. Sci. A-2: Polym. Phys.*, **7**, 1461 (1969).

CHAIN ORIENTATION DISTRIBUTION AND ELASTIC PROPERTIES OF POLY (*p*-PHENYLENE TEREPHTHALAMIDE), A "RIGID ROD" POLYMER

M. G. NORTHOLT and J. J. VAN AARTSEN
Akzo Research Laboratories Arnhem, Corporate Research Department, Arnhem, the Netherlands

SYNOPSIS

On the basis of a single-phase structural model a theory has been developed, which yields a relation between the fiber compliance and the chain axis distribution. It is shown that this relation, viz.

$$S_{33} = e_3{}^{-1} + A\langle \sin^2 \phi \rangle + B\langle \sin^4 \phi \rangle$$

is quantitatively fulfilled for poly(*p*-phenylene terephthalamide) fibers studied in this investigation.

INTRODUCTION

Aromatic polyamide fibers, a new kind of high strength polymeric material, have been developed in the past decade [1, 2]. Poly(*p*-phenylene terephthalamide), abbreviated here as P-*p*PTA, is at present the best known example. This fiber has exceptional mechanical and thermomechanical properties, e.g., high elastic modulus and low elongation at break. The tensile curve in Figure 1 shows a constant modulus up to very high stresses. Another interesting feature is that for fibers with nearly parallel chain orientation the fiber modulus E_{of}, by which here is understood the extensional as well as the sonic modulus, is of the same magnitude as the crystal lattice modulus, E_{ol}, and the estimated theoretical modulus of the polymer chain, E_c, the values lying in between 100 and 220 GN/m^2.

For conventional synthetic polymer fibers, such as the nylons, polyesters, and polyolefins, the situation is very different. Here, E_{of} is nearly one order of magnitude smaller than E_{ol} and E_c, which has been interpreted on the basis of the two-phase model for the micromorphology of these fibers. This model in its simplest form assumes a sandwich structure of crystalline and amorphous domains along the fiber axis. Layers of more or less folded chain crystals are interconnected by amorphous domains containing loosely packed chain segments, ends, and chain folds. Of these two phases it is primarily the amorphous part which determines the tensile properties of the conventional organic fibers.

Evidently this model is not suitable for the interpretation of the physical properties of P-*p*PTA fibers. Investigation of the microstructure of the fiber by

Journal of Polymer Science: Polymer Symposium 58, 283–296 (1977)

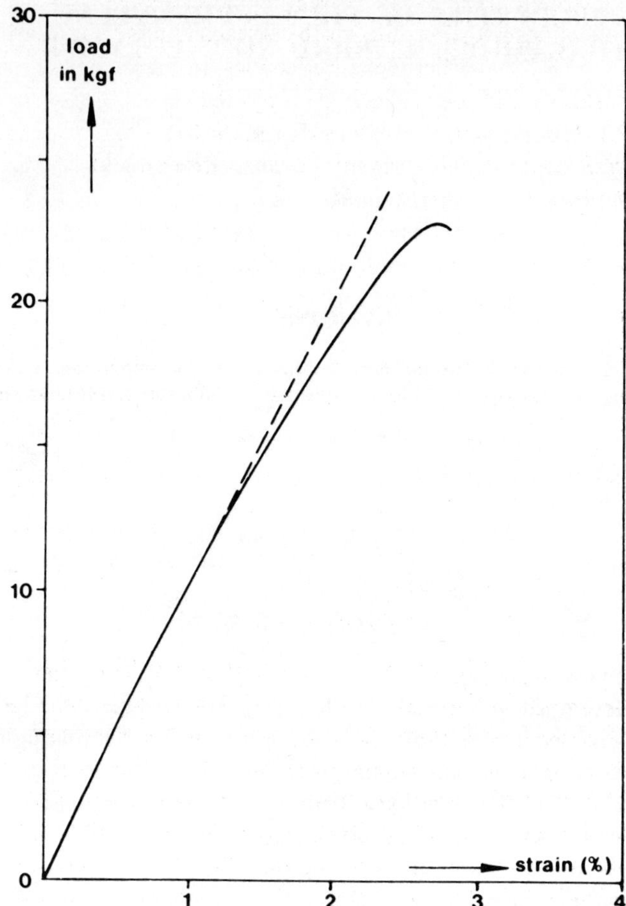

FIG. 1. Tensile curve of poly(*p*-phenylene terephthalamide) yarn with a count of 1737 dtex, strain rate was 8% per min.

means of X-ray diffraction [3, 4] has provided clues which now enable us to understand at least qualitatively its properties. As a result of the intramolecular interactions between sequential phenyl and amide segments in the chain, free rotation around the phenyl-carbonyl and the phenyl-nitrogen bonds is absent. Hence, the polymer chain must be regarded as a rigid rod, and it is this particular chain property that essentially accounts for the structure and properties of the fiber, and in addition, does explain its liquid crystalline behavior in solution. So, chain folding in the wet spinning process is ruled out, and it is unlikely that two distinct phases in terms of crystalline and amorphous domains are created. These considerations have led us to adopt a one-phase paracrystalline model for the microstructure of the fiber. The rigid rod chains are linked together by hydrogen bonds, thus forming crystalline domains which display in the direction of the fiber axis a much greater regularity than in conventional polymeric fibers. The lateral packing order, however, does not deviate very much from the one found in the semicrystalline conventional fibers. These aspects of the structure are clearly demonstrated in the X-ray diffraction patterns of P-pTA, which show equatorial reflections with a profile width corresponding to a crystallite size of about 50 Å and very sharp meridional reflections. In Figure 2 the square of the broadening of the meridional reflections is plotted versus the fourth power of the order of the reflection. According to Hosemann's theory of paracrystallinity [5], a linear relation between these two quantities should be observed for Gaussian profiles if the broadening results only from the crystallite size and the lattice distortions of the second kind. A crystallite size of about 700 Å and a value of 1.7% for the distortion parameter of the second kind are found, which demonstrates the applicability of the concept of paracrystallinity to the microstructure of P-pPTA fibers.

A consequence of this structural model is the dependence of the elastic properties on the orientation distribution of the crystallites. In this report we will first describe the determination of the crystal lattice modulus E_{ol} of P-pPTA and then present the orientation measurements which have yielded the experimental relationship between the fiber compliance and the well-known orientation parameter $\langle \sin^2 \phi \rangle$ of the chain axis distribution. Subsequently it will be shown that a theory of Bartenev and Valishin [6] on the orientational dependence of the compliance for complete crystalline fibers, transformed into a generalized form, gives a proper quantitative description of the elastic properties of P-pPTA fibers. Moreover, the structural concept underlying this theory is found to be in accordance with the paracrystalline model that has emerged from the X-ray diffraction studies so far.

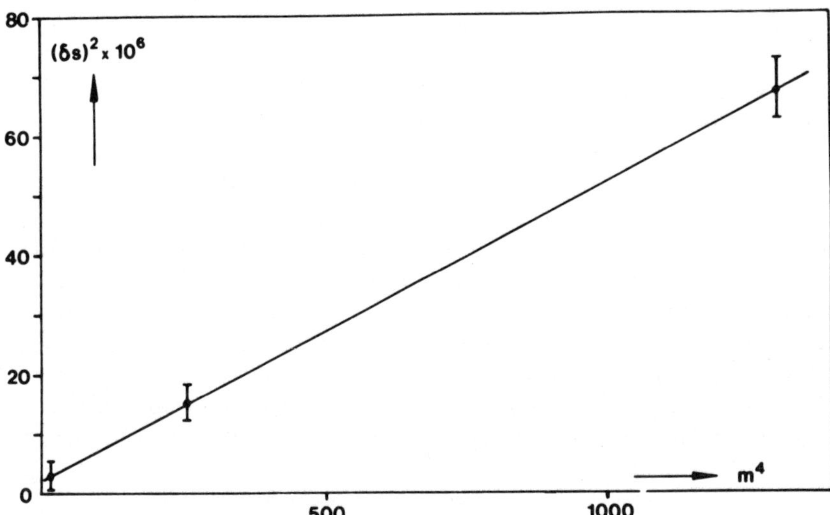

FIG. 2. Profile broadening analysis of meridional reflections of poly(p-phenylene terephthalamide) fibers. The squared half-height width is plotted versus m^4, m being the order of the reflection and S = $(2 \sin \theta)/\lambda$. Crystallite size is 700 Å, distortion parameter of the second kind 1.7%.

EXPERIMENTAL

In this section two experiments are discussed which have provided some valuable data on the tensile properties of P-pPTA fibers, viz., determination of the crystal lattice modulus and measurement of the orientation dependence of the fiber elastic modulus. In both cases the X-ray diffraction technique was employed, use being made of a horizontal Philips diffractometer PW 1380 and CuKα radiation. In the line focus arrangement, the specimen holder was placed in symmetrical transmission position while the diffracted beam was focalized with a curved quartz crystal. The diffracted intensity was recorded with a proportional counter using pulse–height discrimination.

Determination of E_{ol} for P-pPTA Fibers

The elastic modulus of the crystal lattice along the direction of the fiber axis is determined by measuring the stress σ on the lattice and the resulting strain ϵ, defined as the relative change of the distance between the (001) net planes. For the monoclinic P-pPTA fibers this distance, corresponding with the fiber identity period, is 12.8 Å. After a load or an elongation has been imparted to the fiber filaments, ϵ is determined from the shift of the diffraction angle, $\Delta (2\theta)$, of the meridional reflections (004) or (006),

$$\epsilon = \Delta d/d = -\Delta(2\theta)/2tg\theta$$

In case the crystallite size in the fiber axis direction comprises only a few repeat units of the chain, the diffraction angle may shift by a change of the number of repeat units [7]. In P-pPTA fibers the crystallite size is of the order of 700 Å, which implies at least 50 repeat units in the diffracting domain, so any shift of the diffraction angle is brought about only by lattice deformation.

The stress σ on the lattice was assumed to be equal to the macroscopic force on the fiber bundle divided by its cross section. A stretching apparatus with a Statham load cell was mounted on the horizontal diffractometer such that both the fiber bundle and the stress were aligned along the diffraction vector, which implies application of symmetrical transmission geometry. Because of the large value of E_{of} the experimental procedure was to impart an elongation to the fiber bundle, followed by measurement of the force and the shift of the diffraction angle. This shift was determined by scanning the complete profile before and after extension of the fiber bundle. As a result, the data points of the tensile curve are obtained step by step, i.e., E_{ol} is determined on a static way. If extension of the fiber bundle results in a broadening of the diffraction profile to larger angles, it would indicate a nonuniform load distribution over the filaments in the bundle. However, a decrease of the profile width and an increase of the intensity has been observed, which is presumably caused by better alignment and increased perfection of the crystallites. Figure 3 shows the tensile curve of the crystal lattice as measured from the shift of the (006) reflection for a P-pPTA fiber having the following properties: count 403 dtex, density of 1.44 g/cm^3, sonic modulus of 103 GN/m^2, tensile strength of 2.75 GN/m^2 and an elongation at break of 2.9%. The observed E_{ol} moduli, with estimated standard deviation in brackets,

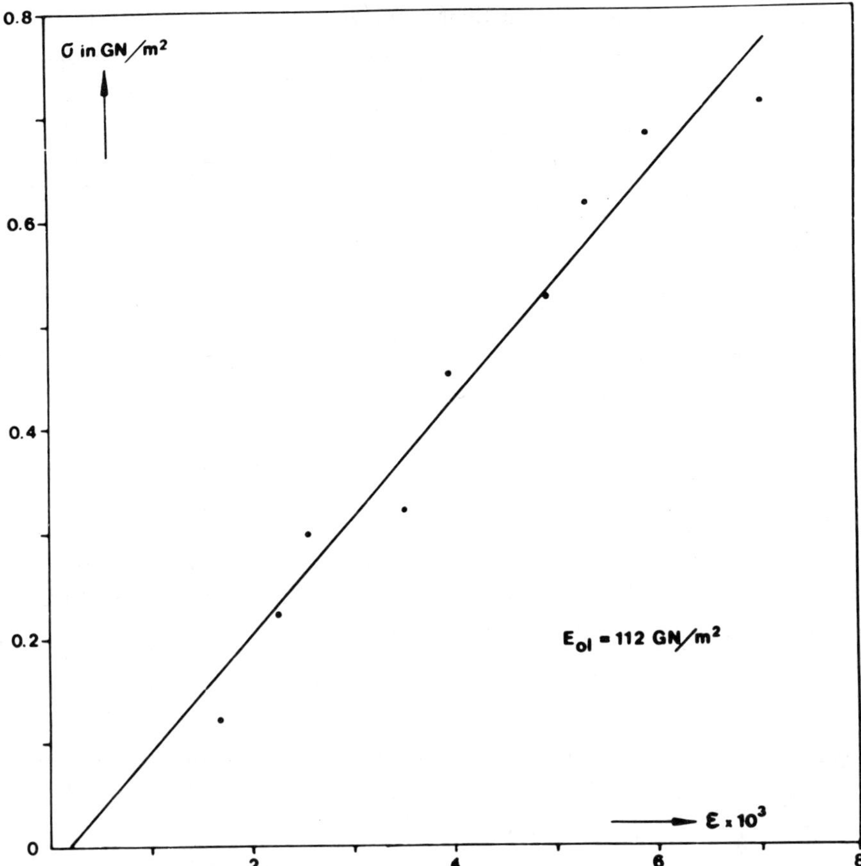

FIG. 3. Determination of the crystal lattice modulus E_{ol} as measured from the shift of the (006) reflection.

determined from the shifts of the (004) and (006) reflection are 112 (16) GN/m² and 112 (9) GN/m².

Orientation and Sonic Modulus Measurements

The chain orientation distribution $I(\phi)$, where ϕ is the angle between the unit cell axis c and the fiber axis z, is given by the orientation profile of the meridional reflections (00l) and can be specified by the parameter

$$\langle \sin^2 \phi_{c,z} \rangle = \int_0^{\pi/2} I(\phi) \sin^3 \phi\, d\phi \bigg/ \int_0^{\pi/2} I(\phi) \sin \phi\, d\phi \tag{1}$$

For medium and highly oriented fibers the parameter was computed from the profile of the (004) reflection at $2\,\theta_0 = 27.86°$, as measured with the method by Decker, Asp and Harker [8]. The fiber specimen was prepared by parallel winding of a single filament around a thin rigid frame, which is set in transmission position on the diffractometer with the fiber axis in the plane formed

288   NORTHOLT AND VAN AARTSEN

by the primary and the diffracted beam. With the counter fixed at 2 θ_0, the θ $- 2\theta$ coupling is disengaged and the orientation profile is measured by slowly rotating the frame around the diffractometer axis. With the specimen in transmission position the method can be applied only for orientation scans of meridional and near-meridional reflections. When the specimen is in reflecting position, also the ϕ profiles of equatorial reflections can be measured but the complete ϕ profile should then be appreciably less than 2 θ_0 degrees. Due to the line-focus optics a high resolution in ϕ is achieved in both cases.

For low oriented P-pPTA fibers the intensity of the meridional reflections is too small to permit measurement of the orientation profile. Therefore, $\langle \sin^2 \phi_{c,z} \rangle$ was determined from the measurement of the orientation profiles of equatorial reflections. The monoclinic symmetry of P-pPTA fiber makes it necessary to determine the orientation parameter $\langle \sin^2 \phi_{hko,\, z} \rangle$ of at least three equatorial reflections in order to calculate $\langle \sin^2 \phi_{c,z} \rangle$. However, the monoclinic angle of 90 degrees and the presence of pseudosymmetry elements [4] give the packing of the chains approximately orthorhombic symmetry, which justifies the measurement of the orientation profile of only two equatorial reflections. After selection of the two strongest equatorial reflections (200) and (110), the chain orientation distribution parameter is then calculated with the relation

$$\langle \sin^2 \phi_{c,z} \rangle = v^{-2}[1 - u^2 + v^2 - \langle \sin^2 \phi_{110,z} \rangle + (u^2 - v^2)\langle \sin^2 \phi_{200,z} \rangle]$$

(2)

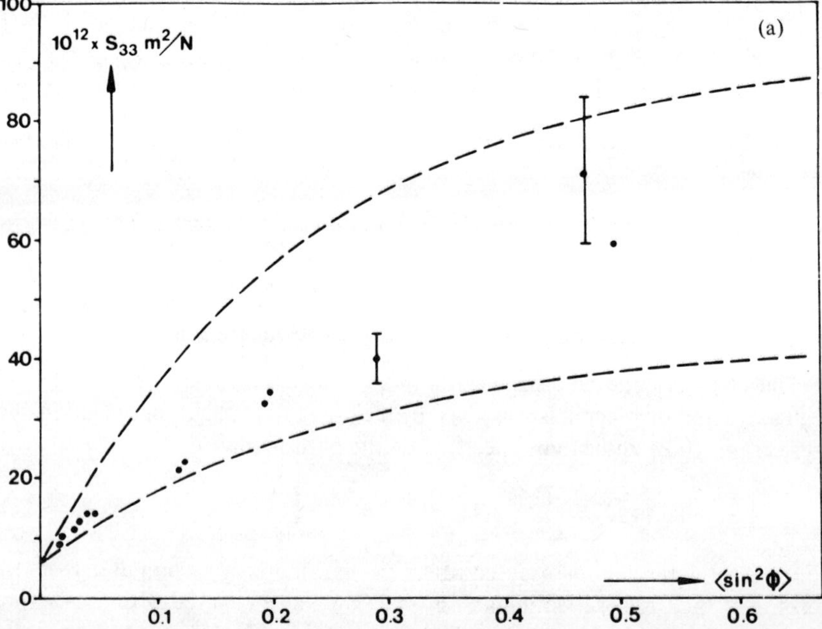

FIG. 4. (a) The compliance S_{33} as a function of $\langle \sin^2 \phi \rangle$. Dashed curves represent the calculated limits. Observed values are indicated by points. Vertical bars give the experimental error in the measured values of S_{33}.

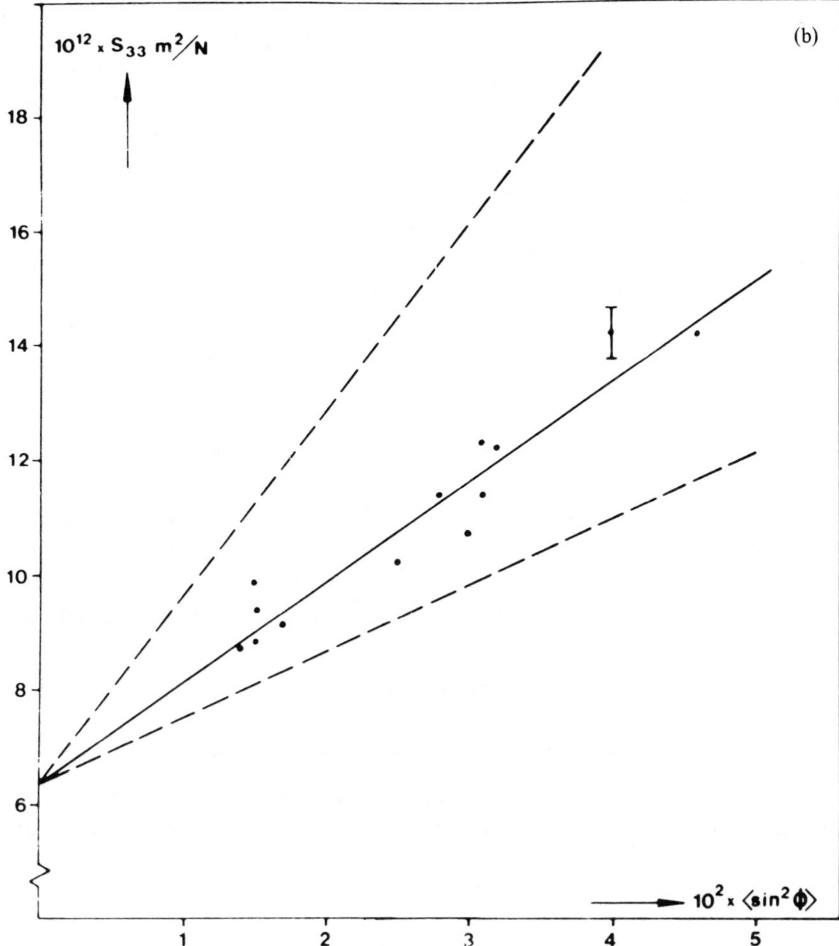

FIG. 4. (b) part of Figure 4(a) on a larger scale. The vertical bar gives the experimental error of S_{33}.

where u and v are the direction cosines of the normal to the (110) plane with the a and b axes. If randomness of the crystallites around the c axis is assumed, we have

$$\langle \sin^2 \phi_{200,z} \rangle = \langle \sin^2 \phi_{110,z} \rangle \tag{3}$$

and (2) reduces to

$$\langle \sin^2 \phi_{c,z} \rangle = 2(1 - \langle \sin^2 \phi_{200,z} \rangle) \tag{4}$$

Measurement of the orientation profile of an equatorial reflection for the low oriented fibers was carried out by rotating the specimen frame in the plane bisecting the primary and diffracted beam. In order to minimize instrumental broadening, a narrow pencil-shaped beam from a point focus was used. With a circular collimator of 0.5 mm diameter a ϕ-resolution of 2.5 degrees at $2\theta_0 = 20°$ was achieved.

For a correct representation of the ϕ profile one should actually measure the integrated intensity $\int I(\phi, 2\theta)d2\theta$ at each point of the ϕ profile, instead of the peak scan $I(\phi, 2\theta_0)$. Comparison of both approaches showed negligible differences in the shape of the distribution, whereupon the less time consuming peak scan measurement has been used for all orientation measurements.

The sonic modulus was measured with the dynamic modulus tester PPM5 of H.M. Morgan Co. For the calculation of the modulus from the pulse propagation velocity a nominal density value of 1.47 g/cm^3 was taken. It should be noted that the density range of the fibers as measured by the submerged cantilever method [9] is 1.44–1.47 g/cm^3, and the crystalline density determined from precise measurements of the unit cell dimensions is 1.525–1.545 g/cm^3.

The results of the sonic modulus and orientation measurements are summarized in Figure 4a and 4b, which present the compliance, $S_{33} = E_{son}^{-1}$, as a function of $\langle \sin^2 \phi_{c,z} \rangle$. For medium and highly oriented fibers a linear relation between the two quantities is observed, the slope being 1.75×10^{-10} m^2/N. An alternative representation of the observations is shown in Figure 5 where the sonic modulus is given as a function of the width at half-height of the orientation profile of the equatorial (200) reflection. By extrapolation a value of about 160 GN/m^2 is obtained for the modulus of a fiber with parallel orientation of the chain axes and values ranging from 10 to 20 GN/m^2 for random orientation.

THEORY

Compliance and Crystallite Orientation Distribution

G. M. Bartenev and A. A. Valishin have developed for completely crystalline polymers a theory which gives a functional relation between the macroscopic

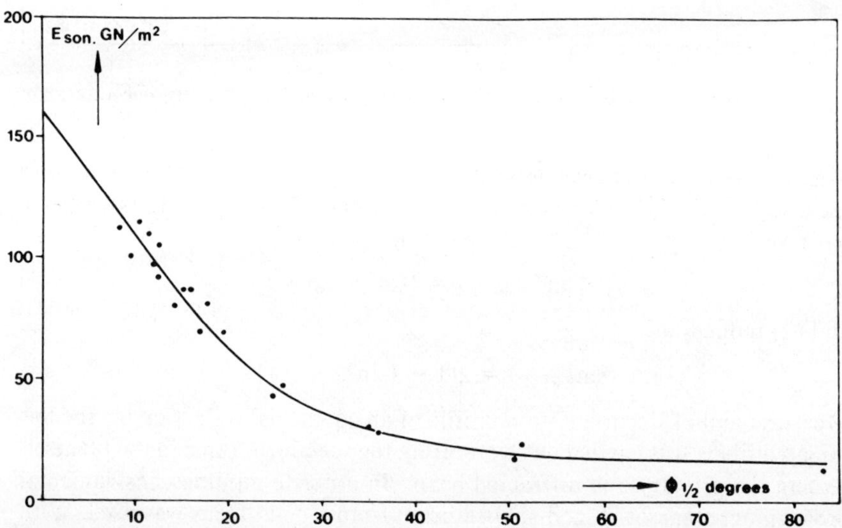

FIG. 5. The sonic modulus E_{son} as a function of the azimuthal breadth $\phi_{1/2}$ at half maximum intensity of the (200) reflection.

compliance and the crystallite orientation distribution in a polymer fiber [6]. The relation was derived by specifying the crystallite orientation distribution, but as will be shown here, a general expression can be obtained for any kind of distribution. It is assumed that during the measurement of the elastic modulus the orientation distribution does not change. The completely crystalline polymer is regarded as an elastically anisotropic medium for which the macroscopic elastic tensor is equal to the microscopic elastic tensor averaged over all crystallite orientations in the fiber. So, interaction between the crystallites is thus not taken into account. The crystallite itself is regarded as an transversally isotropic body with its symmetry axis aligned along the direction of the polymer chains. Its dimensions are small compared with the fiber filament but large enough to contain many unit cells. Although stated as such, perfect crystallinity is not a prerequisite for the application of the theory. The structure of the polymer should rather be devoid of amorphous domains having different mechanical properties, which does not exclude the presence of some structural disorder as, for example, is taken into account by a paracrystalline model.

The four components of the compliance tensor of the crystallite, which need to be considered here, are

$$S'_{11} = 1/e_1, \quad S'_{33} = 1/e_3, \quad S'_{44} = 1/g, \text{ and } S'_{13} = -v/e_1$$

where e_1 and e_3 are the elastic moduli normal and parallel to the symmetry axis, g is the modulus for shear in a plane parallel the symmetry axis and v the Poisson ratio. If ϕ is the angle between the symmetry axis of the crystallite and the fiber axis, the macroscopic compliance of a uniaxially symmetrical specimen with a crystallite orientation distribution $\rho(\phi)$ is given by

$$S_{33} = 4\pi \int_0^{\pi/2} (S'_{11} \sin^4 \phi + 2S'_{13} \sin^2 \phi \cos^2 \phi + S'_{44} \sin^2 \phi \cos^2 \phi$$

$$+ S'_{33} \cos^4 \phi)\rho(\phi) \sin \phi d\phi \quad (5)$$

In terms of the averages $\langle \sin^2 \phi \rangle$ and $\langle \cos^4 \phi \rangle$ this expression becomes

$$S_{33} = \left(\frac{2}{e_1} + \frac{2v}{e_1} - \frac{1}{g}\right) \langle \sin^2 \phi \rangle + \left(\frac{1}{e_1} + \frac{2v}{e_1} - \frac{1}{g} + \frac{1}{e_3}\right) \langle \cos^4 \phi \rangle$$

$$- \left(\frac{1}{e_1} + \frac{2v}{e_1} - \frac{1}{g}\right) \quad (6)$$

Equation (5) can also be expressed in terms of $\langle \sin^2 \phi \rangle$ and $\langle \sin^4 \phi \rangle$ by substitution of $\langle \cos^4 \phi \rangle$, leading to

$$S_{33} = \frac{1}{e_3} + A\langle \sin^2 \phi \rangle + B\langle \sin^4 \phi \rangle \quad (6a)$$

where

$$A = \frac{1}{g} - \frac{2v}{e_1} - \frac{2}{e_3} \quad (7)$$

and

$$B = \frac{1}{e_1} + \frac{2\nu}{e_1} - \frac{1}{g} + \frac{1}{e_3} \tag{8}$$

For highly oriented fibers $\langle \sin^4 \phi \rangle \rightarrow 0$, so the relation becomes linear, viz.,

$$S_{33} = \frac{1}{e_3} + A \langle \sin^2 \phi \rangle \tag{9}$$

and for ideally parallel-oriented crystallite axes along the fiber axis the elastic modulus takes the value

$$E_p = e_3 \tag{10}$$

For completely random orientation of the crystallites in the specimen $\langle \sin^2 \phi \rangle = 2/3$ and $\langle \cos^4 \phi \rangle = 1/5$ which leads to the following relation for the elastic modulus:

$$\frac{1}{E_r} = \frac{1}{15} \left(\frac{8 - 4\nu}{e_1} + \frac{3}{e_3} + \frac{2}{g} \right) \tag{11}$$

Before the theoretical results given by eqs. (6), (9), (10) and (11) can be verified, the elastic constants of the crystallites formed by the P-pPTA chains have to be estimated.

Calculation of the Elastic Constants e_1, e_3 and g

For a number of polymers the calculated estimate of the theoretical elastic modulus of the polymer chain E_c has been reported [10]. Comparison with the observed values for the crystal lattice modulus E_{ol} shows a good agreement for chains with a fully extended configuration. Except for the cellulose chain [11], the intermolecular interactions may be neglected in the calculation of E_c. Hence, it is assumed here that the modulus of the crystallite along the symmetry axis, e_3, equals E_c. We have used the simple valence force field type of calculation, which was first introduced by Treloar [12]. The greatest uncertainty in these calculations arises from the inadequacy of published data for the force constants of valence bond deformation and bond stretching. On the basis of a hypothetical crystal structure Fielding Russell estimated the theoretical modulus of P-pPTA fiber to be 235 GN/m^2 [13].

Defining $E_c = (f/a)/(\delta l/l)$, where f is the tensile force acting along the chain, $\delta l/l$ the relative chain extension, and a the effective cross-section of the chain in the crystal structure, we arrive at the following values, in 10^{-5} m/N, for the contributions to $\delta l/f$: bond stretching 476, extension of the phenylene segment 375, and valence angle deformation 590. The crystal structure determination [4] yields values for a and l of 20.06×10^{-20} m^2 and 6.4×10^{-10} m, which results in an estimated value for E_c or e_3 of 220 GN/m^2. The force constants for valence angle deformation at the nitrogen and carbon atoms of the amide segment, available in the literature and used in this calculation, refer to valence angles at these atoms having the common values of 120 and 115 degrees. The actual angles are larger, viz., about 125 and 117 degrees. Since the direction of the repulsive forces involved in the steric hindrance between the amide and phenylene segments is approximately parallel to the chain axis, it is likely that the actual force constants for valence angle deformation in the amide segment are appreciably lower, which results in a larger contribution to $\delta\,1/f$. If, for example, these force constants are smaller by a factor of 2, a value of 160 GN/m^2 is obtained for e_3. Comparison with values for E_c of polyethylene, nylon 6, and nylon 66, which are 186, 196, and 196 GN/m^2 respectively, shows a good agreement.

In order to arrive at an estimate for the elastic constants e_1 and g, we assume that the hydrogen bonds are the dominant interaction between the chains. For approximately linear hydrogen bonds one can derive that

$$e_1 = \frac{k_s l_H}{a_H} \text{ and } g = \frac{k_b l_H}{a_H} \tag{12}$$

where k_s and k_b are the stretching and bending force constants, l_H the distance between the hydrogen and oxygen in the bond, and a_H the cross-section of the bond. The infrared spectroscopic literature [14, 15] gives for k_s (0.20–0.36) $\times$ 10^5 dyne/cm and for k_b (0.03–0.07) $\times$ 10^5 dyne/cm; from the crystal structure the values l = 2 $\times$ 10^{-10} m and a_H = 24.8 $\times$ 10^{-20} m^2 are obtained, which results for the elastic constants in the approximations 16 $<e_1<$ 29 and 2.4 $<g<$ 5.7, both expressed in GN/m^2.

Estimates for Various Macroscopic Quantities

With the approximations for e_1, g and e_3 given in the preceding paragraph, and a Poisson ratio of 0.3, the range for A, being the limiting slope in a plot of S_{33} versus $\langle \sin^2 \phi \rangle$ for $\langle \sin^2 \phi \rangle \rightarrow 0$, and the range for the modulus E_r of a specimen with random oriented crystallites can be calculated. With eq. (7) we find

$$1.2 < A < 3.7 \quad (10^{-10}\text{m}^2/\text{N})$$

and with (11)

$$11 < E_r < 25 \quad (\text{GN/m}^2)$$

Further evaluation of eq. (6) requires specification of the orientation distribution for which we choose the Gaussian shaped function in the form

$$\rho(\phi) = \frac{pe^{p(2[\cos\phi] - 1)}}{4\pi \sinh p} \tag{13}$$

Parameter p characterizes the distribution: when $p \rightarrow \infty$ the crystallite symmetry axes become oriented parallel to the fiber axis and when $p = 0$ a random orientation is obtained. For distribution (13) the expressions for the averages $\langle \sin^2 \phi \rangle$ and $\langle \cos^4 \phi \rangle$ are

$$\langle \sin^2 \phi \rangle = \frac{e^{2p} - p}{p(e^{2p} - 1)} - \frac{1}{2p^2} \tag{14}$$

$$\langle \cos^4 \phi \rangle = \frac{e^{2p}(p^3 - 2p^2 + 3p - 3)}{p^3(e^{2p} - 1)} + \frac{3}{2p^4} \tag{15}$$

For Gaussian orientation distributions the limits of S_{33} have been computed, using the estimated values for the four elastic constants of the crystallite. Figures 4a and b give a graphic representation of the boundary curves for S_{33} as a function of $\langle \sin^2 \phi \rangle$ together with the experimental values.

DISCUSSION

The determination of the crystal lattice modulus E_{ol} for P-pPTA fibers has yielded a value which is about equal to the modulus E_{of} of highly oriented fibers, but somewhat lower than the theoretical modulus E_c. Presumably this discrepancy is due to the measuring technique employed for the determination of E_{ol}. Under constant elongation of the fiber bundle, stress relaxation occurs and the observed macroscopic stress is likely to be lower than the local stress on the crystallites contributing to the reflection profile. These effects may result in a value for E_{ol} which is lower than would have been obtained by measuring the lattice modulus under constant load. Yet, we conclude that for a large part of the tensile curve the macroscopic strain is mainly brought about by elongation of the crystal lattice through valence angle deformation and bond stretching of the polymer chain. The deviation from linearity, which starts near 1% strain in the tensile curve of Figure 1, may then be attributed to chain breakage and other irreversible processes.

Table I gives a summary of the observed and calculated value for E_p, E_r and the limiting slope A. From this Table and from Figures 4a and b we can conclude that the theory presented in the preceding paragraph provides an adequate explanation for the orientation dependence of the elastic modulus of P-pPTA fibers.

The description of the microstructure in the direction of the fiber axis by means of the paracrystalline model may suggest that in this direction the fiber is composed by an assembly of joined crystallites with dimensions of at least 700 Å. However, the crystallite length as determined by X-ray diffraction is only the maximum distance between two repeat units of the polymer chain for which the phase difference is still close to a multiple of the wavelength. It does not necessarily imply that the boundaires of a crystallite in the fiber axis direction are really separated by a distance equal to the crystallite length obtained by X-ray diffraction. For a first approximation we would rather regard the mi-

TABLE I

Comparison of Observed and Calculated Values for the Moduli E_p and E_r, and the Limiting Slope A

	Observed	Calculated
E_p (GN/m^2)	~ 160	160 - 220
E_r (GN/m^2)	10 - 20	11 - 25
A (10^{-10} m^2/N)	1.75	1.2 - 3.7

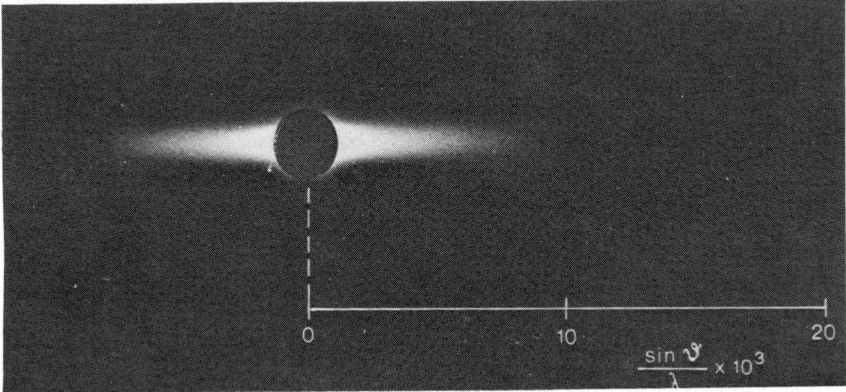

FIG. 6. SAXS pattern of poly(*p*-phenylene terephthalamide) fiber taken with CuKα radiation. Fiber axis vertical. Camera resolution 250 Å.

crostructure along the fiber axis as a continuous linear crystallite composed of parallel but finite chains which are randomly displaced relative to each other. For highly oriented fibers this description is consistent with the approximate agreement between the three moduli E_{ol}, E_{of} and E_c.

Besides the apparent crystallite size of about 50 Å deduced from the wide-angle diffraction pattern, the small-angle X-ray diffraction pattern, shown in Figure 6, provides interesting information on the lateral structure of the fiber. The marked equatorial scattering and the absence of meridional scattering indicate the presence of voids, having their longest dimension along the fiber axis and very small dimensions perpendicular to this axis. Analysis of the diffracted intensity along the equator, applying the method of Debye, Anderson and Brumberger [16] modified for two dimensional inhomogeneous objects, yields a correlation distance of about 20 Å. This distance may be regarded as a measure of the size of the voids in transverse direction.

The authors are indebted to J. L. Salve for the computational contribution and to H. van Hall for his assistance with the experiments.

REFERENCES

[1] W. B. Black, *J. Macromol. Sci.,* **A7,** 3, (1973).

[2] J. Preston, *Polym. Eng. Sci.,* **15,** 199, (1975).

[3] M. G. Northolt and J. J. van Aartsen, *J. Polym. Sci., Polym. Lett. Ed.,* **11,** 333, (1973).

[4] M. G. Northolt, *Eur. Polym. J.,* **10,** 799, (1974).

[5] R. Hosemann and W. Wilke, *Die Makromol. Chem.,* **118,** 230, (1968).

[6] G. M. Bartenev and A. A. Valishin, *Polym. Mech.* **4,** 800, (1968).

[7] L. G. Wallner, *Monatsh. Chem.,* **79,** 279, (1948).

[8] B. F. Decker, E. T. Asp and D. Harker, *J. Appl. Phys.,* **19,** 388, (1948).

[9] H. de Vries and H. G. Weyland, *Text. Res. J.,* **28,** 183, (1958).

[10] L. Holliday and J. W. White, *Pure Appl. Chem.,* **26,** 545, (1971).

[11] P. P. Gillis, *J. Polym. Sci. A2,* **7,** 783, (1969).

[12] L. R. G. Treloar, *Polymer,* **1,** 95, (1963).

[13] G. S. Fielding—Russell, *Text. Res. J.,* **41,** 861, (1971).

[14] T. Shimanouchi and I. Harada, *J. Chem. Phys.,* **41,** 2651, (1964).

[15] S. Meshitsuka, H. Takahashi, K. Higasi, and B. Schrader, *Bull. Chem. Soc. Jpn.* **45**, 1664, (1972).
[16] P. Debye, H. R. Anderson, and H. Brumberger, *J. Appl. Phys.* **28**, 679, (1957).

OBSERVATIONS OF TRIMODAL CRYSTAL TEXTURE IN POLYPROPYLENE BLENDED WITH NYLON 11

K. K. SETH* and C. J. E. KEMPSTER
Department of Polymer and Fibre Science and Department of Pure and Applied Physics, UMIST, Manchester, England

SYNOPSIS

X-ray diffraction techniques, optical microscopy, and scanning electron microscopy have been used to study a series of melt-spun bi-constituent fibers containing isotactic polypropylene and nylon 11. The addition of small amounts of nylon 11 produces bimodal crystal texture in polypropylene with most crystallites having either their c-axes or a-axes oriented nearly parallel to the fiber axis. In blends containing about 50% nylon 11 a third polypropylene crystallite orientation is observed in which the [101] directions are oriented nearly parallel to the fiber axis. The polypropylene chain axes make an angle of about 50° to the fiber axis, and an additional set of layer lines is produced. In blends containing more than 50% nylon 11, most of the polypropylene crystallites adopt the third orientation. It is suggested that this third orientation of polypropylene crystallites is the result of epitaxy on portions of nylon 11 fold surface at the interface between the two components.

INTRODUCTION

The work to be described is part of a general study of the physical properties of melt-spun bi-constituent fibers of isotactic polypropylene and nylon 11 [1]. Previous studies of other bi-constituent fiber systems [2–4] have made little use of X-ray methods, but it is evident that physical properties will depend upon the molecular orientation and crystallization behavior of the component polymers, as well as their state of dispersion in a particular blend. The crystalline texture and morphology of the polyblend systems considered here have been observed using X-ray methods and optical and scanning electron microscopy.

EXPERIMENTAL

A series of blends was prepared from isotactic polypropylene (Carlona P, Shell International Petroleum Co. Ltd.) and nylon 11 (Rilsan BESNO, Aquitain Total Organico S.A.) with melt-flow indices 10.4 and 13.8 g/10 min respectively at 230°C (BS 2782:1970; method 105C). The melting points of polypropylene and nylon 11, as determined by differential scanning calorimetry, were 166°C and 195.5°C respectively (heating rate 20°/min). The two polymers were melt-

---

* Present address: Department of Metallurgy and Materials Science, University of Cambridge, England.

Journal of Polymer Science: Polymer Symposium 58, 297–310 (1977)
© by John Wiley & Sons, Inc.

blended in a Midget Banbury mixer; the vacuum-dried polyblend was then melt-spun at 250°C and air-quenched at room temperature with a jet–stretch ratio of 41 to produce a nine-filament thread.

X-ray diffraction photographs of cylindrical bundles of fifty lengths of thread were obtained using a Searle toroidal-focusing camera and a Rigaki-Denku small-angle scattering camera. A Hilger and Watts four-circle X-ray diffractometer was also used to obtain equatorial and azimuthal scans through selected peaks in the fiber pattern to study crystallite orientation.

RESULTS

X-Ray Diffraction Survey

The wide-angle X-ray diffraction pattern of highly oriented polypropylene is shown in Figure 1a; this was obtained from drawn and annealed fibers of pure polypropylene melt-spun as described above. The diffraction spots can be indexed on a monoclinic unit cell of the α-form, with parameters given in Table I. The c axis is oriented parallel to the fiber axis; the strong reflections on the equatorial layer line have indices (110), (040), (130), (060), and (220) in order of increasing distance from the center and the strong reflection on the first layer line is a combination of (111), (101), and (041).

The diffraction patterns of selected members of the series of blends as-spun are shown in Figures 1c–1e. It will be seen that throughout the series the polypropylene component is well crystallized, with sharp diffraction rings.

In pure polypropylene, as-spun, orientation is poor; Figure 1b shows, however, that (040) and (060) reflections are present close to the equator only, while all other reflections give complete rings. Thus crystallite b axes are oriented approximately perpendicular to the fiber axis, while a and c axes are randomly oriented.

The addition of a small amount of nylon 11 produces much greater orientation of the polypropylene crystallites; Figure 1c for the blend containing 10% nylon 11 (by weight) shows very distinct layer lines corresponding to the c-repeat of α-polypropylene. It will be noted that the (110) and (130) rings show the presence of two distinct orientations of polypropylene crystallites, with either the c axes or the a axes oriented approximately parallel to the fiber axis. Since the a-repeat is only $2\frac{1}{2}$% greater than the c-repeat both sets of reflections lie on the same layer lines. This is the well-documented bimodal texture of polypropylene, recently surveyed by Andersen and Carr [5]; it is produced by the inclusion of as little as 5% nylon 11.

As the proportion of nylon 11 is increased, the diffraction pattern changes. In Figure 1d, for the blend containing 50% nylon 11, the nylon diffraction pattern becomes clearly apparent, showing that the nylon component is fairly well oriented with crystallite c axes roughly parallel to the fiber axis. In addition the (110) ring is thickened in three distinct places. In Figure 1e, for the blend containing 80% nylon 11 and only 20% polypropylene, the third positions of the (110) and (130) peaks are the predominant features of the polypropylene diffraction

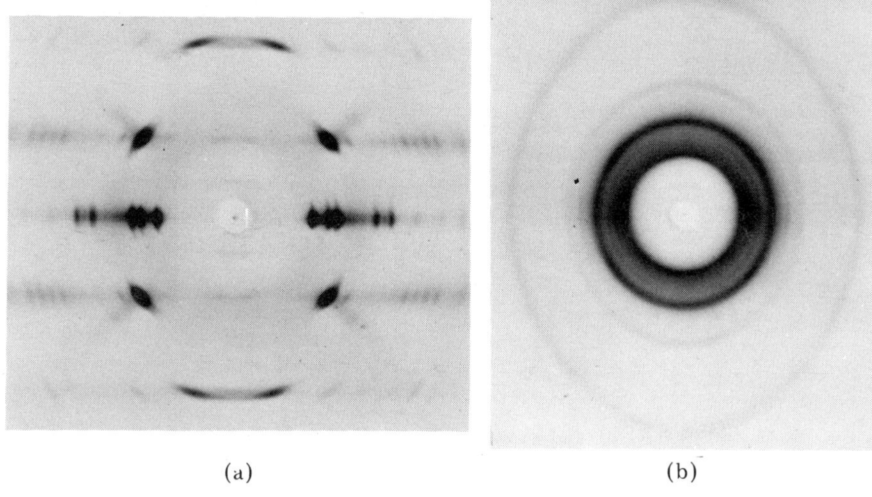

(a)   (b)

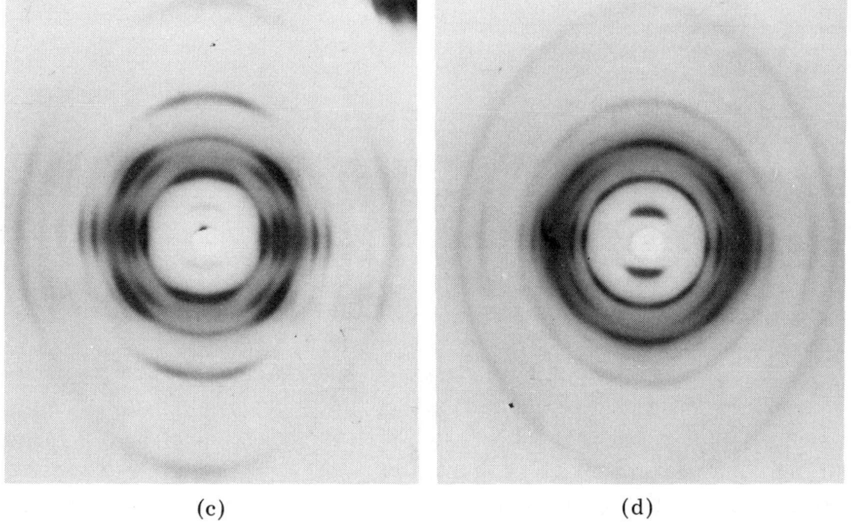

(c)   (d)

FIG. 1. Wide-angle X-ray diffraction photographs from bi-constituent fibers: (a) 100% polypropylene, drawn and annealed; (b) 100% polypropylene, as spun; (c) 90% polypropylene/10% nylon
11; (d) 50% polypropylene/50% nylon 11; (e) 20% polypropylene/80% nylon 11; and (f) 100% nylon
11.

pattern, while (040) remains on the equator unchanged. The layer-lines so
produced correspond to a new repeat distance of about 8.8 Å in the fiber direction. Figure 1f, for pure nylon 11, shows that these additional peaks in the
polypropylene pattern contain no contribution from the nylon component. For

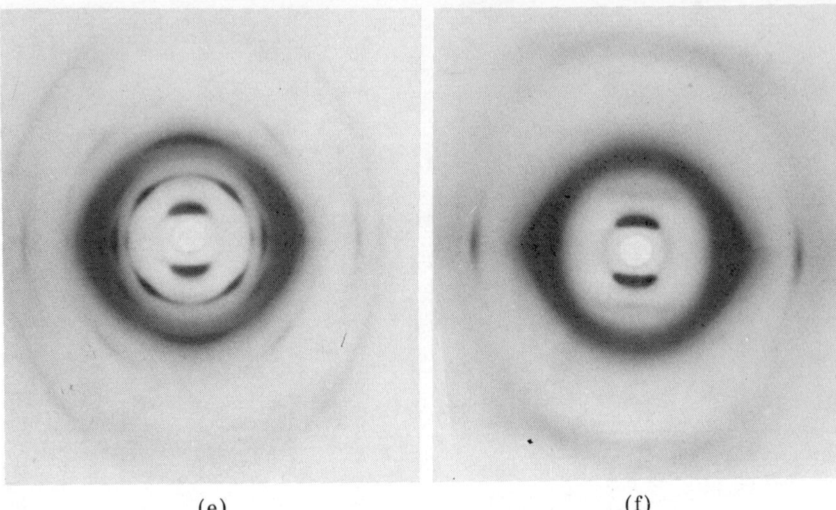

(e)   (f)

FIG. 1. (*continued from previous page*)

TABLE I

Unit-Cell Parameters for the Component Polymers

Isotactic Polypropylene (α-phase)[a]	Nylon 11[b]
a = 6.66 Å	a = 4.9 Å
b = 20.78 Å	b = 5.4 Å
c = 6.495 Å	c = 14.9 Å
β = 99.62°	α = 49°
	β = 77°
	γ = 63°

[a] From reference 6.

[b] From reference 7.

an explanation it is necessary to look in detail at the α-polypropylene unit cell; Figure 2 shows a view of this parallel to the b axis.

The strong reflections in Figure 1e can all be indexed on the polypropylene unit cell provided that an orientation is chosen with the [101]-direction, repeat distance 8.49 Å, parallel to the fiber axis; the polypropylene molecules then make an angle of approximately 51° with the fiber axis. The remaining few reflections correspond to a small amount of the normal bimodal crystal texture.

The intermediate blends clearly contain significant amounts of all three orientations of polypropylene unit cells, as indicated in Figure 3. This trimodal, or three-population, microstructure in isotactic polypropylene has not previously been reported.

To determine the relative proportions of the three crystallite orientations, more detailed measurements were made using a four-circle X-ray diffractometer. Figure 4 shows the results obtained from azimuthal scans round the (110) diffraction ring for each composition. It is apparent that the degree of orientation

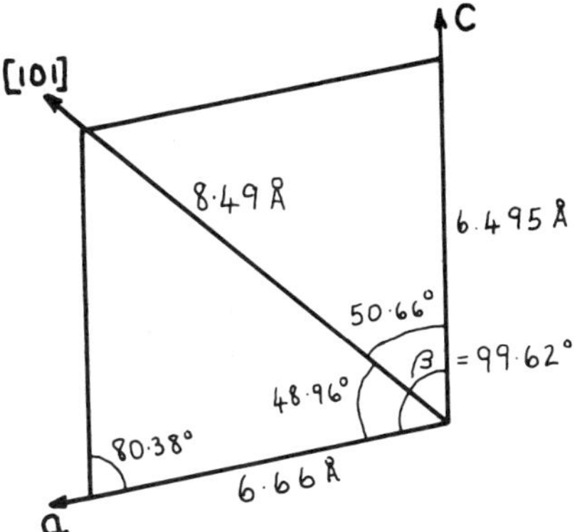

FIG. 2. The polypropylene unit-cell viewed parallel to the b axis.

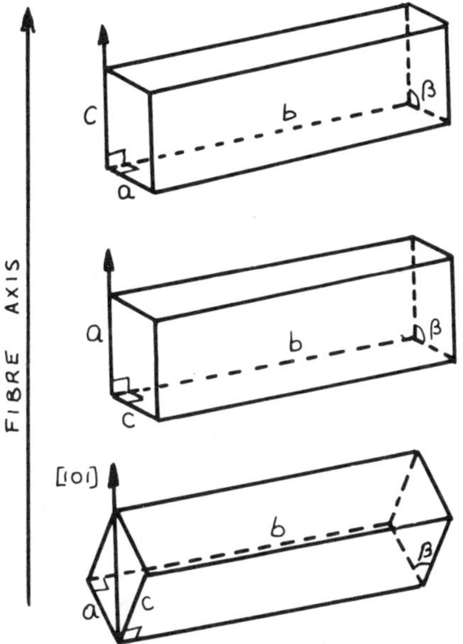

FIG. 3. A schematic representation of the crystallographic relationship between three orientation modes in polypropylene.

of those unit cells with the c-direction roughly parallel to the fiber axis is greatly increased by the addition of only 5% of nylon 11. (The Hermans orientation function for the c-oriented mode increases from 0.29 for pure polypropylene to 0.84 [8]; to obtain these values the azimuthal curves were separated into three components using a procedure analogous to that of Samuels [9].) These c-oriented unit cells predominate in the range 5–40% nylon 11, but the second mode of crystallization with unit cell a-directions parallel to the fiber axis occurs alongside the c-oriented mode. The third mode, with unit cell [101]-directions

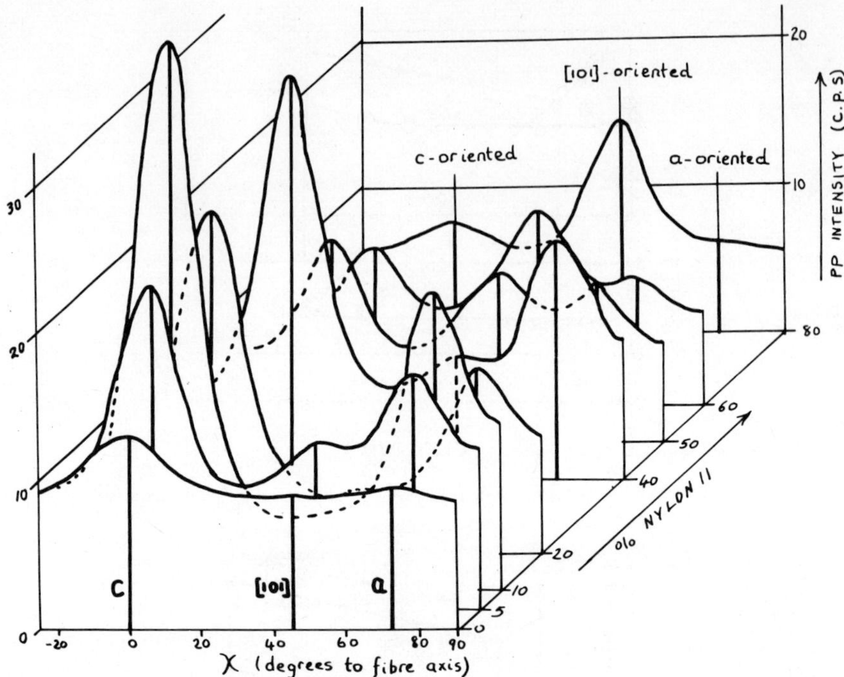

FIG. 4. The variation of intensity with azimuthal angle for the polypropylene (110) diffraction ring plotted as a function of blend composition.

parallel to the fiber axis appears to be present to some extent throughout the series, but only becomes significant in the blends with more than 40% nylon 11; the proportion of the third mode then increases with the amount of nylon 11 present.

Crystallinity changes throughout the series of the three modes of polypropylene crystallization were followed by calculating the observed structure factor F_{110} for each mode from the diffractometer data. The integrated intensity of X-ray scattering was assumed to be proportional to the product (peak height above background) $\times$ (full-width at half-height from equatorial scan) $\times$ (full-width at half-height from azimuthal scan). After correction for the Lorentz factor, and for the proportion of polypropylene present in the sample, the square root was taken to be a measure of F_{110}. A similar procedure was followed to obtain F_{040} as a measure of the total crystallinity.

It will be seen from Figure 5 that F_{110} is practically constant throughout the series for c- and a-oriented crystallites. For [101]-oriented crystallites, F_{110} is small for blends containing small amounts of nylon 11 but increases rapidly for blends with more than 40% nylon 11. The variation of F_{040} and the total of F_{110} for all three modes shows that the crystallinity is constant for blends containing nylon 11 as a minority component, but increases considerably as the proportion of nylon 11 increases above 50%. The increased crystallinity appears to be due almost entirely to the increasing population of [101]-oriented crystalline regions.

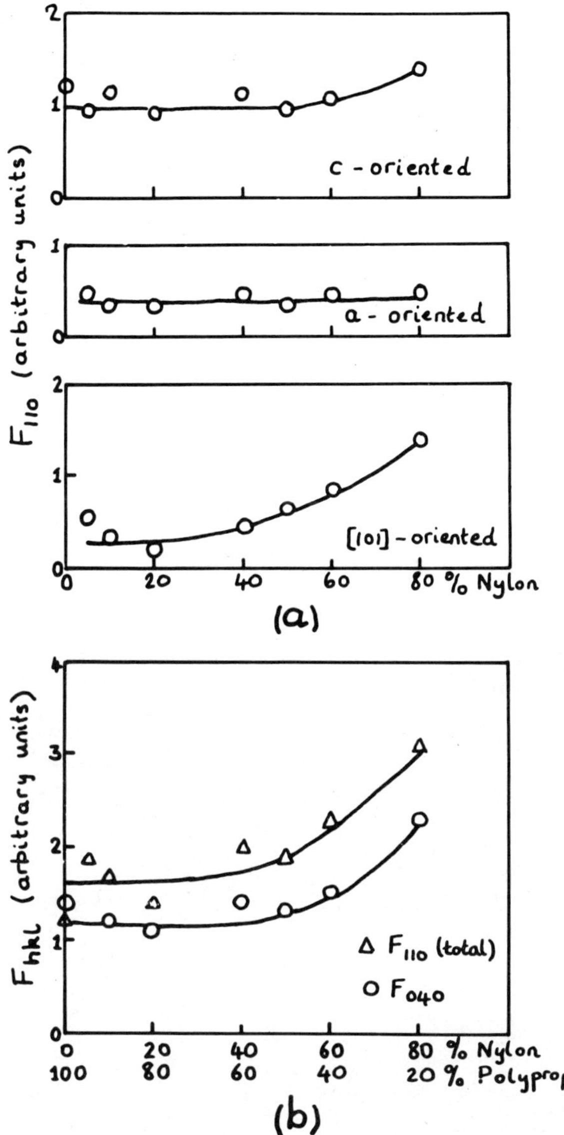

FIG. 5. The variation of the observed polypropylene structure factor with blend composition: (a) for the three orientation modes, and (b) for polypropylene as a whole.

Microscopy

Typical phase contrast micrographs of cross-sections of the bi-constituent fibers are shown in Figure 6; the nylon component, with a higher refractive index, is imaged darker. The fibers containing a small proportion of nylon 11 (Fig. 6a) show that the nylon 11 component is dispersed in domains of average diameter 1.5 μm in a fiber of diameter 45 μm. In fibers containing 50% nylon 11 (Fig. 6b) the dispersion is relatively coarse, with neither component clearly forming the

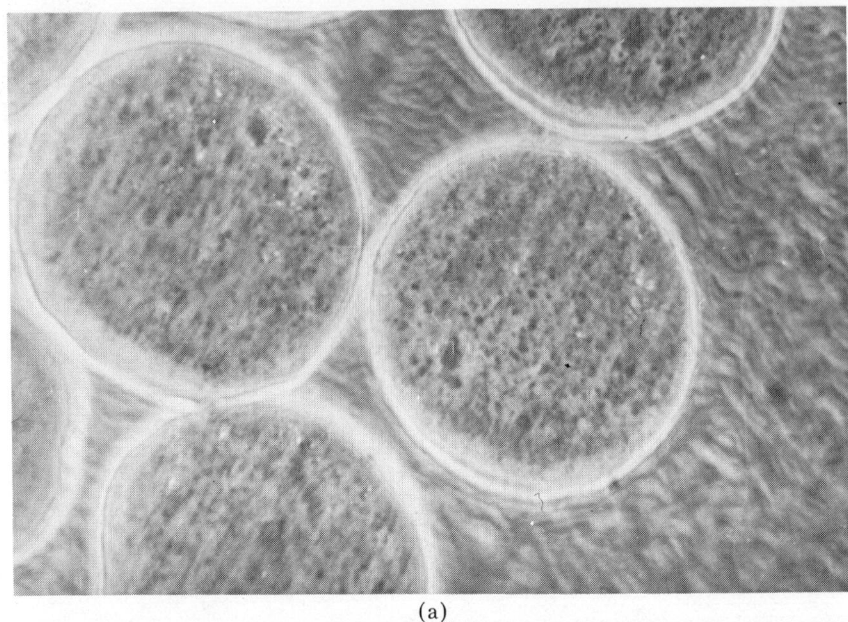

(a)

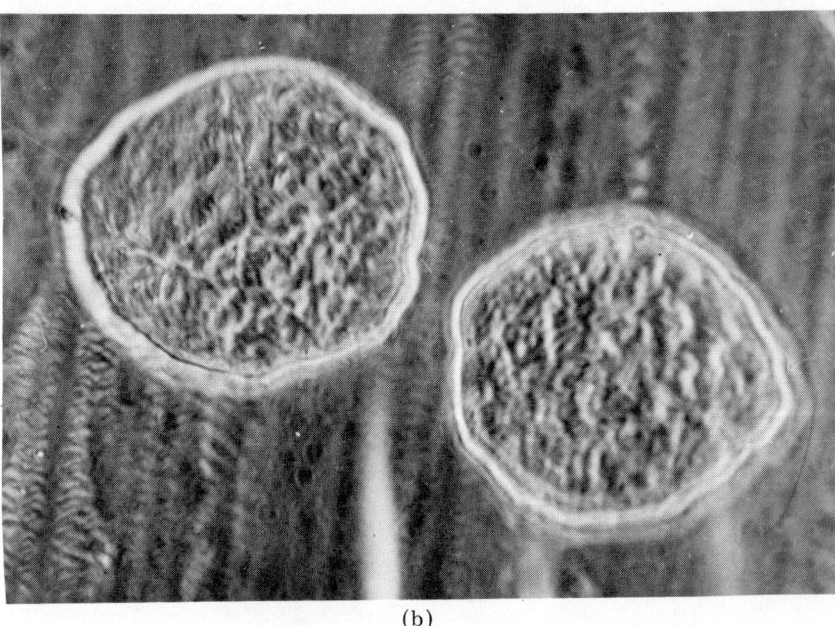

(b)

FIG. 6. Phase contrast micrographs of cross-sections of bi-constituent fibers (×1664): (a) 90%
polypropylene/10% nylon 11; (b) 50% polypropylene/50% nylon 11; (c) continued on next page.

matrix. When the proportion of nylon is increased further, the polypropylene
is dispersed in a matrix of nylon. Fibers with 80% nylon 11 (Fig. 6c) do not show

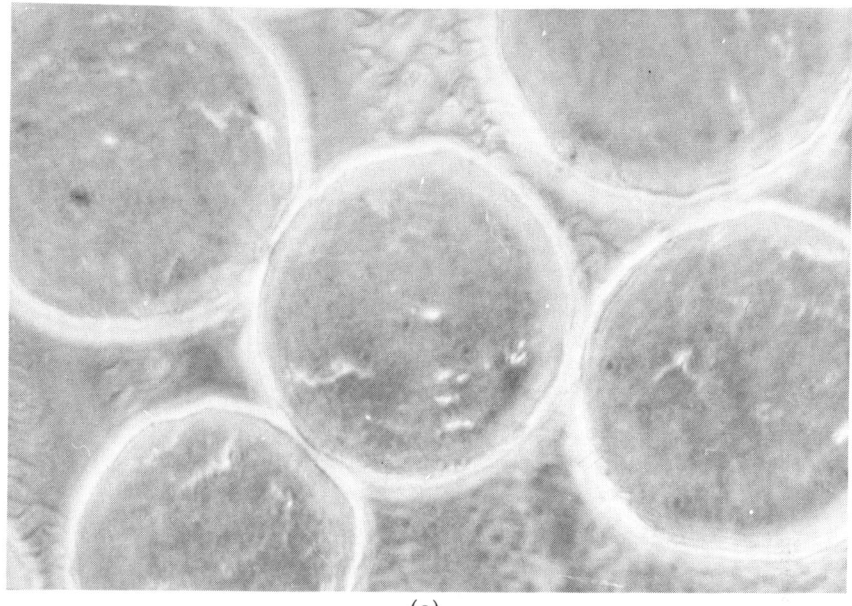

(c)

FIG. 6. (c) 20% polypropylene/80% nylon 11.

up 20% of polypropylene component; it seems likely that the dispersion is on too fine a scale to be observed. This is confirmed by the scanning electron micrographs of Figure 7; longitudinal fiber surfaces are shown before and after leaching out the nylon component in boiling formic acid for 50 hr, drying for 24 hr and coating with silver. The remaining fibrils of polypropylene, lying parallel to the fiber direction, are clearly shown in Figure 7b.

DISCUSSION

It is clear from the results of the last section that the blend containing 20% polypropylene and 80% nylon 11 shows a very good dispersion of fibrils of polypropylene in a matrix of nylon. Consequently the area of interface between the two components is large, about forty times the external surface area of the blend fibers, and the possibility of an epitaxial relationship between the third mode of polypropylene crystallization and the nylon crystallization cannot be ruled out.

During the melt-spinning process the nylon component may be expected to solidify first as the jet of molten polymer is stretched, giving good orientation of the nylon molecules (see Fig. 1f). Further along the thread-line, as the nylon matrix cools to about 166°C, nucleation and crystallite growth will occur in the fine tubes of liquid polypropylene. These will be in intimate contact with the solidified nylon surface, as the fibrils shown in Figure 7 have a diameter of the order of $1/5$ μm, which is only 300 times the polypropylene c-repeat.

The dimensions of the nylon 11 unit cell are given in Table I; the (001) face transverse to the chain axis is inclined to it at an angle close to the angle $\alpha = 49°$. Hearle and Greer [10] suggest that, in fibers of oriented nylon 6.6, the surface

(a)

(b)

FIG. 7. Scanning electron micrographs of the surface of a bi-constituent fiber containing 20% polypropylene/80% nylon 11: (a) before leaching (×5500), and (b) after leaching with formic acid (×4700).

containing the folds must be parallel to the transverse face of the unit cell if the integrity of the underlying crystal lattice is to be maintained. It is reasonable to suppose that nylon 11 would crystallize in the same way in these fibers, although there is some evidence [11] that in bulk nylon 11 the fold surface is not necessarily a simple low-index plane.

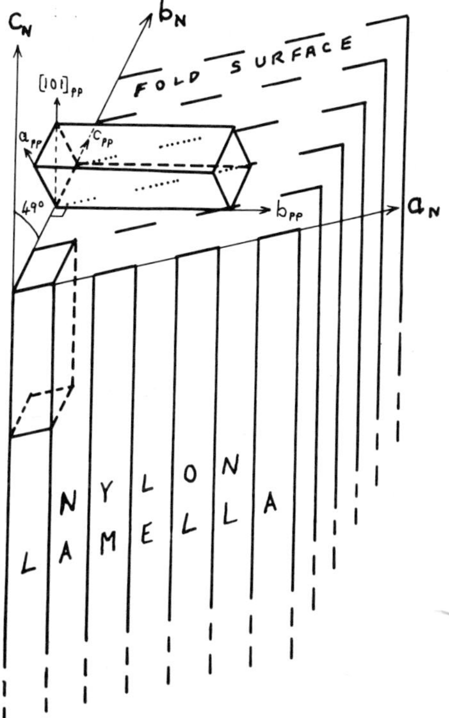

FIG. 8. An epitaxial model for the crystallization of polypropylene on the fold surface of nylon 11.

According to a recent review of epitaxy by Wunderlich [12], the nucleation of a linear high polymer on a plane surface probably occurs by the adsorption of a polymer chain onto the surface. In any nucleation of polypropylene on the nylon 11 fold surface we should therefore expect the polypropylene chain direction to lie parallel to the fold surface, probably along a set of ridges and furrows formed by the array of folds. In Figure 8 it is shown that a unit cell of polypropylene oriented on the nylon 11 fold surface with the chain direction (c) being parallel to the nylon 11 b axis would be within 2° of the [101]-orientation, with the b axis approximately horizontal. This would correspond to a 19% mismatch between the polypropylene c axis and the b axis of nylon 11 and a 16% mismatch between the polypropylene b axis and four times the furrow width in the fold surface of nylon 11. Mismatch of this order has been deduced by Macchi, Morosoff, and Morawetz [13] for the epitaxial growth of nylon 6 on its monomer crystals.

This offers a geometrical explanation of the appearance of the third mode of polypropylene crystallization in blends with nylon. According to Sauer, Morrow, and Richardson [14], polypropylene crystals grown from solution take the form of laths elongated parallel to the a axis; they identify the fold plane with (010) and the fold surface with (001). It is reasonable to assume similar crystallization in molten polypropylene. A small region of nylon 11 fold surface could, therefore, provide a site for a large epitaxial polypropylene crystal growing rapidly out from the nucleus in the a-direction.

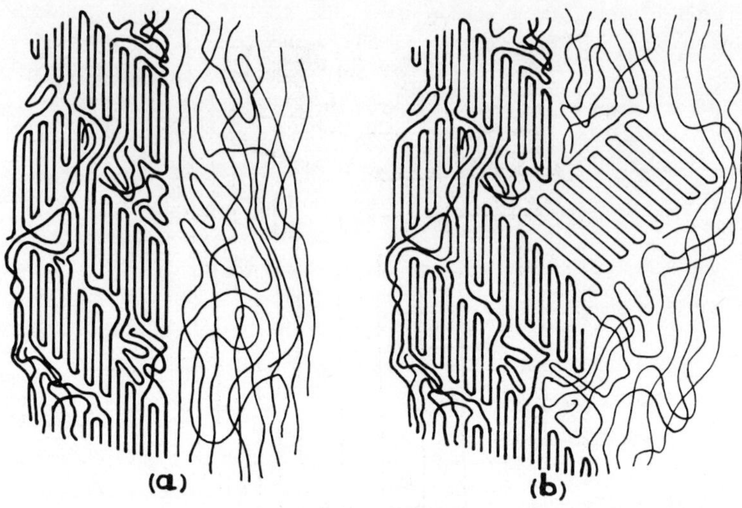

FIG. 9. A schematic representation of lamellar ingrowth: (a) smooth interface between the blend components just below the melting point of nylon 11; and (b) stepped interface formed by the readjustment of nylon 11 molecules; this allows the formation of a polypropylene crystallite in the third orientation mode.

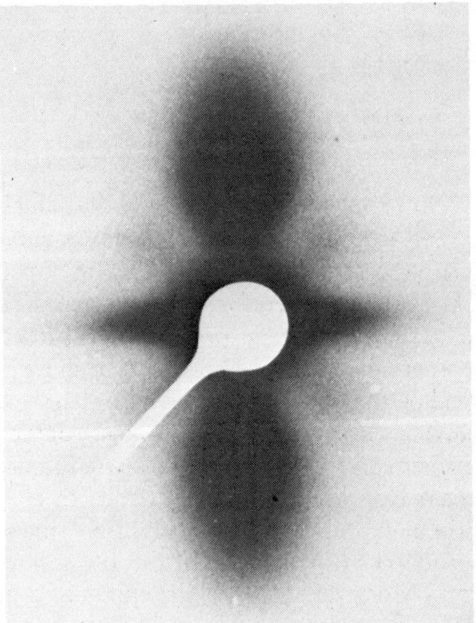

FIG. 10. Small-angle X-ray diffraction pattern from bi-constituent fibers containing 20% polypropylene/80% nylon 11.

While the nylon fold surface cannot be free of disordered material—in the form of cilia, switchboard loops, and tie molecules (see, for example, Keller's recent survey [15]), it is possible that, at the interface between the nylon matrix and the cylinders of liquid polypropylene, at a temperature slightly below the melting point of nylon, nylon lamellae might grow out from the interface into the region occupied by polypropylene. This lamellar ingrowth would give rise to narrow, fairly perfect, ledges on which oriented epitaxial growth might occur. Figure 9 shows a schematic representation of lamellar ingrowth of nylon 11 and subsequent polypropylene nucleation in the [101]-orientation taking place in molten polypropylene.

Further evidence for the oriented growth of polypropylene crystals with chain axes at about 51° to the fiber axis is provided by small-angle X-ray scattering. Figure 10 shows the small-angle scattering by blend fibers containing 80% nylon 11 and 20% polypropylene. The horizontal streak and wide vertical lobes are identical to those for 100% nylon as spun; the shorter lobes, at angles of about 50° to the nylon lobes, are thought to come from the polypropylene component. This suggests that the polypropylene chain axes make an angle of about 50° with the nylon chain axes in this blend.

Having put forward the hypothesis of epitaxial growth of polypropylene on portions of the fold surfaces of nylon 11 lamellae to account for the trimodal texture of nylon-rich blends, it is of interest to see how similar ideas would apply to the bimodal texture of the polypropylene-rich blends. Katayama, Amano, and Nakamura [16] have shown that the secondary crystallization of a-oriented crystallites in melt-spun fibers occurs further along the thread-line, that is, later and at a lower temperature than the formation of the primary c-oriented crystallites. Thereafter the two populations of crystallites grow simultaneously in the later stages of cooling.

It seems possible that the mechanism is as follows. First, thin fibrils of solidified nylon 11 in the molten polypropylene fibers produce increased flow stresses, which increase the orientation of the polypropylene molecules in the direction of the fiber axis. This leads to the primary crystallization of c-oriented crystallites, probably nucleated on a backbone of extended-chain crystallized molecules and forming stacks of lamellae in a shish-kebab array. Continuing flow-stresses give oriented polypropylene molecules at the perimeter of the laterally growing primary lamellae, but in the region between the tips of the lamellae the melt is relatively relaxed and epitaxial growth can occur on the fold surfaces. Nucleation by polypropylene chains adsorbed along furrows parallel to the a axis on small regions of the fold surface forms crystals of the secondary mode exactly in the a-orientation. The mismatch between the a and c axes is only 2½% and the two modes share a common b axis. Both crystallites then continue to grow in their a-directions, as suggested by Figure 9, without interfering with each other. Epitaxy on the (010) face of the lamellae, preferred by Andersen and Carr [5], would not allow both crystallites to continue growing; it also seems less probable since the (010) face is the growth face and is less likely to remain smooth. The introduction of portions of the γ-phase of polypropylene (Padden and Keith [17]) would seem to be an unnecessary complication in the present system.

It should be noted that the proportion of crystallites in the second orientation is always about half of those in the primary orientation (see Fig. 5), which suggests a less efficient crystallization process. Furthermore, in the blend containing 50% of each component the relative amounts of the second and third modes of polypropylene crystallization are approximately equal, as might be expected if each were formed by epitaxial growth on one blend component.

One of the authors (KKS) is grateful to the Ministry of Education and S.W., Government of India for the award of a National Scholarship and to the Department of Polymer and Fibre Science, UMIST for the award of a part-time Teaching-Assistantship.

REFERENCES

[1] K. K. Seth and W. G. Harland, to appear.
[2] P. V. Papero, E. Kubu, and L. Roldan, *Text. Res. J.,* **37,** 823 (1967).
[3] R. Simo and J. Kovac, *Chem. Vlakna,* **22,** 63 (1972).
[4] T. Kitao, H. Kobayashi, S. Ikegami and S. Ohya, *J. Polym. Sci., Polym. Chem. Ed.,* **11,** 2633 (1970).
[5] P. G. Andersen and S. H. Carr, *J. Mater. Sci.,* **10,** 870 (1975).
[6] A. Turner-Jones, J. M. Aizlewood and D. R. Beckett, *Makromol. Chem.,* **75,** 134 (1964).
[7] W. P. Slichter, *J. Polym. Sci.,* **36,** 259 (1959).
[8] K. K. Seth, Ph.D. Thesis, University of Manchester (1976).
[9] R. J. Samuels, *Structured Polymer Properties,* Wiley, New York, 1974, p. 144.
[10] J. W. S. Hearle and R. Greer, *J. Text. Inst.,* **61,** 240 (1970).
[11] J. J. Point, M. Dosiere, M. Gilliot, and A. Goffin-Gerin, *J. Mater. Sci.,* **6,** 479 (1971).
[12] B. Wunderlich, *Macromolecular Physics,* Vol. 1, p. 266, Academic Press, New York, 1973.
[13] E. M. Macchi, N. Morosoff, and H. Morawetz, *J. Polym. Sci., Ar,* **6,** 2033 (1968).
[14] J. A. Sauer, D. R. Morrow and G. C. Richardson, *J. Appl. Phys.,* **36,** 3017 (1965).
[15] A. Keller, *J. Polym. Sci., Symp. No. 51,* **7** (1975).
[16] K. Katayama, T. Amano and K. Nakamura, *Kolloid Z. Z. Polymer.,* **226,** 125 (1968).
[17] F. J. Padden Jr. and H. D. Keith, *J. Appl. Phys.,* **44,** 1217 (1973).

ORIENTATION CHANGES CAUSED BY STRETCHING OF PTE AND PA-6 FIBERS EVALUATED ON THE BASIS OF IR SPECTROSCOPY

GRZEGORZ W. URBAŃCZYK
Instytut Fizyki Włókna i Chemicznej Obróbki Włókna
Politechnika Łódzka, Poland

GENERAL CONSIDERATIONS

The problem of the alteration of molecular orientation raised in PTE and PA-6 fibers by their stretching from the undrawn state has since a long time been the focus of attention of many research centers and has been the subject of numerous publications. Nevertheless, till now, this problem can not be regarded as ultimately cleared-up and wholly recognized.

One of the most essential reasons for inadequate knowledge in respect to this problem has been based on the fact that the process of molecular orientation is accompanied and influenced by a simultanously occurring rearrangement of the ordered polymer fraction. And so, as it is well-known, the increase of molecular orientation in PA-6 fibers is accompanied by a conversion of the ordered fraction from the hexagonal γ-type of arrangement into the monoclinic α-type lattice.

Similarly, in the case of PTE fibers, the increase of molecular orientation caused by fiber stretching brings about initially the creation of a smectic-hexagonal ordered intermediate component and later on, at higher values of draw ratio, a transformation of this component into the triclinic crystalline state or arrangement.

With the view to the above mentioned facts, an attempt has been made to evaluate once again the alteration of orientation caused by stretching of undrawn PTE and PA-6 fibers by means of the IR spectroscopy technique.

In the performed investigations, for the purpose of quantitative characterization of the molecular orientation of the examined fibers, the orientation factor f_d, defined by the general equation of Hermans–Platzek

$$f_d = 1 - \frac{3}{2}\sin^2\theta$$

has been applied.

The values of orientation factors have been calculated on the basis of the previously set-up values for the average orientation angles of the molecule axis, θ. The latter, however, had been calculated according to the formula of Fraser [1], on the basis of the experimentally determined values of the dichroic ratio,

Journal of Polymer Science: Polymer Symposium 58, 311–321 (1977)
© by John Wiley & Sons, Inc.

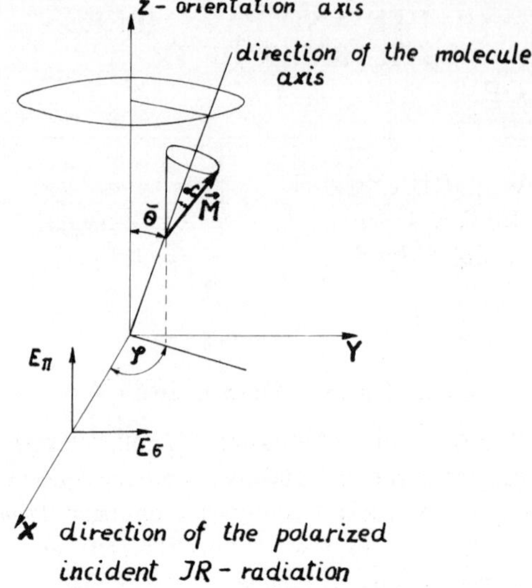

FIG. 1. Scheme of the model of partial axial orientation.

R, of the selected absorption bands and of previously established values of the corresponding transient moments α.*

The used relationship between the measured dichroic ratio R, the established transient moment α, and the sought average orientation angle, has, according to Fraser, the following form:

$$R = \frac{2 \cot^2\alpha \cos^2\theta + \sin^2\theta}{\cot^2\alpha \sin^2\theta + \tfrac{1}{2}(1 + \cos^2\theta)} \tag{1}$$

The calculation of the orientation factors has been founded on the general assumption that the system of the molecular orientation in the examined fibers is due to the model of the partial axial orientation, shown schematically in Figure 1.

In order to evaluate the alteration of the orientation in a complex manner, both the orientation of crystalline aggregates and the orientation of molecules in the noncrystalline (amorphous) fraction of the fibers has been taken into consideration.

For this purpose, in the first case, some of the so-called crystalline absorption bands and, in the second case, some of the so-called amorphous bands have been regarded. As especially suitable for the evaluation, these absorption bands of the IR spectra, summarized in Table I, have been recognized and used.

In order to broaden the scope of the concluding comparative values of orientation, factors corresponding as well to the crystalline as to the noncrystalline fraction have been additionally determined. This has been achieved by applying the X-ray diffraction and light polarization microscopy techniques.

* The determination of the transient moments used by the author in the presented contribution has been described in publications: Faserforschung und Textiltechnik **27**, 4 (1976), 183–187; and Polimery-Tworzywa Wielkoczasteczkowe **21**, 1 (1976), 26–30.

TABLE I

Characteristics of the Considered Absorption Bonds for PTE and PA-6 Fibers

Fiber	Absorbing group and vibration mode	Corresponding "crystalline" absorption band		Corresponding "amorphous" absorption band	
		absorption frequency	type of dichroism	absorption frequency	type of dichroism
PTE	$C_6H_4 - \gamma$ 17B(B_1U)	875 cm^{-1}	σ	1580 cm^{-1}	σ
	$CH_2 - \gamma_w$	1343 cm^{-1}	π	1368 cm^{-1}	π
	$CH_2 - \delta_s$	1473 cm^{-1}	π	–	–
PA-6	$CH_2 - \gamma_r$	729 cm^{-1}	σ	1171 cm^{-1}	σ
	Amide III	1198 cm^{-1}	π	–	–

MATERIALS STUDIED AND THE EXPERIMENTAL TECHNIQUE

In the case of PTE fibers, the examined material consisted of differently stretched filaments obtained from the same source of undrawn multifilament. The following seven draw ratios were selected: 1.5; 2.0; 2.5; 3.0; 3.5; 4.0; 4.5. The applied temperature and stretching speed were 85° and 0.7 m/min, respectively.

In the case of PA-6 fibers, the studied material consisted of variously stretched samples, obtained from the same undrawn multifilament. The drawing was performed at 25°C and the speed was 300–500 m/min. The particular samples are represented by the following draw ratios: 1.0; 1.5; 2.0; 2.5; 3.0; 3.5.

Infrared absorption spectra were obtained in the region between 650 cm^{-1} and 2000 cm^{-1} by using a Specord IR-71 double-beam spectrometer, equipped with a selenium polarizer. The studied materials were examined in the form of parallel coiled monolayer samples.

In order to reduce the light reflection at the filament surface to minimum the sample holder was adjusted between two NaCl monocrystal sheets and filled up with nujol.

RESULTS AND DISCUSSION

The results of the investigations are summarized in Tables II and III for PTE and PA-6 fibers, respectively, and shown in Figures 2–5.

In the case of the investigated PTE fibers, the obtained results lead to the following conclusions:

(1) The curves illustrating the relationship between the orientation factor and draw ratio shown in Figure 2 indicate, that for each of the analyzed crystalline absorption bands there can be noted an increase of the orientation factor with an enhancement of the draw ratio. But as it can be seen, the slope of the curves is not steady. The initial more abrupt course of the curves gives evidence for concluding that of the early stage of stretching, e.g., at low draw ratios (1.5–2.5×), the ordering of crystallites is more effective than at the later phase of drawing.

The values of the orientation factor of the crystalline fraction pertaining to the 1343 cm^{-1} and 1473 cm^{-1} bands respectively as compared to values due to

TABLE II

Orientation Characteristics of the Crystalline and Noncrystalline Fractions in PTE Fibers

Draw Ratio	Birenfringence Δn	Density (g/cm^3)	Crystallinity %	Crystalline fraction orientation indexes								Noncrystalline fraction orientation indexes				
				875 cm^{-1}		1343 cm^{-1}		1473 cm^{-1}		$\sin^2$ /005/	$f_{x,r}$	1368 cm^{-1}		1580 cm^{-1}		$f_{a,o}$
				R	f_{xd}	R	f_{xd}	R	f_{xd}			R	$f_{a,d}$	R	$f_{a,d}$	
1,5 x	0,0143	1,3518	15,0	0,98	0,017	1,21	0,200	1,15	0,157	–	–	1,10	0,090	0,98	0,017	–
2,0 x	0,0449	1,3446	8,7	0,82	0,176	1,29	0,269	1,31	0,309	–	–	1,13	0,116	0,81	0,189	–
2,5 x	0,1090	1,3471	10,9	0,61	0,418	1,76	0,615	1,62	0,566	–	–	1,15	0,133	0,75	0,254	–
3,0 x	0,1128	1,3488	12,4	0,49	0,571	1,88	0,691	1,70	0,626	0,0570	0,914	1,36	0,301	0,63	0,391	0,456
3,5 x	0,1333	1,3555	18,3	0,48	0,585	2,10	0,817	1,81	0,703	0,0547	0,918	1,40	0,332	0,58	0,453	0,536
4,0 x	0,1374	1,3595	21,8	0,39	0,718	2,20	0,870	2,04	0,851	0,0532	0,920	1,12	0,109	0,75	0,254	0,542
4,5 x	0,1485	1,3640	25,8	0,38	0,727	2,30	0,921	2,08	0,875	0,0501	0,925	1,10	0,090	0,73	0,274	0,588

TABLE III

Orientation Characteristics of the Crystalline and Noncrystalline Fractions in PA-6 Fibers

| Draw Ratio | Biren-fringence Δn | Density g/cm^3 | Crystal-linity % | Crystalline fraction orientation indexes | | | | | | Noncrystalline fraction orientation indexes | | |
| | | | | 729 cm^{-1} | | 1198 cm^{-1} | | $\sin^2$ /004/ | $f_{x,r}$ | 1171 cm^{-1} | | $f_{a,o}$ |
				R	$f_{x,d}$	R	$f_{x,d}$			R	$f_{a,o}$	
1,0 x	0,0207	1,1297	34,0	0,79	0,374	2,80	0,560	–	–	0,79	0,374	–
1,5 x	0,0413	1,1320	35,7	0,65	0,656	4,05	0,753	0,0864	0,870	0,75	0,451	0,215
2,0 x	0,0459	1,1321	35,8	0,67	0,614	4,82	0,836	0,0821	0,877	0,78	0,393	0,291
2,5 x	0,0479	1,1350	37,8	0,63	0,698	4,21	0,772	0,0772	0,884	0,80	0,355	0,303
3,0 x	0,0539	1,1360	38,6	0,59	0,786	6,45	0,963	0,0691	0,894	0,74	0,471	0,386
3,5 x	0,0561	1,1415	42,4	0,55	0,876	5,52	0,931	0,0671	0,899	0,66	0,635	0,398

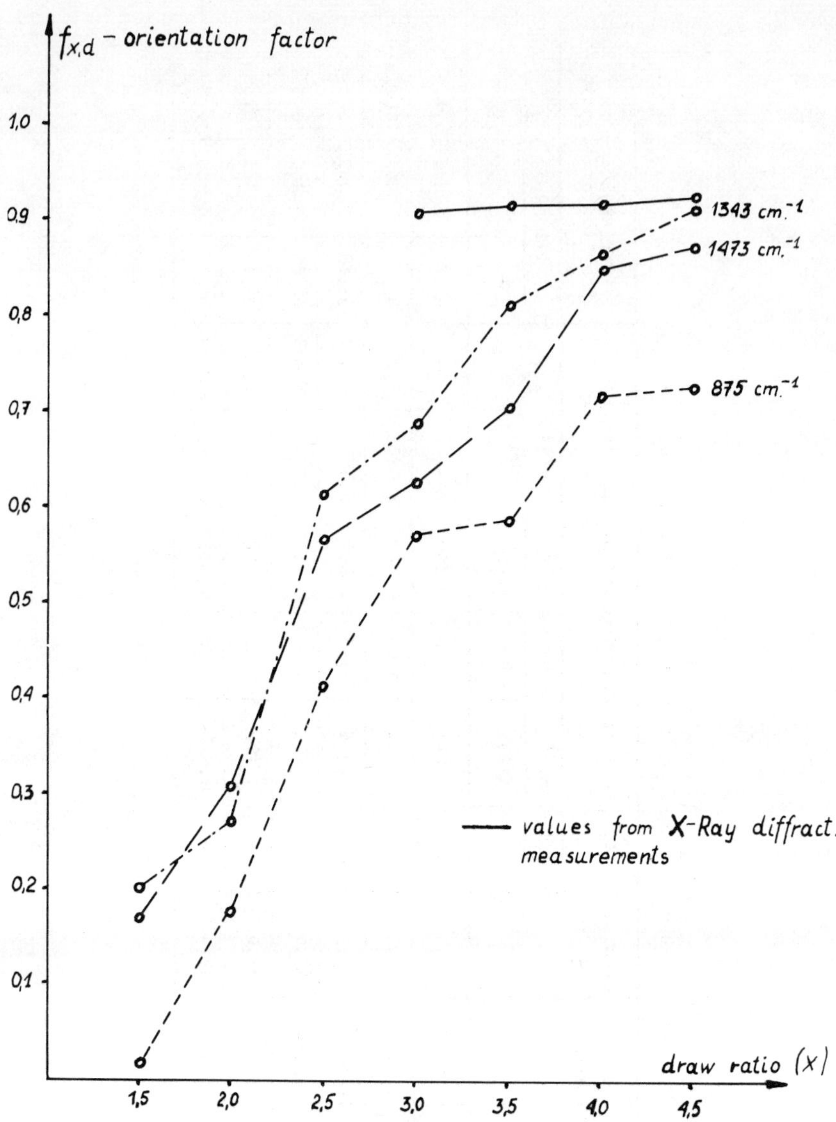

FIG. 2. Orientation changes in the crystalline fraction of a PTE fiber.

the 875 cm^{-1} band are very close to each other and are always bigger than the latter.

Regarding the kind of absorbing groups which are correlated to the analyzed bands, it might imply that in the crystalline state of the molecular aggregation the ordering of the aliphatic chain sections is more perfect than that of the aromatic chain fragments. This might further suggest, that for the considered samples the benzene rings incorporated in the crystalline component take less settled positions then the ethylene glycol residues.

In support of such a conclusion it should be mentioned, that according to the generally accepted opinion, the free rotation of benzene rings in the PTC and the hereby induced possibility for taking the most energetically favorable positions may occur after PTE is heated above 120–130°C.

On the other hand, the rotation of groups which form the aliphatic chain sections proceeds already at temperatures over the range 70–90°C.

In the case of the investigated material the heating temperature was, as has been mentioned, 85°C. Such temperature of heating creates adequate conditions only for a free rotation of groups composing the aliphatic chain section, but is insufficient for an inhindered rotation of benzene rings. These facts could explain the observed differences among the orientation factors determined on the basis

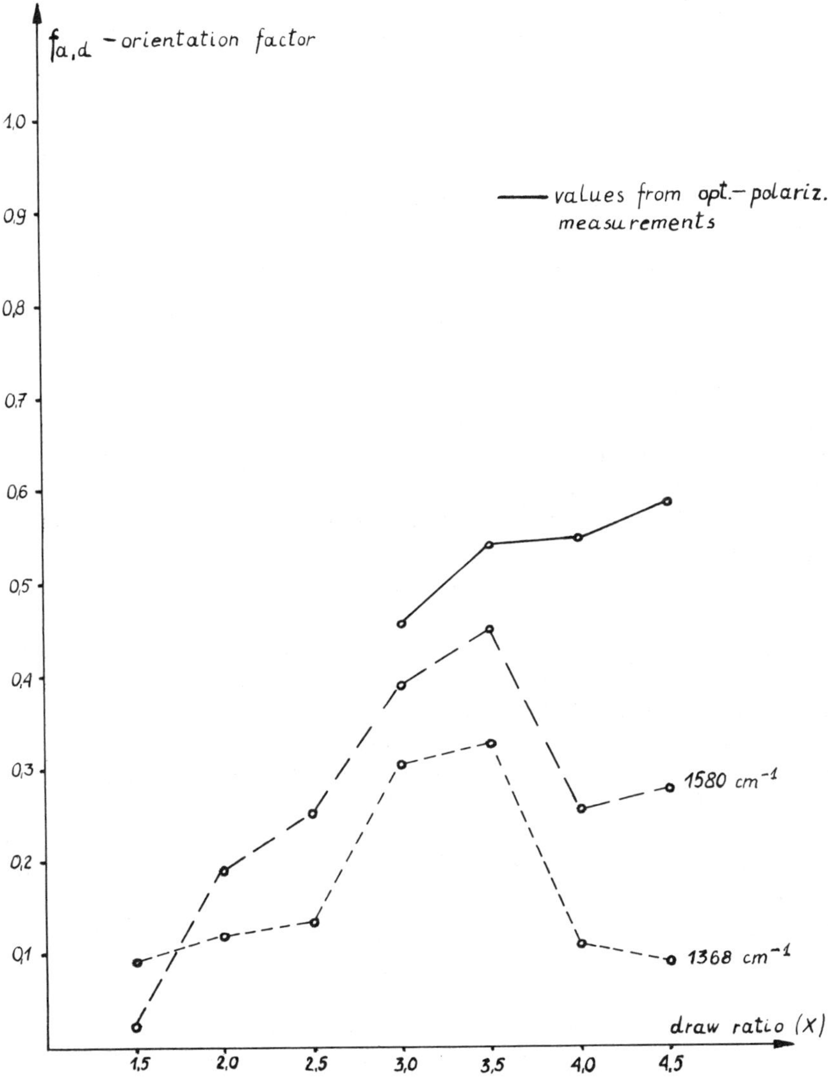

FIG. 3. Orientation changes in the noncrystalline fraction of a PTE fiber.

of the 1343 cm^{-1} and 1473 cm^{-1} bands on one side and on the 875 cm^{-1} band on the other side.

(b) The variation of values of the orientation factors for the noncrystalline matter reflected by the curves shown in Figure 3 reveals the resemblance to both of the analyzed amorphous absorption bands 1368 cm^{-1} and 1580 cm^{-1}. The shape of the curves points out that the ordering of molecules in the noncrystalline component undergoes a specific fluctuation by stretch increasing.

In the range of small draw ratios, below the values of 3.0–3.5, the continuous increase of the orientation factor takes place. After the draw 3.5 is exceeded, a rapid diminishing of the orientation factor occurs.

It seems that this at the first instance unexpected change may be elucidated by the crystallization induced by fiber stretching. As it has been earlier stated by Bonart [2], Asano [3], and Lindner [4], at draw ratios about 3.0–3.5 there proceeds a conversion of the intermediate paracrystalline-smectic polymer fraction into the triclinic crystalline matter.

It is reasonable to assume that the formation of the triclinic crystalline structure must in the first instance proceed at the expense of a part of a noncrystalline matter, a part comprising the best-ordered molecules. As a result, the remaining unconverted noncrystalline matter should be composed of only less ordered molecules and therefore must be characterized by minor values of orientation factor.

In the case of the investigated PA-6 fibers, the obtained results lead to the following conclusions:

(1) As can be seen in Figure 4, the variation of the orientation factor values calculated on the base of the analyzed crystalline absorption bands 729 cm^{-1} and 1198 cm^{-1} brings about evidence that with the increasing stretching the crystallite orientation improves.

Despite some differences in the shape of both curves, from their common features it may be concluded that at the initial stage of stretching a more effective ordering of the crystallites takes place in comparison with the later phase of drawing.

The curves illustrating the change of orientation factors reveal specific "thresholds." These thresholds, or in other words, ranges of diminishing values of the orientation factor, are located differently in both cases. In the case of results obtained on the basis of the crystalline band 729 cm^{-1}, the threshold appears over the range of 1.5–2.0 draw ratio. For the results pertaining to the crystalline band 1193 cm^{-1}, the threshold is noticed for the range of 2.0–2.5 draw ratios.

It seems that the appearance of these thresholds is connected with the polimorphic $\gamma \to \alpha$ transformation which takes place during the stretching of PA-6 fibers in the ordered matter and may be elucidated by it. Additionally, the mechanism of the $\gamma \to \alpha$ transformation might also fully explain the different positions of the "thresholds" in both cases.

According to the prevailing opinions, the transformation consists, among others, in transition of the molecule conformation from the helical into the planar type. It may be therefore assumed that the transformation is in the first order,

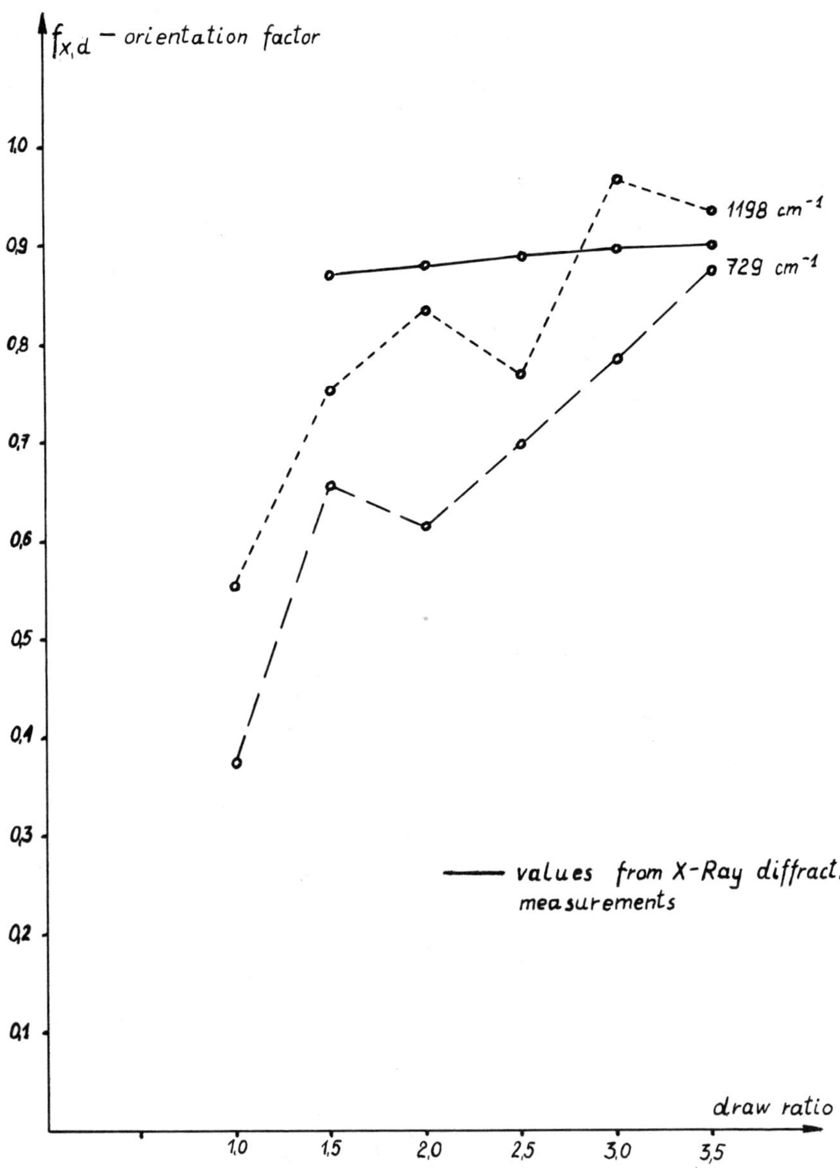

FIG. 4. Orientation changes in the crystalline fraction of a PA-6 fiber.

e.g., for lower draw ratios accomplished by a displacement of CH_2 groups and then later completed by the translocation of CONH groups.

Such a sequence in the translocation of CH_2 and CONH groups agrees well with the ascertained localization of the thresholds. The threshold for the curve corresponding to the 729 cm^{-1} (CH_2 groups) arises at the lower stretch (draw ratios 1.5–2.0), the threshold corresponding to the 1198 cm^{-1} band (CONH groups) arises for the higher stretch (draw ratios 2.0–2.5).

The comparison of orientation factors calculated from the 729 cm^{-1} and 1198

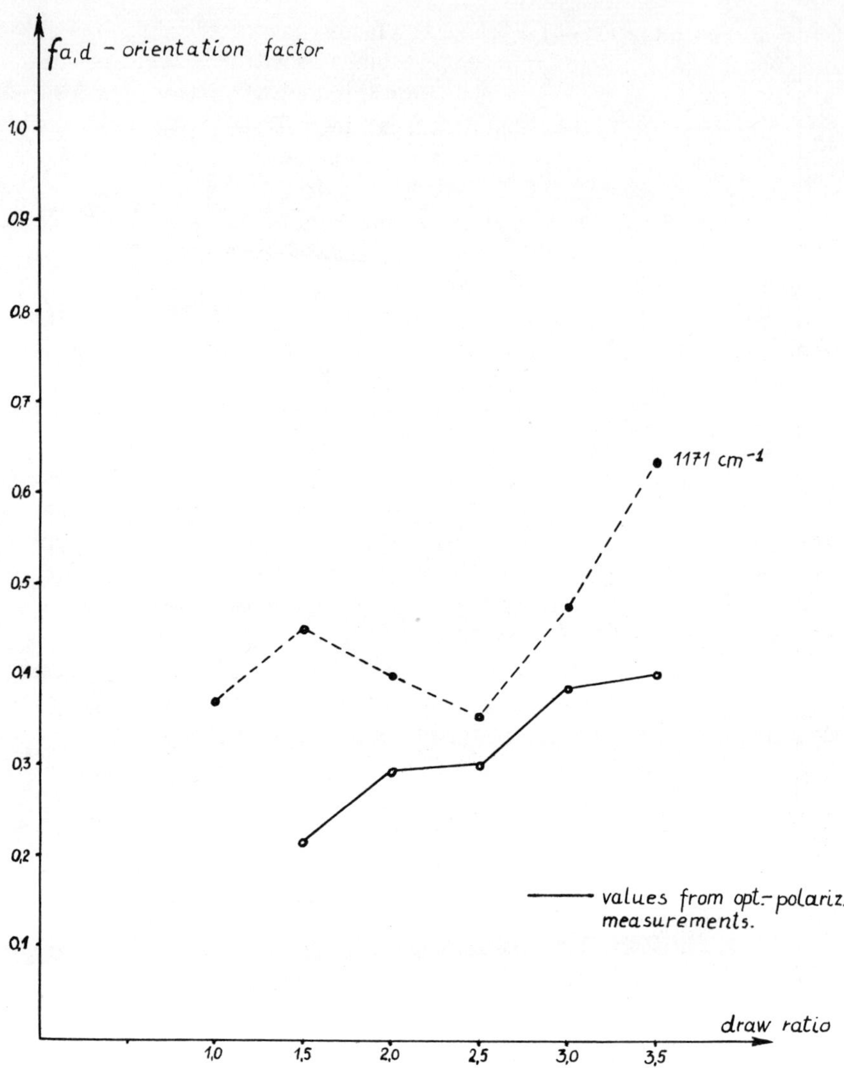

FIG. 5. Orientation changes in the noncrystalline fraction of a PA-6 fiber.

cm^{-1} bands, respectively, indicates that in the crystalline fraction of the fiber, CH_2 groups are worse ordered than CONH groups. Such a relationship seems fully justified.

In view of the fact that the positions of the CONH groups are almost in all cases fixed-up by hydrogen bonds to the adjacent chains it is reasonable to expect that the arranging of these groups must be more regular than that of the CH_2 groups and therefore characterized by higher values of the orientation factor.

On the other hand, the space dislocation of CH_2 groups is described by a relatively high diversity of distribution and therefore, by lower values of the orientation factor. This would arise from the fact that the CH_2 groups are fixed in the lattice only by comparatively weak links of the kind of van der Waals' forces.

(2) The alteration of the orientation factors calculated on the basis of the amorphous 1175 cm^{-1} absorption band exhibits a specific fluctuation. As it can be seen from the curve shown in Figure 5, initially, at smallest draw ratios (1.0–1.5) the values of the orientation factor increase, later, in the range of intermediate stretching (1.5–2.0) the orientation factors decrease, and finally, at the highest stretching (2.0–3.5) increase again.

At the same time it can be noticed that the range of diminishing values of the orientation factors coincides well with the range of decreasing values of orientation factors for the crystalline matter.

It may be supposed that the transient diminishing of the orientation factor has only an apparent but not virtual character. It seems that the ascertained course of orientation factor alteration reflects not only the "pure" change of molecular orientation, but also gives account of the simultaneously occurring crystallization and polymorphic transformation.

One can assume that the crystallization and polymorphic transformation induced by stretching proceeds at the expense of the best ordered molecules of the noncrystalline polymer fraction. As a result, the remaining noncrystalline matter should be composed only of less ordered molecules, which in the course of the matter will further mean that the values of the corresponding orientation factors become lower.

The comparison of the orientation factors obtained from IR spectroscopy measurements with similar data calculated on the basis of X-ray diffraction and light polarization microscopy techniques leads to the conclusion that, as both in evaluating the crystallite orientation and assessing the orientation of the noncrystalline fraction, the first method enables us to characterize the alteration of orientation more precisely than the latter methods.

REFERENCES

[1] R. D. B. Fraser, *J. Chem. Phys.,* **28,** 1113–1115 (1958).
[2] J. H. Bonart, *Kolloid Z. Z. Polym.* **213,** 1 (1966).
[3] T. Asano, *Poly. J.,* **5,** 72–85 (1973).
[4] W. L. Lindner, *Polymer* **14,** 9–15 (1973).

MELT-SPUN POLYPROPYLENE FIBERS WITH HELICAL ORIENTATION. I. MORPHOLOGY AND MECHANICAL PROPERTIES CHARACTERIZATION

T. PAKUŁA, J. MORAWIEC, J. SWIĘTOSŁAWSKI, and
M. KRYSZEWSKI

*Centre of Molecular and Macromolecular Studies, Polish
Academy of Sciences, 90-362 Łódź, Poland*

SYNOPSIS

A new method of melt-spinning of fibers with a twist is presented. The analysis of the dynamics of the spinning process shows differences in the distributions of rotational and longitudinal velocity gradients along the distance from the spinneret. The morphology of polypropylene fibers spun with the twist was examined by various techniques. It was shown that the applied method of spinning produces the distinct helical orientation inside the fibers. The overall and crystalline orientation was observed by birefringence and X-ray diffraction techniques. The light diffraction patterns have shown the periodic helical unhomogenities inside the fibers. Measurements of mechanical properties have shown the dependence of the stress–strain behavior of fibers on the twist rate. It was established that the fibers spun with intermediate rotational speed have the maximal tensile and sonic modulus and the ultimate strength.

INTRODUCTION

The molecular, crystalline, and morphological orientations are important factors determining many physical properties of polymers, first of all the mechanical properties of fibers. The relationship between properties and orientation parameters of man-made fibers is still under study in a great number of laboratories. However much attention is paid, first of all, to uniaxial orientation which is realized in commonly used spinning and drawing operations. Several important and very useful empirical relationships between properties and various aspects of morphology have been established and adequate mechanical analyses have been elaborated for uniaxially oriented fibers.

It is known that a large modification of fiber properties is observed when the overall geometry of orientation is different from uniaxial [1]. An example of special importance is the helical orientation [2]. This type of orientation was found in many natural fibers and can be also achieved in man-made fibers by special drawing techniques.

In this paper we present the primary results of examination of morphology and mechanical properties of polypropylene fibers with internal helical orientation which is generated during the melt-spinning process. Andersen *et. al.* [3] have also started with similar studies.

Journal of Polymer Science: Polymer Symposium 58, 323–338 (1977)
© by John Wiley & Sons, Inc.

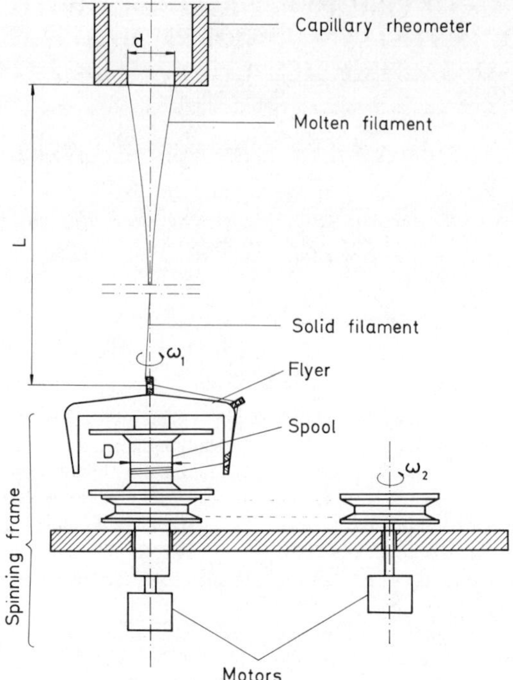

FIG. 1. Scheme of the melt-spinning process with a twist.

The orientation in spinning of polymer melt and that in cold or hot drawing of a solid polymer material differ in many aspects. Therefore, it can be expected that the helical orientation involved during the spinning of polymer melt should give a new morphology and modified properties of fibers.

METHOD OF SPINNING

The commonly used melt-spinning procedure consists of the preparation of the polymer melt, extrusion of the melt through the spinneret, extension of molten filament, and the winding up of the solidified filament by a take-up device.

Our modification of this procedure consists in subjecting the solidified filament to rotation about its own axis. Thus the filament is simultaneously extended and twisted in the molten part. Two types of motion of the filament (rotation and translation) can be achieved by using the well-known ring or flyer spinning frame, as a take-up element. The scheme of such spinning process is shown in Figure 1. The polymer melt is extruded by the simple capillary rheometer and the flyer spinning frame is applied for taking up the solid filament. The spinning frame is powered by two high-speed motors rotating the flyer and the spool independently.

The spinning process is characterized by the same parameters as usual [4] and additionally by the twist rate which is determined by the rotational speed ω_1 of the flyer. The take-up rate is determined by the difference of the rotation speed of the flyer and the spool

$$V = \pi D(\omega_1 - \omega_2)$$

where D is the spool diameter and ω_2' is the rotational speed of the spool.

DYNAMICS OF SPINNING WITH TWIST

The dynamics of melt-spinning were analyzed by several authors [5–8]. A set of partial differential equations describing the heat, force, and material balance has to be considered for full analysis. The problem was solved under various simplifying assumptions.

Some of the results can be adopted in our analysis of the melt-spinning process accompanied with the twist of the filament. To get a rough idea about the dynamics of such spinning we treat the problem with a few simplifications.

The filament is placed in the cylindrical coordinate system as it is shown in Figure 2.

Neglecting the gravitational, inertial, surface tension, and skin-friction forces, the force balance along the filament axis can be expressed as follows:

$$\int_A p_{xx}dA = \text{const.} = F_{ext} \tag{1}$$

where dA is the element of the area and the integral is taken over any cross-sectional area of the filament, F_{ext} is the external tensile force and p_{xx} is the tensile stress along the filament.

In a steady-state the continuity equation takes the form [9]

$$v \left(\frac{\partial A}{\partial x}\right) = -A \left(\frac{\partial v}{\partial x}\right) \tag{2}$$

and after integration it gives

$$G = A \cdot v \cdot \rho \tag{3}$$

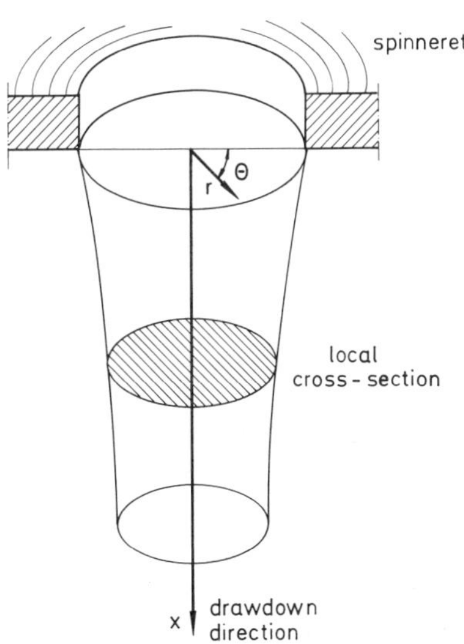

FIG. 2. Illustration of the filament position in the cylindrical coordinate system.

where G is the mass rate of polymer flow, ρ is the density and v is the filament velocity.

Taking into consideration the twist of the molten filament we introduce the momentum balance equation

$$\int_A r p_{\theta x} dA = \text{const.} = M_{\text{ext}} \tag{4}$$

where M_{ext} is the twisting moment applied to the solid part of the filament, $p_{\theta x}$ is the shear stress component. Because of the rotational symmetry of the shear stress distribution it is clear that the resultant force must be zero. The only resultant of the stress distribution is the moment.

The distribution of velocities and stresses in the spinning line is strongly affected by the heat transfer from the filament to the surrounding medium. To consider this effect, the heat balance must be introduced. Assuming only convection-controlled heat transfer one can write [5]

$$v \left(\frac{\partial T}{\partial x}\right) = -\frac{2(T - T_s)\alpha}{\rho R C_p} \tag{5}$$

where R is the local radius of the polymer filament, T is local temperature, T_s is the temperature of the surrounding medium, C_p is the isobaric specific heat of polymer melt, and α is the coefficient of heat transfer at the filament surface.

For further considerations we assume that the melt behaves like an incompressible Newtonian liquid with viscosity η. Then

$$p_{xx} = 3\eta \left(\frac{\partial v}{\partial x}\right) \tag{6}$$

and

$$p_{\theta x} = \eta r \left(\frac{\partial \omega}{\partial x}\right) \tag{7}$$

can be treated as the constitutive equations.

Introducing these equations into eqs. (1) and (4) and assuming the circular cross-section of the filament, we obtain

$$F_{\text{ext}} = 3\eta \left(\frac{\partial v}{\partial x}\right) \int_A dA = 3\pi\eta R^2 \left(\frac{\partial v}{\partial x}\right) \tag{8}$$

and

$$M_{\text{ext}} = \eta \left(\frac{\partial \omega}{\partial x}\right) \int_A r^2 dA = \eta I_2 \left(\frac{\partial \omega}{\partial x}\right) \tag{9}$$

where $I_2 = \int_A r^2 dA$ is called the polar moment of inertia of the cross-sectional area about the axis of the filament. For the cross-section of radius R the integration gives

$$I_2 = \pi R^4/2 \tag{10}$$

Then we can express velocity gradients of the rotational and translational motions as follows

$$\frac{\partial v}{\partial x} = \frac{F_{\text{ext}}}{3\pi\eta R^2} \tag{11}$$

and

$$\frac{\partial \omega}{\partial x} = \frac{2M_{\text{ext}}}{\pi\eta R^4} \tag{12}$$

The viscosity is strongly dependent on temperature. To express this dependence the empirical formula of Williams, Landel, and Ferry [10] can be applied

$$\log_{10}\left(\frac{\eta T}{\eta T_0}\right) = \frac{899.9(T_0 - T)}{(51.6 - T_0 - T_g)(51.6 - T - T_g)} \tag{13}$$

where T_0 is the temperature of the extruded polymer melt and T_g is the glass transition temperature.

Taking into consideration eqs. (3), (5), (11), (12), and (13) as a set with boundary conditions, the solution of the velocity distribution against the distance from the spinneret can be obtained. The boundary conditions are defined by T_s, T_0, d, G, V and ω_1. Such solution contains, of course, many simplifying assumptions. However, not all of them were mentioned here but they are discussed in details by Kase and Matsuo [9]. We have calculated the velocity gradients along the filament using the temperature and cross-section area distributions given earlier by Kase and Matsuo [9] for simple spinning of polypropylene. Figure 3 shows the dependences (replotted from the paper of those authors) of the cross-section area and the temperature on the distance from the spinneret.

Figure 4 shows the velocity gradients calculated using eqs. (11), (12), and (13) and the data of T and R from Figure 3. The velocity gradients in Figure 4 are calculated in arbitrary units and therefore they can not describe any real system. The plots presented can be treated only as an illustration of the difference in the distributions of translational and rotational velocity gradients.

It is seen that the maximum rotational speed gradient is shifted down the filament with respect to the maximum of longitudinal velocity gradient. It means that any volume element of the molten filament is at first stretched and then twisted as it travels from the spinneret to the take-up element. Due to the viscosity distribution against the distance from the spinneret it can be also noticed that the melt is subjected to maximum torsion when it gets the higher value of viscosity than that during the maximum stretching deformation.

Such velocity gradient distributions can have important influence on the orientation effects during spinning.

It is known that the molecular orientation of the molten polymer is proportional to the velocity gradient and to the viscosity-dependent orientation relaxation time. It means that the orientation produced during the torsional deformation can be of major significance in the spinning process, because it takes place in the part of the filament of lower flexibility and in the part which reaches the solidifying zone in shorter time.

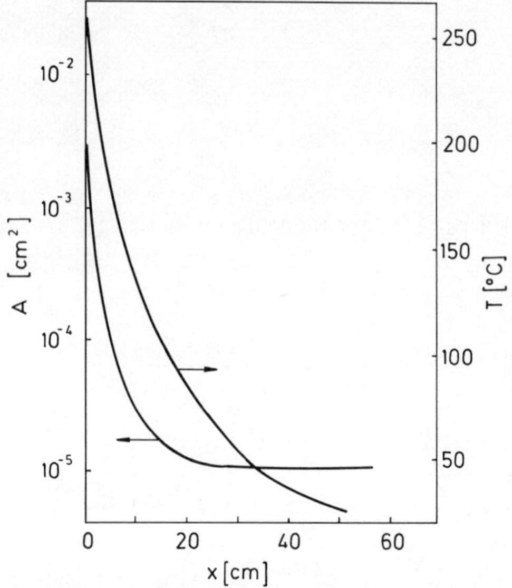

FIG. 3. Calculated cross-section area A and the temperature T of polypropylene fiber as functions of the distance from the spinneret (replotted from: [9]; Fig. 9, curve 4).

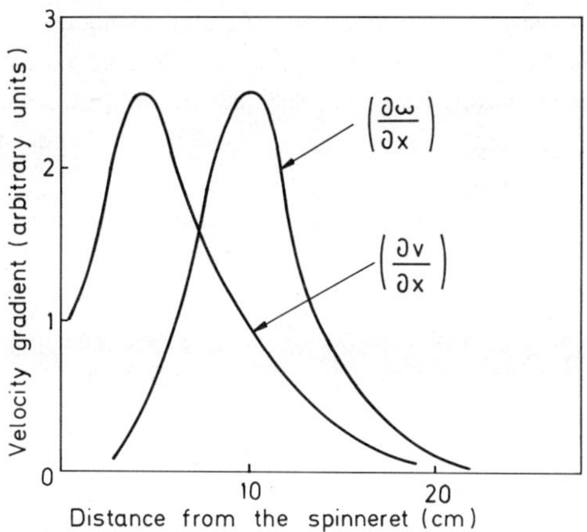

FIG. 4. Calculated rotational and longitudinal velocity gradients as functions of the distance from the spinneret.

SPINNING CONDITIONS

Polypropylene fibers were spun from Moplen S30G/Montecatini (using equipment showed in Fig. 1). In Figure 5 the details at the take-up element are shown.

Polymer melt was extruded with the mass rate 1.8 G/min at 190°C through the capillary having 1 mm in diameter. The solidified filament was taken up by the spinning frame at the distance $L = 1.2$ m from the spinnertt. The fibers were spun with the take-up velocity $V = 72$ m/min and with different rotational speed ω_1 ranging from 10 to 150 Hz. The ambient temperature was 25°C.

FIG. 5. View of the flyer spinning frame.

MORPHOLOGY

The structure of fibers was examined with various techniques. Preliminary observations done using optical microscopy showed that the filament has a constant diameter of about 50 μ independent of the rotational speed. The polarizing micrograph of single fiber twisted with 100 r.p.s. is shown in Figure 6. Figure 7 shows the scanning electron micrograph of the surface of the twisted fiber. The surface of the fiber is smooth and uniform.

The overall orientation of fibers was examined using the interferometer microscope. The birefringence of fibers measured by this technique is plotted in Figure 8. The observations of the thin cross-sections of fibers in polarizing microscope show that the image of untwisted fiber is completely dark while the characteristic Maltese cross appears on the cross-sections of the twisted fibers.

An example for highest rotational speed is depicted in Figure 9. Further observations show that the characteristic dark cross is distinct only in the surface layer for fibers twisted with low rotational speeds, while for fibers twisted with high rotational speeds it is distinct also in the center of the cross-section as it is shown in Figure 9. Such an image proved that the fiber has the cylindrical optical anisotropy. The anisotropy is not constant along the radius of the fiber and it reaches higher values towards the center of the cross-section for fibers obtained at higher rotational speeds.

Comparing these with the results of birefringence measurements one can conclude that the birefringence along the fiber decreases with the increasing optical anisotropy of the cross-section of fibers. It proves that the twisted fibers have the helical orientation and that it can be changed with the twist rate.

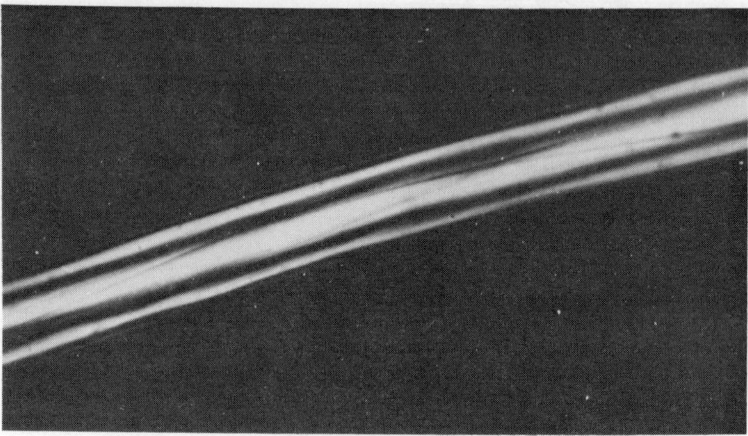

FIG. 6. Micrograph of the single fiber twisted with 100 r.p.s. (polarizing microscope).

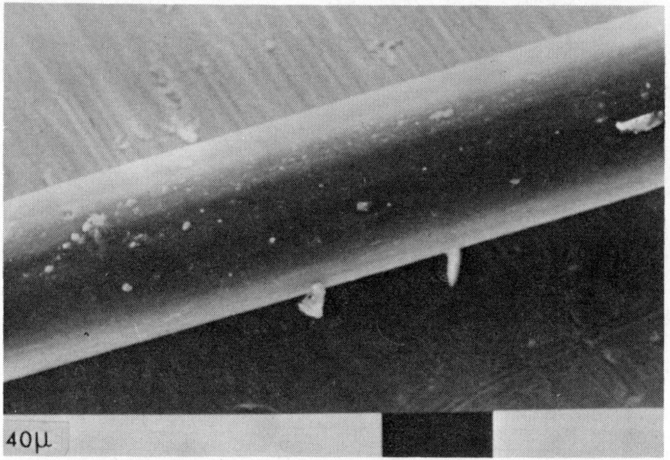

FIG. 7. Fiber surface observed in the scanning electron microscope.

Wide-angle X-ray scattering was applied for the determination of crystalline element orientation. The measurements were done using the DRON1 apparatus (USSR). An example of the diffraction pattern for the untwisted polypropylene fiber is shown in Figure 10. The reflexes taken into consideration in further discussion are marked [11]. It was found that the mentioned reflexes were displaced in azimuthal direction if fibers with different twists were examined. Such displacements are characteristic of changes in the orientation of crystalline elements. An example of the intensity distribution for fibers with various twists is shown in Figure 11. The arrows show the directions of the reflex displacements. The reflexes from (011) planes, first separated for the untwisted fibers, come closer with twist and join together and then for high twist separate again, covering separated reflexes from (040) planes. The orientation function of the plane (110) was calculated from the normalized intensity distribution [12].

In the case of uniaxial orientation, this function has one maximum at the angle

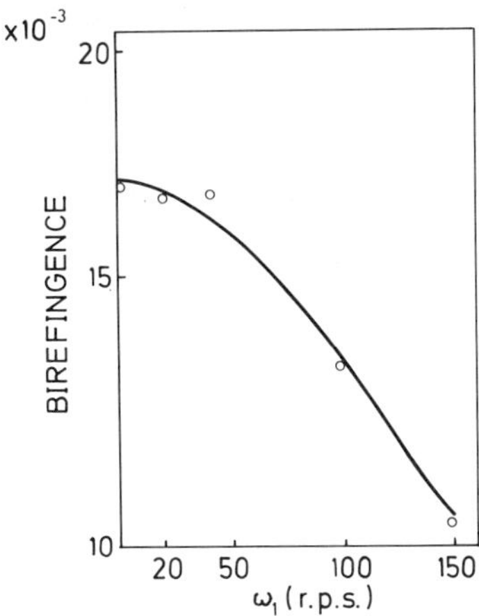

FIG. 8. Birefringence of fibers obtained with various twist rates.

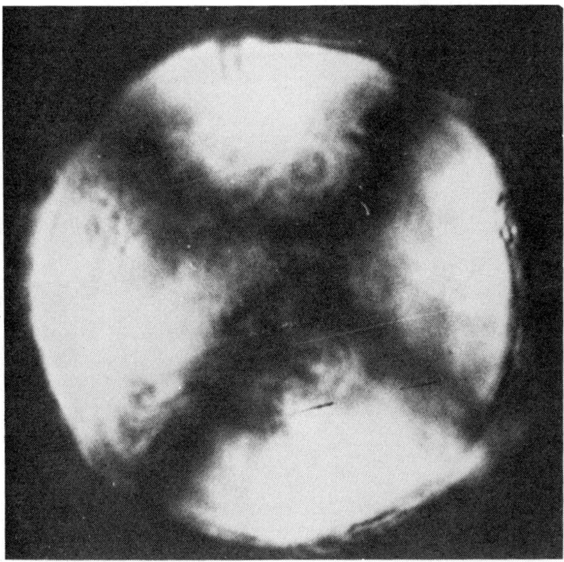

FIG. 9. Cross-section of the twisted fiber observed in the polarizing microscope with crossed polaroids.

of orientation and it has a Gaussian shape if there is statistical deviation from the direction of orientation. The Gaussian shape of the (110) plane orientation function with maximum at 0° was observed for untwisted fibers. The orientation functions for untwisted and twisted fibers are shown in Figure 12. The orientation of the (110) planes for untwisted fibers is shown schematically by means of normal vectors in Figure 13a. For the twisted fibers the shape of orientation distribution of the (110) planes differs from Gaussian. It means that the lowering of the maximum of orientation function is a result of different than uniaxial type of orientation.

As follows from optical observations described above, the central part of

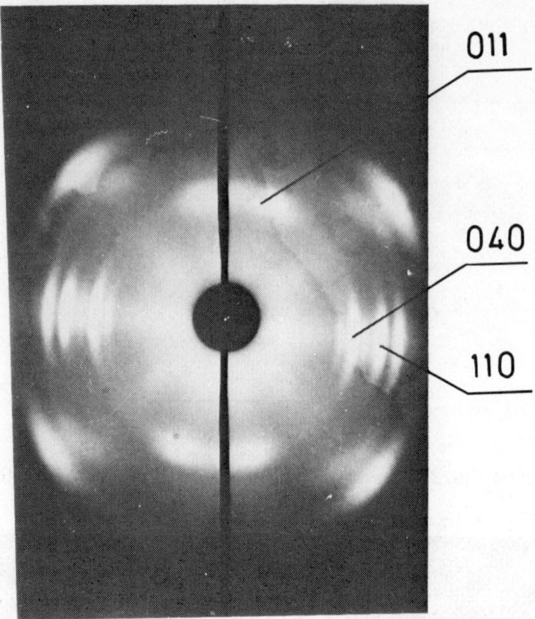

FIG. 10. Wide-angle X-ray diffraction pattern of the untwisted polypropylene fibers.

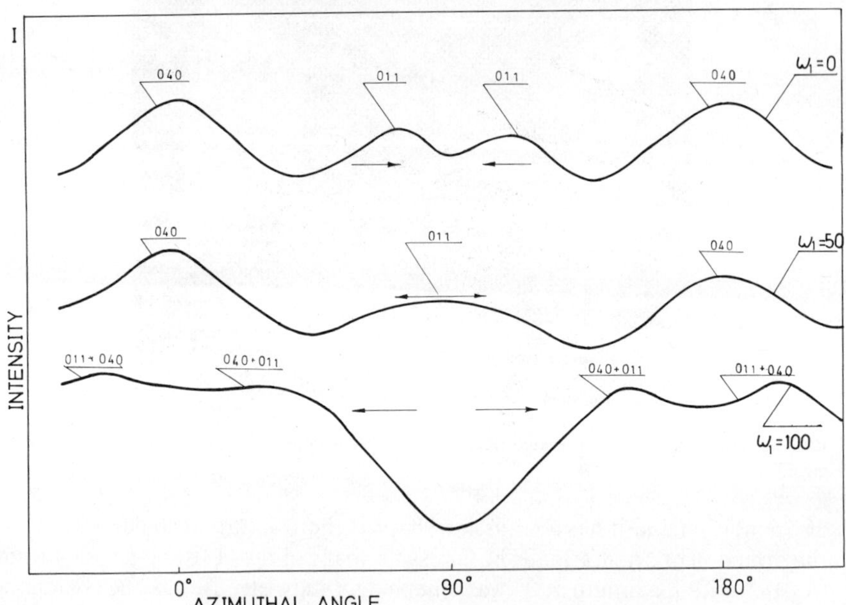

FIG. 11. Azimuthal displacement of the (040) and (011) diffraction reflexes for fibers spun with the twist.

twisted fibers has unchanged orientation. Thus one can accept the Gaussian shape of orientation function of the (110) planes in the central part of twisted fibers.

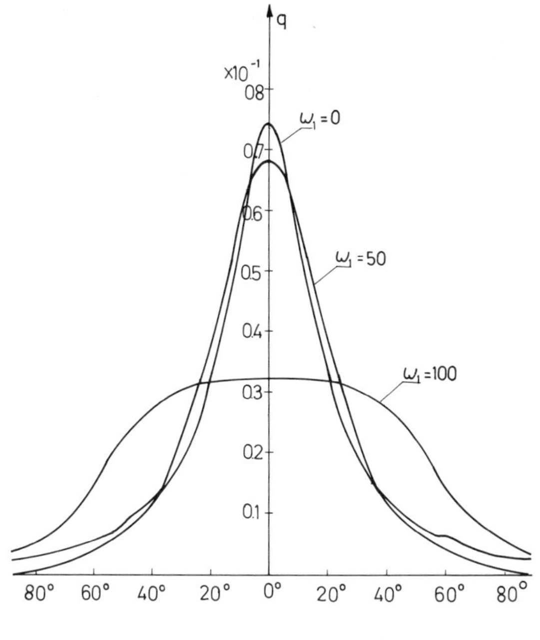

FIG. 12. Orientation functions of the (110) planes for fibers spun with various twist rates.

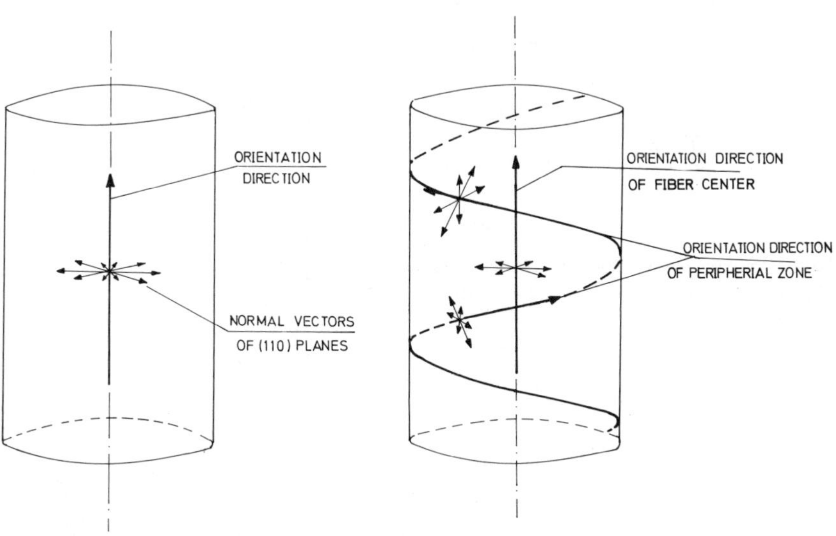

FIG. 13. Illustration of the orientation of the (110) planes by means of normal vectors, (a) untwisted fiber, (b) twisted fiber.

In addition, the orientation distribution function for the (110) planes could be treated as a sum of orientations in the central and peripheral zones of the fiber. Based on that assumption, the orientation function of peripheral zones can be separated from the overall orientation function. It is shown in Figure 14. The function with two distinct maxima, which are symmetric with respect to the fiber axis, is characteristic of helical orientation. Schematic representation of orientation of the (110) planes using normal vectors separately for central and peripherial zones is shown in Figure 13b. The displacements of reflexes for the (011) and (040) planes (Fig. 11) show analogous changes of orientations of these planes and similar calculations can be done.

Using an apparatus schematically shown in Figure 15, the patterns of light diffraction were recorded. The source of light is the He–Ne laser. The polarized light beam passes through the cylindrical lens giving a vertically elongated spot. Fiber in immersion is placed in the beam. The scattered light passes through the analyzer and is observed on the screen or is photographically recorded.

The diffraction patterns for differently twisted fibers are shown in Figure 16. The structure of these patterns is like a projection of a helix on the plane parallel to its axis. It is characteristic that the pattern shifts if the fiber is dis-

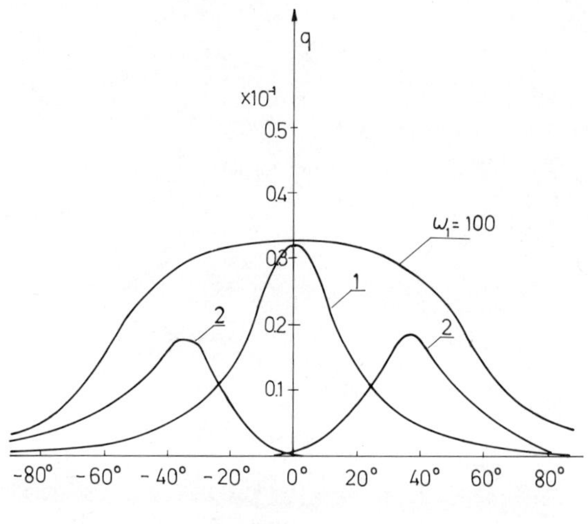

FIG. 14. Separation of the orientation function of (110) planes into the orientation functions of the central (1) and peripheral (2) zones of the twisted fiber.

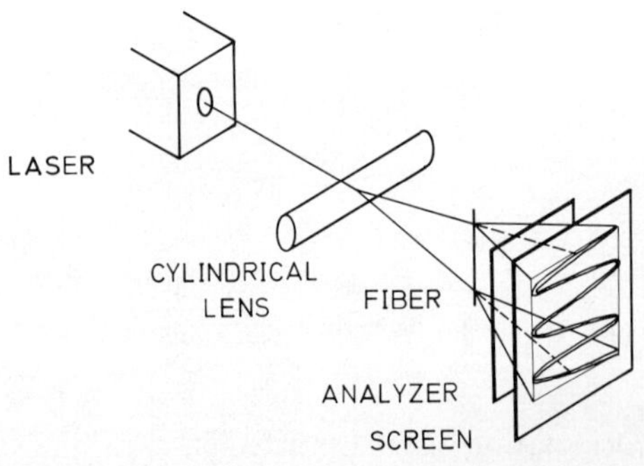

FIG. 15. Scheme of the light diffraction experiment.

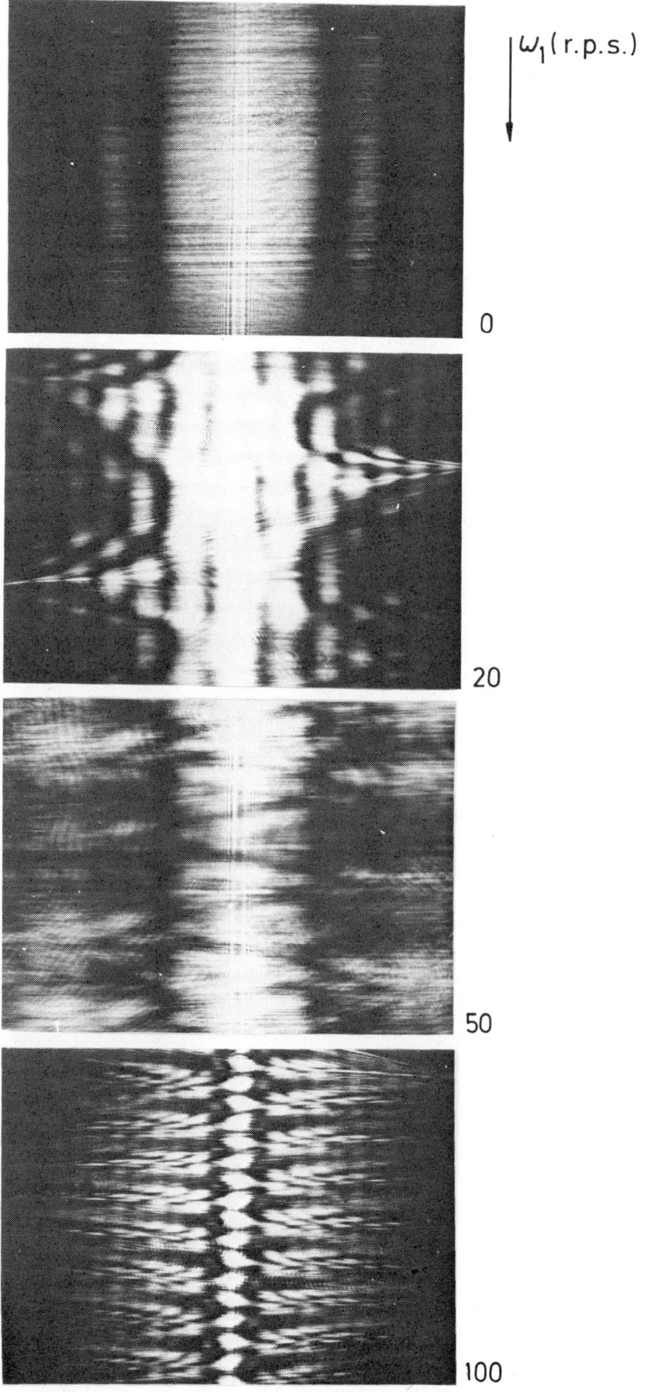

FIG. 16. Light diffraction patterns for the untwisted and twisted fibers.

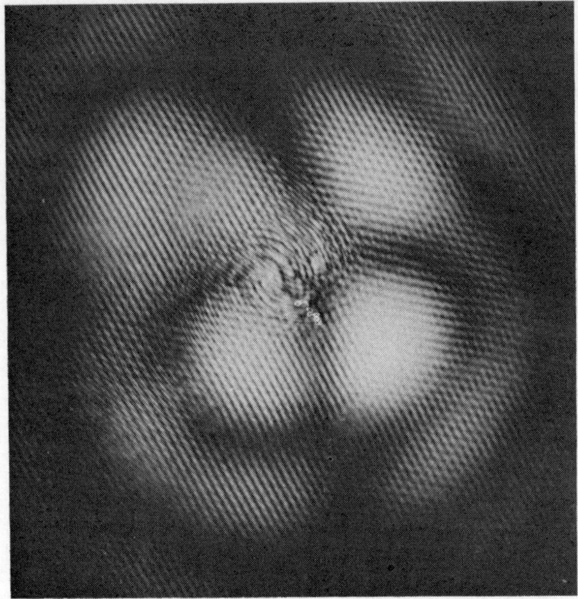

FIG. 17. Small-angle light scattering from the thin cross-section of the twisted fiber.

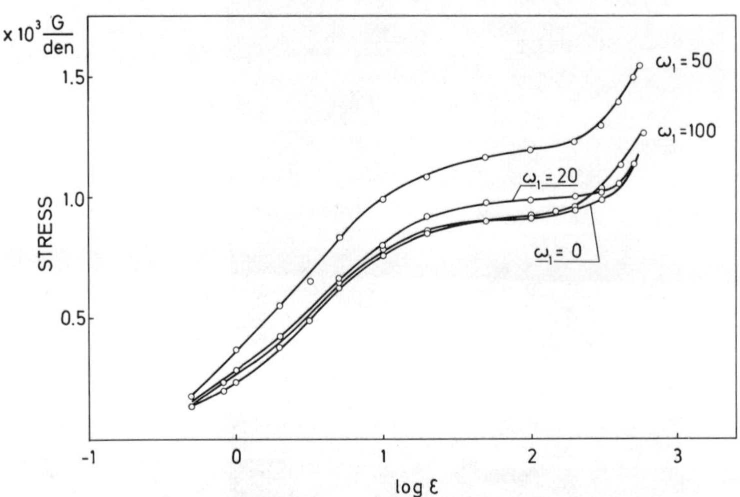

FIG. 18. Stress–strain curves for the fibers obtained with various twist rates.

placed along its axis and shifts also when rotating the fiber. It seems that the
fiber has the regular periodic macrostructure, and the observed light diffraction
pattern results from the periodic unhomogeneities of the density or optical anisotropy. The period in such a diffraction pattern results from the periodic unhomogeneities of the density or optical anisotropy. The period in such a diffraction pattern is exactly the same as the pitch of helix calculated from the ratio
of the take-up rate and rotational speed. The shape of the diffraction pattern
depends markedly on the refractive index of the immersion liquid. The immersion

having the refractive index equal to the mean refractive index of the fiber gives the best results.

The small-angle diffraction pattern from the cross-section of the fiber for crossed polaroids is shown in Figure 17. The obtained four-leaf pattern proves additionally the previous conclusion about the cylindrical symmetry of the optical anisotropy of the twisted fiber [13].

MECHANICAL PROPERTIES

The stress–strain measurements of the fibers with various twists were conducted using the Instron Testing Machine with elongation rate 60%/min. The results are shown in Figure 18. The elastic modulus of fibers was calculated from these results in the range of 1% deformation and then was plotted against the rotational speed as is seen in Figure 19. In addition, the sonic modulus was measured using the Morgan Sonic Modulus Tester at 10 kHz and the results are also drawn in Figure 19.

Both the tensile and sonic modulus reach the maximum values for fibers spun with intermediate rotational speed. It can be also seen from Figure 18 that higher stress is needed to deform the fibers obtained with rotational speeds of 20 and 50 Hz than that needed for untwisted fiber deformation. The ultimate strength of twisted fibers is also extreme for fibers with $\omega_1 = 50$ Hz. The above results show that the measured mechanical properties of twisted fibers are clearly affected by a rotational speed of spinning.

CONCLUSIONS

The simplified analysis of the dynamics of the melt-spinning with the twist has shown that the shear deformation of the spun melt can have significant influence on the orientation due to the shift of the maximum of rotational velocity gradient down the filament. The morphological examination of the fibers spun with the twist has shown the distinct helical orientation as well as the periodic

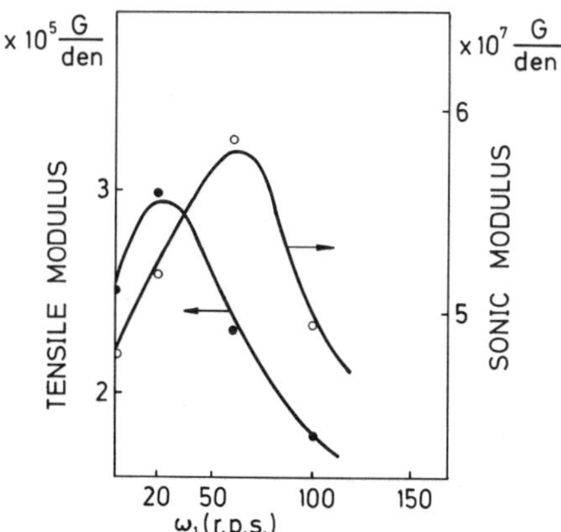

FIG. 19. Dependence of the tensile and sonic modulus on the rotational speed of spinning.

macrostructure inside the fibers. The observed morphology modification causes the changes of mechanical properties of the twisted fibers with respect to fibers obtained without twist. It was observed that the tensile and sonic modulus and the ultimate strength are maximal for fibers spun with intermediate rotational speed.

REFERENCES

[1] I. W. S. Hearle, *J. Polym. Sci.,* **C20,** 213 (1967).
[2] I. W. S. Hearle, *J. Appl. Polym. Sci.,* **7,** 1207 (1963).
[3] P. G. Andersen, S. Test, S. R. Deo, and S. H. Carr, *J. Appl. Polym. Sci.,* **19,** 1735 (1975).
[4] A. Ziabicki, *Man-Made Fibers—Science and Technology,* Eds. H. F. Mark, S. M. Atlas, and E. Cernia, Interscience, New York, 1967.
[5] S. Kase and T. Matsuo, *J. Polym. Sci.,* **3A,** 2541 (1965).
[6] A. Ziabicki and K. Kędzierska, *Kolloid-Z.,* **171,** 51 (1960).
[7] A. Ziabicki, *Kolloid-Z.,* **175,** 14 (1961).
[8] E. H. Andrews, *Br. J. Appl. Phys.,* **10,** 39 (1959).
[9] S. Kase and T. Matsuo, *J. Appl. Polym. Sci.,* **11,** 251 (1967).
[10] M. L. Williams, R. F. Landel and J. D. Ferry, *J. Am. Chem. Soc.,* **77,** 3701 (1955).
[11] R. J. Samuels, *Structured Polymer Properties,* John Wiley, New York, 1974.
[12] M. Kakudo and N. Kasai, *X-Ray Diffraction by Polymers,* Kodansha Ltd., Tokyo, 1972.
[13] R. S. Stein and M. B. Rhodes, *J. Appl. Phys.,* **31,** 1873 (1960).

A COMPARISON OF THE SUPERMOLECULAR STRUCTURE OF WHOLLY-AROMATIC AND ALIPHATIC POLYMER FIBERS

L. I. SLUTSKER, L. E. UTEVSKII, Z. YU. CHEREISKII, and K. E. PEREPELKIN

Leningrad Branch of the All-Union Scientific Research Institute of Artificial Fibres, Leningrad, 195030, U.S.S.R.

SYNOPSIS

This paper considers specific features of the supermolecular structure of wholly-aromatic polymer (WAP) fibers as compared with that of aliphatic polymer (AP) fibers revealed by small- and wide-angle X-ray diffraction and acoustic methods. Long periods of 150–230 Å in size were observed in many WAP fibers. Special studies showed these periods to consist of crystalline and intercrystalline amorphous regions alternating along the fiber axis. These intercrystallite regions represent, just as in AP fibers, "weak" spots in the structure with respect to deformation and are responsible for the sorption properties. Acoustic measurements permitted the evaluation of the fraction of load-holding chains in intercrystallite regions from the number of chains in the crystallite per the same section. It can reach 60–80% for WAP, while only 10–15% for the strongest AP fibers. Besides, in WAP fibers all tie molecules in intercrystallite regions are load-holding, while only a small fraction of them hold the load in AP fibers. A study of poly-*p*-benzamide fibers showed their structures to be close to homogeneous. This accounts for the unusually high elastic modulus of this fiber (16,000–17,000 kg/mm^2) which is close to that of crystallites (18,600 kg/mm^2) and to the theoretical modulus of the molecule. Such a homogeneous supermolecular structure of oriented polymers is observed for the first time.

INTRODUCTION

According to modern concepts, the supermolecular structure of AP fibers has a characteristic feature consisting in the existence of the so-called long periods, i.e., in a regular alternation along the fiber axis of more dense, more ordered crystalline regions and of less dense, less ordered amorphous regions. These periods may be a few tens or hundreds of Å. In most cases the existence of long periods brings about the appearance of small-angle meridional X-ray reflections (Fig. 1). There are, however, cases where for AP fibers (e.g., highly-oriented fibers of polyethylene, polyvinyl alcohol (PVA), polyoxymethylene) as well as for fibers of hydrated cellulose (HC) such a reflection is not observed. Special studies carried out on PVA [1, 2] and HC [3, 4] showed such long periods to exist in these specimens; however they do not reveal themselves in diffraction. This is possibly a combined result of a simultaneous action of such factors as a small difference in the density of amorphous and crystalline regions, a large dispersion in the size of the long periods, and the small diameter of the fibrils [1, 2].

Journal of Polymer Science: Polymer Symposium 58, 339–358 (1977)
© by John Wiley & Sons, Inc.

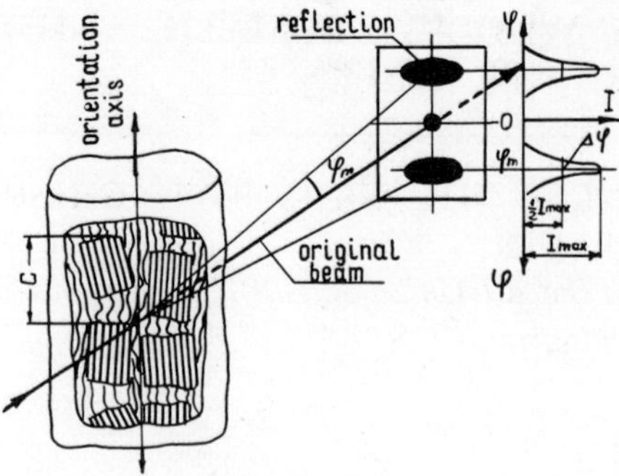

FIG. 1. Small-angle X-ray scattering from oriented crystalline polymers (schematic).

One of the most convincing arguments for this is the fact that the relative deformation of high-strength "no-reflection" PVA specimens exceeds by an order of magnitude that of crystallites [2] just as it is the case usually. Thus one may suggest that all AP fibers including the strongest ones have a heterogeneous supermolecular structure characterized by long periods. Most authors consider the model of Hosemann–Bonart with folded chains in the crystallites to be the most acceptable model for the supermolecular structure in aliphatic polymers. A study of the effect of longitudinal elastic extention [2, 5–8], iodine sorption [9–12], heat [3, 4], and hydrolysis [13] on the supermolecular structure of various APs and HC showed the intercrystallite amorphous layers to be "weak" spots of the structure since it is in them that the processes of deformation and rupture, thermal destruction, sorption, and chemical reaction primarily occur.

Thus the presence of the amorphous layers and their structure proper affect many essential properties of the fibers.

At the same time practically no attention has been paid up to now to studying the heterogeneity in the supermolecular structure of oriented specimens of wholly-aromatic, heat-resistant polymers, the presence of "weak" spots in their structure. Such studies are necessary for the understanding of the variety of properties of these fibers which find at present an ever increasing use. Therefore the purpose of the present work was to investigate various features in the supermolecular structure of fibers of WAPs as compared with those of APs, and to study the effect of these features on the fiber properties.

EXPERIMENTAL

As subjects for the study we chose fibers of polymers specified in Table I.

The fibers of WAPs of interest were prepared in the form of a bundle of monofibers by wet-spinning from solution with subsequent plastification and heat drawing. The conditions of preparation of the specimens were reported elsewhere [14–18]. As subjects for the study of the AP fibers we chose the corresponding fibers obtained in the usual way. The strength σ and the breaking

TABLE I
Polymers Studied

Item No.	Name	Notation	Monomer link
1	Poly(4,4′-diphenyloxide)-pyromellitimide	PI-PM	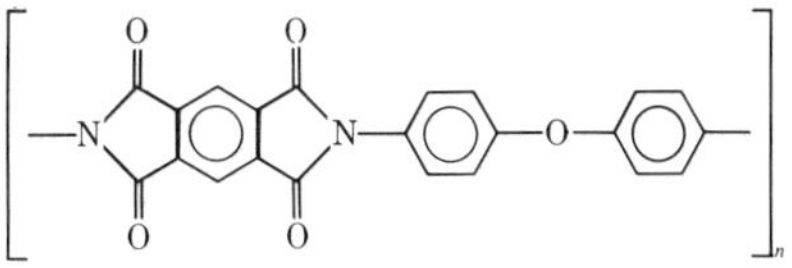 (structural formula)
2	Poly-p-phenyleneterephthalamide	PPTA	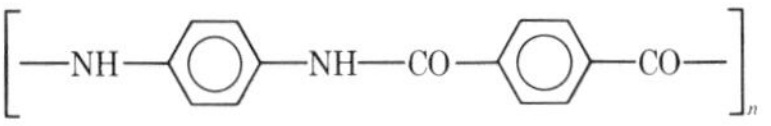 (structural formula)
3	Poly-m-phenyleneisophthalamide	PPIA	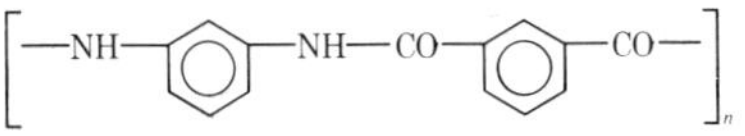 (structural formula)
4	Polyarylene-1,3,4-oxadiazole	POD	copolymer 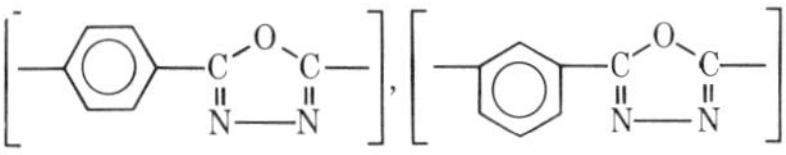 (structural formula)
5	Poly-p-benzamide	PBA	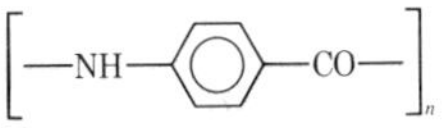 (structural formula)
6	Polyvinyl alcohol	PVA	$[-CH_2-CHOH-]_n$
7	Polyethylene	PE	$[-CH_2-CH_2-]_n$
8	Polyoxymethylene	POM	$[-CH_2-O-]_n$
9	Polycaproamide	PCA	$[-NH-(CH_2)_5-CO-]_n$

TABLE II
Mechanical Properties of the Ultimate-Drawn Fibers

Item No.	Polymer	Strength, σ, kg/mm^2	Breaking elongation, ε, %
1	PI-PM	72	5.6
2	PPTA	85	2.6
3	PBA	170	2.4
4	PPIA	70	11.2
5	POD	66	5.3
6	PVA	128	3.5
7	PE	73	25.0
8	POM	106	8.0
9	PCA	90	20.0

elongation ε of the ultimate-drawn fibers, i.e., of the fibers which cannot be drawn still more under the given conditions of orientation without rupture, are presented in Table II.

The structural studies were carried out by small- and wide-angle X-ray diffraction methods and by acoustic techniques.

We used in the X-ray studies nickel-filtered CuK_α-radiation ($\lambda = 1.54$ Å). The ionization measurements were performed on the KRM-1 (small-angle),

$\overline{\text{U}}$RS-50$\overline{\text{I}}$M and DRON-1 (wide-angle) equipment made in the USSR. Slit collimation (collimation by Kratky in KRM-1) and scintillation detection were used throughout.

The wide- and small-angle diffraction was studied on unloaded fibers and under longitudinal elastic extention. The loading was carried out on a proof-ring dynamometer with the specimen drawn in the form of a bundle of parallel fibers (a few tens of filaments) 8 mm wide and $\sim$0.2 mm thick. The relative deformation of the specimen ϵ_{sp} was derived from the change in the distances between two sets of points traced on the fiber 20 mm apart measured with a microscope. The relative error in determining ϵ_{sp} was $\pm0.5\%$. The load applied to the bundle was determined from the deformation of the dynamometer spring-loaded ring. The relative deformation of the long period ϵ_c was evaluated from the change in the angular position of the small-angle reflex [5, 6]. The relative deformation of the crystal lattice ϵ_{cr} along the fiber axis was derived from the angular displacement of the wide-angle meridional reflexes [19, 20] lying at double Bragg angles: 10°54′ for PI-PM, 24°14′ for POD, 42° for PPTA and PBA, 75° for PVA. The relative error in the determination of ϵ_{cr} constituted 0.1–0.5%. Experiments with loading permit the evaluation of both the elastic modulus of the specimen proper E_{sp} and the modulus of its crystal lattice E_{cr} [19, 20]. To do this, we assumed, just as in refs. [19, 20], that the stress on the crystallites in a loaded specimen is equal to the average stress on the specimen. This is equivalent to assuming that crystallites occupy all the cross-section of the specimen.

We carried out structural studies of AP fibers subjected to iodine sorption. Iodine was from the vapor phase [9, 10]. Specimens with different iodine content were obtained by maintaining fibers in iodine vapor at 90–120°C for different times. Next, they were kept at the same temperature but without iodine to ensure a more uniform distribution of iodine over the bulk of the fiber. Reference specimens were maintained under the same conditions (temperature, time) as the ones under study but without the action of iodine vapor. A differential technique was employed in small- and wide-angle X-ray measurements [9] providing for a comparison of the scattering intensity from specimens with different absorption.

The acoustic measurements permitted us to determine and calculate the following quantities [21, 22]: the dynamic elastic modulus of specimens, the orientation of tie and load-holding molecules in the intercrystallite amorphous regions, the fraction of the load-holding chains in these regions of the number of chains in a crystallite for the same cross-section. For this purpose, we measured the change in the speed of sound propagation in the fibers at both room and liquid nitrogen temperatures at the transducer frequency of 50 kHz and pulse repetition frequency of 2.10^4 cps. The accuracy of the some speed measurements was 2.5–3%.

EXPERIMENTAL RESULTS AND DISCUSSIONS

Wide-angle X-ray patterns obtained on fibers of WAPs showed that we had well-oriented crystallizing polymers (Fig. 2). In view of the purpose of our work

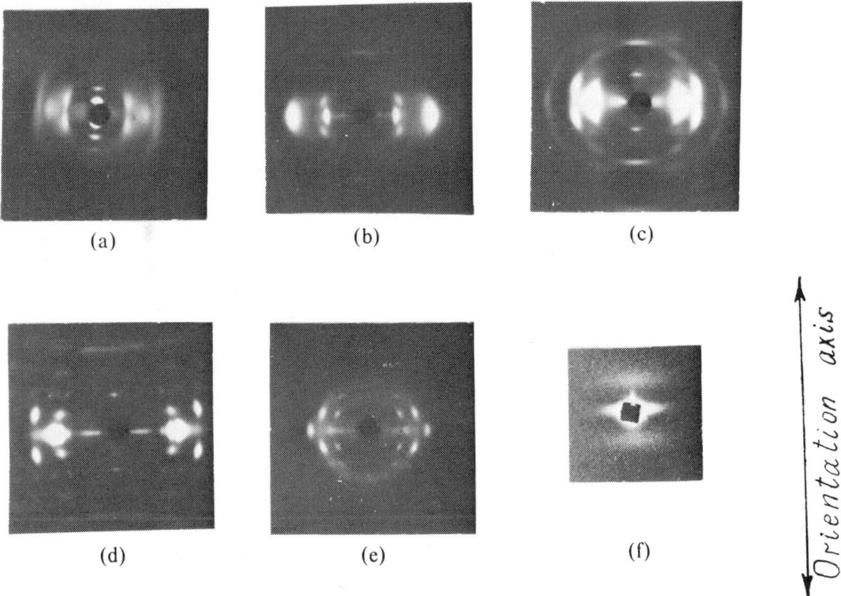

FIG. 2. Diffraction pattern for wholly-aromatic polymer fibers. Wide-angle: (a) PI-PM, (b) POD, (c) PPTA, (d) PBA, (e) PPIA; Small-angle: (f) POD.

it was of interest to see whether these fibers have regions differing in their structure and properties from the crystalline ones. To research this, we first studied small-angle X-ray scattering. It turned out that small-angle meridional reflections having a line-shaped pattern can be observed for all WAPs examined (for example, POD in Fig. 2f) except for PBA. Small-angle X-ray reflections for PI–PM and POD appear only at a thermal treatment or drawing temperature of ~500°C or higher, no reflections being present at lower temperatures. (The conditions at which small-angle reflections appear for PPTA and PPIA were not studied in this work.)

The results of small-angle X-ray diffraction measurements carried out on aromatic polymers and PVA (the example of APs) are presented in Figure 3. After separation of the reflections with smooth curves as shown in Figure 3, the angular position of the reflections yielded the average size of the long periods which is given in Table III. Also presented here are data for PCA and POM.

The absence of small-angle reflections for some PI-PM and POD fibers, as well as for PBA, does not indicate the absence of density heterogeneity (long periods) in them, which has been already pointed out in the case of APs. A question arises of the nature of the long periods in WAP fibers for which one observes small-angle reflections, and whether heterogeneity in density exists in "no-reflection" fibers.

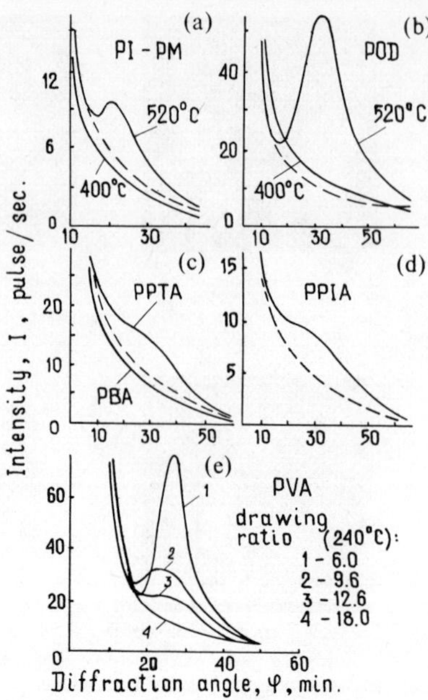

FIG. 3. Small-angle meridional X-ray scattering from different fibers. Temperature of heat drawing specified.

TABLE III

Data on Long Periods of Ultimate-Drawn Fibers

Item No.	Polymer	Long period, C, Å	Size dispersion of long periods, $\frac{\Delta c}{c}$
1	PI–PM	230	0.36
2	PPTA	190	0.40
3	PPIA	160	–
4	POD	155	0.34
5	PVA	200	0.30 [a]
6	PCA	105	0.37 [a]
7	POM	150	0.40 [a]

[a] Data from [22].

Small-Angle Reflections Are Observed

Iodine Sorption

As seen from Figure 4 presenting data for **PI-PM** and **POD** fibers, introduction of iodine reduces drastically the intensity of small-angle reflections. Small- and wide-angle X-ray measurements on reference specimens showed

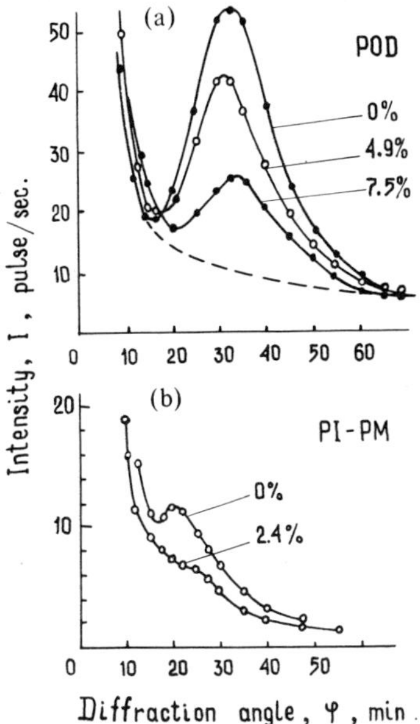

FIG. 4. Variation in small-angle X-ray reflection intensity versus iodine weight content (in per cent).

the temperature itself not to affect the fiber structure. It turned out also that the equatorial and meridional wide-angle reflections for the specimens with iodine and reference specimens coincide in shape, intensity, and angular position. Thus the change in the intensity of small-angle reflections is connected only with iodine sorption. Note that iodine does not enter the crystallites and concentrates in intercrystallite layers, thus reducing the difference in the density of the crystalline regions and intercrystalline layers. The results obtained agree with the data on selective iodine sorption by the intercrystallite amorphous layers for polyvinyl alcohol [10, 11], polycapromide [9], and polyethylene [12].

The fact that iodine sorption slightly changes the intensity of diffuse scattering at small angles (Fig. 4) indicates, just as in the case of PVA fibers [10], that this diffuse scattering is caused mainly by the heterogeneity of a different nature than the observed long periods, possibly by submicrocracks.

Loading

Figure 5 presents data on the effect of loading on small-angle diffraction for the PI-PM and POD fibers. We see here qualitatively the same pattern: indeed, the reflections increase in intensity with increasing load and restore to the original level when unloaded. In the case of PI-PM one can readily see the reversible nature of the reflection shift to smaller angles (Fig. 5a). For POD fibers this effect is very small because of the small magnitude of the relative specimen deformation ϵ_{sp} (Fig. 5b). (For PPTA fibers the effect of loading on small-angle diffraction was not studied because of the very small magnitude of ϵ_{sp} below 1%.) For PI-PM we estimated the relative deformation ϵ_c of the long period from the

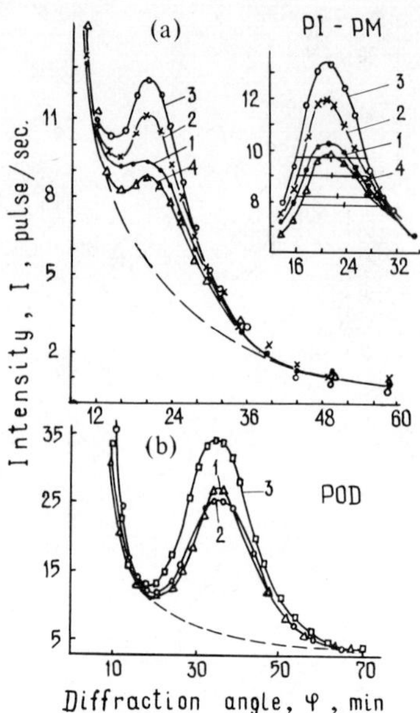

FIG. 5. Variation in small-angle X-ray reflection intensity versus relative deformation of fiber in %. (a) PI-PM: 1–0 (before loading), 2–2.8, 3–4.4, 4–0.2 (after removal of load); (b) POD: 1–0 (before loading), 2–0.4 (after removal of load), 3–2.0.

change in the angular position of reflections. Because of the small magnitude of ϵ_c the accuracy of its determination is naturally lower than that of ϵ_{sp}. It turned out that at specimen extensions of 2.8 and 4.4% the long period deforms by ~3 and ~6%, respectively. Thus one may assume, in a first approximation, that ϵ_{sp} and ϵ_c for PI-PM are close, as this is the case for most APs [5–7].

In order to establish the reasons for the enhanced intensity of the small-angle reflections in loaded fibers and to reveal the structural regions which are the weakest with respect to deformation, we compared relative deformations of the specimens ϵ_{sp} (macrodeformation) and of the crystallites ϵ_{cr} (microdeformation) at different loads. These data are given in Figure 6 for the fibers of PI-PM, POD and PPTA. Also presented for comparison are the data for the PVA fibers where small-angle reflections are observed (Fig. 3e, curve 2). In all cases, the pattern is essentially the same, i.e., ϵ_{sp} exceeds by far ϵ_{cr} throughout the loading range covered. Among aromatic polymers, the strongest effect is revealed in the case of POD where ϵ_{sp} and ϵ_{cr} differ by a factor of ~8. Such a strong difference is typical for such aliphatic polymers as PVA (Fig. 6d), PE, and PCA [19] whose molecules in the crystallites are arranged in a planar zigzag pattern. In this respect PI-PM and PPTA differ from the above mentioned aliphatic polymers. For PPTA, ϵ_{sp} exceeds ϵ_{cr} by a factor of ~2, and for PI-PM, by a factor of ~1.7. Note that in crystallites PI-PM molecules also have the shape of a planar zigzag [23] and in PPTA molecules the small fixed angle between the planes of phenylene rings is the source of more rigidity of the chains [24].

The crystallites behave here in an elastic way. PI-PM and PPTA fibers likewise exhibit an elastic behavior while POD fibers show permanent deformation.

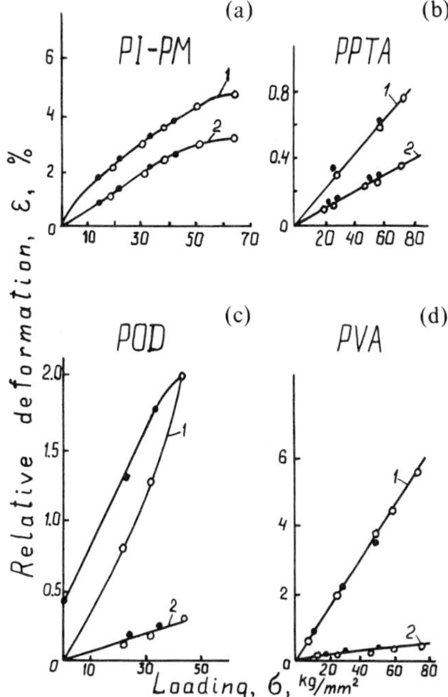

FIG. 6. Loading dependence of relative deformation for various fibers (1) exhibiting small-angle X-ray reflections, and for their crystallites (2). ○, loading; ●, removal of load.

From the slope of the $\epsilon = f(\sigma)$ graphs in Figure 6a, b, and c, we derived the elastic moduli E_{sp} and E_{cr} (see Table IV). Also given in Table IV are the dynamic elastic moduli for the specimens E_d calculated from the acoustic data, and theoretical values for the molecular moduli taken from various papers, together with some data for APs. We did not calculate E_{sp} for POD fibers since they reveal inelastic deformation (Fig. 6c). If the assumption on the equality of stress in a specimen and in crystallites (see above) is not valid and the stress in the crystallites is higher, then the true value of E_{cr} will be still higher which means that the true value of E_{cr} should be not less than the figures presented in Table IV. It is of interest that in the case of aromatic polymers, the values of E_{sp} and E_d lie much closer to the values of E_{cr} and E_{theor} than for aliphatic polymers. This indicates more perfect structure of the intercrystallite regions in WAP fibers.

PI-PM fibers reveal an increase of E_{sp} and E_{cr} with load. This feature will be discussed later together with other mechanical properties.

The observed relation between the quantities $\epsilon_{cr}-\epsilon_{sp}$ and $E_{cr}-E_{sp}$ indicates that intercrystallite layers are more yielding with respect to deformation. In the course of extension these more yielding regions possibly become looser, thus increasing the density difference within a long period which results in the observed increase in the intensity of small-angle reflection. A qualitatively similar situation is observed in APs [5, 6].

Thus the experiments on iodine sorption and loading permit us to conclude that the observed long periods in PI-PM, POD and PPTA fibers are made up

TABLE IV
Elastic Moduli for Ultimate-Drawn Fibers

Item No.	Polymer	E_{sp}	E_d	E_{cr}	E_{theor}
				kg/sq.mm	
1	PI–PM	1250–1900	1200	1600–4000	1950 /33/
2	POD	–	5000	14000	–
3	PPTA	9000	12300	18600	20400 /30/ 18200 /31/ [a] 23700 /31/ [b]
4	PBA	16000	17000	18600	20400 /30/ 18200 /31/ [a] 23700 /31/ [b]
5	PVA	3500	9100	22000 25500 /19/ 22000 /7/	19400– –21000 /37/
6	PE	1000	3800	24000 /19/	18600 /38/ 22500–25000 /37/

[a] Calculated for *cis*-conformation.
[b] Calculated for *trans*-conformation.

of alternating more dense, more ordered crystalline regions and less dense, less ordered intercrystallite layers. The latter may probably be considered amorphous because of their lower density and less ordered nature. They are more yielding with respect to deformation and are responsible for the sorption properties.

The system of long periods in WAP fibers exhibits a considerable dispersion in period size about the mean. This is evidenced by the large diffraction width of the small-angle reflections. The ratio of the angular width of the reflections at their half-maximum to the angular position $\Delta\varphi/\varphi_m$ (Fig. 2) constitutes 0.5–0.6. This value permits the estimation of the relative standard deviation of long period size Δ_c/C [25, 26]. The results are given in Table III. The values of Δ_c/C are seen to be close, 0.34–0.40, and they are close to the corresponding figures for ultimate-drawn aliphatic polymer fibers (Table III).

The aromatic polymer fibers examined are likewise ultimate-drawn in the conditions of orientation chosen, since they can no more be drawn until rupture. The fact that they can no more be drawn, in accordance with the concept outlined in refs. [27, 28], can be attributed to their having reached a high dispersion in the size of the long periods and amorphous layers resulting in the appearance of strong local overstress.

We evaluated the diameter of the fibrils for POD fibers exhibiting a strong small-angle reflection. This was done by studying the effect of the specimen tilting angle relative to the incident X-ray beam on the small-angle reflection intensity [29]. It turned out to be ~46 Å. In ultimate-drawn fibers of PVA and polycaproamide, the fibril diameter is about 60 Å [34] which is close to the result obtained.

Although the above figures characterizing the long period structure of WAPs (the long period size, its dispersion, the fibril diameter) should be considered only as approximate, nevertheless they indicate that this structure is close in its geometrical parameters to the long-period structure of oriented APs. At the same time there are essential differences in the structure of intercrystallite regions in polymer fibers of the two types which will be discussed later.

Small-Angle Reflections Are Absent

Figure 7 presents data to compare the micro- and macrodeformations for POD fibers drawn at 400°C, PBA fibers, and ultimate-drawn PVA fibers which do not exhibit small-angle reflections. In the case of PVA we see that ϵ_{sp} exceeds by about an order of magnitude ϵ_{cr} and hence, as this was pointed out above, these fibers have usual long periods with "weaker" intercrystallite regions. The same probably is true also for POD fibers for which ϵ_{sp} exceeds ϵ_{cr} by a few times (Fig. 7a). The value of E_{cr} calculated for POD fibers turned out in this case to be 7600 kg/mm^2 which is much smaller than E_{cr} for POD fibers considered above (Fig. 6c, Table IV). This may be attributed to the fact that in the case of heat drawing at 400°C the degree of crystallinity is much smaller than at 500°C (noticeable crystallization proceeds in the 400–500°C range), so that the crystallites do not occupy all the cross-section of the specimen and the stress in them is larger than the average stress in the specimen. Since we do not take this into account, our value of E_{cr} is underestimated. In this case the true discrepancy between ϵ_{sp} and ϵ_{cr} should be still larger than seen from Figure 7a.

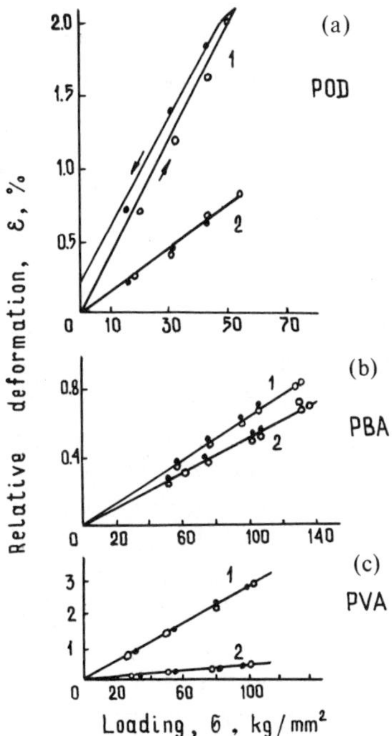

FIG. 7. Loading dependence of relative deformation for various fibers (1) which do not exhibit small-angle X-ray reflections, and for their crystallites (2). O, loading; ●, removal of load.

The pattern for PBA fibers is completely unusual (Fig. 7b). Indeed, ϵ_{sp} exceeds ϵ_{cr} by only ~16%, and the calculated elastic moduli E_{sp} and E_{cr} likewise differ by ~16% constituting 16,000 and 18,600 kg/mm². These data are shown in Table IV whence it is seen that the values of E_{cr} for PBA and PPTA are equal while the values of E_{sp} are strongly different. The equality of E_{cr} agrees well with that of the theoretical molecular moduli E_{theor} for these polymers [30, 31] (see Table IV).

Since both relative deformations ϵ_{sp} and ϵ_{cr} are very small (<1.0%), data from several loading–unloading cycles were used to improve accuracy. In addition, prior to measuring ϵ_{sp} and ϵ_{cr} the specimen was subjected to one such cycle to enhance elasticity of the specimen proper. Consider now how reliable is the dependence of ϵ_{cr} on σ obtained, i.e., whether the assumption on the equality of stress in the specimen and the crystallites is valid. In favor of the validity of this assumption are the following facts: (1) equal values of E_{cr} for PBA and PPTA specimens obtained in different conditions and having different super-molecular structure but the same elastic properties of the molecules (equal E_{theor}); this conclusion is similar to the one drawn for PVA fibers of different supermolecular structure in ref. [7]; (2) the closeness of E_{cr} to E_{theor} calculated in ref. [30] and for the cis-conformation in ref. [31] (see Table IV).

Thus the closeness of ϵ_{sp} and ϵ_{cr}, E_{sp} and E_{cr} for PBA appears to be a reliably established fact. With PBA fibers, such a close agreement between these quantities is observed for the first time for crystallizing polymers. One may assume the supermolecular structure of PBA fibers to be close to homogeneous, which means that we deal here either with imperfect monocrystals or with a partially crystalline long-period structure whose intercrystallite regions are characterized by a high density and high degree of molecular order.

On the Structure of Intercrystallite Regions

It is well known that the sections of the molecules in the intercrystallite amorphous regions of oriented aliphatic polymers are of different length. A fraction of the molecules are probably totally extended (or extended even at low loads) while some of them are bent to a lesser or greater extent. The elastic modulus of extended molecules is the same as that of molecules in the crystallites while the modulus of bent molecules in APs differs probably by two to four orders of magnitude from that of extended chains, depending on the actual degree of bending [7]. Thus the elastic modulus of amorphous layers and of the specimen as a whole is mainly determined by extended molecules, the contribution of the bent molecules to the total modulus being not large because of the smallness of these moduli. This indicates that the extended molecules hold the predominant part of the load, thus acting as overstressed load-holding chains, while bent molecules practically do not hold the load. The fraction of the load-holding chains in amorphous layers from the number of chains in crystallites per unit cross section, n/N, was estimated by means of mechanical and X-ray diffraction measurements for some aliphatic polymers [7, 32]. It turned out to be moderate, 0.05–0.15.

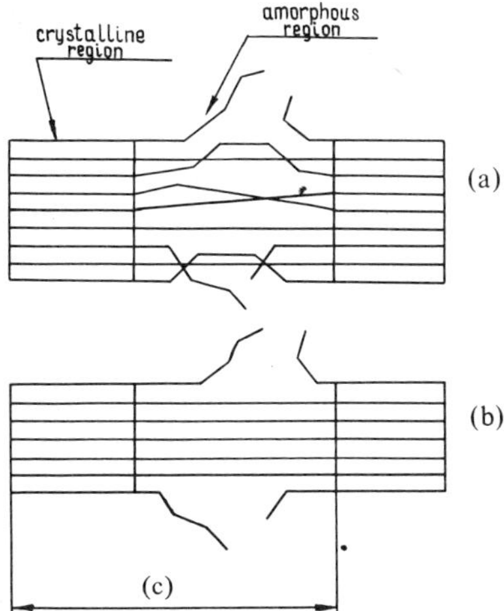

FIG. 8. A structural model for intercrystallite amorphous regions in wholly-aromatic polymer fibers (a) and its idealized version (b) which is equivalent in the elasticity modulus.

It would be of interest to establish the structure of the amorphous regions in WAPs. The sections of the tie molecules in these regions can be expected to be also of different length, so that some of them are bent while the others are extended (Fig. 8a). However, in contrast to APs the elastic modulus of bent molecules in aromatic polymers does not differ so strongly from that of extended molecules. The rigidity of the macromolecules in aromatic polymers should result in the modulus of bent sections being smaller than that extended ones by a few times rather than by a few orders of magnitude, as is the case with APs. For example, conformational differences in the molecules at PPIA (which is the least rigid aromatic polymer) may account for a decrease of the modulus by a factor of 3 [31] (Fig. 9). The elastic moduli of the extended and strongly bent conformations are 15,500 and 5,600 kg/mm^2, respectively.

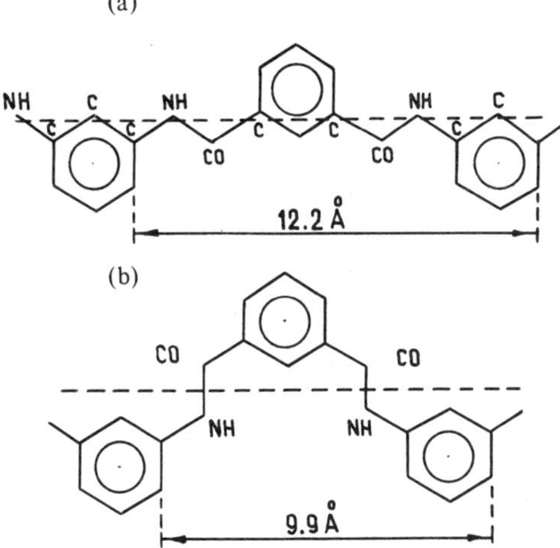

FIG. 9. Two different conformations of PPIA molecule in a crystallite.

Thus the elastic modulus of amorphous layers in WAPs is determined not only by the contribution of the extended chains but rather, to a considerable extent, by those of the bent chains. Thus not only extended but also bent chains in amorphous layers are load-holding. Note that the stress distribution of the holding chains in aromatic polymers should be considerably broader than that in aliphatic polymers since in the latter only extended molecules are considered as load-holding.

To check the above considerations, we carried out acoustic measurements [22]. We estimated separately the orientation of the tie and load-holding molecules in the intercrystallite amorphous regions of aliphatic and wholly-aromatic polymers (Table V). It is of interest that, in contrast to the APs, in aromatic polymers the orientation of the tie and load-holding chains is the same. This confirms the above suggestion that in WAPs all the tie molecules in the intercrystallite amorphous regions are load-holding already at room temperature.

Acoustic measurements [21] permitted the estimation of the fraction of the load-holding chains in amorphous layers from the number of chains in crystallites per unit cross section, n/N. The corresponding data are listed in Table V. A simple formula was proposed for the determination of the quantity n/N:

$$\frac{n}{N} = \frac{E_d(1 - \varkappa)}{E_{cr} - E_d\varkappa}$$

where $\varkappa$ is the fibril crystallinity.

This formula defines the quantity n/N within a model equivalent in the elastic modulus, in which all the molecules are completely extended and all are load-holding (Fig. 8b). The elastic modulus for such a model is equal to that for a model of a real polymer with tie chains comprising both bent and extended

TABLE V
Structural Characteristics of Intercrystallite Layers in Various Polymer Fibers

| Polymer | Orientation angle, deg. | | $\dfrac{n}{N}$ at 20°C | $\dfrac{n}{N}$ at -180°C |
	of tie chains in intercrystallite regions	of load-holding chains in intercrystallite regions		
PE	16.5	4.1	0.06	
PVA	26	18	0.14	
POM	39	14	0.05	
PCA	31.5	21.1	0.09	
PPIA	32	32	0.11	–
PI–PM	16	16	0.56	0.70
PPTA	15.3	15.3	0.50	0.60
PBA [a]	–	–	0.80	0.80

[a] For PBA we assumed $\varkappa = 0.70$ which can not be checked experimentally because of the absence of small-angle reflections.

molecules. Since in APs, as this was pointed out before, practically only extended molecules are load-holding, the value of n/N determined by us corresponds to the real number of load-holding molecules in the amorphous regions. As seen from Table V, in APs the value of n/N varies from 0.05 to 0.14 which agrees satisfactorily with the results of other works presented above.

For WAPs where bent molecules are also load-holding, the value of n/N determined by us will be an underestimation since bent load-holding molecules may be considered as replaced in the equivalent model by a smaller number of extended molecules (Fig. 8). Presented in Table V are the values of n/N for aromatic polymers (see n/N at 20°C). We see that these values can both equal those for aliphatic polymers (PPIA) and exceed them by far reaching 0.6–0.8 (these values of n/N are not smaller than the real ones and can in fact be still larger on the grounds of what has been said before).

In our opinion, for WAPs one should obtain more realistic values of n/N when determining n/N at low temperatures (−180°C), since in this case the modulus of bent molecules increases so that the same number of bent molecules will be replaced in the equivalent model by a larger number of extended molecules. The values of n/N obtained at −180°C are listed in Table V. For PI-PM and PPIA one observes an increase of n/N with decreasing temperature, while in PBA such a phenomenon is not observed. The latter agrees well with the concept of a comparatively homogeneous structure of PBA where bent molecules probably do not exist at all.

Thus WAP fibers have a more ordered structure of intercrystallite regions with a larger number of load-holding molecules.

What is the mechanism of intercrystallite layer formation in WAPs and why there is the lack of polymer material between the crystallites? While in APs this occurs as a result of the folding of molecules in crystallites and escape of some molecules in the lateral direction from the amorphous layers (model of Hosemann–Bonart), in the case of aromatic polymers having rigid molecules the reasons for the heterogeneity cannot be associated with the folding of the molecules (with the only possible exclusion of PPIA). We believe the most probable model for the supermolecular structure in aromatic polymers to be that of the model of Hearle where polmyer material deficiency in amorphous regions results

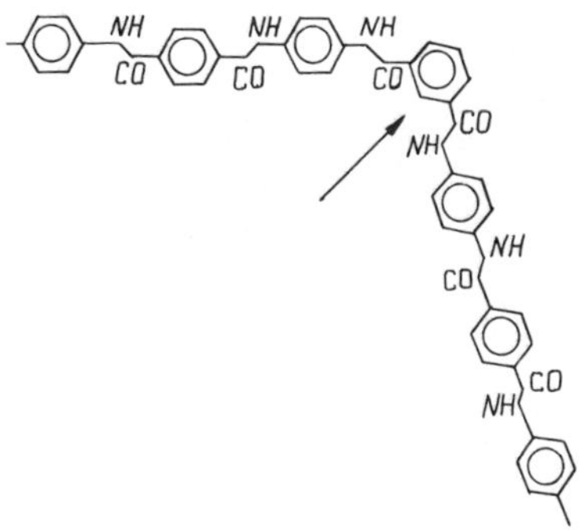

FIG. 10. An example of bending of PPTA molecule resulting from a meta-link in the chain ("genetic" defect).

from lateral escape of the molecules (Fig. 9a). Two reasons may account the bending of rigid chains. One is the "errors" in the synthesis [35] where, for example, one meta-link can appear per a few tens of "normal" para-links which eventually can result in a bending of the molecular chain (Fig. 10). Such "genetic" defects producing eventually "weak" spots at the next level of order which is the supermolecular structure can undoubtedly occur in synthetic polymers. Indeed, even in biological systems characterized by a much higher degree of perfection, accidental mutations representing errors in the genetic code invariably take place.

The second reason may lie in the so-called "conformational" defects consisting in a combination of different possible conformations in the same molecule (Figs. 11 and 12). Figure 11 presents a case of a cis-link entering the trans-conformer in PPTA. A similar situation is shown in Figure 12 for POD.

Now which of the above reasons for the bending of molecules are more dangerous? "Genetic" defects appear practically uncorrectable since even in biopolymers they frequently do not heal. At the same time the occurrence of "conformational" defects can be reduced by purely technological means, e.g., by properly choosing the solvent, the conditions of spinning, heat treatment, and drawing.

It should be noted in conclusion to this section that the case with PBA illustrates a real possibility of producing an oriented crystalline polymer with a nearly homogeneous supermolecular structure. We consider, however, that this homogeneity, which is the main condition of high mechanical qualities, not necessarily should be associated with the crystalline state but can rather be reached also in highly oriented amorphous polymers with molecules of high rigidity, since it is high rigidity of the molecules which, besides a strong molecular interaction, is the most important factor in producing homogeneous structures with high orientation while the property of crystallization is not a necessary condition.

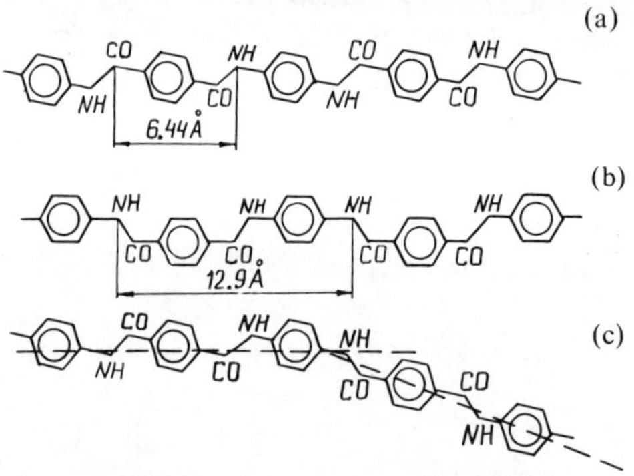

FIG. 11. An example of bending of PPTA molecule (c) resulting from merging of cis- (b) and trans-conformation (a).

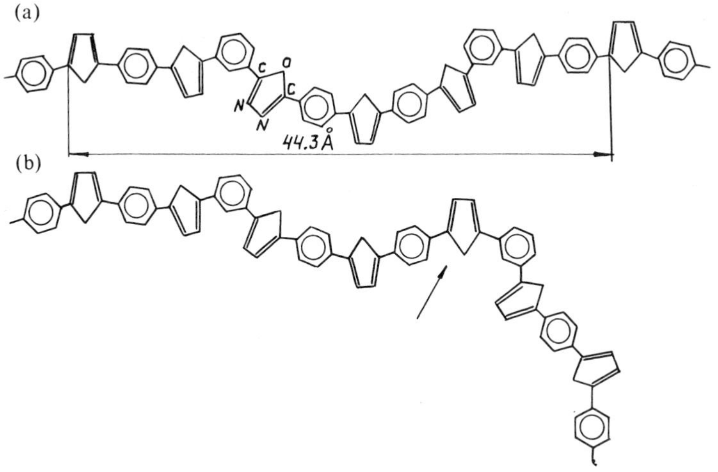

FIG. 12. An example of bending of POD molecule resulting from conformation merging; (a) extended molecule, (b) bent molecule.

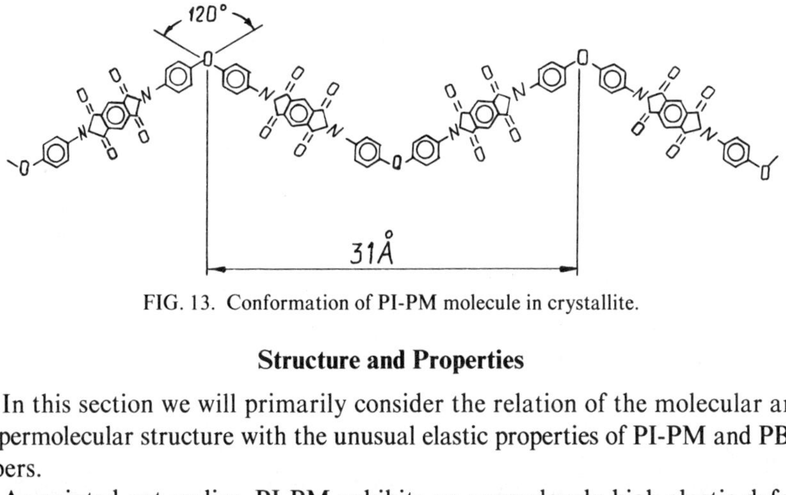

FIG. 13. Conformation of PI-PM molecule in crystallite.

Structure and Properties

In this section we will primarily consider the relation of the molecular and supermolecular structure with the unusual elastic properties of PI-PM and PBA fibers.

As pointed out earlier, PI-PM exhibits an anomalously high elastic deformation of the crystal lattice (Fig. 6a) constituting $\sim$3% at a specimen deformation $\sim$5%. This is quite unusual for polymers whose molecules in crystallites are arranged in a planar zigzag, to which also PI-PM belongs [23] (Fig. 13). Apart from that, the lattice modulus turned out to be extremely small while increasing with load from 1620 to $\sim$4000 kg/mm^2 (Fig. 14). (The data in Fig. 14 were derived from Fig. 6a.) Such a modulus is closer in magnitude to that of crystallites with molecules of helical or bent shape (polypropylene, isotactic polystyrene, polycaproamide, etc.) [19]. The reason for such a low value of E_{cr} and a high deformability of crystallites in PI-PM lies in the conformation feature of the molecule proper which is the presence in the chain of an oxygen "hinge" and a large moment of forces around it (Fig. 13). Calculations [33] confirmed this suggestion yielding for the theoretical modulus of the molecule a value E_{theor}

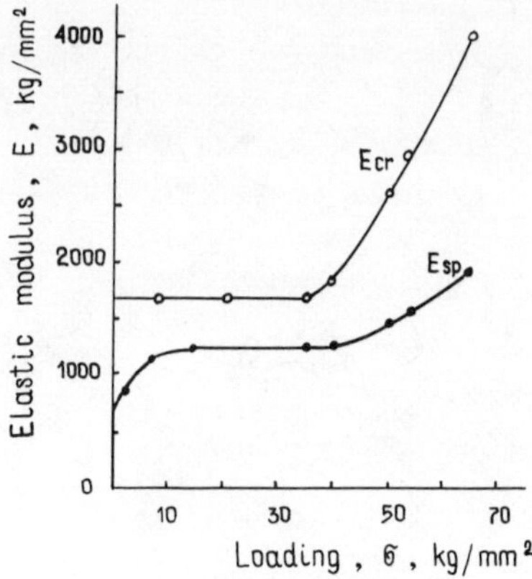

FIG. 14. Loading dependence of longitudinal elastic moduli of PI-PM fiber (E_{sp}) and of its crystallites (E_{cr}).

≈ 1950 kg/mm^2 which agrees well with our experiment. The increase in the crystallite modulus at loads close to rupture (Fig. 13) results from increasing modulus of deformation of the valence angle at the oxygen atom.

The fact that crystallites contribute essentially to the total specimen deformation accounts for the unusually high elasticity of the PI-PM fiber proper. Total elastic restoration of the fiber occurs immediately after the removal of load even in cases where the specimen was maintained at loads up to 0.7–0.9 of the breaking limit for 48 hr. Total restoration of length occurred in a few seconds after load removal. Besides, this fiber does not loose its high elasticity even at liquid nitrogen temperature ($-180°$C). It should be noted that a less crystalline fiber (drawn at 400°C) exhibits poorer elastic properties.

The unusually high modulus of PBA fibers (16,000–17,000 kg/mm^2) is explained by two facts, viz., the supermolecular morphology being close to a homogeneous structure and a high modulus of the PBA molecule proper. On the other hand, despite high values of n/N, the PBA as well as PI-PM and PPTA fibers exhibit comparatively low strengths which lie considerably below the theortical values (Table II). The strength of these fibers is probably limited by the existence of defects (e.g., submicrocracks) or a nonuniform load distribution among the holding molecules (it is obtained for PI-PM in ref. [36]) resulting in their nonsimultaneous rupture.

CONCLUSIONS

1. WAP fibers reveal long periods of 150–230 Å made up of crystalline regions and intercrystallite amorphous layers. The latter represent "weak" spots of the structure with respect to deformation, sorption and, probably, strength.

2. The distinctive features in the structure of intercrystallite amorphous regions in WAP fibers as compared with the amorphous layers in AP fibers are

as follows: (a) all the tie molecules are load-holding; (b) the fraction of the load-holding molecules of the number of chains per the same cross-section in a crystallite is larger by a factor 5–6 and can reach 0.6–0.8.

3. A real possibility of the existence of fibers with a nearly homogeneous supermolecular structure was illustrated for the first time in the case of PBA fibers. This structure either represents imperfect monocrystals or is partially crystalline with a high density and high degree of order of intercrystallite amorphous regions.

We are greatly indebted to A. S. Semenova, V. D. Kalmykova, T. S. Sokolova, A. V. Volokhina, V. N. Kuz'min and A. P. Verkhovets for assistance in the work, fruitful discussions, and for the specimens submitted for the study.

REFERENCES

[1] L. I. Slutsker, A. V. Savitskii, L. E. Utevskii, I. M. Stark, and O. S. Lelinkov, *Vysokomol. Soedin.*, **A13**, 2785 (1971).

[2] L. I. Slutsker, in *New Chemical Fibres for Industrial Use* (in Russian), Khimiya, Leningrad, 1973, p. 120.

[3] W. Ruland, *J. Polym. Sci., C.*, **28**, 143 (1969).

[4] I. P. Dobrovolskaya, L. I. Slutsker, L. E. Utevskii, and Z. Yu. Chereiskii, *Vysokomol. Soedin.*, **A17**, 1555 (1975).

[5] S. N. Zhurkov, A. I. Slutsker, and A. A. Yastrebinskii, *Dokl. Akad. Nauk USSR*, **153**, 303 (1963); *Fiz. Tverd. Tela*, **6**, 3601 (1964).

[6] V. S. Kuksenko and A. I. Slutsker, *Fiz. Tverd. Tela*, **10**, 838 (1968).

[7] V. S. Kuksenko, A. V. Ovchinnikov, and A. I. Slutsker, *Mekh. Polim.*, **6**, 1002 (1969).

[8] A. I. Slutsker and V. S. Kuksenko, *Mekh. Polym.*, **1**, 81 (1975).

[9] V. A. Marikhin, A. I. Slutsker, and A. A. Yastrebinkii, *Fiz. Tverd. Tela*, **7**, 441 (1965).

[10] L. I. Slutsker, Z. Yu. Chereiskii, L. E. Utevskii, and K. E. Perepelkin, *Vysokomol. Soedin.*, **B11**, 594 (1969).

[11] Sh. Tuichiev, N. Sultanov, B. M. Ginzburg, and S. Ya. Frenkel, *Vysokomol. Soedin.*, **A12**, 2025 (1970).

[12] E. W. Fisher, H. Goddar, and G. Schmitt, *Makromol. Chem.*, **119**, 170 (1968).

[13] G. Kiessig, *Das Papier*, **22**, 5 (1968).

[14] L. I. Slutsker, L. E. Utevskii, Z. Yu. Chereiskii, I. M. Stark, and N. D. Minkova, *Vysokomol. Soedin.*, **A15**, 2372 (1973).

[15] L. I. Slutsker, L. E. Utevskii, Z. Yu. Chereiskii, V. N. Kuz'min, V. D. Kalmykova, T. S. Sokolova, A. V. Volokhina, and G. I. Kudryavtsev, *Vysokomol. Soedin.*, **A17**, 1569 (1975).

[16] L. I. Slutsker, Z. Yu. Chereiskii, L. E. Utevskii, V. N. Kuz'min, A. S. Semenova, and A. V. Volokhina, *Vysokomol. Soedin.*, **A17**, 2080 (1975).

[17] A. S. Semenova, N. P. Okromchedlidze, A. B. Raskina, and A. V. Volokhina, *Preprints of Int. Symp. on Man-Made Fibres*, Kalinin, USSR, **4**, 25 (1974).

[18] G. I. Kudryavtsev and A. M. Shchetinin, *Khim. Volokna*, **6**, 2 (1968).

[19] I. Sakurada, T. Ito, and Y. Nukushina, *J. Polym. Sci.*, **57**, 652 (1962).

[20] V. S. Kuksenko, V. A. Ovchinnikov, and A. I. Slutsker, *Vysokomol. Soedin.*, **A11**, 1953 (1969).

[21] L. E. Utevskii, *Vysokomol. Soedin.*, **B14**, 308 (1972).

[22] L. E. Utevskii, *Vysokomol. Soedin.*, **A16**, 2339 (1974).

[23] L. G. Kazaryan, E. T. Lur'e, and L. A. Igonin, *Vysokomol. Soedin.*, **B11**, 779 (1969).

[24] M. G. Northolt and J. J. Van Aartsen, *Polymer Lett.*, **11**, 333 (1973).

[25] L. I. Slutsker, *Vysokomol. Soedin.*, **A17**, 262 (1975).

[26] L. I. Slutsker and V. N. Kuz'min, *Vysokomol. Soedin.*, **A17**, 490 (1975).

[27] L. I. Slutsker, V. N. Kuz'min, A. V. Savitskii, L. E. Utevskii, and K. E. Perepelkin, Preprints of *Int. Symp. on Man-Made Fibres,* Kalinin, USSR, **1,** 39 (1974).

[28] L. I. Slutsker, Thesis, 1973.

[29] M. A. Gezalov, V. S. Kuksenko and A. I. Slutsker, *Vysokomol. Soedin.,* **A12,** 1787 (1970).

[30] G. S. Fielding-Russell, *Text. Res. J.,* **41,** 860 (1971).

[31] K. E. Perepelkin and Z. Yu. Chereiskii, *Prepr. Int. Symp. on Man-Made Fibres,* Kalinin, USSR, **1,** 142 (1974).

[32] A. Peterlin, *J. Polym. Sci.,* **7, A-2,** 1151 (1969).

[33] Z. Yu. Chereiskii, Abstracts of Reports submitted to the Meeting on Heat-Resistant Fibres (Russian), Mytishchi, USSR, p. 28, 1972.

[34] V. S. Kuksenko, O. D. Orlova, and L. I. Slutsker, *Vysokomol. Soedin.,* **A15,** 2517 (1973).

[35] V. V. Korshak, *Vysokomol. Soedin.,* **A15,** 298 (1973).

[36] N. P. Leksovskaya, V. I. Vettegren', and K. Yu. Fridlyand, *Vysokomol. Soedin.,* **A17,** 811 (1975).

[37] K. E. Perepelkin, *Faserforsch. Textiltech.,* **25,** 251 (1974).

[38] L. C. G. Treloar, *Polymer,* **1,** 95 (1960).

PECULIARITIES OF SUPERMOLECULAR STRUCTURE OF POLYIMIDES AND POLYESTER IMIDES

A. V. SIDOROVICH, YU. G. BAKLAGINA, A. V. KENAROV, YU. S. NADEZHIN, N. A. ADROVA, and F. S. FLORINSKY

Institute of High Polymer Compounds, Bolshoi 31, 199004, Leningrad, U.S.S.R.

SYNOPSIS

Supermolecular structures of fourteen heterocyclic thermally stable polymers of the polyimide (PI) and polyesterimide (PEI) class have been investigated by X-ray scattering, optical, dilatometric and thermomechanical methods. Quasi-crystalline structure of these polymers in the initial amorphous state of poly(amic acid) and poly(ester amic acid) and also after chemical imidization has been revealed. The quasi-crystalline structure is characterized by a biaxial polarizability ellipsoid with a negative sign. This is indicated by distinct conoscopic figures recorded with a polarizing microscope. The effect of the number of oxyphenylene groups on the crystallizability of PI and PEI was investigated. For PEI the effect of self-orientation of macromolecules induced by conformational transition was observed. Discrete small-angle x-ray scattering pattern (a long period) was also revealed for PEI. The appearance of long periods is caused by polymorphism of PEI uniquely determined for one of its types (PEI-II). X-ray scattering data confirmed by calculations of the lattice energy show that dianhydride pyromellitimide chain portion is the element determining the crystalline structure of PI. The structure is determined by van der Waals' forces and obeys the principle of dense packing. This leads to the appearance of layer structure with alternating layers of diamine and dianhydride portions. It was found that, depending on the rigidity of the diamine portion, PI and PEI exhibit different abilities of forming axial-planar or axial textures in nonoriented films.

Preparation of polymers in an oriented state is of great practical and scientific interest. Self-oriented systems receive increasing attention in this respect [1]. Flory [2] has shown that when the rigidity of the polymer chain changes, the requirement of the minimum free energy on attaining a critical concentration leads to the appearance of the nematic phase, i.e., of self-ordering. Effects of self-ordering appear in conformational transition for polypeptides [3] and in spontaneous elongation for cellulose diacetate [4]; they are responsible for specific properties of solutions of aromatic polyamides [5]. Self-ordering of the polymer chain can appear only if it contains both flexible and rigid portions [1]. From this standpoint, heterochain polymers of the polyimide (PI) and polyester imide (PEI) classes are of great interest.

PIs investigated by us (Table I) were prepared from pyromellitic dianhydride and different diamines representing both rigid portions: para-phenylene, benzidine, meta-phenylene (group I, 1–4) and more flexible oxyphenylene units (group II, 5–9). In PEI (group III, 10–14) the dianhydride component is more complex owing to the introduction of an oxyphenylene ring with a planar rigid

Journal of Polymer Science: Polymer Symposium 58, 359–367 (1977)

TABLE I

The Chemical Composition and Crystalline States of Studied Materials

PA GROUP		POLYMER DESIGNATION AND CHEMICAL FORMULA	CRYSTALLOGRAPH DATA			
			CELL PARAMETERS $\overset{o}{A}$	Z	X-RAY g/cm^3	P_{exp} g/cm^3
I	1	PMPF	$a=8,58$; $b=5,48$; $c=12,3$; $\alpha=\beta=\gamma=90°$	2	I,66	I,56
	2	PMB	$a=8,58$; $b=5,48$; $c=16,6$; $\alpha=\beta=\gamma=90°$	2	I,56	I,48
	3	PMPF	$a=8,58$; $b=5,48$; $c=20,9$; $\alpha=\beta=\gamma=90°$	2	I,49	I,45
	4	PMMF	AMORPHOUS	–	–	I,42
II	5	PM	$a=6,3$; $b=3,97$; $c=32,0$ $\alpha=\beta=\gamma=90°$	8	I,56	I,42–I,45
	6	PFG	$a=8,30$; $b=5,64$; $c=22,0$; $\alpha=\gamma=90°$ $\beta=98°$	2	I,55	I,40
	7	PMR	$a=8,76$; $b=5,47$; $c=43,4$; $\alpha=\beta=\gamma=90°$	4	I,5I	I,40
	8	PM-4	$a=8,44$; $b=5,66$; $c=53,4$; $\alpha=\beta=\gamma=90°$	4	I,50	I,40
	9	PM-5	$a=8,44$; $b=5,66$; $c=31,3$; $\alpha=\beta=\gamma=90°$	4	I,45	I,42
III	10	PEI-2 (A)	$a=5,56$; $b=7,94$; $c=59,0$; $\alpha=\gamma=90°$ $\beta=86$	4	I,59	I,44
	11	PEI-II (B)	$a=13,12$; $b=7,84$; $c=57,0$; $\alpha=\beta=\gamma=90°$	8	I,44	I,43
	12	PEI-III	$a=5,64$; $b=8,95$; $c=34,5$; $\alpha=\beta=\gamma=90°$	2	I,48	I,44
	13	PEI-IV	$a=5,68$; $b=8,09$; $c=78,6$; $\alpha=\beta=\gamma=90°$	4	I,48	I,43
	14	PEI-V	$a=5,66$; $b=8,14$; $c=43,5$; $\alpha=\beta=\gamma=90°$	2	I,45	I,4I

ester group, whereas the diamine component is similar to that in PI (group II, 5–9).

The preparation of PI and PEI is a two-stage process. In the first stage, corresponding poly(amic acids) (PAA) and poly(esteramic acids) (PEAA) were obtained. This was followed by chemical or thermal imidization [6]. In the initial state PAA and PEAA were amorphous substances as shown by X-ray diffraction patterns.

However, all the polymers investigated, both in the initial state of PAA or PEAA and after chemical imidization up to the beginning of intense crystallization, exhibit a considerable specific order of the quasi-crystalline character. This uniquely follows from microscopic studies. Identification of structures from conoscopic observations clearly showed that the polymer structure on an optical level may be characterized by a biaxial polarizability ellipsoid with a negative sign, i.e., it is similar to biaxial crystals. The conoscopic figure in the initial state is characteristic of a section surface of biaxial crystal perpendicular to the bisectrix of the acute angle of optic axes which coincides with the minor semi-axis of the polarizability ellipsoid. The film surface normal to the optic axis of the microscope is the section surface. In accordance with this, when the microscope stage is rotated by 45° to either side, the characteristic cross is separated into hyperbolas diverging to opposite sides (Fig. 1 a, b, and c). The sign of the polarizability ellipsoid shows that the polymer chains are oriented predominantly parallel to the film surface.

Investigations of the effect of temperature on the structure of quasi-crystallites showed that conoscopic figures of biaxial crystals of PAA and PEAA are stable up to 75–100°, i.e., until the beginning of imidization, and chemically imidized samples up to 200°, until the beginning of crystallization. Changes in conoscopic figures over these ranges of temperatures show that the polarizability ellipsoid rotates to the position corresponding to the section surface of a biaxial crystal perpendicular to the optic axis.

The appearance of ordering of quasi-crystalline character explains the fact that PI and PEI readily crystallize: probably their macromolecules would not be able to crystallize if an entangled network were formed in the amorphous state.

Figure 2 shows dilatometric data for PAA with a rigid PMB group, for PAA PM-4, and PEAA PEI-II (curves 1, 2, 3, respectively).

FIG. 1. Conoscopie figures for PI and PEI. The position of the microscope stage: (a) 0°, (b) 45°, (c) 135°.

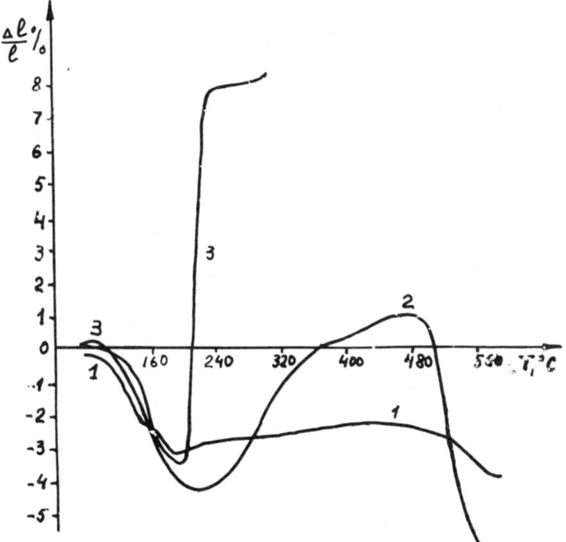

FIG. 2. Dilatometric curves: (1) PAA PMB; (2) PAA PM-4; (3) PEAA PEI-II.

At temperatures up to 220° they all contain a contracted portion resulting from the imidization process which is almost complete at 200–210° [7]. X-ray diffraction patterns and infrared spectra show that at higher temperatures the process of crystalline ordering begins; moreover, different dilatometric behavior is observed. For PAA PMB with a rigid diamine portion the curve is shallow and PAA PM-4 exhibits a relatively pronounced elongation up 480° (decrease in size of higher temperatures is caused by degradation effects).

For PEI-II an anomalous abrupt elongation of up to 6–8% is observed when the polymer is in the amorphous state according to X-ray diffraction patterns. Thus, depending on the structure of diamine and dianhydride components, on the combination of rigid and flexible portions, different orientational effects appear. In all PEIs investigated, the effect of self-orientation appears both in the PEAA state and in PEI after chemical imidization when the temperature of 200–210°C is attained. This is similar to conformational helix–coil transition [3] and to spontaneous elongation in cellulose diacetate [4]. The self-orientation

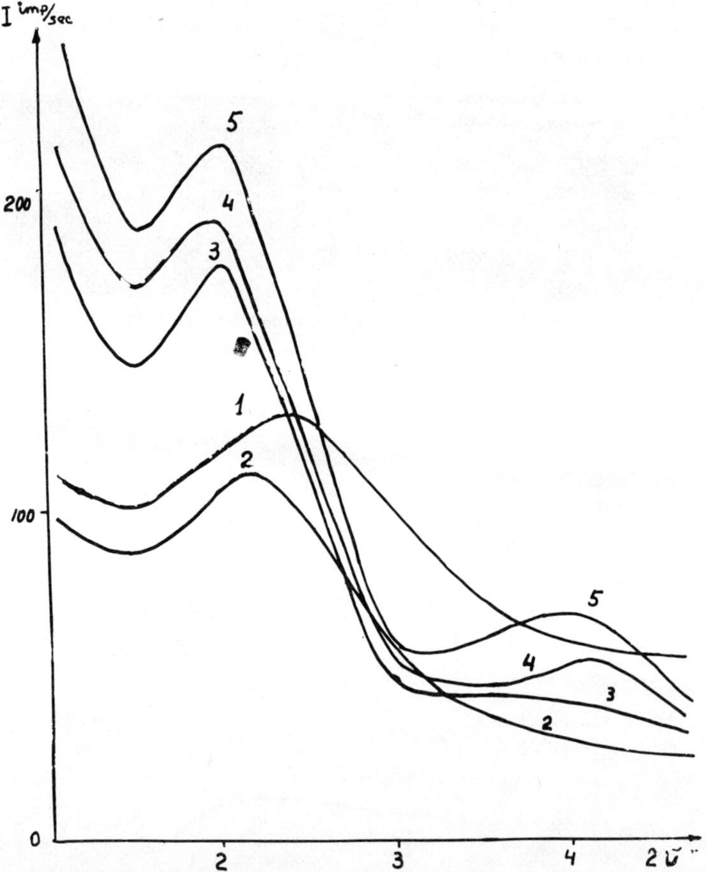

FIG. 3. Diffraction, patterns of PEI-V, unoriented film after heating to different temperatures: (1) the initial state the temperatures: (2) 170°; (3) 200°; (4) 220°; (5) 230°. Cu Kα radiation nickel filter.

effect is caused by conformational transition in the amorphous state in the range of glass transition temperature.

Figure 3 shows X-ray diffraction patterns of PEI-V confirming this conclusion. Amorphous films: an initial film and a film heated to 170°C (curves 1 and 2) exhibit a broad reflection at $2V = 2°16'$ due to longitudinal ordering of chains (002 reflections). After heating to 170° the reflection becomes sharper but its position does not change. Heating to 200°C (curve 3) causes a change in the identity period from 39 to 43 Å and the reflection shifts from $2°16'$ to $2°03'$. Further heating from 220 to 230°C (curves 4 and 5) leads only to increasing intensity of reflections but does not shift them. The appearance of a second maximum at 4° is due to the beginning of crystallization which proceeds with the highest intensity at 350–380°C.

Determinations of the identity period of PEI in the crystalline state after conformational transition showed that it corresponds to the most extended and thermodynamic equilibrium conformation. Depending on the number of phenylene oxide groups in the repeat unit, the thermodynamic equilibrium conformation varies from a straight (PEI-II and PEI-V) to a curved conformation which contains the axis of the second order (PEI-II and PEI-IV, Fig. 4). Apart from the most extended and thermodynamic equilibrium conformation A, the formation of a slightly coiled conformation B in PEI-II in the crystalline state is also possible, i.e., polymorphic forms of crystallites may exist (Table I, 10 and 11). Systems of reflections for each conformation could be resolved for PEI-II; their schemes are shown in Fig. 5b and d. A monoclinic cell with $\rho cr_A = 1.59$ g/cm^3 and the identity period along the "C" axis $l_c = 59 \pm 0.5$ Å corresponds to the thermodynamic equilibrium form A (Fig. 5b). For the B conformation $l_c = 57 \pm 0.5$ Å and $\rho cr_B = 1.44$ g/cm^3 (Fig. 4d). The predominance of one or the other conformation is determined by kinetic factors, i.e., by conditions of precrystallization ordering of chains in the amic acid or the imide form. Probably the broadening of reflections usually observed indicates that polymorphism also exists in other PEIs.

PEIs exhibit discrete small-angle X-ray scattering (SAXS) pattern. When PEI-II crystallizes, long periods appear in the direction of the drawing axis. For initial amorphous samples with conformations A (Fig. 5a) and B (Fig. 5c) the SAXS intensity—curves 1 monotonically decreases (diffuse scattering) and when films are heated to 250°C, small-angle reflection appears—curves 2. When samples undergo further annealing up to 350–380°C, the intensity of SAXS

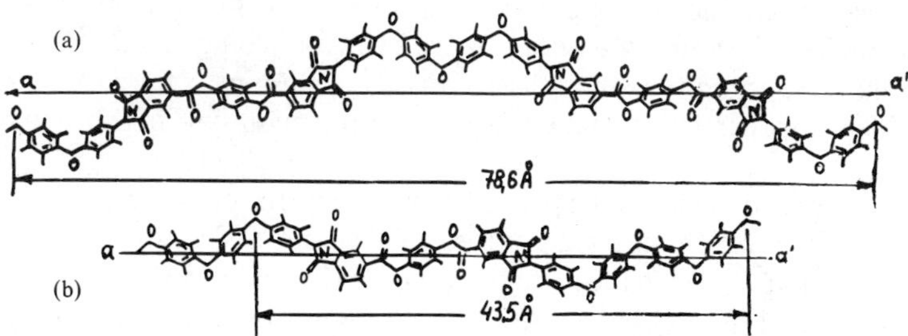

FIG. 4. Conformations of chains in the crystalline state: (a) PEI-IV, (b) PEI-V.

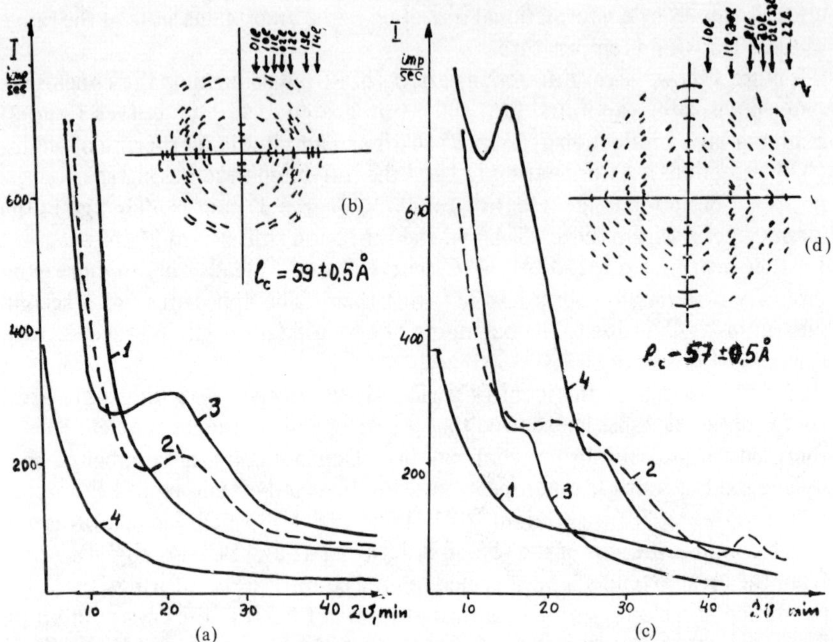

FIG. 5. Schemes of X-ray diffraction patterns: PEI-II(A)-(b); PEI-II(B)-(d); Curves of SAXS: PEI-II(A)-a; PEI-II(B)-c; Initial samples—I; samples heated to the following temperatures: 2–250°C; 3–380°C; 4–425°C in Figure (a) and 390°C in Figure (c).

reflections increases and the long period increases to 260 Å for the A conformation (Fig. 5a, curve 3) and to 380 Å for the B conformation (Fig. 5c, curve 3). For the B conformation at $2V = 27'$ reflection of the second order is observed (Fig. 5c, curve 4). This indicates a high degree of regularity and low dispersity of size for chain portions with higher and lower electron density in the direction of the texture axis. When films are heated to 400–425°, polymers become amorphous and SAXS reflections disappear (Fig. 5a, curve 4).

Thus, it may be concluded that the appearance of long periods is observed only in the range of the crystalline state of PEI-II; the crystallinity maximum corresponds to the highest intensity of SAXS. The polymorphism that we observed in PEI-II is of the conformational character [9]. It leads to the formation of crystallites greatly differing in density. It is natural to assume that just the ordered portions of oriented macromolecules of different polymorphic forms cause the appearance of long periods in SAXS in contrast to flexible-chain polymers for which the long period is due to chain folding [10].

In the first group (polymers 1–4, Table I) containing rigid portions in the diamine component, torsional vibrations are the main type of molecular motion.

The addition of nuclei in the para position ensures the preparation of polymers readily crystallized in the course of thermal treatment. The change of addition from the para position to the meta position (PMPF and PMMF) leads to great conformational difficulties owing to which the polymer does not crystallize. Values of X-ray diffraction patterns and experimental densities of PMPF, PMB, and PMTF show that the increase in the number of phenylene rings in the diamine component leads to decreasing density. The peculiarity of crystallization of this group of polymers is the formation of an axial-planar texture in nonoriented films, and the formation of an amorphous texture in oriented films and fibers in the PAA stage. Heating of samples with an amorphous texture leads to the formation of a crystalline and distinctly axial texture.

Crystalline cells of PMPF, PMB, and PMTF belong to the rhombic syngony and differ only in the value of the "C" period by 4.3 Å. This corresponds to the sum of the length of the benzene ring (2.8 Å) and the distance between the rings (1.5 Å). The section of the elementary cell remains unchanged, i.e., the structure exhibits homologous isomorphism [11]. X-ray diffraction patterns were used to calculate the packing of macromolecules (Fig. 6). They were found to have the following packing: densely packed layers of pyromellitimide portions (Fig. 6a) alternate with layers of phenylene rings rotated by 60–90° with respect to the dianhydride component (Fig. 6c,d). The data reported show that the PI structure obeys the principle of dense packing [12] and is governed by van der Waals' forces. This is confirmed by the calculation of the lattice energy [13]. Hence, the pyromellitimide portion determines the packing of macromolecules in the crystalline lattice.

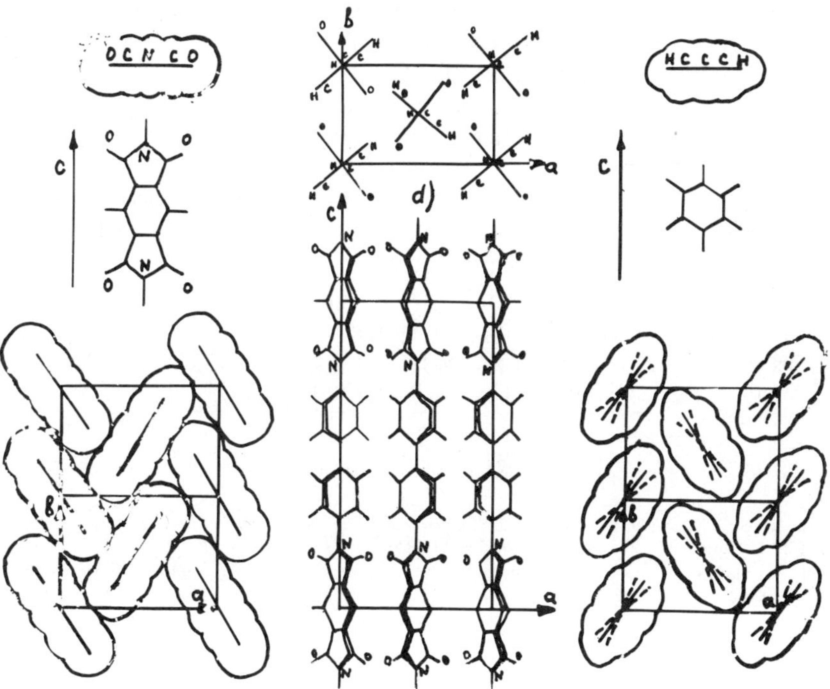

FIG. 6. Structure of PMB in the crystalline state. Projection of the "ab" section of the lattice—the pyromellitimide portion—(a); the diamine portion—(c); projection of the structure: along the "b" axis—(b); along the "c" axis—(d).

Polymers II of the PI-II group (Table I, 5–9) contain oxygen as a mobile link. No crystallization of PM in film occurs on heating. Other polymers beginning with PFG readily crystallize. Not all the polymers of group II form a texture in the unoriented state and an amorphous texture on drawing of PAA films and fibers, but after crystallization in the oriented state they all exhibit a distinct axial texture which permits a reliable determination to be made of crystallographic parameters of cells and of chain conformations (Table I). The parameters of crystalline cells of PM-4 and PM-5 differ only in their size along the "c" axis. Thus, the introduction of an oxygen atom does not lead to great conformational hindrances and facilitates crystallization of PI; even the meta position does not hinder crystallization (PMR).

Investigations of PEI showed that they readily crystallize on heating, and in unoriented films yield axial or axial-planar textures. Drawing of amorphous PEI leads to the formation of a distinct amorphous texture; crystalline lattices of some PEIs exhibit homologous isomorphism.

The following conclusions may be drawn from the results of these investigations. Data of X-ray diffraction patterns confirmed by calculation of the lattice energy show that it is the rigid pyromellitimide portion that determines intermolecular packing of PI chains. Intermolecular structure obeys the principle of dense packing and is mainly induced by van der Waals' forces. The greater size of this portion and its greater contribution to the interaction energy are responsible for the quasi-crystalline structure of polymers in the amorphous state and their layer structure in the crystalline state. The introduction of oxyphenylene groups leads to considerable flexibility of the polymer chain. Consequently, in some cases PIs in the unoriented state do not exhibit an axial or an axial-planar texture and their dilatometric curves differ markedly from those of PIs of group I.

The introduction of a rigid planar ester group directly attached to the dianhydride portion greatly affects the processes of ordering and leads to the self-orientation effect because of the transition of the macromolecule from a coiled to an extended and thermodynamic equilibrium conformation. This transition takes place in the range of the glass transition temperature in which chains exhibit considerable mobility in the amorphous state. These phenomena can qualitatively be interpreted by the Flory theory [2].

REFERENCES

[1] S. Ya. Frenkel, *Fizika segodnya i zavtra,* Nauka, Leningrad, 214–219, 1973.

[2] P. I. Flory, *Proc. R. Soc., London,* **234,** 60 (1956); *J. Polym. Sci.,* **49,** 105 (1961).

[3] E. Blout, R. Karlson, P. Doty, and B. Hargity, *J. Am. Chem. Soc.,* **76,** 4492 (1954); **78,** 947 (1956).

[4] T. G. Majury, H. I. Wellard, *Simposio internazionale di chimica macromoleculare,* Milano-Torino, **26,** 354, S-06 (1954).

[5] V. N. Tsvetkov, G. I. Kudryavtsev, E. J. Rjumtsev, V. Ya. Nikolaev, V. D. Kalmykova, A. V. Volokhima, *Dokl. Akad. Nauk, SSSR,* **224,** 398 (1975); V. D. Kalmykova, G. I. Kudriavtsev, S. P. Papkov, A. V. Volochina, U. U. Iovleva, L. P. Milckova, V. G. Kulichihin, and S. I. Bandurjn, *Vysokomol. Soedin.,* **A15,** 310 (1973).

[6] M. M. Koton, *Vysokomol. Soedin.,* **A15,** 310 (1973).

[7] A. V. Sidorovich and E. V. Kuvshinsky, *Vysokomol. Soedin.*, **A10,** 1401 (1968); E. V. Kuvshinsky, V. N. Nikitin, N. A. Adrova, Yu. G. Baklagina, and N. V. Michailova, *J. Polym. Sci.*, **12,** 1375 (1974).

[8] Ya. G. Baklagina, N. V. Mikhailova, V. N. Nikitin, A. V. Sidorovich, and L. N. Korzhavin, *Vysokomol. Soedin.*, **A15,** 2738 (1973).

[9] V. G. Dachevskj, J. O. Murzatinay, *Itogi Nauki, Ser. Khim., Collektion, Vysokomol. Soedin.* (1969).

[10] R. Bonart and R. Hoseman, *Makromol. Chem.*, **39,** 105 (1960).

[11] A. J. Kitaigorodsky, *Molekulyarnye Kristally,* Nauka, Moscow, p. 55, 1971.

[12] A. I. Kitaigorodsky, *Organicheskaj Kristallokhimija,* ed. A. N. USSR, 1955.

[13] Yu. G. Baklagina, I. S. Melewskaj, A. V. Sidorovich, and L. N. Korzhavin, *I Vsesojsnoe soveschanie po organicheskoj Kristallokhimii* (*tesisi dokladow*) Znanie, Riga, 1975, p. 122.

PROPERTIES OF ORIENTED BLOCK COPOLYMERS

A. SKOULIOS

*Centre de Recherches sur les Macromolécules, 6, rue Boussingault,
Strasbourg, France*

SYNOPSIS

In this paper, we shall be concerned with single crystals of block copolymers. As we shall have occasion to note later, these are highly oriented specimens of block copolymers in the mesomorphic state. The present condensed review of the subject is based mainly on published evidence worked out conjointly at Bristol and Genoa; it also refers, however, to unpublished work carried on in Strasbourg at Ecole d'Application des Hauts Polymères and at Centre de Recherches sur les Macromolécules. It is proper to specify immediately that the only copolymers found so far to produce "single crystals" are formed of polystyrene and polydiene blocks.

MESOMORPHIC STATE OF BLOCK COPOLYMERS

Block copolymers are usually in the mesomorphic state [1–4]. Owing to their fundamental incompatibility, blocks segregate and locate themselves in distinct and intimately interdispersed microdomains, the size of which is comparable to the dimensions of the macromolecules. When the molecular weight distribution of the blocks is sufficiently sharp, the segregation microdomains rise to surprisingly regular and periodic arrangements in space.

Three types of structure have been found to occur. The lamellar structure [5] consists of a set of two kinds of layers alternately piled up over one another, each of them containing only one kind of blocks. The cylindrical structure [5] consists of a set of parallel cylinders made up of one of the blocks, embedded in a continuous matrix formed of the other blocks, and packed together according to a two-dimensional hexagonal lattice. The last structure, which is cubic in symmetry, corresponds to the packing in space of spherical microdomains [6]. The cell parameters which characterize the structures, that is, the layer thicknesses or the distance between neighboring cylinders and spheres, are typically in the range from 100 to 1000 Å.

The segregation of the blocks and the organization of the polymer take place spontaneously, provided the mobility of the chains is sufficiently high. They happen in the absence as well as in the presence of small amounts of solvent. Many factors determine the type of structure that will be adopted by a given system. Two of these will be mentioned here for the sake of reference. The first is the relative lengths of the blocks, the second is the degree of swelling of the sample. The more pronounced the inequality of the volumes of the (swollen) blocks, the more marked will be the curvature of the interfaces between microdomains. As the ratio of the volumes of the two blocks increases from unity, the lamellar structure turns into the cylindrical, then into the spherical one.

Journal of Polymer Science: Polymer Symposium 58, 369–380 (1977)

This type of organization is identical to that of lyotropic mesophases [7]. Now, these are characterized by their texture as revealed by the distribution in space of their optical birefringence. With block copolymers, the overall texture is generally deprived of those specific geometrical features which typically characterize the classical liquid crystals [8]: It is rather diffuse and cloudy in appearance. However, it shows quite clearly that the orientation of the macrolattice varies continuously from place to place throughout the sample.

FREEDOM OF BLOCKS, THERMOPLASTIC ELASTOMERS, AND SINGLE CRYSTALS

Although the segregation microdomains are microscopic in size, they still are very large compared to the polymer segments viewed on a local scale. As a consequence, blocks preserve their own properties and habits to a large extent. Depending upon their chemical nature and the experimental conditions, they can crystallize or stay in the amorphous state; they can be liquid or glassy in character.

This freedom is at the origin of the thermoplastic and elastic properties of a new class of moldable rubbers that attracted recent considerable attention: the three-block styrene–butadiene–styrene (SBS) or styrene–isoprene–styrene (SIS) copolymers. At room temperature, the styrene blocks are in the glassy state, while the diene ones form a liquid. Now, if the structure and morphology of the polymer is such that the former are located in cylinders or spheres embedded in a continuous matrix of the latter, it is clear that the sample will behave as a typical elastomer. The styrene microdomains will act as vulcanization sites for the liquid matrix. However, the strength of the rigid nodules is not permanent. Above the glass transition temperature of polystyrene, these become soft and mobile themselves, and cease to prevent flow of the material. The sample can thus be molded in new shapes.

The structure of this family of copolymers has extensively been studied by low-angle X-ray diffraction and electron microscopy [9–13]. Because of the physical and technological properties which could be predicted for them, research in the field has unwearyingly been developed, aiming at enriching the list of available engineering materials.

In such a study, Keller, et al. [14] focused their attention on Shell's Kraton 102, a 75,000 molecular weight SBS three-block copolymer, containing about 20% in volume of styrene ($M_S = 10^4$; $M_B = 5.5 \times 10^4$) and presenting the cylindrical structure. Using a sample in the form of the extruded rod 9 mm in diameter, they were able to observe discrete spot-like low-angle X-ray diffraction effects indicative of the presence in the specimen of macroscopic individual "crystals." Apparently, the extrusion of the polymer results in the orientation of the hexagonal axis of the macrolattice parallel to the axis of the plug. Similar observations were made simultaneously by Terrisse [15, 16] who also succeed in orienting three-block SIS copolymers with cylindrical and lamellar morphology.

TECHNIQUES OF ORIENTATION

To develop the orientation of the macrolattice many experimental techniques may be used. All those already tested with SBS and SIS block copolymers call upon the shear flow of the material.

The technique used by Pedemonte et al. [17] consists in gently extruding the copolymer from a steel cylinder heated in the range from 100 to 120°C, that is, slightly above the glass transition temperature of the polystyrene blocks, into a glass tube maintained at the lowest possible temperature ($\sim$50°C) compatible with a reasonable rate of flow of the material. When the tube is cylindrical, heat is applied symmetrically around it, and when it is rectangular, heat is supplied on only one side. Ordinarily, a period of 2–3 hr is required to prepare a plug of about 10 cm long and 9 mm in diameter. It is believed that the orientation results from the combined effects of shear flow and solidification of hot material on the relatively cold wall of the glass tube. It is important to note that, depending upon the chemical composition of the copolymer, much higher extrusion temperatures (up to 220°C) [18, 19] may be required to carry out the experiment. The orientation process is achieved by annealing the plug at about 120–200 °C over a long period of time scaling from a few hours up to one month; this thermal treatment allows for the improvement of the organization of the material.

A variation of this technique has also been used very recently in the case of block copolymers with lamallar morphology [20]. Instead of extruding the hot material into a glass tube, the technique consists of injecting it between two circular brass plates and squeezing it therein so that it flows outwards radially. As previously, the best oriented samples were obtained when one of the plates was cooled.

The technique of orientation used by Terrisse [15] calls upon the velocity gradient set up in the molten polymer placed in a Couette viscometer type apparatus. The sample is located in the annular space between two concentric cylinders; the outer cylinder is fixed while the inner one rotates to create a shearing rate of 10^{-2}–10^{-1} sec.$^{-1}$. The polymer is kept at about 170°C and is generally sheared over a travel of about 300 times its thickness. It is worth stressing that orientation is here achieved in the absence of any temperature gradient.

The orientational device that we are testing in our laboratory [21] is very similar to the previous one. Shear is however obtained by the backward and forward gliding at constant speed of two metal plates, placed over one another at a distance of 2–3 mm, between which is located the polymer to be oriented. The sample is first molded in vacuum in a disc form; it is then introduced between the two plates and sheared long enough at about 130–170°C. Specimens are thus found to be uniformly oriented throughout their whole volume. Completely flat and relaxed (the annealing process is operated in situ after the motion has stopped), they are furthermore ready to be cut up in the form of the standard probes required for the mechanical tests.

TEXTURE OF THE ORIENTED SPECIMENS

It is of interest to consider now the quality of the orientation and the texture of the block copolymer samples submitted to the above mentioned techniques. Two experimental methods have been used to deal with this problem: low-angle X-ray diffraction [14, 15, 21] and electron microscopy [18, 19, 22–24].

The texture of the samples oriented by extrusion reveals it not to be perfectly uniform throughout the plug: However, it is possible to get high yield of "single crystalline" material by choosing the appropriate conditions of fabrication [17–19]. With cylindrical probes, the quality of orientation varies from the core to the outer annulua. In fact, it is from the latter area that samples are found to be the best oriented.Likewise, the orientation of rectangular plugs changes from one edge to the other. At any rate, in the best oriented regions of the probes, the axes of the cylindrical microdomains in the case of a cylindrical structure [17–19] and the planar interfaces in the case of a lamellar one [22] are found to lie parallel to the extrusion direction.

Samples with cylindrical structure cut perpendicular to the rod axis and observed by electron microscopy reveal the hexagonal packing of the cylinders, while samples cut parallel to the extrusion direction reveal the parallel arrangements of these [24]. It is important to note that the orientation of the hexagonal lattice about the axis of the plug is not the same everywhere in the sample: "Grain" boundaries are visible that separate domains of different orientations. Likewise, samples with lamellar structure cut perpendicular to the extrusion direction reveal the presence of circonvolutions and defects; they clearly show that the lamellae, viewed edge on, change direction from place to place. Undoubtedly, the ordering of the "single crystals" is far from being perfect; but this is consistent with the symmetry that the extrusion process is capable of imposing upon the material.

As for the specimens obtained by mechanical shearing, their texture seems to be more uniform. Samples presenting the lamellar structure cleave very easily over their whole extension parallel to the shearing plane [15–21]. Moreover, the distribution of the intensity of the (OOl) X-ray diffraction peaks measured as a function of the angle between the direct beam and the flat surface of the sample is very sharp, indicating that the orientational dispersion of the lamallae is not important [21]. Likewise, in samples presenting the cylindrical structure, not only the cylinders are actually all parallel to one another, but the (100) lattice-planes, that is, those planes which correspond to the highest packing density of the cylinders, tend to lie significantly parallel to the shearing plane [21].

SYMMETRY OF THE PROPERTIES OF ORIENTED
BLOCK COPOLYMERS

Materials such as those we have just described, presenting such a high degree of orientation combined with a very good quality of organization, provide a unique opportunity for studying the macroscopic properties of macromolecular systems in close connection with their intimate structure and the conformation of the polymer chains.

Before going further on, it is of interest to ask the following question. Can specimens of perfectly organized and oriented block copolymers be considered as true single crystals in the conventional sense? The answer is clearly no. First of all, the material is completely amorphous and disorganized on a local scale. Thus, the structural elements that intervene in their formation are very likely not all of strictly the same size; although their distribution is probably extremely sharp, the thickness of the lamellae and the diameter of the cylinders can vary from place to place owing to the dispersion of the molecular weights and to the "liquid" character of the matter on a local scale. Finally, the material does not result from the periodical juxtaposition of the structural elements following an infinitely extended three-dimensional crystal lattice. Instead, it results from the one- or the two-dimensional periodical repetition of unit patterns, infinitely extended themselves in two or one directions. As a consequence, the macroscopic orientation observed is not the natural outcome of a three-dimensional translational order. It is merely due to the planarity of the lamallae and the linearity of the cylinders, which are attained in practice only in an artificial way by submitting the material to an appropriate thermomechanical treatment.

At any rate, whatever their exact nature and definition, much can be gained in the analysis of their macroscopic properties in considering these systems as being true single crystals. Quite fortunately, the physical properties of crystals are qualitatively independent of the relative positions of the symmetry elements; they are only concerned with their relative orientation [25]. In other words, in the formal study of a crystal we have only to take into account the point group symmetry of the system, that is, the combination of its symmetry elements (center, mirror planes, rotation or inversion axes) passing through a single point. In this sense, oriented block copolymers can be safely compared with true single crystals.

A quick inspection of their symmetry shows that block copolymers with cylindrical or lamellar structure must exhibit macroscopic physical properties, the symmetry of which includes all the elements of the 6/mmm class of crystals [25]. The most salient features of this class are the absence of piezo- and pyroelectric effects due to the presence of a center of symmetry, the uniaxial character of their second-rank tensor properties such as dielectric, diamagnetic properties, electrical and thermal conductivities, and the strong anisotropy of the elastic compliances and photoelastic coefficients.

We shall now discuss very briefly those anisotropic properties of oriented block copolymers which have been studied so far from an experimental standpoint.

OPTICAL BIREFRINGENCE

As expected from symmetry considerations, the optical birefringence of oriented samples of SBS block copolymers has been found to be uniaxial in character, with the optic axis perpendicular to the layers for the lamellar structure [20], or parallel to the 6-fold rotation axis of symmetry for the cylindrical one [26–28]. With the indicent beam parallel to the extrusion direction for the lamellar structure, or perpendicular to it for the cylindrical one, a sharp extinction can be observed with crossed polars when the plane of polarization of light is either parallel or perpendicular to these directions. When the indicent beam is directed perpendicularly to the layers or parallel to the cylinders, the material appears completely isotropic.

Now, birefringence in polymers is normally associated with molecular orientation, which determines the anisotropic distribution in space of the polarizabilities of the monomer writs. With organized block copolymers, the situation is more complex. Owing to the presence in the sample of microdomains of differing refractive indices, which have at least one dimension smaller than the wavelength of light, another optical effect, the so-called form birefringence, may intervene [29] and contribute noticeably to the overall birefringence of the system.

This effect depends solely on the volume fractions v_1 and v_2, and the refractive indices n_1 and n_2 of the two microphases. With samples consisting of parallel cylinders, form birefringence is expected to be positive with a value [26] of

$$\Delta n = \frac{v_1 v_2 (n_1^2 - n_2^2)^2}{2[(1 + v_1)n_2^2 + v_2 n_1^2][v_1 n_1^2 + v_2 n_2^2]^{1/2}}$$

With samples consisting of parallel layers, it is expected to be negative with a value [20] of

$$\Delta n = \frac{-v_1 v_2 (n_1^2 - n_2^2)^2}{2[v_1 n_2^2 + v_2 n_1^2][v_1 n_1^2 + v_2 n_2^2]^{1/2}}$$

The sign and the magnitude of the birefringence of oriented samples of SBS block copolymers have been measured experimentally in two particular cases. With Shell's Kraton 102, which contains about 20% polystyrene in volume and displays the cylindrical structure, the birefringence was found to be positive with a value of $+4.9 \times 10^{-2}$ [26]. On the other hand, with Shell's TR-41–1649, which contains about 43% polystyrene in volume and presents the lamellar structure, it was found to be negative with a value of -27.0×10^{-4} [20].

These experimental values were compared with those of form birefringence that can be calculated from the known volume composition of the samples and the measured value of the refractive index of homopolystyrene ($n_1 = 1.59$) and homopolybutadiene ($n_2 = 1.52$). The excellent agreement observed in the case of Kraton 102 was interpreted to mean that the contribution of molecular orientation to the total birefringence of the system is negligible [26]. In other words, either there is no molecular orientation, or the molecular birefringence of the two blocks, which are opposite in sign, cancel each other fortuitously. Infrared dichroism [30] and pulsed NMR [31] experiments substantially support the first interpretation. In the case of TR-41-1649, the observed value of the birefringence is not fully accounted for by form birefringence alone [20]. This was interpreted to imply the existence of an additional birefringence due to molecular orientation. Measurement of the infrared dichroism indicated indeed a slight orientation of the polystyrene blocks along [20].

ANISOTROPY OF SWELLING

When incorporated in an organized block copolymer [3, 4], solvents concentrate in those segregation microdomains which are formed of the more soluble blocks. The value of the geometrical parameters which characterize the structure vary with concentration in a continuous way. Beyond a certain degree of swelling, it frequently happens that the type of structure of the system changes into a new polymorphic form.

Before going on, it is of interest to briefly report here the conclusions of an X-ray study of the swelling of a two-block styrene–isoprene copolymer with a lamellar structure in the presence of dodecane used as a solvent [32]. This system is indeed similar to those we shall now consider from the standpoint of anisotropy. Contrary to all expectations, dodecane proved to be only a preferential solvent of polyisoprene blocks with a partition coefficient of about 75%. Up to a solvent content of about 25% by weight, the thickness of the lamellae increase on dilution by only a few percent, while the polymer layers expand laterally by more than 50%.

The anisotropy of swelling of block copolymer "single crystals" has specifically been studied recently with Shell's Kraton 102 [28] and TR-41-1649 [20]. Put in the presence of hydrocarbon vapor, these two samples were observed from the standpoint of the macroscopic change of their external shape. On swelling, the change of sample dimensions revealed highly anisotropic. In a first stage, which was interpreted as being reversible, only that size of the sample which corresponds to the thickness of the lamellae, or to the lateral distance between neighboring cylinders, was found to increase. In a second stage, which was found to be irreversible, this type of swelling levels off while the size of the sample starts to drastically increase in the other directions. Severe changes of the structure were then suspected to occur: The lamellar structure was supposed to become cylindrical, and the cylindrical to change into a spherical one. These observations were interpreted to mean that the solvent absorbed contributes to swelling only the butadiene matrix until a point is reached where the swelling of the system ceases and the structure undergoes a polymorphic change.

In our opinion, this behavior simply shows the anisotropy of diffusion of the solvent through the sample. Indeed, hydrocarbons are not exclusive solvents of only the diene blocks [32]; they do actually dissolve much better this type of polymer chains, but they also dissolve the polystyrene blocks to an appreciable extent.

As a result, they diffuse quickly into the butadiene matrix contributing to producing the dimensional changes observed during the first stage of swelling, but they also diffuse, although very slowly, into the styrene microdomains. It is the relative rate of diffusion through the two types of phases that determines the general turn of the swelling process, and to be specific, the moment where the dilation of the sample in one direction ceases letting the dilation proceed in the other direction: The lower the molecular weight of the hydrocarbon, that is, the higher the diffusion coefficient, the more rapidly the change in the swelling regime is approached [20–28].

STRUCTURAL DEFORMATION UNDER MECHANICAL STRESS

In the next section, we shall be concerned with the mechanical properties of block copolymer "single crystals." But in order to analyze usefully the stress–strain tensorial responses of the material, it is necessary to briefly describe first how the structure changes as a function of stress.

From what we know of their organizational characteristics, we can very easily guess how SBS or SIS block copolymers will behave from the structural standpoint when submitted to an external mechanical stress. Located within distinct microdomains, blocks preserve their intrinsic properties to a large extent. At room temperature, the styrene blocks are in the glassy state, while the diene ones are liquid-like in character. The styrene microdomains will thus be many orders of magnitude stiffer than the domains which are formed of the diene chains.

To simplify, let us assume that only the diene microdomains are capable of changing their shape, while the styrene ones cannot get deformed at all. Although schematic, this must very likely be the case at room temperature where the time dependence of the properties of polystyrene is negligible. The consequence is as follows. In the case of a lamellar structure, the styrene layers can only slide over one another in a pure shearing process, the interlamellar distance remaining constant. In the case of a cylindrical structure with a continuous matrix of styrene, no appreciable deformation is likely to occur, at least for stresses remaining weak. Finally, in the case of a cylindrical structure with a continuous matrix of diene blocks, all structural deformations are possible, provided they keep constant the surface of unit cell of the two-dimensional crystal lattice.

Some preliminary experimental observations that we have done [21] tend to support these ideas. The test probes that we used had been cut in two carefully oriented SIS specimens, the first of which was lamellar in structure, and the second cylindrical with styrene rods embedded in an isoprene continuous matrix. They both had the shape of a rectangular parellelopiped, the edges of which were respectively parallel and perpendicular to the lamellae or to the axes of the cylinders. Using X-ray diffraction, we noticed that a tensile stress, whatever its direction, results very easily in the destruction of the lamellar organization: The X-ray diagrams fade off and vanish for strains exceeding a few percent. The same occurs with the cylindrical structure when the stress acts parallel to the axis of the cylinders. Clearly, this is due to the rupture of the styrene microdomains which are stiff and break down easily.

When the tensile stress is applied perpendicularly to the cylinders, the quality of the organization remains intact. A simple visual inspection of the shape of the probes shows that the sample does not change in size in the direction of the axis of the cylinders, that it lengthens in the direction of the stress, and that it shrinks in the direction perpendicular both to the stress and to the axis of the cylinders (see next section). During this deformation, which can attain values of more than 300%, the volume of the sample remains constant. It is of interest to note that, in this process, the crystal lattice, as observed with low-angle X-ray diffraction, changes in an affine way with respect to the external shape of the material.

MECHANICAL ANISOTROPY

To test for the mechanical anisotropy of block copolymer "single crystals," Keller, Folkes and co-workers [26, 27, 33, 34] focused their attention on Shell's Kraton 102, the behavior of which was repeatedly been commented upon in the previous sections. The structure is cylindrical, and the continuous matrix is formed of the elastic diene blocks.

For small strains, the mechanical properties of an elastic single crystal are fully described by the corresponding generalized Hooke's law $\epsilon_i = S_{ij}\sigma_j$ which relates the stresses σ_j to the strains ϵ_i through the compliance matrix S_{ij} [25]. For transversely isotropic crystals:

$$\begin{Bmatrix} S_{11} & S_{12} & S_{13} & 0 & 0 & 0 \\ S_{12} & S_{11} & S_{13} & 0 & 0 & 0 \\ S_{13} & S_{13} & S_{33} & 0 & 0 & 0 \\ 0 & 0 & 0 & S_{44} & 0 & 0 \\ 0 & 0 & 0 & 0 & S_{44} & 0 \\ 0 & 0 & 0 & 0 & 0 & S_{66} \end{Bmatrix}$$

where $S_{66} = 2(S_{11} - S_{12})$, S_{ij} contains only five independent elastic constants.

To determine the value of these constants for Kraton 102, three series of experiments were carried out. The longitudinal Young's modulus E_θ was first measured as a function of the angle θ between the direction of stress and the axis of the cylinders [26, 27, 33], and compared to the modulus which can be calculated theoretically from the matrix S_{ij} by rotating the axes defining its direction:

$$1/E_\theta = S_{33}\cos^4\theta + (2S_{13} + S_{44})\sin^2\theta\cos^2\theta + S_{11}\sin^4\theta$$

An excellent fit was obtained, verifying the internal consistency of the data and leading to the value of:

$$S_{11} = (2.3 \pm 0.3) \times 10^{-8} \text{ dyn}^{-1} \text{ cm}^2$$

$$S_{33} = 2.4 \times 10^{-10} \text{ dyn}^{-1} \text{ cm}^2$$

$$2S_{13} + S_{44} = (6.6 \pm 0.7) \times 10^{-8} \text{ dyn}^{-1} \text{ cm}^2$$

The high mechanical anisotropy of the material was thus established quantitatively: The sample is two orders of magnitude stiffer along the cylinders than it is in the transverse direction.

In a second series of experiments, the two Posson's ratios

$$\nu_{12} = -S_{12}/S_{11}; \nu_{13} = -S_{13}/S_{11}$$

were considered. The first represents the ratio of the strains measured across the cylinders in two mutually perpendicular directions; the second is the ratio of the strains along the cylinders and perpendicular to them. Owing to the high anisotropy of the system, ν_{12} is very close to unity, while ν_{13} is almost zero [26]: $S_{12} \simeq -S_{11}; |S_{13}| \ll S_{11}.$

This is in agreement with the relations:

$$S_{11} + S_{12} + S_{13} = 0$$

$$2S_{13} + S_{33} = 0$$

that can be deduced from the matrix, assuming the material to be incompressible. Using the data obtained from the first series of experiments, the following values can be estimated:

$$S_{44} \simeq S_{44} + 2S_{13} = 6.6 \times 10^{-8} \text{ dyn}^{-1} \text{ cm}^2$$

$$S_{12} \simeq -2.3 \times 10^{-8} \text{ dyn}^{-1} \text{ cm}^2$$

$$S_{66} = 2(S_{11} - S_{12}) \simeq 9.2 \times 10^{-8} \text{ dyn}^{-1} \text{ cm}^2$$

In a third series of experiments [33, 34], using a free oscillation torsion pendulum at a frequency in the range from 0.1 to 1.0 Hz, the torsional modulus G of the material was measured. When torsion takes place about an axis parallel to the cylinders, G provides a value of S_{44}, and when it takes place about an axis perpendicular to the cylinders, G provides a value of S_{66}:

$$S_{44} = (3.1 \pm 0.2) \times 10^{-8} \text{ dyn}^{-1} \text{ cm}^2$$

$$S_{66} = (5.9 \pm 0.3) \times 10^{-8} \text{ dyn}^{-1} \text{ cm}^2$$

Although they are unexpectedly large, these values are not very different from those derived previously from the Young's modulus combined with the Poisson's ratios. This was associated with the extreme anisotropy of the samples; but it might very well be only due to the quality of the single crystals used which is not perfect.

After the five independent elastic constants in the compliance matrix S_{ij} had been determined experimentally, the mechanical behavior of the system was analyzed on the ground of some current theories of fiber reinforcement [33]. It is not proper to enter here into the details of the discussion; it suffices to only point out that reasonable agreement can be obtained by assuming the Kraton 102 single crystalline sample to be a miniature fiber reinforced rubbery material. For instance, the experimental values of S_{11} and S_{33} are close to those which can be calculated considering a simple model of series and parallel arrangements of polystyrene and polybutadiene. However, this problem needs further examination, in particular the observed departure of the behavior of the butadiene matrix from this of normal rubber.

CONCLUDING REMARKS

We have reached the end of our review of the properties of oriented block copolymers. Our principal aim has been to expose the experimental evidence that is available in the field, without entering details and undue theoretical developments. It is only fair to pause now a moment for the sake of contemplating the subject as a whole.

We have seen that block copolymers are as a rule capable of spontaneously organizing in well-defined mesomorphic structures. The nature of these is closely related to the segregation of the blocks and the creation of interfaces between two types of distinct microphases. In the case of polystyrene–polydiene block copolymers, the mesomorphic phases are able to orient themselves in what may very conveniently, though improperly, be designated as "single crystals." The physical properties of these can then quite profitably be commented upon and interpreted in the usual tensorial terms of crystal physics; and this is how their optical, swelling, and mechanical properties were actually discussed.

The information that we gave about the subject has, however, been deficient in many respects. Several aspects of the problem are left to be elucidated hopefully in the near future. Among these let us mention the following.

What is, first, the mechanism of the orientational thermomechanical process that is used to bring about the formation of the "single crystals?" Is it only related to the anisotropic rheological properties of the medium in the molten state, or is it also affected, as claimed, by the temperature gradient in the system? Could it not happen that, during the mechanical shearing of the material, only those crystallites which are correctly oriented with respect to the shear direction stand efficiently the treatment, while the others are destroyed and subsequently reformed?

How can block copolymers different from those containing styrene and diene blocks be oriented in the form of single crystals, and what will be their specific properties? It is clear that no answer will be given this question as long as the orientational mechanism is not elucidated.

What actually is the chain conformation of the blocks within the segregation microdomains? Optical birefringence and infrared dichroism observations suggest that blocks are real random coils; yet, a weak anisotropy can be detected in the case of the lamellar structure. What is the scale of the random distribution of the monomer units? Apparently only neutron-scattering measurements of the radius of gyration of deuterated blocks, about an axis normal to the interfaces, can help clarify this problem.

Finally, how do the macroscopic properties of single crystals depend upon the molecular weight of the blocks and the structural parameters of the system? What is their time dependence, and how fast do the glassy microphases and the overall mesomorphic structure of the medium relax under stress?

The author would like to thank Drs. A. Mathis, A. Thierry, and D. Guillon for very valuable discussions.

REFERENCES

[1] M. J. Flokes and A. Keller, "The Morphology of Regular Block Copolymers," in *Physics of Glassy Polymers,* R. N. Haward, Ed., Applied Science Publishers Ltd, London 1973.

[2] M. J. Folkes and A. Keller, "Morphology of Block Copolymers and its Consequences," in *Block and Graft Copolymers,* J.J. Burke and V. Weiss, Eds., Syracuse University Press, Syracuse, New York, 1973.

[3] A. Skoulios, "Organizational and Structural Problems in Block and Graft Copolymers," in *Block and Graft Copolymers,* J. J. Burke and V. Weiss, Eds., Syracuse University Press, Syracuse, New York, 1973.

[4] A. Skoulios, "Mesomorphic Properties of Block Copolymers," in *Advances in Liquid Crystals: Vol 1,* G. H. Brown, Ed., Academic Press, New York, 1975.

[5] A. Skoulios and G. Finaz, *C. R. Acad. Sci., Paris,* **252,** 3467, (1961).

[6] G. Tsouladze and A. Skoulios, *J. Chim. Phys.,* **60,** 626 (1963).

[7] P. Ekwall, "Composition, Properties and Structures of Liquid Crystalline Phases in Systems of Amphiphilic Compounds," in *Advances in Liquid Crystals: Vol. 1,* G. H. Brown, Ed., Academic Press, New York, 1975.

[8] G. Friedel, *Ann. Phys., Paris,* **18,** 273, (1922).

[9] B. Gallot, R. Mayer, and C. Sadron, *C. R. Acad. Sci., Paris,* **C 263,** 42 (1966).

[10] K. Kato, *Polym. Lett.,* **4,** 35 (1966).

[11] E. Vanzo, *J. Polym. Sci.,* **A1-4,** 1727 (1966).

[12] H. Hendus, K. Illers, and E. Röpte, *Koll. Z.Z. Polym.,* **216,** 110 (1967).

[13] E. Fischer, *J. Macromol. Sci., Chem.,* **2,** 1285 (1968).

[14] A. Keller, E. Pedemonte, and F. M. Willmouth, *Nature (London),* **225,** 538 (1970); *Koll. Z.Z. Polym.,* **238,** 385 (1970).

[15] J. Terrisse, Ph.D. Thesis, Strasbourg (1973).

[16] J. Le Meur, J. Terrisse, and C. Schwab, Evian Meeting of French Physical Society (1971).

[17] M. J. Folkes, A. Keller, and F. P. Scalisi, *Koll. Z.Z. Polym.,* **251,** 1 (1973).

[18] E. Pedemonte, S. Cartasegna, P. Devetta, and A. Turturro, *Chim. Ind., Milan,* **55,** 861 (1973).

[19] E. Pedemonte, G. Dondero, G. Alfonso and F. de Candia, *Polymer,* **16,** 531 (1975).

[20] M. J. Folkes and A. Keller, *J. Polym. Sci., Phys.,* Ed. to appear.

[21] A. Mathis and A. Skoulios, unpublished.

[22] J. Dlugosz, A. Keller, and E. Pedemonte, *Koll. Z.Z. Polym.,* **242,** 1125 (1970).

[23] J. Dlugosz, M. J. Folkes, and A. Keller, *J. Polym. Sci., Phys.,* **11,** 929 (1973).

[24] J. A. Odell, J. Dlugosz, and A. Keller, *J. Polym. Sci., Phys.,* in press.

[25] J. F. Nye, *Physical Properties of Crystals,* Oxford University Press, 1969.

[26] M. J. Folkes and A. Keller, *Polymer,* **12,** 222 (1971).

[27] A. Keller, J. Dlugosz, M. J. Folkes, E. Pedemonte, F. P. Scalisi, and F. M. Willmouth, Evian Meeting of French Physical Society (1971).

[28] M. J. Folkes, A. Keller, and J. A. Odell, *J. Polym. Sci., Phys.,* in press.

[29] H. Ambronn and A. Frey, *das Polarisationmikroskop,* Leipzig p. 119, 1926.

[30] M. J. Folkes, A. Keller, and F. P. Scalisi, *Polymer,* **12,** 793 (1971).

[31] G. E. Wardell, D. C. Douglass, and V. J. McBrierty, *Polymer,* **17,** 41 (1976).

[32] M. L. Ionescu and A. Skoulios, *Makromol. Chem.,* **177,** 257 (1976).

[33] R. G. C. Arridge and M. J. Folkes, *J. Phys. D: Appl. Phys.,* **5,** 344 (1972).

[34] M. J. Folkes and R. G. C. Arridge, *J. Phys. D: Appl. Phys.,* **8,** 1053 (1975).

STYRENE-DIENE TRIBLOCK COPOLYMERS: ORIENTATION CONDITIONS AND MECHANICAL PROPERTIES OF THE ORIENTED MATERIALS

A. WEILL and R. PIXA

E.A.H.P., 4, rue Boussingault, F-67000, Strasbourg, France

SYNOPSIS

The organization of the hexagonal structure of some block copolymers has been obtained directly from the melt. We have pointed out the temperature and shear conditions which give a well-organized and oriented structure. In order to represent the viscous and elastic behavior of the melt, we have drawn flow curves and have measured the die swell ratio. A flow mechanism, corresponding to grain structure, is proposed. The results obtained enabled us to build a molding device. Fracture tests under tension on ABA poly(styrene–b-isoprene) samples exhibiting hexagonal structure showed a highly anisotropic behavior: If the crack propagates in a plane perpendicular to the isoprene cylinder axes or parallel with it, the fracture is either brittle or ductile accordingly. In the latter case, branching also occurs on both sides of the main crack. By using optical and scanning electron microscopy, the corresponding fracture surface morphologies have been studied. The brittle fracture surface shows a pattern of alternating rough and smooth bands which is characteristic of pure polystyrene. On the other hand, the ductile fracture surface is composed of an array of elongated "chips." They are disposed practically parallel to the fracture propagation direction. Moreover, indentations made with a spherical steel penetrator led to residual ellipsoidal impressions. By measuring their dimensions, the anisotropy of the mechanical properties was displayed.

INTRODUCTION

The understanding of the block copolymers properties in connection with their organization and orientation possibilities is increasing. Contrary to the orientation of homopolymers which is due to an anisotropic distribution of the molecules, it is the geometry of the microphases of the block copolymers and their regular arrangement which provides, in certain cases, anisotropic properties. Hexagonal structure is the subject of most papers [1] and represents the greatest industrial production of block copolymers. Concerning the orientation process from the melt, there is no generally accepted model. Therefore we are interested in rheological studies of block copolymers yielding anisotropic material. The copolymers we have chosen are of the poly(styrene-b-diene) type, exhibiting hexagonal structure, the matrix being either rubbery or rigid. Mechanical properties of oriented block copolymers have already been examined [2]. To complete these previous works we have carried out some tensile fracture experiments and indentations with a spherical steel indentator.

Journal of Polymer Science: Polymer Symposium 58, 381–394 (1977)

TABLE I
Materials Used in Experimental Work

Materials	Molecular weights $(\overline{M}_w)$	Styrene content (wt %)
Anionic Polymerisation :		
SIS 75	47000–36000–49000	76
SIS 309	53000–33000–36000	73
SI 317	75000–25000	75
Commercial Polymer :		
Cariflex 4113 (SBS)	15000–50000–15000	38

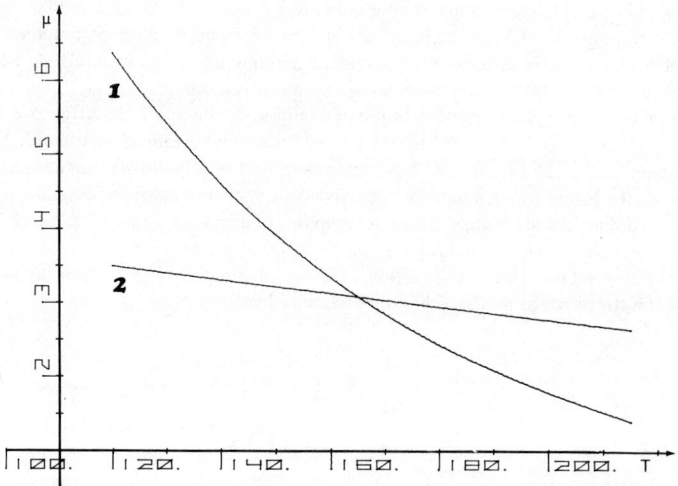

FIG. 1. Log of the zero-shear rate viscosity (Poise) versus temperature (°C) for the two blocks of a SBS triblock copolymer (15000-50000-15000). 1: polystyrene, $\overline{M}_w$ = 15000; 2: polybutadiene, $\overline{M}_w$ = 50000. The curves are calculated after [19].

MATERIALS

The block copolymers we have studied were prepared by anionic polymerization [3]. The syntheses were carried out in three stages using n-butyllithium as catalyst and benzene as solvent. After complete polymerization of each block, a sample of solution was taken for the analyses.

By gel permeation chromatography, the molecular weights of the first blocks A were determined. The mole percentages of styrene in the block copolymers AB and ABA were then obtained by UV absorption measurements of the phenyl groups. The values for all the block copolymers studied are shown in Table I. After completion of the polymerization an antioxidant (Nonox) was added. The solution was then frozen at 233°K and dried in vacuum.

RHEOLOGY

Different Block Viscosities

Every molecule is made up of two or more blocks. Each block presents a different average molecular weight and glass transition temperature. As a result of these differences, each segment forming the block copolymer has a different viscosity. Therefore, locally, the material is not homogeneous.

Let us, for instance, discuss the case where only two different blocks are present in the molecule (Fig. 1). If one of them is far more fluid than the other, the former will support the total deformation imposed by shearing, whereas the latter will practically be unaffected in its form, although being subjected to the mechanical strain. If one considers the case where the matrix is more fluid than the cylinders, not only will the latter be undeformed during shear, but they will also align themselves parallel to the direction of flow. This is practically the case of rigid rods within a free flowing mass.

The degree of structural organization of the cylinders within the annealed samples is given by the ratio R of the Young's modulus measured in the direction of flow to that in the perpendicular direction. We have carried out orientation experiments at 120°C and 180°C. R was respectively 50 and 18. This result is in good agreement with Figure 1 showing clearly the existence of a transition between the state of fluid cylinders–viscous matrix (more than 160°C) and that of viscous cylinders–fluid matrix (less than 160°C). When the cylinders are more viscous than the matrix, the degree of orientation is higher than on the contrary.

On the other hand, if the matrix is more viscous than the cylinders, the structural organization risks being perturbed by shear. Copolymers with a cylindrical structure and a high percentage of styrene represent always the case of viscous matrix surrounding fluid cylinders, in the whole range of temperature. The melt cannot tolerate high shear without destroying its phase structure. Weccan suppose that there is a balance between lattice building (by heat) and lattice destruction (by shear). Furthermore, the higher the molecular weight of the copolymer, the bigger is the thermic activation energy, and, due to the

TABLE II

Experimental Conditions for Obtaining Well-Organized and Oriented Plaques from SIS Block Copolymers with 20 to 30% Polyisoprene at 180°C

AVERAGE MOLECULAR WEIGHT OF THE COPOLYMERS SIS	SHEAR RATE (s^{-1})	TIME (HOURS)	ROUND PER HOUR
$\leq$ 100 000	10^{-1}	2	6
200 000	10^{-2}	24	0.6
400 000	$< 3.10^{-3}$	> 70	0.2

viscosity, the more difficult are the internal motions. Because of the oxydative degradation of the polyisoprene, the temperature cannot be too high. Therefore the samples need to be sheared very slowly, in order to be well organized and oriented. The experimental work has been made by Beaudouin [4] and his results are given in Table II. When the samples are submitted to a shear rate higher than the limit, form birefringence and mechanical anisotropy are not observed.

Flow Curves

The curves plotted in Figure 2 were established by means of a constant speed ZWICK capillary rheometer. Since a die of 60 mm in length and 0.5 mm in radius is used, the Rabinowitsch equation is applied without end and barrel corrections. By comparing the flow curves of diblock and triblock copolymers of about the same total molecular weight and the same percentage of polyisoprene (see Fig. 2), one notes that the non-Newtonian behavior of the triblock is much more marked than that of the diblock copolymer. This corresponds logically to the fact that almost all the cylinders are anchored to each other in a triblock, whereas they are not in a diblock. This anchorage of the cylinders acts on the non-Newtonian behavior of the polymer in the same way as does the density of entanglements. It is also interesting to note that the compared flow curves tend to be superimposable for high shear. One may interpret this observation remembering the disappearing of the phase structure at high shear.

Die Swelling

Measurements of swelling made either at the exit of a rheometer or an extruder tend to show that the grain structure (observed for several years on extruded rods and on films made from evaporated solutions) already exists in the molten state.

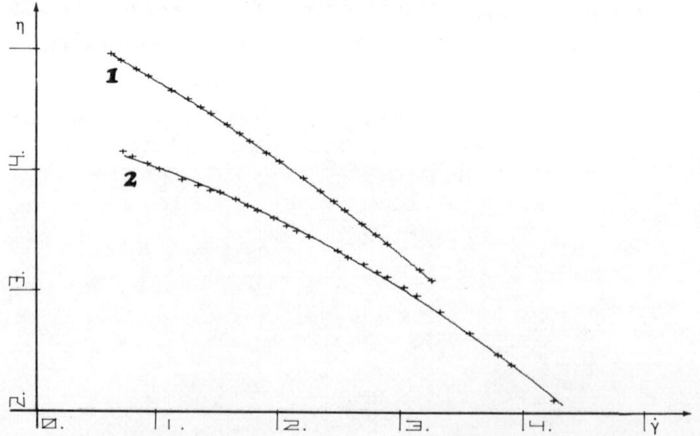

FIG. 2. Flow curves: Log of the apparent viscosity (Poise) versus log shear rate (s^{-1}) at 210°C. 1: SIS 309; 2: SI 317. The data are obtained with the capillary rheometer.

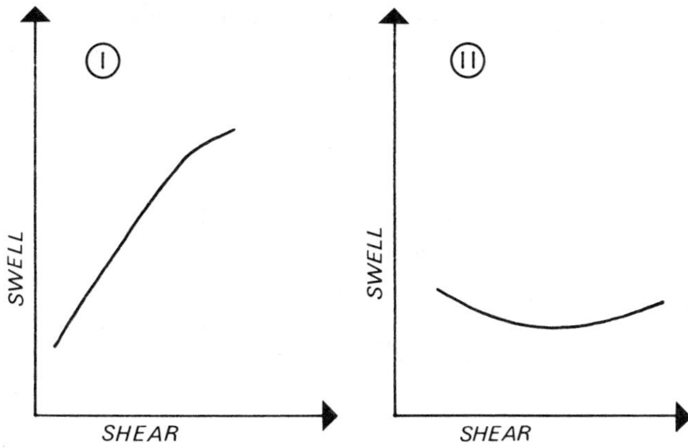

FIG. 3. Die swell plotted against rate of shear. I: General form (after [6]); II: Results for emulsion polymerized PVC (after [5]).

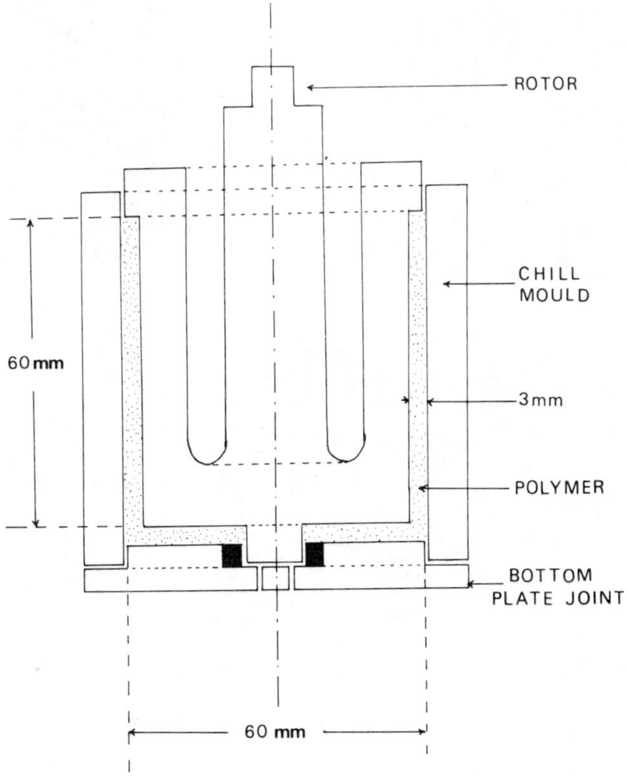

FIG. 4. Schematic representation of the device yielding oriented plaques.

The die swell plotted against rate of shear leads us to observe a certain analogy between triblock copolymer flow and the flow of homopolymers prepared by emulsion polymerization such as polyvinyl chloride [5] or polystyrene. In our experiments, the curves of triblock copolymers seem to belong to the type II of Figure 3. Furthermore, the polymer beads synthetized in emulsion have an approximate size of some tenths of microns, and the grain structure remains partially in the melt if the temperature is not too high. Now, the size of the domains (0.2–0.6 μm) found in triblock copolymers by different authors [7, 8] is in good agreement with our hypothesis of the existence of a granular structure within the melt. In spite of the anchorage of the cylinders, the phase lattice orientates itself in the hydrodynamic field by slippage and rotation of the grains past each other rather than by homogeneous deformation of the melt.

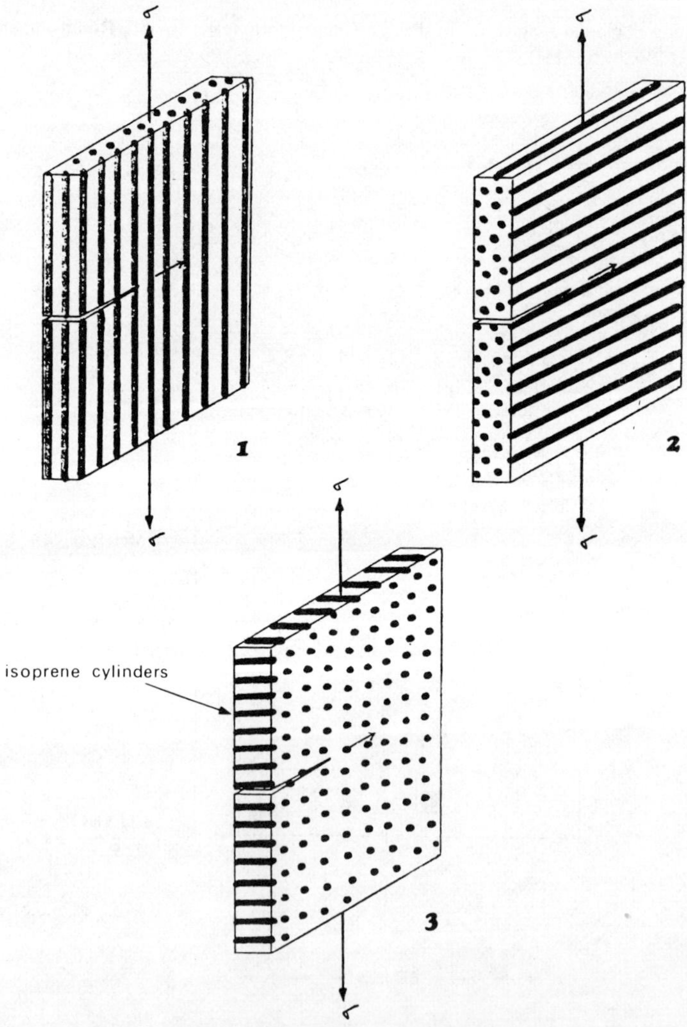

FIG. 5. Disposition of the internal structure in the samples.

ORIENTATION DEVICE

The device yielding plaques of well-oriented material was built at the EAHP [4, 9] and is shown in Figure 4. Like a rotating cylinder viscometer, it essentially consists of two concentric cylinders and a bottom plate joint. The rotating part is the inner cylinder. When experiments are carried out, the different parameters are: the temperature of the melt, the total duration of the rotary motion, and the number of revolutions per hour. After each operation the apparatus is allowed to cool, the bottom plate is removed and the outer cylinder is opened. The polymer molded in the form of a cup has to be cut along a generating line and the bottom of the cup has to be removed. By annealing above T_g, the remaining hollow cylinder is unrolled and the residual stresses are released. By means of low-angle X-ray scattering and transmission electron microscopy the existence of hexagonal symmetry is verified.

MECHANICAL TESTS

Fracture Tests

Specimens of rectangular shape are cut out from oriented plaques of SIS 75, their edges being parallel to the symmetry planes. The area of the samples is about 10 cm^2. An edge-crack several mm long is introduced with a knife-edge in a plane normal to the long side. In order to ensure rectilinear fracture propagation normal to the applied tensile stress, the long side is chosen greater than three times the short one [10]. There exist three possibilities of obtaining a symmetrical disposition of the edge-crack with regard to the internal structure (see Fig. 5). We have studied the first two of them. The fracture tests are carried out at 23°C on a Zwick tensile machine. The strain rate is about 10^{-4} s^{-1}. The fracture surfaces are examined using a Zeiss optical microscope and a Cameca scanning electron microscope. Specimens for examination in the scanning electron microscope were coated with a thin layer of gold–palladium alloy by vapor deposition in vacuum.

Fracture Normal to the Isoprene Cylinders

In this case, we observe a sudden and complete rupture normal to the applied tension when a critical stress is reached. This behavior occurs generally in brittle materials where cracks are growing along a path normal to the direction of maximum tensile stress.

Figure 6 shows a part of the corresponding fracture surface. One sees two families of markings with different distances between. Immediately behind the precrack a first system of fine ribs develops and slightly farther appear more irregular rough bands. The spacing between the rough markings increases with increasing distance from the precrack. It varies between about 130 μm at the beginning to about 210 μm at a distance of 15 mm from the precrack. On the other hand, the smooth ribs are regularly spaced with an interval of about 30 μm (Fig. 7).

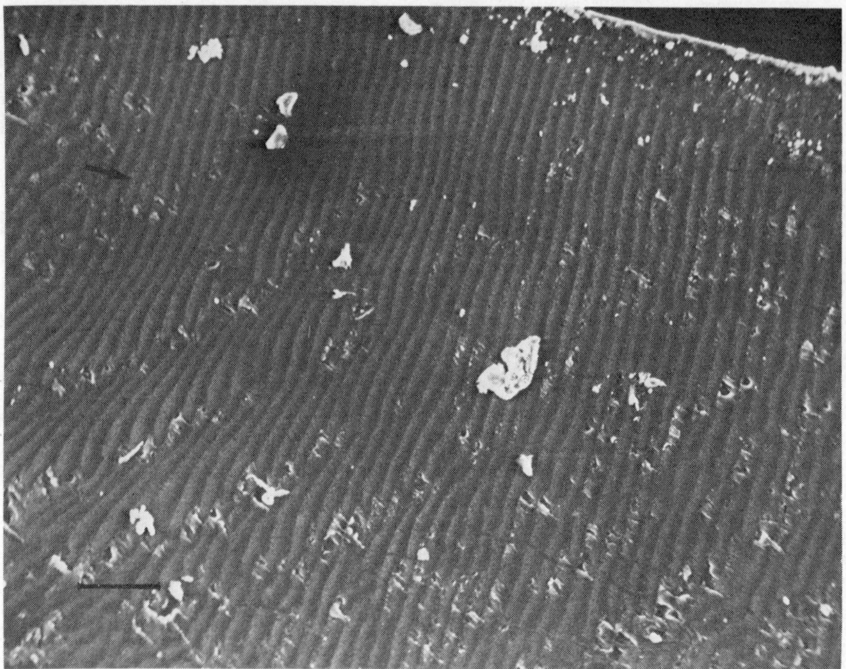

FIG. 6. Type 1 fracture surface, the arrow indicates the propagation direction. Scale represents 0.1 mm. Scanning electron micrograph.

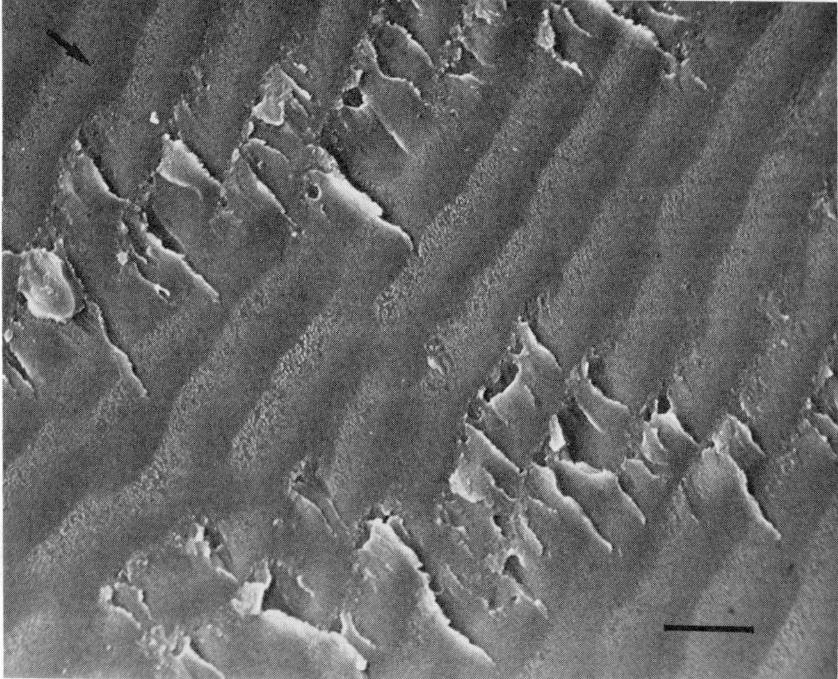

FIG. 7. Type 1 fracture surface, the arrow indicates the propagation direction. Scale represents 20 μm. Scanning electron micrograph.

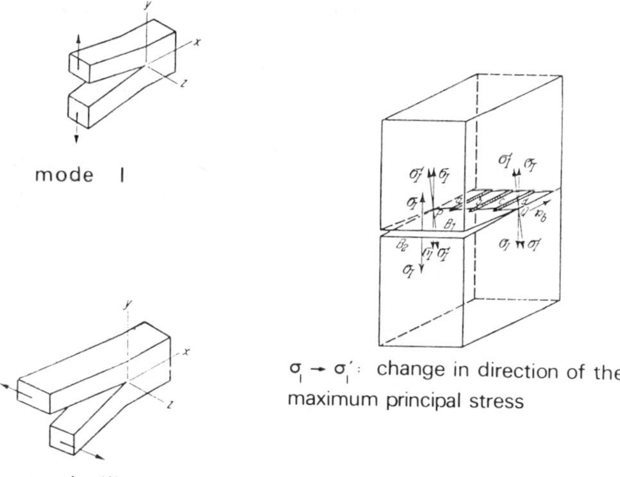

FIG. 8. Scheme of the formation of fracture lances by the superposition of mode I and mode III loading (after [13]).

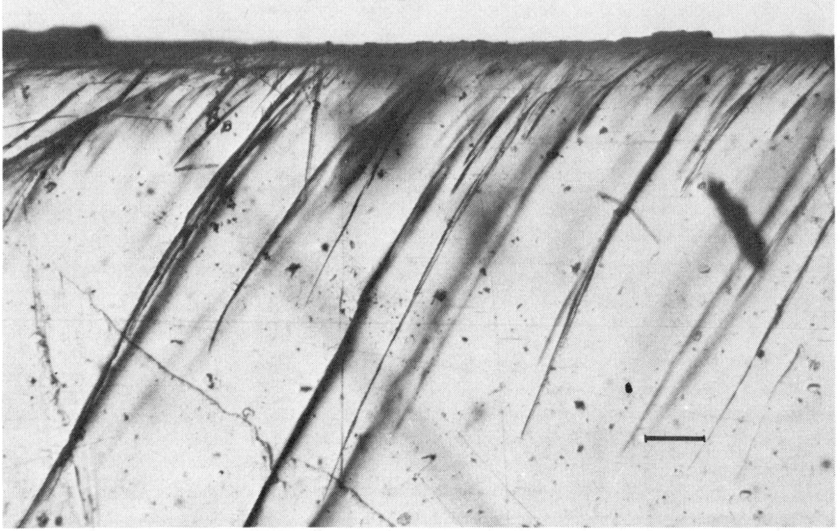

FIG. 9. Side-view of a fractured specimen. Secondary crazes in type 2 fracture. Propagation from right to left. Scale represents 0.2 mm.

We may also note that the two rib systems interact mutually. They intersect forming an acute angle. This may indicate that they have not the same basic origin.

The rough bands are formed out of separated markings disposed practically parallel to the propagation direction. The separations delimit partial fracture planes turned through a small angle with respect to the main fracture plane. The axis of rotation of the partial fracture surfaces is parallel to the fracture direction. These markings are probably fracture lances [11]. They are due to a superpo-

FIG. 10. Type 2 fracture surface, the arrow indicates the propagation direction. Scale represents 10 μm. Scanning electron micrograph.

sition of tensile (mode I) and anti-plane shear (mode III) loadings (see Fig. 8) [12]. The periodicity might result from a periodical concentration of shear strain due to the interference of transverse waves with the propagating fracture front. The transverse waves are certainly created by reflection on inhomogeneities within the sample. The markings are therefore corresponding to Wallner lines [14]. The smooth bands exhibit the same micromorphology as the "patch pattern" described in polystyrene by others [15].

Fracture Parallel to the Isoprene Cylinders

When the sample is subjected to a tensile load according to case two (Fig. 5), first an array of crazes appears perpendicularly to the tensile stress. Then the sample breaks in a manner similar to case one.

A particular feature is the appearance of an array of many secondary crazes on both sides of the crack plane (see Fig. 9). They form an angle of about 60° to the fracture plane. A similar phenomenon was observed in polystyrene containing a set of preformed crazes parallel to the crack plane [16]. Probably, the presence of the isoprene cylinders parallel to the fracture path modifies the stress field around the crack tip. In this manner, the maximum principal stress direction is no longer coinciding with the tensile stress direction.

On the corresponding fracture surface (see Fig. 10) one observes striations and "chips" disposed parallel to the propagation direction. The striations oc-

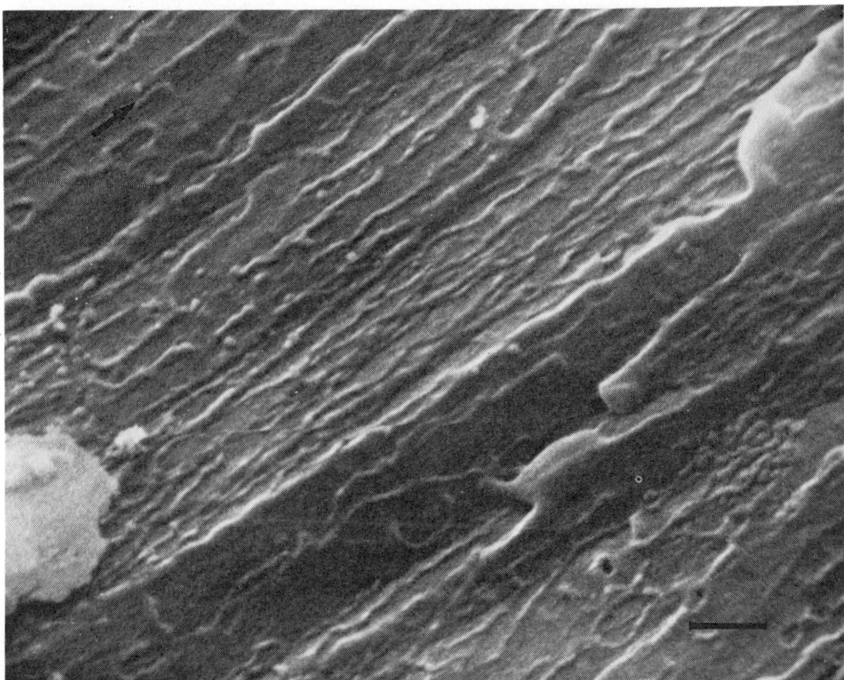

FIG. 11. Type 2 fracture surface, the arrow indicates the propagation direction. Scale represents 1 μm. Scanning electron micrograph.

casionally tend to incurve and to open fan-wise (e.g., in A). The general appearance corresponds to a viscous deformation of the matter induced by adiabatic heating during fracture propagation. This is more noticeable in Figure 11. The feature appearing like striations (B in Fig. 10) is a result of the subdivision of the fracture surface in facets. Adjacent facets are slightly rotated one in relation to another. The axis of rotation is parallel to the propagation direction. The microstructure of these facets is formed by a "patch pattern" like in the previous case. The particular features of the fracture surface in this fracture mode are interpreted as induced by the internal structure of the specimens, i.e., the crack propagates parallel to the isoprene cylinders.

Indentation Tests

Another way to show the anisotropy of these samples is the Brinell hardness test. In this test, a hard spherical indenter is pressed under a fixed normal load on to the surface under examination [17]. At first, the ball and the material will both deform elastically. For isotropic materials, these deformations are described by the classical equations of Hertz [18]. As the load on the indenter is increased, the material deforms plastically until equilibrium is reached. When the load is removed, elastic recovery of the indentation occurs, with a corresponding change in its shape. Therefore the residual indentation is produced only by plastic deformation. When considering viscoelastic materials, it is necessary to take

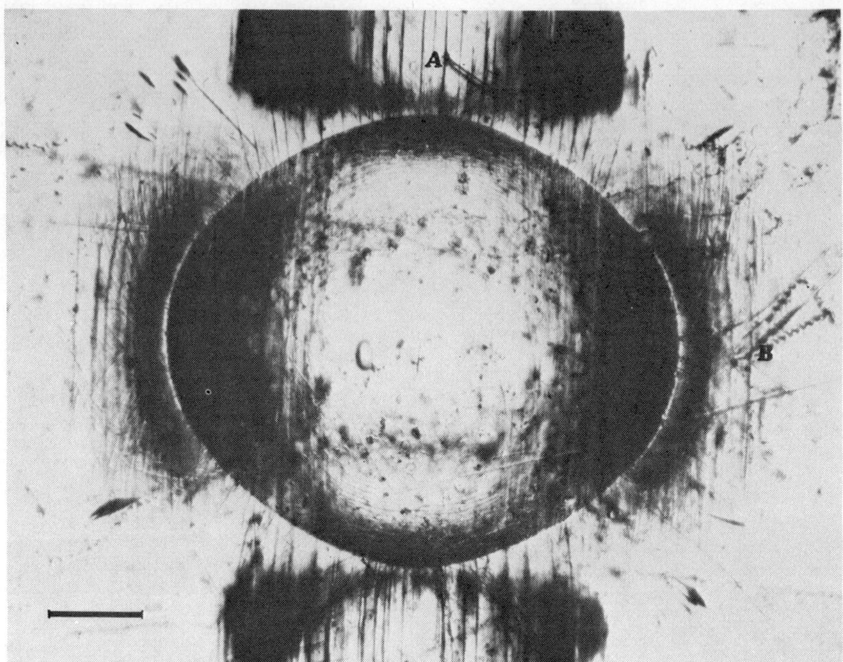

FIG. 12. Residual indentation, obtained with an indenter of diameter 5 mm, loaded with 40 daN during 1 min. Scale represents 0.5 mm.

into account the time dependence of the deformation. Under certain conditions it is possible to calculate the yield stress and Young's modulus by measuring the parameters of the indentation with and without load.

For an anisotropic material, a quantitative interpretation is more critical. As a matter of fact, there exists no analytical solution for the contact problem of an anisotropic body. Figure 12 shows a top-view of such a residual indentation obtained in an oriented sample. Figure 13 explains schematically the disposition of the indentation with regard to the internal structure of the polymer. It shows also the two cross-sections AA′ and BB′ along the symmetry planes. The material is displaced more strongly upward in the directions normal to the cylinder axes ("piling-up"). Along the AA′ section, we observe rather a "sinking-in" effect. This behavior is connected with the anisotropy of the plastic deformation. Obviously the yield stress is lower along the BB′ direction.

In Figure 12 we observe another interesting phenomenon: the formation of crazes exhibiting unusual orientations. Indeed, a first array of crazes has developed parallel with the isoprene cylinders. The density of these crazes is very high around the points B and B′. At two areas near the points A and A′ and extending symmetrically to the two sides, we see also a high craze density. A second system of crazes has radial direction. It has developed more weakly than the first one.

When the indenter is pressed with an increasing normal force P onto an isotropic brittle solid, a cone crack develops when P exceeds a critical value. This fracture is due to the tensile stresses which are maximum in an area surrounding

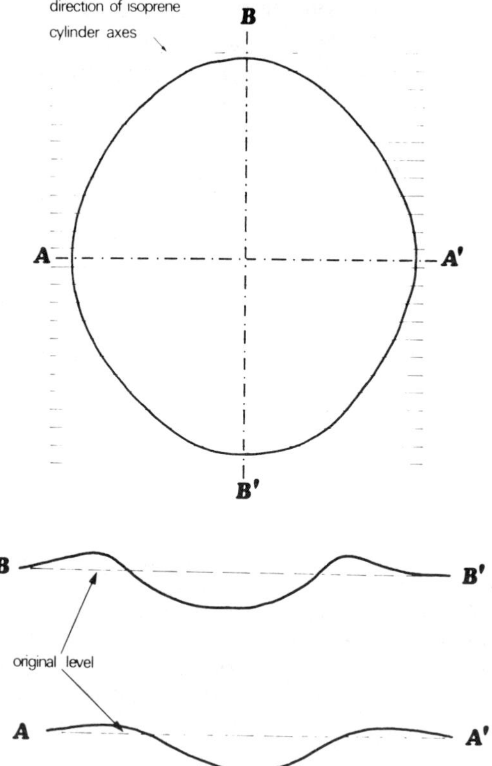

FIG. 13. Schematic representation of a residual indentation. Top-view and two cross-sections.

the contact surface [18]. For a circular contact surface the tensile stresses act radially with respect to its center.

In the case considered here, the stress field around the contact surface is not of axial symmetry. However, it is also surrounded by an area bearing tensile stresses. On the other hand the stress σ_c needed to initiate crazes is also anisotropic. The preferential formation of crazes parallel to the isoprene cylinders indicates that σ_c has a lower value normal to this direction. If we assume only tensile stresses acting radially like in the isotropic case, the crazes in the regions A and A' cannot be accounted for. This fact indicates a more complicated stress distribution around the contact area.

CONCLUSION

The rheological behavior of the block copolymers we studied bears resemblance to that of some polymers synthesized in emulsion. It has been shown [5] that the melt flow of such polymers involves a particle-slippage flow mechanism. We propose to interpret the flow behavior of our block copolymers by an analogous phenomenon. This provides also an explanation for the observation of grain boundaries in not perfectly oriented solid samples [7, 8].

The fracture tests of the anisotropic glassy block copolymers behave very differently in relation to the propagation direction. The polyisoprene rods does almost not influence a crack propagating normal to them: the fracture surface bears strong resemblance to that of pure polystyrene. On the contrary, the propagation parallel to the polyisoprene rods gives rise to a more ductile fracture and to unusual fracture surfaces.

Finally, the hardness tests on the same oriented samples give us information about the anisotropy of the yield stress and the craze initiation process.

The results reported in this paper are necessarily incomplete. A further investigation of these phenomena is currently in progress.

We are particularly indebted to C. Goett and J. Journé for the synthesis of the copolymers. We must also acknowledge the help of Miss I. Peterschmidt for examining the samples in SEM. The work reported here has been carried out in part under contracts 72-7-0760 and 74-7-0157 from the Delegation à la Recherche Scientifique et Technique.

REFERENCES

[1] M. J. Folkes and A. Keller, "The Morphology of Regular Block Copolymers," in *The Physics of Glassy Polymers,* R. N. Haward, Ed. Applied Science Publishers Ltd., London, 1973.

[2] R. G. C. Arridge and M. J. Folkes, *J. Phys. D,* **5,** 344 (1972).

[3] C. Goett and J. Journé, *Caoutch. Plastiques,* **53,** 558, 47 (1976).

[4] L. Beaudouin, Thesis, University of Strasbourg, 1973.

[5] A. R. Berens and V. L. Folt, *Polym. Eng. Sci.,* **9,** 27 (1969).

[6] J. A. Brydson, *Flow Properties of Polymer Melts,* The Plastics Institute, London, 1970, p. 65.

[7] G. Kämpf, H. Krömer and M. Hoffmann, *J. Macromol. Sci. Phys.,* **B 6,** 167 (1972).

[8] C. Price, T. P. Lally, A. G. Watson, D. Woods and M. T. Chow, *Br. Polym. J.,* **4,** 413 (1972).

[9] J. Terrisse, Thesis, University of Strasbourg, 1973.

[10] P. C. Paris and G. C. Sih, *ASTM Spec. Techn. Publ.* **381,** 30 (1965).

[11] A. Smekal, *Oesterr. Ing. Arch.,* **7,** 49 (1953).

[12] E. Sommer, *Eng. Fract. Mech.,* **1,** 539 (1969).

[13] F. Kerkhof, *Bruchvorgänge in Gläsern,* Glastechnische Gesellschaft, Frankfurt, 1970.

[14] H. Wallner, *Z. Phys.,* **114,** 368 (1939).

[15] P. Beahan, M. Bevis, and D. Hull, *Proc. R. Soc. London A,* **343,** 525 (1975).

[16] D. Hull, *J. Mater. Sci.,* **5,** 357 (1970).

[17] D. Tabor, *The Hardness of Metals,* Clarendon Press, Oxford, 1951.

[18] S. P. Timoshenko and J. N. Goodier, *Theory of Elasticity,* 3rd Ed., McGraw Hill Kogakusha Ltd, Tokyo, 1970, p. 409.

[19] D. W. van Krevelen, *Properties of Polymers, Correlation with Chemical Structure,* Elsevier, Amsterdam, 1972, p. 259.

A CURRENT ACCOUNT OF CHAIN EXTENSION, FIBROUS CRYSTALLIZATION, AND FIBER FORMATION

A. KELLER

H. H. Wills Physics Laboratory, University of Bristol, Bristol BS8 1TL, U.K.

SYNOPSIS

A broad, rather personal survey is presented on chain alignment and fibrous crystallization. First, attention is drawn to the fact that fibrous crystal entities can be of fundamentally different origin and certain methods for their identification are proposed. Following this, the largest part of the survey is concerned with fibrous crystallization from pre-extended molecules. As a first step, methods for extending chains and for observing their extension are outlined. Elongational flow is receiving special attention: conditions by which it can lead to ultrahigh chain extension are briefly defined as high-lighted both by a priori considerations and by most recent experiments, methods of realizing these conditions being indicated both in solutions and melts. Following this, attention is given to the mode of crystallization of extended chains and to the structures which arise. Amongst the many factors involved, the part played by the local flow field arising from the presence of the growing crystal entity within the general flow field is outlined, leading up to the basic mechanism by which fibrous crystals grow. It is shown how the recognition of these principles can be utilized for producing crystalline fibers. The unusual thermal and mechanical properties of the products with special emphasis on melting behavior and modulus are indicated. A short section is devoted to the achievement of very high chain extensions by means of deforming material crystalline to begin with and the resulting products are contrasted with those obtained by means of crystallizing pre-extended chains as regards thermal and mechanical behavior.

PREFACE

The present article is only a brief survey of a very extensive subject. It does not claim to be comprehensive in its coverage, nor to be analytical in its presentation. Its purpose is merely to outline the field and to indicate some general guidelines. It is not intended to serve as a first announcement of new information which has not yet passed through the recognized publication channels. Yet it is the nature of knowledge that once known it cannot be undone. Thus I cannot avoid taking cognizance of, and being influenced by, what I happen to know at this time. Even so I shall refrain from providing specific documentation of items which have not yet appeared in publication before. I only hope that by merely pointing out what there is to know and the source from where it is awaited or which, if so desired, may be approached, I am not going too far toward making this article an original source reference, nor am I holding back too much to leave the reader totally unenlightened.

Journal of Polymer Science: Polymer Symposium 58, 395–422 (1977)
© by John Wiley & Sons, Inc.

CLASSIFICATION OF FIBROUS CRYSTALLINE POLYMERS AND METHODS FOR DISTINGUISHING BETWEEN THEM

Crystallization in fibrous form is the one which is intuitively expected from long chain molecules, as indeed such molecules form the fibrous materials in nature and are being used for the manufacture of synthetic fibers in technology. In spite of this, laboratory experiments have shown that crystallization of long chain molecules leads to lamellae containing the chains in folded configuration [1]. Even so, fibrous crystal entities can be observed under the following rather special circumstances:

(1) When crystallization takes place concurrently with the formation of the molecules, i.e., in the course of the polymerization reaction. This is often referred to as the nascent crystalline state. Not only is fibrosity the native state of many important polymers (polyethylene, polypropylene) prior to further processing, but this state is finding increasing utilization as such, e.g., as pulp material for paper products. There is very little basic knowledge of the formation and structure of such nascent fibers, hence they will not feature further in this article (for some particulars, see recent textbook by Wunderlich [2]).

(2) When the chains are stretched first and crystallization is allowed to occur subsequently from the extended chains. This phenomenon will be the main subject of the paper.

(3) When polymer already crystallized in the lamellar (spherulitic) form is deformed by uniaxial tension as in conventional fiber drawing. This subject will only be briefly touched upon mainly to contrast it with material obtained along route 2.

First it needs stating that the mere observation of crystalline fibrous entities in itself does not reveal its mode of origin, e.g., whether they were obtained along routes 1, 2, or 3, a point not always realized when assignments are made to one or the other source in the literature. It is often not adequately appreciated that in spite of comparable degree of chain alignment different kinds of fiber can be of fundamentally different origin: those formed along route 2 are the products of primary crystallization while those along route 3 result from deformation of material crystalline to begin with. This difference offers a ready test for distinguishing between them. Namely class 3 represents a strained system with locked-in restoring forces. Accordingly, when heated close to the melting range the fiber contracts often up to its undeformed dimension even from elongations as large as 30X. In contrast, fibers of class 1 and 2 are thermally stable, in fact can even substantially superheat. This "thermal retraction test" [3] applies equally to macroscopic samples and to fibrils on the electron microscope scale where they might have formed through genuine flow-induced crystallization or through intentional or accidental microdrawing due to sample handling or through the effect of the electron beam. Application of this test to distinguishing between classes 2 and 3 together with examples have been reported elsewhere [3]. Here merely a qualifying statement will be added: The internal stresses in fibers of class 3 can relax on heating the fiber at a fixed length. Such a heat-relaxed fiber will not retract any longer even when held freely, hence the above criterion for identifying fibers obtained along route 3 becomes inapplicable. Distinction between classes 1 and 2 is usually self evident from the macroscopic consistency or from the circumstances under which the fibrous material has been obtained. Even so conditions can arise when the origin of fibers, particularly on the electron microscopic scale, can be in doubt, i.e., whether route 1 or 2 pertains specially when some flow is present during the polymerization reaction in solutions [4]. In view of the scarcity of knowledge on nascent polymers there is no ready answer to this problem.

CRYSTALLIZATION FROM PRE-EXTENDED MOLECULES

Principle and Subject Matters

Crystallization of long chains from the pre-extended state yields fibrous crystal entities both from the melt and solutions (class 2 of the preceding section). While this may appear self evident the converse is not so: Thus unless the chains are extended, crystallization will yield chain folded lamellae. As a further general point it will be stated at this stage that in practice chain extension is never uniform. In fact the usual orienting effects result in a bimodal distribution: Some chains are very highly stretched while others are practically unaffected. The former will form fibrous crystals with extended chains while the latter will use these resulting fibers as nuclei and crystallize onto them by chain folding, giving rise to transversely oriented platelets, the whole composite assembly resembling a series of platelets strung on a central string termed "shish-kebabs."

The subjects to be discussed individually comprise the following items: methods of chain elongation, modes of crystallization, structure of the crystals, and properties of the crystals.

Methods of Chain Elongation

The materials and their methods of stretching fall essentially into two categories according to whether the initially random chains are in a crosslinked form or whether they are individual entities, the presence of physical entanglements giving rise to transition stages between the two.

Crosslinked random chains form the class of materials known as elastomers. As is well known, molecular networks can be readily extended under static conditions, sufficient extension at appropriate temperatures resulting in crystallization. While the crystallization itself leads to many novel features (to be commented on later) the process of chain extension involves no further issue beyond that encompassed by the subject of rubber elasticity.

The second process, namely the stretching of polymer molecules in their liquid state, however, forms a special branch of rheology which is not as familiar to the general polymer public. While on the other hand it has been subject to theoretical studies of considerable sophistication (e.g., refs. [5–8] to quote only a few) there has been little rapport between theory and experiment in as far as there have been few experiments, if any, by which the consequences of theoretical predictions could be assessed. Conversely, the scientific literature and technological practice abound with mention of effects which have their origin in chain extension by flow, however, under circumstances which does not make them amenable to evaluation in terms of existing theories or even allow their connection with flow induced chain extension to be readily recognized. The work on which the present section relies is essentially experiment based and has been designed specifically so as to bridge the gap just indicated. In the course of the work, centered on the researches of M. R. Mackley with continual lead given by F. C. Frank, first simple flow systems, capable of producing ultrahigh chain extentions, were created in a way so as to permit observations on the chain extending effect of the flow. The observation thus made usually called for a reversion to further aspects of the underlying fundamentals. This feed-back between experiment and theory eventually led to a clearcut and simple formulation of the requirements for extending chains which will now be listed.

Conditions for Extending Chains in the Liquid Form

In general, any shearing flow, i.e., flow containing velocity gradients, will possess an extensional and a rotational component by which a given fluid element will be stretched and rotated respectively. As has been recognized for some time, it is the former that has the ability to extend chains. For the ultrahigh alignment we are interested in, the extensional component has to dominate and in addition the chain molecules have to comply with specific criteria. (This is in contrast to the traditional flow birefringence studies, say in a Couette apparatus, where in view of the equality of the rotational and extensional components merely a distortion of the random coil is observed).

The conditions the flow field has to satisfy are as follows (Frank [9]):

(i) The fluid element in question must be persistently extending. According to Frank, in the simple two dimensional situation this condition can be expressed in terms of the persistent strain rate σ

$$\sigma^2 = S^2 - \omega^2; \quad |S| \geq |\omega|$$

$$\sigma = 0; \quad |S| \leq |\omega|$$

where S is the principal extension rate and ω the rotation rate. When $\omega = 0$ the flow is rotation free and entirely extension dominated. In two-dimensional flow it corresponds to a flow with velocity components which increase and decrease in two mutually perpendicular directions according to

$$v = (x, -y, 0)S$$

i.e., to stretching of a fluid element having constant width (pure shear) while in three dimensions to a flow with a velocity

$$v = \left(x, -\frac{1}{2}y, -\frac{1}{2}z\right)S$$

representing stretching of a cylindrical fluid element, i.e., uniaxial extension. $\omega = S$ corresponds to simple shearing flow

$$v = (y, 0, 0)\dot{v}$$

which is the case of the familiar Poiseuille flow and is the limiting condition for persistent extension.

(ii) The extension of the molecule within the fluid element can at most be equal to and is usually smaller than that of the fluid element itself. Hence the persistent extension under (i) must itself persist for a sufficiently long time for the fluid element to attain this required extension. The extensions in question are large. For an isolated chain molecule it will correspond to the ratio between its random coil and its outstretched dimensions which, dependent on the degree of polymerization, can amount to several magnitudes.

(iii) Molecular characteristics—relaxation time. If a chain is placed in an elongating fluid element, its own elongation will be determined by the balance of two forces: a) friction which extends the chain (characterized by friction coefficient f), and b) entropic retraction which opposes this extension, (in first approximation defined by force constant k). The quantity characterizing this balance is the configurational relaxation time given by

$$\tau = \frac{f}{k} \propto M^{1.5} \text{ to } M^2$$

(M = molecular weight) where the exact value of the exponent depends on the extent to which the coil is free draining [48].

Most past treatments refer to the case where an equilibrium extension has already been reached, i.e., rate of stretching and retraction of the chain balance in spite of the fact that the fluid element containing it continues to extend, e.g., [5, 6, 8]. The familiar theories, to which we have referred to in past works [10, 11], relate to this equilibrium situation. Broadly, all such treatments are based on exponential relaxations and contain a factor

$$\epsilon \propto \dot{\epsilon}\tau/(1 - \dot{\epsilon}\tau)$$

relating strain ϵ to strain rate $\dot{\epsilon}$. From this a catastrophic upswing of ϵ as $\dot{\epsilon}\tau \rightarrow 1$ will be apparent. This condition can be achieved by increasing $\dot{\epsilon}$ (i.e., the velocity gradient of the flow field) or by increasing τ accomplished by altering the nature of the molecule, solvent, or temperature. With a given molecular species, for a given solvent and temperature, this involves changing the molecular weight. If there is a spectrum of molecular weights a catastrophic upswing in ϵ should occur at and beyond a critical M value such as makes $\dot{\epsilon}\tau$ approach unity. This last consideration applies not only to isolated molecules but also to interacting groups of molecules where the relaxation time will be higher than that of the single component molecules. Thus entangled and/or associated molecules will extend at lower strain rates than the isolated chain.

Experimental Realizations and Observations

The underlying experiments consist of the creation of predesigned flow geometries and observation of the happenings at the relevant portion of the flow field by optical microscopy. Two kinds of polymer systems were examined: solutions and melts. In both cases examinations were conducted under two sets of conditions: first, where the polymer remained uncrystallized and second, where crystallization could also set in, due to appropriate choice of temperature and solvent. In the first case when crystallization was absent, the appearance of pronounced optical retardation, as seen between crossed polars, indicated the region of ultrahigh chain extension within the flow field while the magnitude of the birefringence gave a measure of the chain extension. Under conditions when crystallization also occurred, the crystalline regions and their formation could be observed directly and could be made available for subsequent examinations of their fine structure and properties. The flow lines were characterized by microscopic observations of tracer particles and the local velocities, required for defining the strain rates, by the track length of these particles obtained by means of suitably chosen photographic exposures. The fundamental flow studies on which the preceding conclusions were based were all on solutions. The observations on the melt certainly fell in line with these conclusions; the ultimate purpose of the melt experiments, nevertheless, was the prospect of utilizing the principles involved for melt processing.

The elongational flow fields created were essentially of two kinds:

(1) Three-dimensional flow systems. These embrace uniaxial extension and compression obtained by sucking and blowing solutions through pairs of impinging jets (Fig. 1) [10, 11] (an appropriately modified version of the apparatus has been used for melts [12]), the uniaxial extension case having been given more prominent attention;

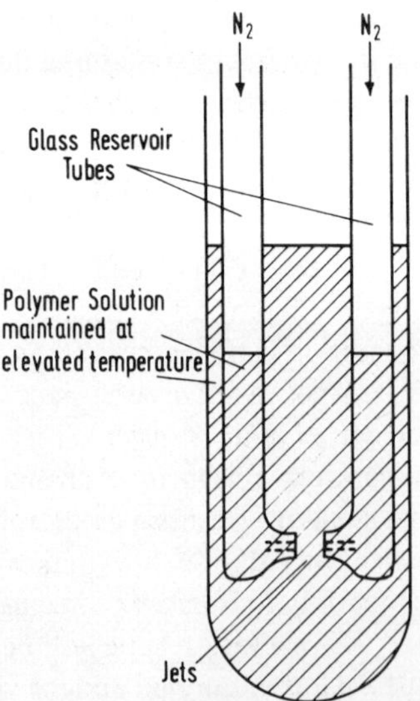

FIG. 1. Essential features of the "impinging jet" apparatus (ref. [16], based on ref. [10]).

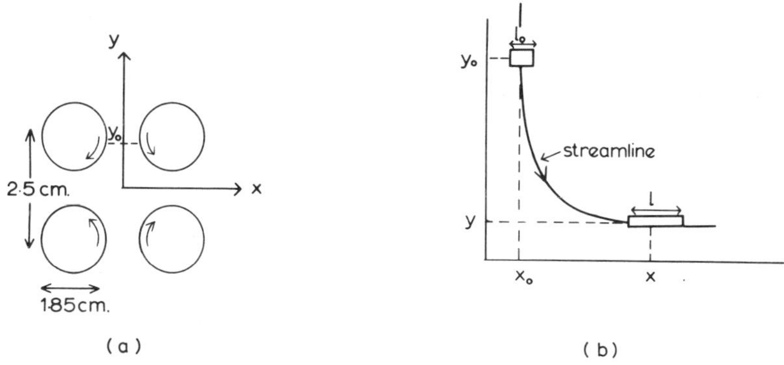

FIG. 2. Schematic diagram of the "four roll mill"; (a) the four counter rotating cylinders as viewed along their axes, (b) extension of a volume element along a streamline [17].

(2) Two-dimensional flow systems. These were achieved by a set of rollers. The most symmetrical case, representing pure shearing flow, (see i) in preceding section) was obtained by the use of four counter rotating rollers (Fig. 2) [13] (the "four roll mill" of G. I. Taylor [14] while more general forms of shear embodying a predesigned superposition of extension and rotation rates were obtained by means of a pair of rollers [15].

The salient results of the work under 1 have already been presented [10, 11, 16] while those under 2 i.e., [13], [15], and [17] were still in the press at the time this lecture was delivered and the paper submitted. For this reason only some general statements could be made to support the contentions of the preceding section at the time of writing.

A highly localized region of pronounced birefringence was observed in those localities where persistent extensional flow was expected. Figure 3 should serve as an illustration for the double jet case for solutions of polyethylene at temperatures too high for crystallization to occur. As reported previously [11], this birefringence increased with strain rate up to a limiting value (Fig. 4), which

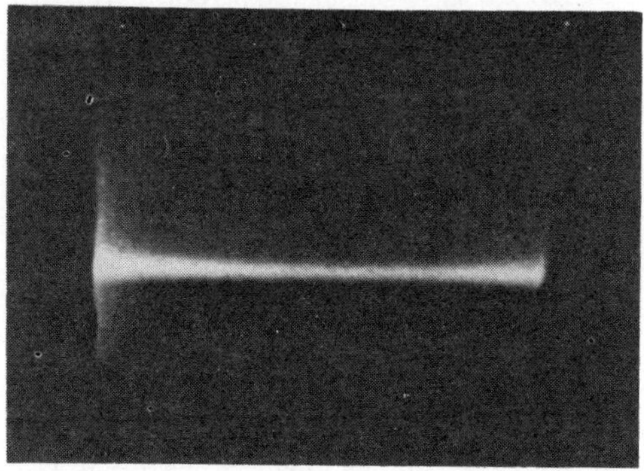

FIG. 3. Highly localized birefringence seen along the center line between two jets in the "suck" mode of operation in a polyethylene solution at 124°C. Crossed polaroids; polariser directions at 45° [11].

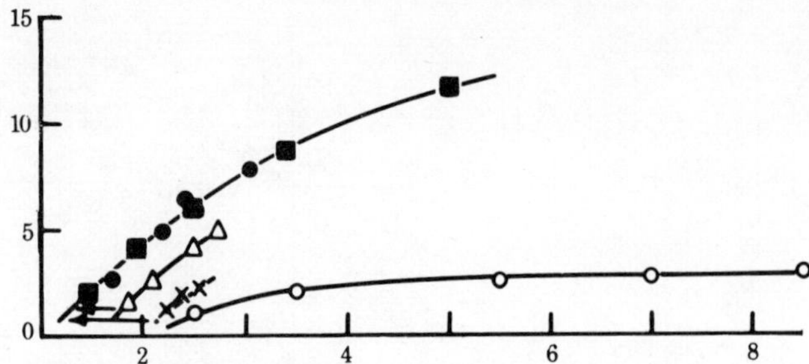

FIG. 4. Birefringence plotted against strain rate in jet experiments on polyethylene as in Figure 3 for different jet separations. The curve most relevant for the present article is the one defined by the solid squares and circles corresponding to jet separations 1.4 and 2.0 mm, respectively [11].

when allowing for the concentration (Fig. 5), correspond to complete or near complete alignments of a representative amount of chains in the solution. In addition it is to be noted that the region of high birefringence is confined to a narrow filamental region along the central axis of the system. Previously we attempted to account for this by finding reasons why the persistently elongational flow should be localized along this axis [11]. However, analogous observations on the localization of birefringence were made in subsequent experiments on the two-dimensional flow systems (type 2 above) where the uniformity of the strain rates could be guaranteed over extended areas by a priori considerations [13, 15, 17]. For example, in the situation represented by Figure 2, the strain rate is uniform in the area between the rollers, yet the orientation observed was confined to a narrow line, corresponding to a sheet seen edgewise along the x axis. This has led to the recognition of condition (ii) in the preceding section, namely that this chain extending effect of the flow has to persist for long enough time, or what is equivalent, over long enough distances, to produce the elongation required and that this condition will only be satisfied in narrowly confined regions of the flow field. This remains true even when in other parts of the flow field the velocity gradient is of the appropriate kind and magnitude to produce chain extension if the particular stream line (with the appropriate velocity gradient along it) persisted over a longer distance than it does in the actual system. It will be mentioned that the degree of the localization of the orientation itself has further information to convey about the relaxation time, hence the mass of the entity which is being extended, for given parameters of the flow field [17] which in turn permits a distinction to be made between, say, small extensions of a large fraction and large extension of a small fraction of the molecules in a given system producing a particular maximum value of the birefringence.

It is felt that the recognition of the confinement of chain extension to narrow regions within a flow field, which by the criterion of persistent strain rate (i) in preceding section, should be chain extending throughout, is of wider consequence for studies of macromolecules in elongational flow fields. It implies amongst others that methods which sample the entire flow system such as viscosity and light, X-ray or neutron scattering, as normally used will be inadequate for obtaining readily interpretable results; for this purpose the appropriate localized region of the birefringence would need to be probed.

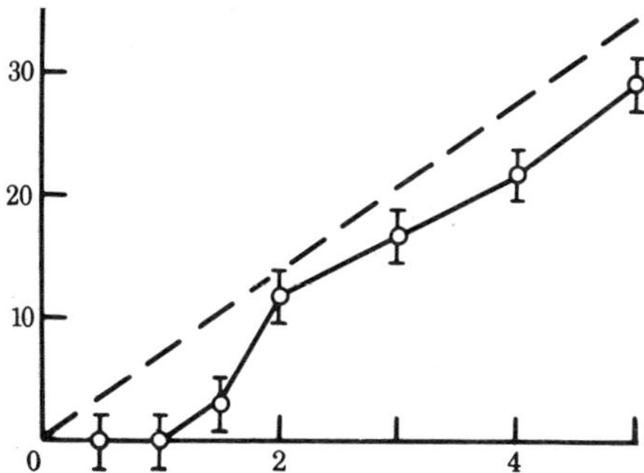

FIG. 5. Observed maximum birefringence against concentration from experiments as in Figure 4. (The dashed line represents 1:1 proportionality) [11].

Of the numerous issues arising, only one will be commented on further. Figure 5 reveals that the birefringence between the jets is broadly proportional to the polymer concentration, which means that chain extension is independent of concentration. However, it drops to zero abruptly below a critical concentration (1–2% for $M_w \sim 150,000$ in Fig. 5; the cut-off is at a lower concentration for higher M_ws). This indicates that the units which are being extended in these experiments are not individual molecules but assemblies of them, which in the instance quoted possibly arise through entanglements. (That individual chains, possibly of the highest end of the distribution, can extend is suggested by the fact that fibrous crystals do form from much higher dilutions when the experiments are conducted at lower temperatures where crystallization is possible [11]. However, the number of these molecules is too small for the birefringence caused by them to become observable). Results using the rollers also indicate that orientation of molecular assemblies rather than individual units are involved particularly at the much lower strain rates achievable in these experiments [13, 17]. In some instances (experiments with polystyrene) the rollers produced orientation at unexpectedly low strain rates, at much lower strain rates than for some other polymers of comparable molecular weight and concentration, which amongst a host of other effects (e.g., unusual persistence of the birefringence after the movement of the rollers has been arrested) strongly indicates the existence of specific associations between the chains [17]. Such effects are the more conspicuous as they vary from one source of polystyrene (even if all nominally atactic) to another suggesting chemical, possibly isomeric, differences between the different products such as may not be readily detectable otherwise.

Wider Implications

The last point emphasizes that observations of localized chain orientations in persistently elongational flow promises to have implications which go beyond the original, more restricted objectives of such studies. As just stated they may serve as indicators of molecular interactions, and in particular of molecular homogeneity in certain systems. Further, the existence of sharp localization of birefringence can be an aid for the identification of singularities in flow fields [9, 15]. Accordingly, chain elongation can be placed at the service of more general hydrodynamical enquiries, polymer interests apart. In particular, its potential in the study of vorticity in general has been indicated with such diverse applications as meteorology and drag reduction due to polymeric additives in turbulent flow [9]. Observations relating drag reduction to the formation of Taylor vortices, such as normally produce chain extension via elongational flow, have in fact been recently placed on record [18] thus indicating a link between a macroscopically observed hydrodynamic and a molecular phenomenon. The use of a particular combination of rollers has even been suggested as a means for investigating the applicability of theorems in the mathematics of discontinuous changes (catastrophe theory); Berry quoted in ref. [19] a line which being explored here (and published since the submission of this review [20]). When listing these numerous ramifications of our subject we have not even touched upon its diverse and profound relevance to methods of technological processing of polymers and to its effect on the properties of the resulting products, the principal motivation of this research, to be enlarged upon further below. Also, we have not yet mentioned that the principles involved are relevant not only to the orientation of macromolecules but to that of any orientable particle on any scale, such as suspended microscopic or macroscopic fibers or platelets. In fact particle orientations are more readily realized because of the larger sizes involved and in some ways more readily interpreted as no entropic restoring forces are operative. Indeed the four roll mill can be used for the study of the orientation of macroscopic fibers, as relevant, e.g., to paper making [21] and the double jet, as prompted by our polymer experiments, to orient asbestos and mica crystals in molten silicate glass leading to glass composites of unusual, possibly advantageous properties [22, 23], while isotropically shaped objects (droplets) can be deformed [14] or broken up [21] with potential use for making dispersions and emulsions. We see that the sphere of relevance of the subject matter outlined in this chapter could hardly be wider.

THE CRYSTALS: STRUCTURE AND GROWTH

Observation of Structures

As indicated in the beginning, crystallization from oriented molecules results in a composite structure consisting of a nucleating fibrous central thread with transverse lamellar overgrowth. In the case of melt crystallization where such crystal entities had been first inferred, they were referred to as "row nucleated structures", while in the analogous units of crystallization from solution they acquired the name of "shish-kebabs". Before proceeding along these lines, first an alternative approach will be quoted.

An Alternative View

According to one group of authors the structures in question have an entirely different origin from what has been implied above [24, 25]. They claim that the shish-kebab type structure is the result of deformation of chain folded single crystals in the course of their growth, the consecutive spiral terraces being pulled out in an accordion fashion. While taking note of this point of view, I shall revert to the picture of a flow-induced central thread with lamellar overgrowth in what follows, which I think is the relevant line of approach.

"Row Structures" in Elastomeric Materials

The history of this subject has been summarized in a previous review [16]. It essentially relates to crystallization of stretched polyisoprene (rubbers) or crosslinked polyethylene melts. The essential point is that the most conspicuous features are columns of parallel lamellae. This is certainly consistent with the general scheme presented here for crystallization from preoriented chains. Nevertheless it raises the following points.

Existence of a Central Thread

As summarized in ref. [16] this is often debated. While in certain instances a central thread can definitely be [26, 27], in others its presence may be questioned. Some of the reasons for the latter are self evident: The structure is more compact being completely filled out with material. The central thread would represent only a small fraction of the total material in any case, hence it is not surprising if it were less apparent than in the loosely stacked platelet structures of the solution grown shish-kebabs. Further, there are indications that the central thread is initially present but may disappear as the lamellar overgrowth develops, not only because it is obscured, but also because it may melt out as the stress relaxes, presumably when the lamellar overgrowth becomes interlocked and takes on the load [26]. Finally and chiefly, the central thread is becoming less regular and continuous where chemical crosslinks are introduced, which is frequently done in these experiments in order to enable the maintenance of the load under static conditions [26, 27]. It is possible that in the comparatively dense networks of such crosslinked systems, fiber continuity is ensured only by isolated stretched chains or groups of chains instead of complete fibrous crystal entities [28].

Nature of Lamellar Overgrowth

As implied above, these are far the dominating features in the structures discussed in the present sections. As mentioned repeatedly before, they may twist as in spherulites or may be all straight and parallel according to whether the stress is low or high, with intermediate stages in between. As the lamellae dominate, the overall texture of the macroscopic sample will be determined by this overgrowth orientation. When these lamellae are fully twisted (e.g., complete randomization around the b axis in polyethylene) the overall structure is that of a stack of two-dimensional spherulites; when they are fully aligned it becomes indistinguishable from that of a drawn fiber [29]. Composite textures containing a small amount of the highly oriented fiber texture, together with a texture appropriate to two-dimensional spherulites (or the appropriate variant of it) have also been observed by X-ray diffraction and considered as evidence for the model consisting of a central thread (giving the drawn fiber structure) and of the nucleated transverse lamellar overgrowth [30].

In polyethylenes crystallized during melt extrusion or from slightly stretched melts, sometimes a texture is obtained which is different both from that attributable to a c axis oriented drawn fiber and from that of a fully or partially randomized transverse lamellar overgrowth. Here a crystallographic direction (a), which is lateral to the chains, is dominantly along the direction of the orienting influence, the so-called a axis orientation. As pointed out in the previous review [16] this is best regarded as a variant of the row structure where the transverse lamellae are not randomized around the growth direction, (b in polyethylene) but have a preference to align with their lamellar planes along that of the orienting influence. A very realistic cause for such a situation has been suggested recently [31] relating the strength of the a axis orientation to the density of the nucleating threads. More precisely, when the lines of nuclei are close, the lamellae do not have a chance to randomize completely before impinging on others from adjacent lines of nuclei. It is shown that the tendency towards a axis orientation can be related to incomplete randomization due to this source. It follows that this a axis type orientation, where it occurs, is not in conflict with the general scheme of row nucleation; on the contrary it is one of its variants and thus falls in line with it.

Parallel lamellae in bulk or thin film samples offer a convenient opportunity to observe lamellae edge-on. Isoprene based elastomers are particularly useful in this respect. In combination with selective staining techniques they enable many interesting studies on the fine structure of the lamellae and also on the kinetics of crystal growth, both lamellar and fibrillar (e.g., refs. [32], [33], and [27]). Finally, stacks of parallel lamellae forming from melts under stressed conditions give rise to a new class of technological material the "hard elastic fibers" [34, 35] where the high degree of reversible elasticity associated with the high modulus is linked with reversible separation and accompanying bending of lamellae. The formation of this type of structure is likely to be related to "row nucleation" although other factors may also play part in it.

Shish-Kebabs

It is the solution-grown shish-kebabs which are most readily related to the basic work on chain extension discussed in the earlier parts of this paper. The principal features and their classification have been summarized in the previous review [16]; here only the most essential points will be recapitulated.

The primary flow induced structure is not a smooth fiber but is a shish-kebab itself, as shown by Figure 6. This is so even when the chains have been completely aligned prior to crystallization by the birefringence criterion in the double jet experiment, i.e., when we have approached the asymptotic maximum in Figure 4, and when all the molecules which are unattached to the fibers have been removed by exchange of solvent before cooling down the system from the crystallization temperature (which in itself is too high for platelets to form sepa-

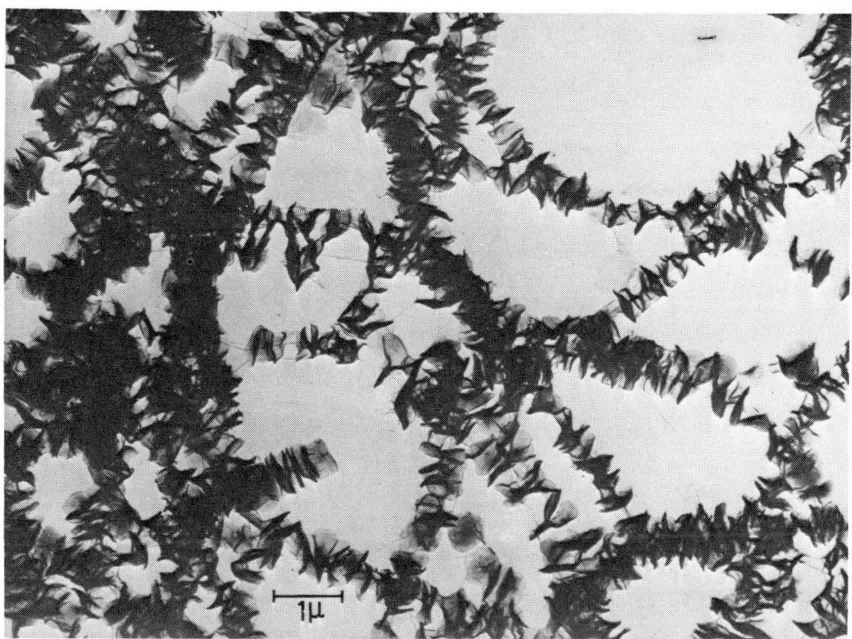

FIG. 6. Pure backbone shish-kebabs produced by elongational flow from 1.5% solution of polyethylene at 103°C between two jets during suction. The strain rate applied produces maximum birefringence (by Fig. 4) prior to crystallization. All unattached material was removed by washing at 103°C [11].

rately). Previously this type of shish-kebab structure has been termed microshish-kebabs [59]. Here the platelets are connected by veils to varying extents and are not detachable by subsequent washing with hot solvents, hence are intrinsically connected to the central fibers. This whole assembly has distinctive properties of its own. In contrast, when the system is cooled without adequate exchange of solvent the separate molecules which are left uncrystallized can precipitate onto these microshish-kebabs, producing a much more massive lamellar overgrowth structure. The latter can be removed by selective dissolution, hence represent pure overgrowth which is not molecularly connected to the central microshish-kebab. This overgrowth has the same properties as the usual chain folded single crystals. As pointed out in the previous review [16] in any real situation it is important to identify the true nature of the shish-kebab, namely whether it is a microshish-kebab, i.e., an intrinsic product of crystallization from extended chains, or a composite consisting of such units plus the usual chain folded crystals deposited onto them.

Method of Growth

In all the earlier descriptions, shish-kebabs were grown by means of stirring solutions. The resultant crystalline material was collected as a coherent aggregate onto the stirrer; for fine structural studies this had to be dispersed into its constituents, or alternatively stripping methods had to be used. As pointed out

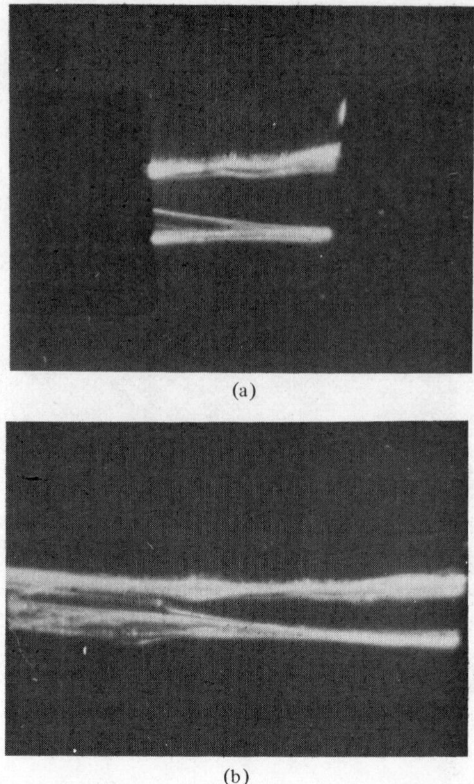

(a)

(b)

FIG. 7. Crystalline polyethylene filaments as formed from solution between jets on suction at 100°C, (a) as seen after formation; (b) as (a) but jets separated showing that the filaments extend into jets. Polars crossed at 45° [11].

previously [59], these operations can give rise to artefacts. Also in the jet method the fibers, as extracted from the flow field, are in a more or less coagulated form which makes them directly visible under the low power optical microscope (Fig. 7). A way to avoid preparative manipulations and obtain the basic fibers in isolation ready for observations under the electron microscope has been achieved by Mackley [36], based on the realization that solid obstacles placed in any flow field (even if uniform) create a velocity gradient down-stream due to the fact that the velocity immediately behind the obstacle is zero but must reach the average velocity of the flow further down-stream. This fact, that solid obstacles in a flow field can be the source of fibers, was first recognized in experiments carried out on melts. Figure 8 shows the effect of a grid, Figure 9 that of a needle (in combination with the much weaker fiber forming effect of the elongational flow created by the confluence of the stream lines at a nearby single jet orifice where the fibers arise subsequent to those at the needle tip). Both cases have useful applications and analogies in the fundamental studies on solutions. In particular, when a microscopic grid is placed edgewise like a paddle onto a stirrer immersed in an appropriately supercooled solution it gives rise to shish-kebab fibers for reasons just stated even on *slow* stirring speeds which, without the grids,

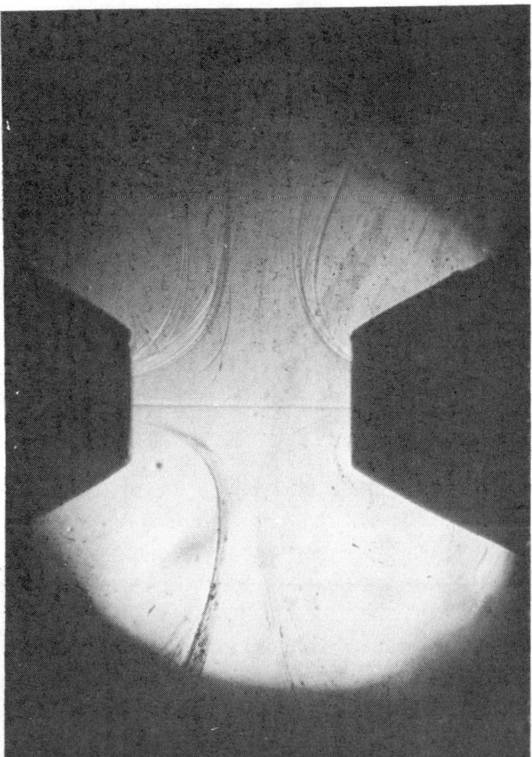

FIG. 8. Formation of crystalline polyethylene filaments in an impinging jet experiment on the melt at 140°C. In this case the melt was forced through the two jets from above through a copper gauze (outside the field of view). First a filament forms along the center line between the jets with fibers starting to emanate from the gauze afterwards. Unpolarized light [12].

would create no longitudinal velocity gradients in themselves [36]. Some of the fibers collect on the grids and are available for fine structural studies as such on the same grids where they have formed (Fig. 10).

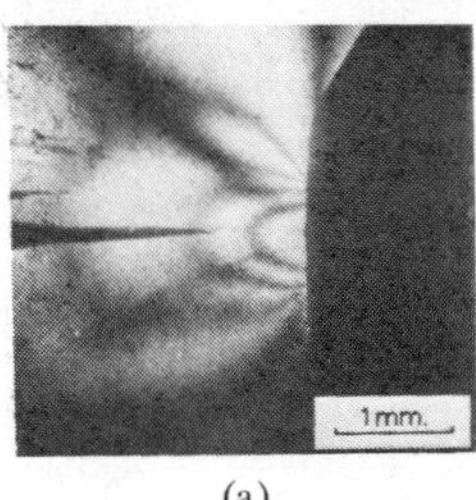

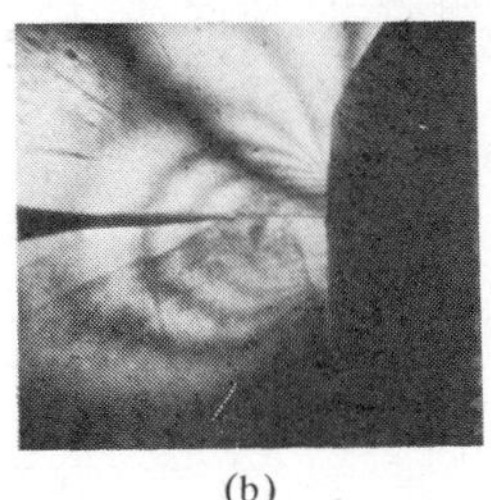
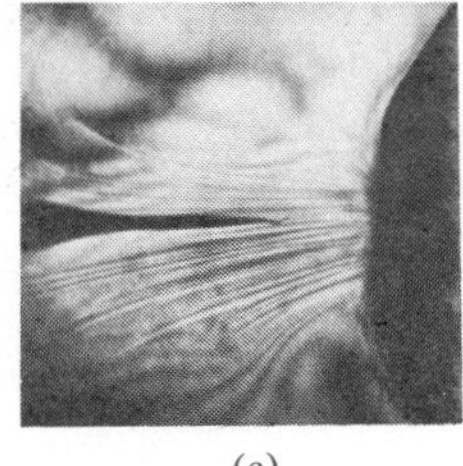

(a)	(b)	(c)

FIG. 9. Formation of crystalline filament for molten polyethylene flowing into an orifice with a stationary needle-like obstruction upstream at 140°C. The flow velocity increased in the sequence a)b)c). For low velocities (0.8 cm sec^{-1} and 3.3 cm sec^{-1}) fiber formation is confined to the needle tip—the main purpose of the illustration—but for higher velocities (8.0 cm sec^{-1}) the orifice itself generates fibers growing in the opposite direction [39].

Local Flow Fields

Several puzzling observations have been noted. One already referred to is that the basic flow-induced entity contains lamellar overgrowth in itself even when arising from molecules that are likely to be fully extended prior to crystallization. Another one closely related to it is that the intrinsic overgrowth has a limiting width, continued growth of the filaments occurring not in the transverse but only in the length direction [24]. This phenomenon has been used to grow filaments by immersing seed fibers into supercooled streaming solutions. The flow did not have to be elongational; it could be even simple shear flow, the seed being induced to grow largely lengthwise [37]. In the first place the seed was the polymer itself obtained beforehand in the fibrous form by the familiar stirring induced crystallization method. It was found nevertheless that foreign fibrous objects like a thread of linen could serve the same purpose. The major puzzle at this stage was the fact that this seed-induced growth did not require elongational flow fields.

All these observations fall in place when considering the effect of local flow fields. The importance of the fact that the presence of a solid object influences the flow field around it has been mentioned above in connection with Figures 8 and 9. The underlying idea can now be carried over to the fibrous crystal itself. Once formed, it will modify the flow field around it. It should create a simple shear field along its side surface which will impede chain alignment there, and hence limit lateral growth in accord with observations, as suggested first in [11] and [16]. But perhaps even more importantly, just as in the case of the macroscopic analogy of Figure 9, it will enhance the elongational gradient at the fiber tip promoting chain alignment and consequently growth there [38]. This can occur even when the overall flow field itself has no longitudinal velocity gradient, even when the flow is uniform; in this case the tip creates its own longitudinal velocity gradient, which provides an explanation for Pennings' seeding experi-

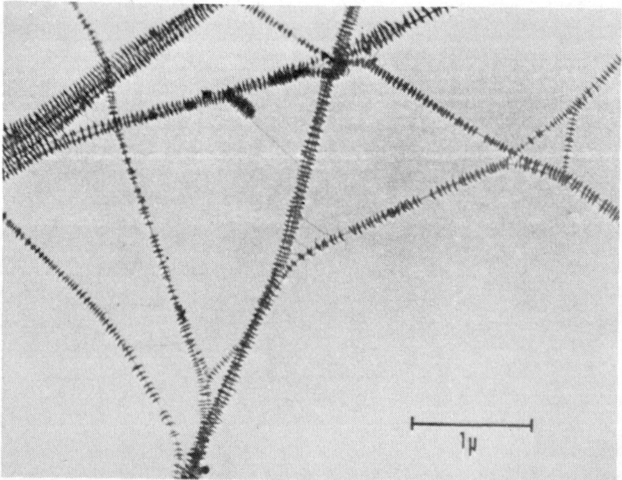

FIG. 10. Solution growth shish-kebabs of polyethylene as produced on a grid in situ. Formed and washed at 106°C. Electronmicrograph [36].

ment [37] and provides a general explanation for the preferred longitudinal growth of shish-kebabs [24] under all circumstances. The feasibility of these ideas has been demonstrated by the mathematical analysis of flow around elongated ellipsoidal objects closely simulating fibrous crystals [38]. In summary, once a fiber has formed, its continuing fibrillar growth will occur at the tip due to the velocity gradient created by the fiber itself, while growth will be arrested along the sides due to the simple shear field arising there.

The idea of tip growth through self-generated velocity gradients has potentially important consequences for producing fibrous crystal entities by design. The seeding experiment from solution referred to above [37] has been adapted for the production of fibers by a continuous process which consists of withdrawal of the fiber upstream at the same rate at which it grows at the tip, the tip remaining in a stationary position within a flowing solution [37]. The flow is capillary (shear) flow, the withdrawal being countercurrent. An analogous method of producing a fiber was developed independently in the course of experiments from the melt [39]. This consists of the melt being forced through a single jet orifice with fibrous crystallization occurring in the hot zone before the melt enters the jet (Fig. 11). By suitable adjustment of conditions the fiber tip can be kept at a stationary position while the fiber is being wound up outside

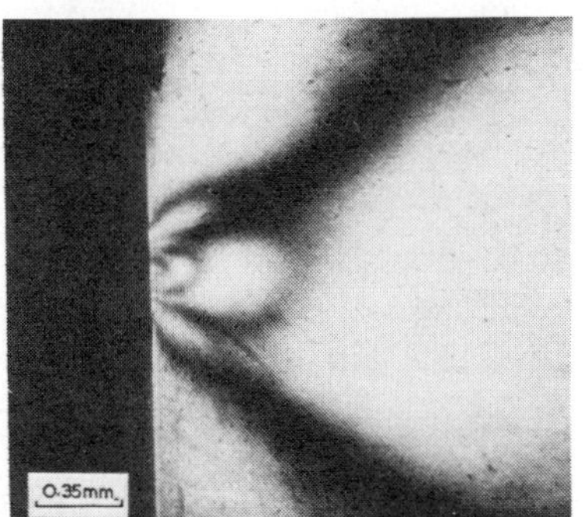

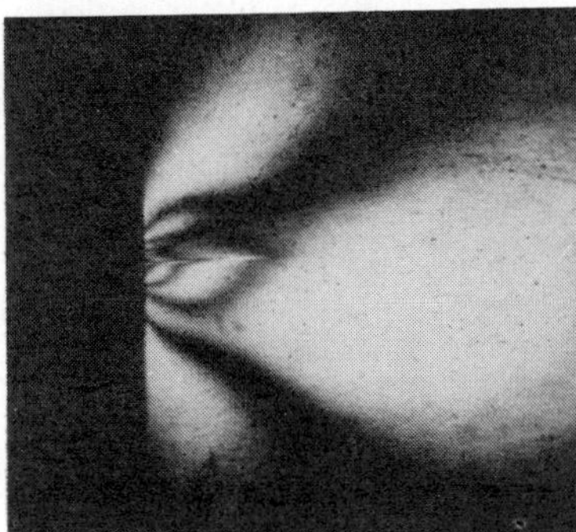

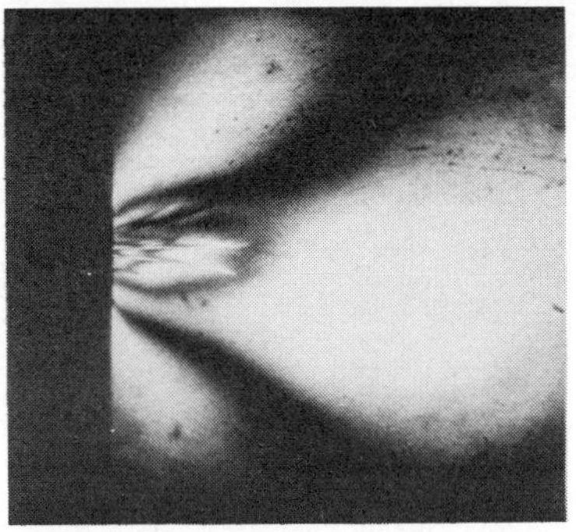
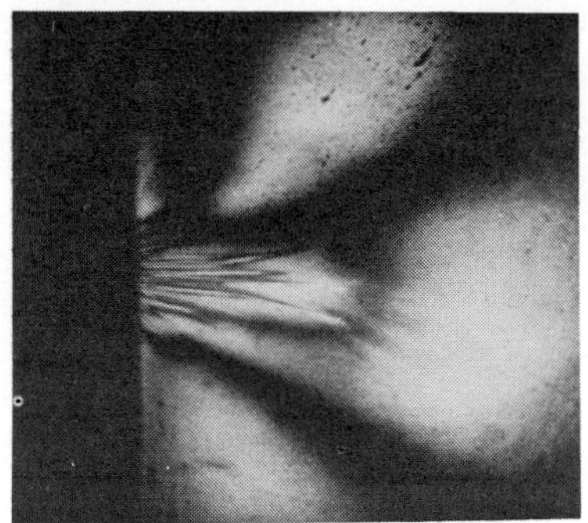

FIG. 11. Four sequences in the development with time of fibrous crystals from polyethylene melts at 140°C at the entrance of a jet orifice. Flow is from right to left. Diagonally crossed polaroids [39].

beyond the downstream end of the jet, the whole operation being a one-stage fiber producing spinning process. There are two influences operating: a velocity gradient towards the jets due to convergence of stream lines, an effect which increases as the melt approaches the jet orifice. The fibrils formed will be carried outwards through the capillary by the surrounding melt with the velocity appropriate to the melt flow within the capillary. This gives rise to the second influence, a velocity gradient at the fiber tip located further inside the melt interior where the melt flow velocity is lower than within the capillary. This will induce crystal growth upstream with an effectiveness which will increase as the growth proceeds towards the melt interior. The growing tip will reach a stationary position corresponding to the dynamic equilibrium due to the two opposing effects. These conditions are illustrated schematically by Figure 12 with the actual observation of fiber formation by Figure 11.

The Origin of the Oriented Overgrowth

In the preceding section it has been stated that the limited lateral growth can be attributed to a reduction of chain alignment due to the development of simple shear flow along the fiber surfaces. While this should account for the eventual arrest of lateral growth, in itself it does not explain the structural features actually seen, i.e., the chain folded overgrowth which is molecularly attached to the fiber, the apparent periodicity of this overgrowth and the veils connecting them. This subject of current topicality will only be touched upon in its broadest generality. The alternative possibilities of whether these connected overgrowth features form at the same temperature as the fiber or whether later on cooling has been raised previously [11]. There appears to be strong support for the latter case [40] namely, that the overgrowth originates during cooling from molecules which are partly attached to the fibers and partly dangling in solution, a picture for which an increasing amount of new evidence is accumulating also in our own laboratory. It would follow that the fibers when formed are "hairy" [40], a

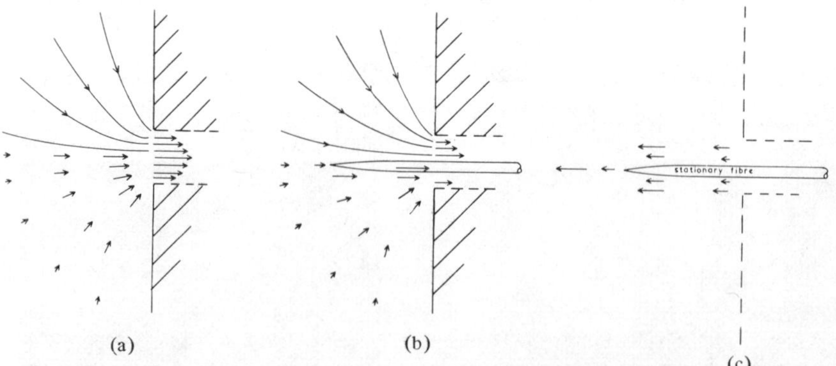

FIG. 12. Schematic representation of fiber formation as in Figure 11, (a) flow into single orifice; (b) flow field after a growing fiber has appeared. The fiber is being swept out towards the right while it keeps growing towards the melt interior at the tip; (c) local flow field around the growing fiber considering the fiber to be stationary [39].

picture also implied in some form or other by other works [41, 42] where these hairs give rise to chain folded overgrowth subsequently. When the same hairs become incorporated in several lamellae simultaneously they produce the connecting veils. Further, the hairs located at the fiber tip can be visualized as forming loose bundles trailing behind the tip and aligning in the local elongational flow field. In addition they could also "hook" further molecules which after extension in the local flow field could become attached to the fiber ensuring continuing growth at the tip [40]. (For particulars see [62] which appeared since submission of present paper.)

New Issues

Under this heading two items which are likely to come into their own in the nearest future will merely be mentioned.

Surface Effects

The pinning effect of surfaces was amongst the first speculations about the origin of ultrahigh chain extensions once the latter was recognized as the source of fibrous crystals. It was proposed [1] that if part of the chain is absorbed along a solid stationary surface, while the rest is freely dangling in a flowing solution (or conversely for a moving surface in a stationary solution), the unattached portion of the molecule will have a chance to become highly extended. Such speculations, however, became superseded by the recognition of the role played by elongational flow. Nevertheless, recent developments and ideas are focusing attention again to the possible chain extending effect of partial chain attachment to solid surfaces within flowing systems. These are arising amongst others in current work in our own laboratory prompted by private information from Pennings relating to the same topic, namely that adsorbed layers along solid walls may play a part in extending chains. In fact the extension of "hairs" referred to above is a manifestation of the same class of phenomenon. At this place only that much will be said that partial attachment along surfaces, whether they are foreign surfaces such as walls of vessels and stirrers, or along surfaces of the same polymer itself, are likely to be contributory factors—in addition to that of elongational flow—to the development of shish-kebabs and possibly also to row structures in the melt [43].

The Scale Factor

It will only be noted here that fibrous structures feature on various dimensional levels throughout the literature of the subject as they have done in the present article. Thus the fibers seen in Figures 7, 8, 9, and 11 are of the order of hundreds of microns in diameter and so are the fibers grown from solutions by Pennings' seeding methods [37]. The fibers in Figures 6 and 10 are in the range of 100 Å (central thread) and 500 Å–1μ (overgrowth). Clearly the finer scale structures represent the primary entities. Nevertheless, the former do not appear to be accidental aggregations of the latter but seem to be coherently developing structure units. The reason for this and the relation of the electron microscopic basic unit to the fiber observed by low power optical microscopy is probably of significance requiring further attention.

Some Comments on the Melt

In the foregoing, no distinction was made between experiments from solutions and melts, examples for both have been used interchangeably. The basic principles of chain extension and the ensuing crystallization in fact were first established in the case of solutions. It was very gratifying to note that subsequent experimentation on melts complied with the same principles. For example, the fibers formed exactly in the same way between the jets at least as far as visible with the optical microscope, both in solutions and melts, i.e., within the central localized regions of the elongational flow field (Figs. 7 and 8). Also, continuous fiber production could be achieved in both cases based on the same principle of dynamic equilibrium at the fiber tip as described earlier. Of course the technicalities of the experimentation were different and less straightforward in the case of the melt, however, this will not be enlarged upon at this place. Here only two additional comments will be made in connection with the melt.

First, from the point of view of the fundamentals there must be significant difference between the melt and solution cases. In the melt the chain should be highly entangled. This means that the configurational relaxation time should be much longer, hence much lower strain rates should suffice to extend the chain. This is in accord with observation [12]. Further, due to this intimate entanglement, rotation of individual coils cannot be readily envisaged. This means that the rotational component of shearing flows should be less of an impediment to chain extension than in the case of moderately dilute solutions. This is again consistent with observation, namely that in the case of melts row nucleated structures are readily observed even as a result of simple shear. (The latter was assessed from the way the flow was induced macroscopically, e.g., by sliding past two surfaces in mutually opposed directions with the melt in between, which does not necessarily mean that the flow itself corresponds to simple shearing flow or even that the flow field is homogeneous. For example, in the experiments being referred to, melt adhesion to the solid surfaces was observed [43].)

Second, the self evident technological implications of chain elongation in melt flow may deserve a single mention. The importance of jets, orifices, solid obstacles—accidental or deliberate—in various aspects of melt processing needs no elaboration. The possibility of creating localized regions of ultrahigh chain extension and ensuing fibrous crystallization will be obvious. This could occur unintentionally say in injection moulding. On the other hand the effect could be utilized for deliberate production of oriented fibers and films (e.g., the one-stage spinning referred to earlier).

PROPERTIES

The flow-induced crystals have distinctive properties. As in the previous sections we shall be only concerned here with the primary flow-induced products and not with the overgrowth formed by unattached molecules, wherever such a distinction can be made as, e.g., in solution grown shish-kebabs, where the latter can be detached or prevented from forming. Two kinds of property will be considered: thermal and mechanical.

Thermal Properties

It was amongst the earliest observations that solution grown shish-kebabs have unusually high melting points and, in addition, that they superheat. There are numerous works on the problem of how the two effects, namely genuinely high melting point and the superheating effects can be separated (e.g., ref. [59]). In fact the two are not readily separable even conceptually. The main source of the high melting point is not so much the high melting enthalpy due to the extensive development of the lattice along the chain direction (which can be partly offset by the large lateral surface if the fibrous entity is thin) but by the small entropy change which would pertain if the extended chains remain oriented even after melting. However, the latter can only be temporary as the chain extension in the molten state should eventually relax. The time scale of this relaxation, however, may be long and according to the time scale of the observation super-heating or genuinely high melting point may be inferred. There are numerous works on this issue. The experimental fact to concern us here is that the solid state can be preserved up to quite exceptionally high temperatures. Two instances will be specifically quoted for the case of polyethylene.

The first relates to a particular calorimetric work on solution-grown shish-kebabs [44]. In the course of it, just as in a number of preceding studies, several melting peaks are observed above what is considered normal for polyethylene, namely in the range of 145–155°C. In addition, however, a small melting peak is reported at the exceptionally high temperature of 180°C and a corresponding exothermic peak at 165°C on subsequent cooling. Clearly a small portion of the material, possibly the innermost central core of the assembly, persists 40°C above the usual melting point and can recrystallize at temperatures still far above the equilibrium melting temperature. This effect must clearly be due to the fact that the chains remain extended even after the lattice has broken down.

The second class of observation relates to hot stage polarizing optical microscopy of melt crystallized polyethylene and polypropylene [45]. Here fibrous entities produced by elongational flow could be seen to persist as birefringent lines in the otherwise isotropic material up to ~170°C and proved to be sites of nucleation on subsequent cooling; even when heated above this temperature, when they become practically undetectable, these sites retained their ability to nucleate crystals on subsequent cooling to lower temperatures. Rather remarkably, isolated birefringent lines of this kind could be seen also in material not deliberately exposed to elongational influences. Apparently, elongational flow arising accidentally in local areas during the usual handling of the melt, or around heterogeneities such as contracting bubbles, can produce the fibrous crystallization products in question which become only detectable on visual inspection when the sample is heated to temperatures high enough to melt out all other types of crystal, spherulites in particular, which mask them. This is an issue which should be of obvious consequence for all melt processing.

Mechanical Properties

Possibly the greatest significance of fibrous crystals lies in the area of mechanical properties. It is the intrinsic nature of the fibers that they are "stiff" and "strong" along the fiber direction and this on account of the covalent connections along this direction achieved through alignment of the chain molecules. In spite of this, it is well known that the mechanical properties expected on this basis are far from being achieved. Thus the modulus of conventionally aligned synthetic fiber is 50 to 100 times lower than the theoretical value. I have pointed out soon after the recognition of chain folded crystals that a high degree of chain orientation is not synonymous with high chain extension [46]; e.g., a stack of parallel chain folded platelets could be indistinguishable from a system of fully stretched chains by the usual criteria of orientation (wide-angle X-ray diffraction, birefringence) a point brought home since, by the recognition of chain folded lamellar structures even in drawn fibers (e.g., ref. [47]). From the mechanical point of view the folded chain may not be directly load bearing which would then be the source of the lower than theoretical modulus. Much of the interest in the achievement of improved extension—as opposed to improved chain orientation—has been triggered off by such considerations [48]. In order to give a feel for the numbers involved, the following modulus figures will be quoted. The modulus of unoriented polyethylene is about 5 Kbar, that of a conventionally oriented fiber about 20–50 Kbar, the theoretical modulus about 2300 Kbar.

Results on solution grown shish-kebabs are certainly encouraging in the above respect. Pennings observed that the modulus could be increased systematically with increasing crystallization temperature when producing oriented sheets of shish-kebabs by deposition on a stirrer [49], and by application of the recent fiber growing method, in fact quite a remarkable increase well on towards the theoretical value (about 1000 Kbar) could be achieved [40]. The origin of the increase in modulus is related to the increase in the ratio of central thread to lateral overgrowth.

As compared to solution grown fibers, fibers produced from the melt by crystallization via elongational flow-induced chain extension showed only modest improvement in modulus so far [50]. The reason for this is indicated by structural studies. From these it appears [60] that the continuous fibrous structure is only a small fraction of the total material, the rest consisting of transverse overgrowth lamellae which are parts of the flow-induced fibrous unit, together with further material which crystallized independently and is apparently unrelated to the flow-induced structure. Clearly while structures such as in Figures 8, 9, and 11 conform to expectation based on the effect of elongational flow, there is much scope for improvement as regards mechanical properties. Some reference to the possibilities will be made in the last section.

FIBER DRAWING

Everything said so far was about fibrous crystals formed from pre-extended molecules. As mentioned in the first chapter this is only one of the methods of forming fibers even if the one to which most of the present paper is devoted. Another method (no. 3 in the introductory section), the more conventional one in fiber technology, is the deformation of solid polymers crystallized under quiescent conditions without external orienting influence. This is the commonly termed "cold drawing." There will be no attempt made here to cover this large subject; only one aspect will be touched upon relevant to the achievement of ultrahigh chain extension and high modulus.

It has been stated earlier that the mechanical properties, modulus in particular, of conventional synthetic fibers do not even approach what could be ultimately expected from them and that this is attributed to the fact that in spite of high molecular orientation they do not contain the chains in a sufficiently stretched out form: The conventional "natural" draw ratios of 4–10 are insufficient to pull out large proportions of the folds. Higher draw ratios have featured occasionally in the literature but not much attention has been given to their production, neither to the properties or to the underlying structure of the resulting material, until the recent work of Ward and collaborators [51–53] who have shown that draw ratios of about 30X can be systematically achieved and that the resulting modulus can be as high as one-third of the theoretical ($\sim$700 Kbars). It is not the purpose of this section to describe these advances, merely a few points will be quoted from our own material on this subject primarily for comparison with the oriented structures obtained along the different route of the preceding sections.

Our own work [54] on ultrahigh drawing was commenced at a stage when it appeared from the literature that a combination of very specific conditions have to be satisfied to achieve it. The usual limitation to the draw ratio is the breakage of the fiber; this has to be overcome, or rather the elongation where breakage occurs to be pushed up, if ultrahigh draw ratios are to be achieved. It has emerged that while there are certain guidelines (foremost amongst them is the need for elevated draw temperature) ultrahigh draw ratios can be achieved under a fairly wide variety of circumstances. Chiefly, however, it has emerged that the resulting modulus is a unique function of draw ratio, i.e., the principal condition for high modulus is that the sample should be capable of the necessary high degree of extension, or conversely in the case of a given sample, a method should be found by which this extension can be achieved. This is illustrated by Figure 13, arrived at independently from the similar examples in the work by Ward and collaborators [53].

One may expect that in such highly drawn samples the chains are largely stretched out, and that the high modulus is attributable to the high degree of chain continuity along the fiber. Nevertheless, in line with what was said in the first section, samples obtained by deformation are of basically different origin from those obtained by crystallization from pre-extended chains in as far as they preserve the memory of the state they have come from. Normally on heating they shrink back near to their original length, a phenomenon which is remarkable

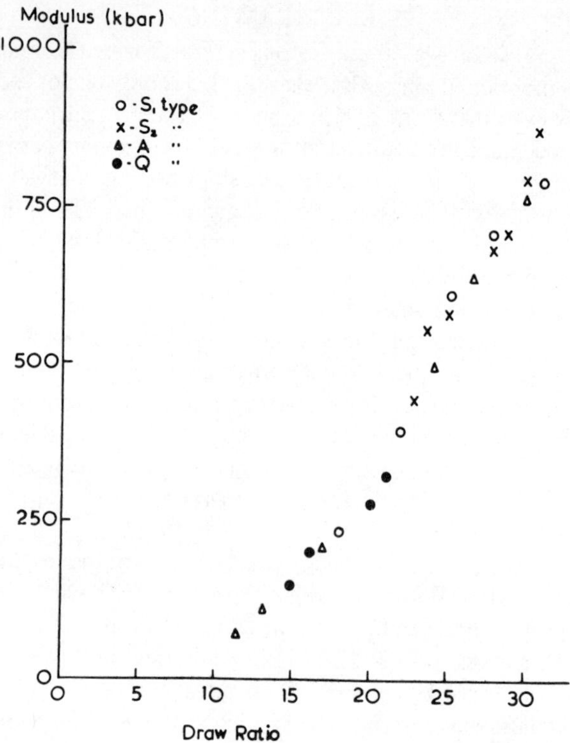

FIG. 13. Modulus versus draw ratio for drawn polyethylene sheets for differently treated starting materials (not to be itemized here) [54].

in itself and also serves to distinguish such fibers from those obtained by crystallization from pre-extended chains. This retraction criterion holds also for the ultrahigh extensions in question as shown by Figure 14.

Even more remarkable than the enormous shrinkage itself is the behavior of these highly drawn samples after heating at constant length [61]. During such heating, contractile stresses develop but if shrinkage is prevented they eventually relax. At this stage the crystal orientation is still maintained but the modulus is reduced to that of a conventionally drawn fiber ($\sim$50 Kbar). Now if such samples are stored at room temperature the modulus starts to increase spontaneously and the original value of say 600 Kbar is regained or closely approached over a period of 1 or 2 days, the resulting high modulus sample becoming dimensionally stable on subsequent heat treatments [61].

The drawing mechanism, the special mechanical properties, the thermal retraction behavior and the underlying structural model raise many further questions. At this point we confine ourselves to the statement that by "cold" drawing a very high degree of chain extension can be achieved which in terms of chain continuity rivals the best that can be obtained by crystallization from pre-extended chains. In fact at present it seems to offer a readier route to high modulus (for qualifications, see next section). Nevertheless the dimensional instability of the product on heating contrasts strikingly with the superheatability

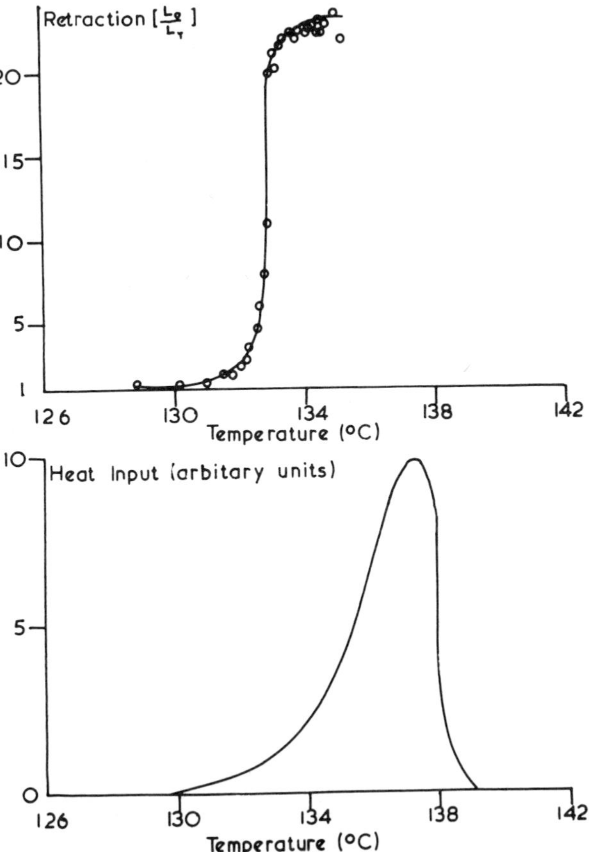

FIG. 14. (a) Graph showing retraction of same specimen used in Figure 13 after 30 sec at the elevated temperature [54]. (b) Thermogram of same sample as in (a) [54].

of materials crystallized from pre-extended chains which however can be eliminated by the treatment leading to the self hardening procedure as outlined above.

The self-hardening phenomenon itself raises entirely new perspectives both as regards structural knowledge and application. At this early stage only that much will be said that most probably a composite structure consisting of permanent fibrous crystals created by the original drawing process and a variable, initially contractile matrix is involved. If so this would create a link between permanent and contractile oriented structures produced by crystallization and deformation respectively.

SOME FURTHER METHODS FOR ACHIEVING HIGH CHAIN EXTENSION

It follows from a priori considerations [48] and from the example of ultra-highly drawn fibers that the achievable modulus is determined by the chain continuity along the macroscopic fiber, hence by the extension of the chains, and that in the first approximation it makes no difference in what way this extension had been obtained. The thermal stability and the possibility of a one-stage extrusion–drawing operation would make the approach via crystallization from pre-extended chains clearly an attractive alternative to cold drawing, not to speak of its interest in its own right. (Pennings' latest announcement of solution grown shish-kebabs with ultrahigh moduli certainly proves that this is a realistic objective). It follows from what has been said that an increase in the amount of central thread relative to lateral overgrowth should be favorable in this respect. However, in contrast to this self evident criterion the increase in the amount of the connecting veil as opposed to "serrated" type overgrowth should have an effect in the same direction by enhancing molecular connectedness along the fiber. Veils are promoted by closely spaced nucleation from a rather dense hair cover along the fiber surface. Thus, if the above suggestion has any validity at all, the overgrowth may also be put to useful service if appropriately designed. Current results not to be detailed here, indicate that this orientation may well hold and lead to useful results [57].

In addition, ultrahigh modulus has been achieved by other conceptually related methods. One of them involves the blockage of melt flow when molten material is forced through capillaries. In this case the soldified plugs, extracted or extruding from the capillaries, were found to possess exceptional properties including high modulus which is within reach of the theoretical [55]. While undoubtedly crystallization induced by elongational flow is involved, Porter et al. claim that extrusion of the solid material subsequent to crystallization is responsible for the high modulus [56]. Current work nevertheless has shown [57] that about one-third of the theoretical modulus can be achieved by a variant of this method where the absence of postsolidification extrusion can be guaranteed, hence when crystallization alone is involved. Other methods leading to ultrahigh modulus, possibly associated with very high chain extension, involve hydrostatic extrusion of solid material [58] a kind of solid state forming process which no doubt is also part of the capillary blockage method in Porter's version just quoted. Squeezing solid material through an orifice is one way of achieving very high chain extension as here, in principle at least, the draw ratio is unlimited because the fiber does not break as in tensile elongations depending only on the ratio of barrel to capillary diameter. Even so, the "solid state forming" is as a rule carried out at elevated temperatures, in fact within or near the melting range of the polymer. Here deformation of crystalline material may be coupled with crystallization of molten material containing pre-extended chains, hence the whole process is likely to be a combination of the two principal categories under discussion [3].

To end with I shall be reverting to the introductory paragraph, namely that fiber formation and the underlying chain elongation is an intrinsic consequence of the long chain nature of polymer molecules. This itself of course is well established and self evident. However, what is comparatively novel and possibly unexpected, is the truly enormous variety of issues, theoretical, experimental, and practical which have emerged in the course of examining the principles involved, and the structures and properties which arise. As so often the case with long chain molecules the reality is more varied and interesting than one may have guessed a priori. Even now, it is not yet certain that we are at the end of the road in discovering new classes of effects, and that no further surprises are lying in store. It is on this note of the open-endedness of the subject, which may well have appeared more or less closed already decades ago, that I wish to conclude the present account.

REFERENCES

[1]   A. Keller, *Rep. Prog. Phys.,* **31,** part 2, 623 (1968).

[2]   B. Wunderlich, *Macromolecular Physics*, Vol. 1, Academic Press, New York (1973); Vol. 2 (1975).

[3]   P. J. Barham and A. Keller, *J. Polym. Sci., Lett.,* **13,** 197 (1975).

[4]   A. Keller and F. M. Willmouth, *Makromol. Chem.,* **121,** 42 (1969).

[5]   A. Ziabicki, *J. Appl. Polym. Sci.,* **11,** 14 (1959).

[6]   A. Peterlin, *Pure Appl. Chem.,* **12,** 563 (1966).

[7]   G. Marucci, *Polym. Eng. Sci.,* **15,** 229 (1975).

[8]   P. G. de Gennes, *J. Chem. Phys.,* **60,** 5030 (1974).

[9]   F. C. Frank, Lecture given at 50th Anniversary of the Canadian Pulp and Paper Research Institute, Montreal, November, 1975.

[10] F. C. Frank, A. Keller, and M. R. Mackley, *Polymer,* **12,** 467 (1971).

[11] M. R. Mackley and A. Keller, *Phil. Trans. R. Soc. London,* **278,** 29 (1975).

[12] M. R. Mackley and A. Keller, *Polymer,* **14,** 16 (1973).

[13] D. G. Crowley, F. C. Frank, M. R. Mackley, and R. G. Stephenson, *J. Polym. Sci. Phys. Ed.,* **14,** 1111 (1976).

[14] G. I. Taylor, *Proc. R. Soc. London,* **146,** 501 (1934).

[15] F. C. Frank and M. R. Mackley, *J. Polym. Sci. Phys.,* **14,** 1121 (1976).

[16] A. Keller and M. R. Mackley, *Pure Appl. Chem.,* **39,** 195 (1974).

[17] D. P. Pope and A. Keller, *Colloid Polym. Sci.,* **255,** 633 (1977).

[18] A. Keller, G. Kiss, and M. R. Mackley, *Nature, (London),* **275,** 304 (1975).

[19] I. Stewart, *New Sci.,* **68,** 447 (1975).

[20] M. V. Berry and M. R. Mackley, *Philos. Trans. R. Soc. London,* **287,** 1 (1977).

[21] S. V. Kao and S. G. Mason, *Nature, (London),* **253,** 619 (1975).

[22] K. H. G. Ashbee, *Nature, (London),* **252,** 469 (1974).

[23] K. H. G. Ashbee, *J. Mater. Sci.,* **10,** 911 (1975).

[24] T. Nagasawa, T. Matsumura, and S. Hashino, *Appl. Polym. Symp.,* **20,** 275 (1973).

[25] T. Nagasawa and Y. Shimomura, *J. Polym. Sci. Phys.,* **12,** 2291 (1974).

[26] E. H. Andrews, Lecture given at the Conference of the Royal Microscopical Society, Oxford (1968), also private communication.

[27] G. K. L. Davies and Ong Eng Long, private communication, 1972; also Ong Eng Long Ph.D. Thesis, Queen Mary College, 1972.

[28] H. Jenkins, Ph.D. Thesis, Bristol 1974.

[29] A. Keller and M. J. Machin, *J. Macromol. Sci.* **B1,** 41 (1967).

[30] M. J. Hill and A. Keller, *J. Macromol. Sci.* **B3,** 153 (1969).

[31] T. Nagasawa, T. Matsumura, and S. Hashino, *Appl. Polym. Symp.,* **20,** 295 (1973).

[32] E. H. Andrews, P. J. Owen, and A. Singh; *Proc. R. Soc. London, A* **324,** 79 (1971).

[33] E. H. Andrews and B. Reeve, *J. Mater. Sci.,* **6,** 547 (1971).

[34] R. G. Quynn and H. Brody, *J. Macromol. Sci. Phys.,* **B5,** 721 (1971).

[35] E. S. Clark, *Structure and Properties of Polymer Fibers,* R. W. Lenz and R. S. Stein, Plenum Publishing, New York, 1974, p. 267.

[36] M. R. Mackley, *Colloid Polym. Sci.,* **253,** 393 (1975).

[37] A. Zwijnenburg and A. J. Pennings, *Colloid Polym. Sci.,* **253,** 452 (1975).

[38] M. R. Mackley, *Colloid Polym. Sci.,* **253,** 373 (1975).

[39] M. R. Mackley, F. C. Frank, and A. Keller, *J. Mater. Sci.,* **10,** 1501 (1975).

[40] A. J. Pennings, Lecture given at "Colloque les Polymères en Papeterie," Grenoble, Oct. 1975.

[41] D. Krueger and G. S. Y. Yeh, *J. Macromol. Sci. Phys.*, **B6,** 431 (1972).

[42] A. J. McHugh and E. H. Forrest, *J. Macromol. Sci. Phys.*, **B11,** 219 (1975).

[43] D. T. Grubb and A. Keller, *J. Polym. Sci., Letters,* **12,** 419 (1974).

[44] A. J. Pennings and J. M. A. A. van der Mark, *Rheol. Acta,* **10,** 174 (1971).

[45] D. T. Grubb, J. A. Odell, and A. Keller, *J. Mater. Sci.,* **10,** 1510 (1975).

[46] A. Keller, *Polymer,* **3,** 393 (1962).

[47] A. Peterlin, *J. Mater. Sci.,* **6,** 490 (1971).

[48] F. C. Frank, *Proc. R. Soc. London* **A314,** 127 (1970).

[49] A. J. Pennings, C. J. H. Schouteten, and A. M. Kiel, *J. Polym. Sci. C,* **38,** 167 (1972).

[50] M. R. Mackley, *Polymer Rheology and Plastics Processing,* Ing. Conf. Loughborough (Br. Soc. Rheology and Plastics and Rubber Inst.), 250 (1975).

[51] G. Capaccio and I. M. Ward, *Polymer* **15,** 233 (1974).

[52] J. B. Smith, A. J. Manuel, and I. M. Ward, *Polymer,* **16,** 51 (1975).

[53] G. Capaccio, T. J. Chapman, and I. M. Ward, *Polymer,* **16,** 468 (1975).

[54] P. J. Barham and A. Keller, *J. Mater. Sci.,* **11,** 27 (1976).

[55] J. H. Southern, N. Weeks, and R. S. Porter, *Makromol. Chem.,* **162,** 19 (1972).

[56] T. Niikuni and R. S. Porter, *J. Mater. Sci.* **9,** 389 (1974).

[57] J. A. Odell and A. Keller, *Polym. Symp.,* (Fraser P. Price Memorial Symposium, Amherst, 1977) in press.

[58] A. G. Gibson, I. M. Ward, B. N. Cole, and B. Parsons, *J. Mater. Sci.,* **9,** 1045 (1974).

[59] A. Keller and F. M. Willmouth, *J. Macromol. Sci. Phys.,* **B6,** 493 (1972).

[60] D. T. Grubb, J. Dlugosz, A. Keller, and M. R. Mackley, to appear.

[61] R. G. C. Arridge, P. J. Barham, and A. Keller, *J. Polym. Sci., Polym. Phys. Ed.,* **15,** 389 (1977).

[62] A. Zwijnenburg and A. J. Pennings, *Colloid Polym. Sci.,* **254,** 868 (1976).

Author Index

Journal of Polymer Science: Polymer Symposium 58, 143 (1977)
© 1977 by John Wiley & Sons Inc.

Published Polymer Symposia

1963   No. 1.   First Biannual American Chemical Society Polymer Symposium
                        Edited by H. W. Starkweather, Jr.
        No. 2.   Fourth Cellulose Conference
                        Edited by R. H. Marchessault
        No. 3.   Morphology of Polymers
                        Edited by T. G. Rochow

1964   No. 4.   Macromolecular Chemistry, Paris 1963 (Published in 3 parts)
                        Edited by M. Magat
        No. 5.   Rheo-optics of Polymers
                        Edited by Richard S. Stein
        No. 6.   Thermal Analysis of High Polymers
                        Edited by Bacon Ke
        No. 7.   Vibrational Spectra of High Polymers
                        Edited by Giulio Natta and Giuseppe Zerbi

1965   No. 8.   Analysis and Fractionation of Polymers
                        Edited by John Mitchell, Jr. and Fred W. Billmeyer, Jr.
        No. 9.   Structure and Properties of Polymers
                        Edited by Arthur V. Tobolsky
        No. 10.   Transport Phenomena in Polymeric Films
                        Edited by Charles A. Kumiss
        No. 11.   Fifth Cellulose Conference
                        Edited by T. E. Timell

1966   No. 12.   Perspectives in Polymer Science
                        Edited by E. S. Proskauer, E. H. Immergut, and C. G. Overberger
        No. 13.   Small Angle Scattering from Fibrous and Partially Ordered Systems
                        Edited by R. H. Marchessault
        No. 14.   Transitions and Relaxations in Polymers
                        Edited by Raymond F. Boyer
        No. 15.   U.S.–Japan Seminar in Polymer Physics
                        Edited by Richard S. Stein and Shigeharu Onogi

1967   No. 16.   Macromolecular Chemistry, Prague 1965 (Published in 8 parts: Parts 1–5,
                        1967; Parts 6 and 7, 1968; Part 8, 1969)
                        Chairmen: O. Wichterle and B. Sedláček
        No. 17.   Electrical Conduction Properties of Polymers
                        Edited by A. Rembaum and R. F. Landel
        No. 18.   The Meaning of Crystallinity in Polymers
                        Edited by Fraser P. Price
        No. 19.   High Temperature Resistant Fibers
                        Edited by A. H. Frazer
        No. 20.   Supramolecular Structure in Fibers
                        Edited by Paul H. Lindenmayer

1968   No. 21.   Analytical Gel Permeation Chromatography
                        Edited by Julian F. Johnson and Roger S. Porter
        No. 22.   Macromolecular Chemistry, Brussels-Louvain 1967 (Published in 2 parts:
                        Part 1, 1968; Part 2, 1969)
                        Chairman: G. Smets
        No. 23.   Macromolecular Chemistry, Tokyo–Kyoto 1966 (Published in 2 parts: Part
                        1, 1968; Part 2, 1969)
                        Chairmen: I. Sakurada and S. Okamura
        No. 24.   Polymer Reactions
                          Edited by E. M. Fettes

No. 51.   International Symposium on Macromolecules in Honor of Professor Herman F. Mark
Edited by E. F. Casassa, T. G Fox, H. Markovitz, C. G. Overberger, and E. M. Pearce

No. 52.   Contributions from Students and Friends of Professor Champetier on the Occasion of his 70th Birthday
Edited by P. Sigwalt, T. G Fox, C. G. Overberger, and H. F. Mark

No. 53.   Crosslinking and Networks
Edited by K. Dušek, B. Sedláček, C. G. Overberger, H. F. Mark, and T. G Fox

1976   No. 54.   Polymer Science: Achievements and Prospects (In Honor of P. J. Flory)
Edited by H. Markovitz and E. F. Casassa

No. 55.   Proceedings of the Eighth Australian Polymer Symposium
Edited by I. C. Watt

No. 56.   Fourth International Symposium on Cationic Polymerization
Edited by J. P. Kennedy

No. 57.   Degradation and Stabilization of Polyolefins
Edited by B. Sedláček, C. G. Overberger, H. F. Mark, and T. G Fox

1977   No. 58.   Orientation Effects in Solid Polymers
Edited by G. Bodor

No. 59.   Recent Advances in the Field of Crystallization and Fusion of Polymers
Edited by J. P. Mercier and R. Legras

All the above symposia can be individually purchased through the Subscription Department, John Wiley & Sons.